Nanocarrier Vaccines

Scrivener Publishing
100 Cummings Center, Suite 541J
Beverly, MA 01915-6106

Publishers at Scrivener
Martin Scrivener (martin@scrivenerpublishing.com)
Phillip Carmical (pcarmical@scrivenerpublishing.com)

Nanocarrier Vaccines

Biopharmaceutics-Based Fast Track Development

Edited by

Vivek P. Chavda

Department of Pharmaceutics and Pharmaceutical Technology, L. M. College of Pharmacy, Ahmedabad, India

and

Vasso Apostolopoulos

Institute for Health and Sport, Immunology and Translational Research Group, Victoria University, Melbourne, Australia

Scrivener Publishing

WILEY

This edition first published 2024 by John Wiley & Sons, Inc., 111 River Street, Hoboken, NJ 07030, USA and Scrivener Publishing LLC, 100 Cummings Center, Suite 541J, Beverly, MA 01915, USA

For more information about Scrivener publications please visit www.scrivenerpublishing.com.

Wiley Global Headquarters
111 River Street, Hoboken, NJ 07030, USA

For details of our global editorial offices, customer services, and more information about Wiley products visit us at www.wiley.com.

Library of Congress Cataloging-in-Publication Data

ISBN 978-1-394-17468-3

Cover image: Pixabay.Com
Cover design by Russell Richardson

Set in size of 11pt and Minion Pro by Manila Typesetting Company, Makati, Philippines

Printed in the USA

10 9 8 7 6 5 4 3 2 1

Dedicated to the 75th year anniversary of the LM College of Pharmacy, Ahmedabad Gujarat India.

Vivek P. Chavda also wants to dedicate this book to his wife Disha and his parents

Contents

Preface

In nanomedicine and nano-delivery systems, materials in the nanoscale range are used as diagnostic instruments or to administer therapeutic compounds to particular targeted regions in a controlled manner. By delivering precise medications to specified locations and targets, nanotechnology provides several advantages in treating chronic human illnesses. The use of nanomedicine (including chemotherapeutic medicines, biological agents, immunotherapeutic agents, etc.) in the treatment of various diseases has recently seen many notable applications. This book aims to be a single source material for understanding all the current and novel advancement in the field of nanotechnology.

Chapter 1 discusses the history and constantly evolving field of nanoparticles. Chapter 2 describes the overall composition of a nanoparticle. The next chapters explain the formulation strategy and the influencing factors in therapeutic approaches, such as vaccine development (Chapter 3), biodegradable and non-biodegradable formulation, and properties such as size, shape, charge, inertness, efficacy, morphology, and more (Chapters 4 and 5). Different nanoparticles, such as lipid-based, viral vector-based, and metal, uphold very significant properties individually, which suggests their applicability in various management tactics, as described in Chapters 6 and 7.

Chapters 8 and 9 examine how genetic information carrying entities are becoming the norm for evacuating tedious diseases. Furthermore, Chapters 10, 11, and 12 gather an exhaustive amount of information on routes of administration for the same, such as the oral route, mucosal immunity, intramuscular, subcutaneous, and intradermal. This treatment has had an astonishing effectiveness in veterinary disease management, as described in Chapter 13. Finally, Chapter 14 explores the legal regulatory for nanotechnology-based approaches.

We hope this book will help to bolster your knowledge on this vastly changing and expanding subject. Our thanks go to the prestigious Wiley and Scrivener Publishing for their continuous kind support and guidance.

Editors
Vivek P. Chavda and Vasso Apostolopoulos

Part 1
GENERAL

1

History of Nanoparticles

Keshava L. Jetha[1,2], Arya Vyas[1], Divya Teli[3], Amit Chaudhari[3], Riyansi Satasiya[3], Vishwa Patel[3], Shailvi Soni[4], Shail Modi[4] and Vasso Apostolopoulos[5*]

[1]Department of Pharmaceutics and Pharmaceutical Technology, L. M. College of Pharmacy, Navrangpura, Ahmedabad, Gujarat, India
[2]Gujarat Technological University, Chandkheda, Ahmedabad, Gujarat, India
[3]Department of Pharmaceutical Chemistry, L. M. College of Pharmacy, Ahmedabad, Gujarat, India
[4]Department of Pharmaceutical Sciences, Massachusetts College of Pharmacy and Health Sciences, MA, USA
[5]Institute for Health and Sport, Victoria University, Immunology and Translational Research Group, Melbourne, VIC, Australia

Abstract
Nanoparticles (NPs) have become a widely researched area in modern medicine due to their unique properties and potential applications. This article provides an overview of the field of NPs in healthcare, starting with a brief introduction to NPs and their history. The article then delves into modern developments in the field of NPs, including their production and various applications. It also covers the different types of NPs that have been studied, along with their properties and advantages. Furthermore, the article discusses the importance of NPs in various healthcare areas, such as drug delivery, medical imaging, and diagnostics. Finally, the article concludes with a summary of the current state of the field and the future prospects for NPs in healthcare. Understanding the properties and potential applications of NPs can contribute to the development of innovative medical therapies and advance the field of healthcare.

Keywords: Nanoparticles, history, medical imaging, diagnosis, future prospects

**Corresponding author*: vasso.apostolopoulos@vu.edu.au

Vivek P. Chavda and Vasso Apostolopoulos (eds.) Nanocarrier Vaccines: Biopharmaceutics-Based Fast Track Development, (3–24) © 2024 Scrivener Publishing LLC

1.1 Introduction

Nanoparticles (NPs) have emerged as a potential research area in nanotechnology that frequently appears in materials science, biomedical engineering, and nanomedicine sectors. A NP is the fundamental unit in the fabrication of a nanostructure with one or more nanometric dimensions ranging in size that may differ from the bulk material. The word "nano" represents a nanometer (nm, 10^{-9} m), an International System of Unit for length. In principle, NPs are materials with lengths ranging from 1 to 100 nm. There are numerous examples from ancient times where nanostructures or NPs have been used for various purposes. The varieties of glorious colors of glass windows of medieval cathedrals are an ancient witness to the utilization of metal oxide NPs [1]. NPs evolved in different eras and from different regions such as hand stencils of Sulawesi cave in Indonesia and hair dyes with lead sulfide NPs in Egypt. Mesopotamia and Egypt produced glassware using inorganic NPs in the fourteenth century BC [2].

The different types of NP classifications, functionalization techniques, various types of synthesis approaches, and growth-related mechanisms are evolved. NPs may be classified into various groups related to dimensionality space; morphology major groups are organic polymeric NPs, inorganic NPs, ceramic NPs, and bionanoparticles [3]. NPs can be synthesized either from a simple material or using a range of multiple composite objects. The synthesis methods of NM are generally classified into "top–down" and "bottom–up" approaches. In the top–down approach, a solid material is broken into smaller particles by external forces, while in the bottom–up approach, nanostructures are synthesized through the buildup of molecules or atoms. These synthetic approaches can be further differentiated by chemical, physical, and biological processes that, through improvement over time, also emerged as including mechanochemical and physiochemical processes. In the current scenario, tremendous metallic nanomaterials are being synthesized in bulk using titanium, copper, zinc, magnesium, alginate, aluminum oxide, silica, gold, and silver.

NPs are widely used to improve the pharmaceutical properties of medicines including penetration, plasma distribution, half-life, and target site accumulation. Size, shape, charge, and elasticity are physical properties of NPs and play a role to provide desired pharmacokinetic properties for implementation in drug delivery systems. Mechanical properties of NPs include elasticity; these overcome biological barriers from the site of application to the site of solid tumor and provide superior cancer drug delivery [4]. NPs have diverted physical and chemical characteristics from bulk material and show a wide range of applications in a multitude of fields, such as medical treatments; use in various industry departments; the manufacture of oxide fuel; and solar batteries for energy storage, cosmetics, and clothes. Nanotechnology can also enhance the properties of construction materials, where recycling concrete with NPs support sustainability [5]. Nanotechnology-based products used for the control of disease in the healthcare system are referred to as "nanomedicine." In recent years, nano-sized compounds such as liposomes, polymers, and virus-sized NPs become attractive development as targeted delivery vehicles for viral antigens. NPs provide similar size distribution as the viruses and therefore NPs loaded with viral antigens can enter the virus-targeted cells. In recent years, the utilization of NPs has also expanded toward vaccine delivery with high bioavailability, elevated immunogenicity, and controlled release profiles [6].

1.2 History of Nanoparticles

Although NPs have been studied for centuries, the term was not coined until the late 1970s. In 1959, Nobel Prize–winning American physicist Richard Feynman proposed the concept of nanotechnology. "There's Plenty of Space at the Bottom" was the title of a lecture Feynman gave at the California Institute of Technology at the American Physical Society's annual meeting (Caltech). The question "Why can't we write the full 24 volumes of the Encyclopaedia Britannica on the head of a pin?" was posed by Feynman in this lecture, and he also sketched out a vision of utilizing machines to build smaller machines all the way down to the molecular level [7]. Refer to Figure 1.1.

Feynman's position as the father of modern nanotechnology was finally cemented by the validation of the ground-breaking idea he offered. Approximately 15 years later, in 1974, Japanese scientist Norio Taniguchi used the word "nanotechnology" for the first time. He defined nanotechnology as the manipulation of materials at the atomic or molecular scale by procedures like separation, consolidation, and deformation [8].

NPs have been utilized for centuries before Feynman put forward the concept of nanotechnology. More than 4,500 years ago, humans used natural asbestos nanofibers to reinforce ceramic matrix materials. More than 4,000 years ago, the ancient Egyptians used nanomaterials (NM) as well. They produced PbS NPs with a diameter of about 5 nm for use in hair dye. When Egyptians and Mesopotamians began employing metals to make

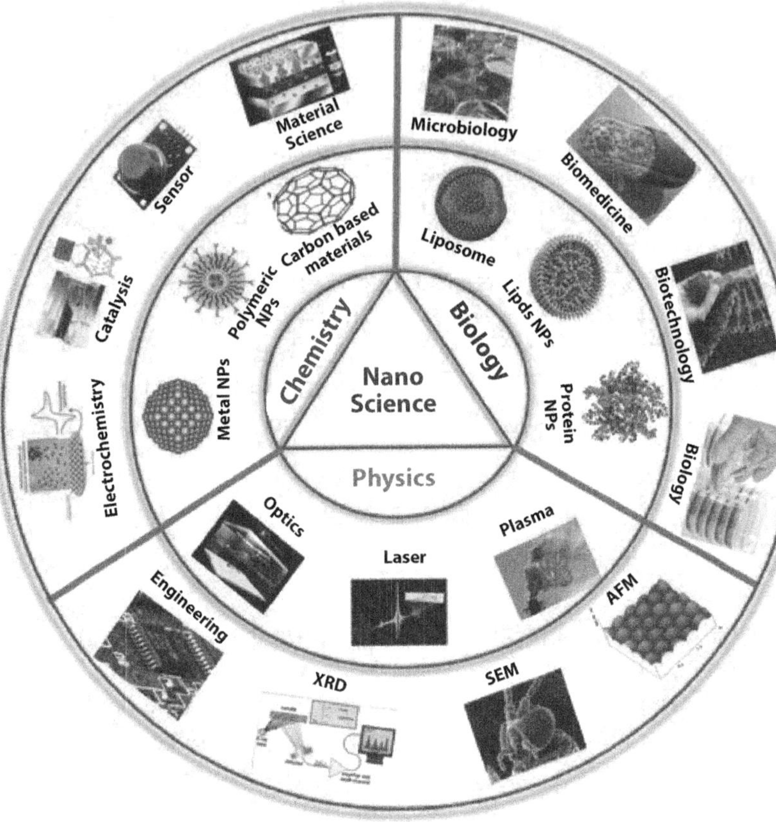

Figure 1.1 The evaluation of NPs on the basis of broad terms including chemistry, biology, and physics. Adopted under CC BY 4 from [8].

glass in the 14th and 13th centuries BC, it may be said that the period of metallic NPs began. Since then, metallic NPs have been synthesized via chemical processes. One of the most intriguing examples of nanotechnology in the ancient world was presented by the Romans in the fourth century AD, who employed NPs and structures. Yet, the most well-known use of ancient metallic NPs is on a piece of Roman glass. The Lycurgus cups are dichroic glass cups from the fourth century AD that show different colors depending on the direction of the light: red when it comes from behind and green when it comes from the front. To understand the dichroism phenomena, scientists examined the cup in 1990 using a transmission electron microscope. The presence of NPs with a diameter of 50–100 nm is what causes the dichroism (two colors) that has been seen. These NPs were identified by X-ray analysis as silver–gold (Ag-Au) alloy with an Ag:Au ratio of approximately 7:3 and 10% copper (Cu) distributed in a glass matrix. The Au NPs absorb light at a wavelength of about 520 nm, which gives them a red tint. The green hue is a result of colloidal dispersions of Ag NPs with a size > 40 nm scattering light, while the red-purple color results from the absorption by the larger particles. One of the oldest artificial nanomaterials is recognized as the Lycurgus cup. The fusion of Au and Ag produces a similar effect in late medieval church windows, which shine dazzling red and yellow colors [9, 10].

Glowing, glistening "luster" ceramic glazes were used in the Islamic world between the ninth and seventeenth centuries and later in Europe. These glazes contained silver (Ag), copper (Cu), or other NPs. Around the 16th century, Italians used NPs to create Renaissance ceramics. They were influenced by the 13th–18th century techniques that were utilized in the production of "Damascus" blades, cementite nanowires, and carbon nanotubes, which were used to provide strength, resilience, and the ability to hold a keen edge [11, 12].

Michael Faraday conducted research on the preparation and characteristics of colloidal suspensions of "ruby" gold by the reduction of an aqueous solution of chloroauric acid ($AuCl_4$) using phosphorus in carbon disulfide [CS_2] in 1857. This work involved the use of gold NPs and marked the first scientific investigation of NP preparation. As part of his research, Michael Faraday examined the optical properties of thin films that were created from dried colloidal solutions. During his experiments, he observed that the color of these films could be altered reversibly by applying mechanical compression, with the films changing from bluish purple to green when pressurized. His study provided the initial scientific description of gold NP solutions, which played a significant role in the history of nanomaterials in the scientific community [13, 14].

In the IBM Zurich Research Laboratory, physicists Gerd Binnig and Heinrich Rohrer created the scanning tunneling microscope (STM) in 1981 as nanotechnology advanced. STM analyzes the surface properties by moving a sharp tip so closely to a material's conducting surface that the atoms' electron wave functions overlap with the surface atom wave functions. For their creation of the STM, Binnig and Rohrer were awarded the 1986 Nobel Prize in Physics. Due to this discovery, the atomic force microscope and scanning probe microscopes (SPMs), which are currently the preferred tools for nanotechnology researchers, were developed [8, 15].

Xu *et al.* discovered a new type of carbon nanomaterials called carbon dots (C-dots) in 2004. These particles had a size of less than 10 nm and were accidentally discovered during the process of purifying single-walled carbon nanotubes [8]. Nanoscience has significantly advanced a number of different scientific disciplines over time, including computer science, biology, and engineering. Nanotechnology has been used in computer science to transform computers from bulky, room-sized equipment to highly portable laptops. The interest in and progress achieved in the subject of nanoscience and nanotechnology increased as the twenty-first century began. The enormous potential of nanotechnologies in biomedicine, notably in the detection and treatment of a number of human diseases, has recently come to light in a number of studies [16].

1.3 Modern Development of Nanoparticles

NPs are small particles with diameters ranging from 1 to 100 nm. NPs have distinct features that make them helpful in a variety of applications ranging from electronics to healthcare. Significant improvements in the production of NPs have occurred throughout the years, resulting in a growing interest in their possible applications [17]. NPs contribute significantly to the field of pharmaceutical sector, advancing drug delivery and facilitating novel therapeutic approaches [18].

NPs have various benefits for drug delivery because of their smaller size and large surface area. NPs can encapsulate drugs, which increases their shelf-life and can deliver drugs to a specific targeted site in the body. Liposomes, polymeric NPs, lipid NPs, and inorganic NPs are different kinds of NPs that are developed in enhancing the drug delivery for different diseases, including cancer, infectious diseases, and neurological disorders. One of the most significant advances was the development of the sol–gel process, which allows for the creation of NPs with a high degree of uniformity and control over their properties [19].

NPs can be especially designed to target infectious cells or tissues, enabling targeted treatment. The addition of ligands, antibodies, peptides, or other targeting moieties with NPs allows for selective binding to targeted cells or tissues, resulting in increased drug accumulation and effectiveness at the site of action. This allows for customized treatments in which NPs may be designed for individual patients or conditions, resulting in more effective and safer therapies. NPs can be formulated for the controlled release of the drug, allowing for sustained and prolonged drug delivery [20]. This can reduce dose frequency, increase patient compliance, and reduce adverse effects. NPs may be designed to respond to a variety of stimuli such as pH, temperature, enzymes, or external triggers, allowing medications to be released at rates depending on concentration, offering adequate control over drug release kinetics. NPs are also used in the combination treatment, in which different drugs or therapeutic agents are coencapsulated within NPs and administered to the targeted region at the same time [14]. This can lead to synergistic effects, better therapeutic results, and decreased medication resistance. Combination therapy using NPs has shown potential in cancer treatment, allowing numerous medicines with distinct modes of action to be combined to improve anticancer efficacy [21]. NPs are being developed for use as theranostic platforms, which combine treatment and diagnostics in a single system [22]. Theranostic NPs can deliver therapeutic medicines while simultaneously imaging the illness site, enabling real-time monitoring of treatment efficacy [23]. This can help with early detection of the disease state, providing feedback on therapy response, and enabling individualized approaches to therapy. NPs are also being investigated for their potential use in the delivery of novel therapies such as nucleic acids (DNA and RNA), peptides, and proteins. NP-based delivery methods can shield these fragile compounds from degradation, improve cellular absorption, and improve therapeutic effectiveness [24, 25]. This has created new opportunities for gene therapy, protein therapy, and other modern therapeutic strategies [26, 27].

The therapeutic agent must be delivered to its site of action to exert its efficacy. There are numerous delivery methods; however, each of them has challenges associated with them. NP-based delivery systems including dendrimers, liposomes, peptide-based NPs, carbon nanotubes, quantum dots, polymer-based NPs, inorganic vectors, lipid-based NPs, hybrid NPs, and metal NPs are the advanced forms of NPs [28]. Dendrimers are radially symmetric, nanoscale molecules having well-defined, homogenous, and monodispersed structures with arms or branches that resemble trees. These are polymeric macromolecules that is ideal to study the effects of polymer size, charge, and composition on biologically relevant properties such as lipid bilayer interactions, cytotoxicity, internalization, blood plasma retention time, and biodistribution [29]. Lipid-based NPs include liposomes, solid lipid NPs and nanostructured lipid carriers. Liposomes are tiny, spherical artificial vesicles that can be made using natural phospholipids. Liposomes have hydrophobic and hydrophilic components forming a bilayer which enhances their specific targeting properties [30]. Peptide-based NPs are conjugation of self-assembled peptides with nucleic acids that is widely used as gene therapy [31]. Carbon nanotubes are nanosized tube-shaped structures that are either single-walled or multi-walled. One of the major concerns with carbon nanotubes is health and occupational safety. Exposure to carbon nanotubes results in toxicity to the ecosystem. It is also classified as a carcinogen [32]. Polymer-based NPs are Nano-capsules or nanospheres that contain the drug or therapeutic agent within the polymeric matrix. It helps in controlled release, enhancing bioavailability, protecting the drug and other molecules in the biological environment [33]. Metal NPs are among the most widely studied and used NPs due to their unique optical, electronic, and catalytic properties. Gold and silver NPs are particularly popular due to their biocompatibility and low toxicity, making them ideal for use in biomedical applications such as drug delivery and imaging.

One of the key challenges in the development of NPs is ensuring that they are safe for use in humans and the environment. This requires a deep understanding of how NPs interact with biological systems and the environment, as well as rigorous testing to ensure that they do not have adverse effects. To address these challenges, researchers are exploring a range of approaches, including the use of biocompatible materials, the development of targeted delivery systems, and the use of NPs with controlled release properties. Overall, the modern development of NPs has led to a wide range of new materials and technologies that have the potential to revolutionize fields ranging from healthcare to energy. While there are still many challenges to overcome, the rapid pace of development in this field suggests that we are likely to see even more exciting advances in the years to come [34, 35].

In conclusion, the modern development of drug NPs represents a promising area of research and development in the pharmaceutical industry. Advances in NP synthesis and characterization, as well as our understanding of how NPs interact with biological systems, have resulted in new therapies and improved treatment options for a range of diseases. While there are still challenges to be overcome, the potential benefits of NP-based drug delivery make this a field of research that is likely to see continued growth and innovation in the years to come. However, further research, regulatory compliance, and careful evaluation of safety and efficacy are necessary for the responsible development and translation of NP-based pharmaceuticals.

1.4 Type of Nanoparticles

NPs can be of different types depending on their size, texture, and properties. Generally, they are classified into the three basic types as shown in Figure 1.2.

1. **Organic-Based NPs:** These NPs are the most prevalent of all sorts and are frequently employed in the biomedical industry and different cancer treatments. Many lipids,

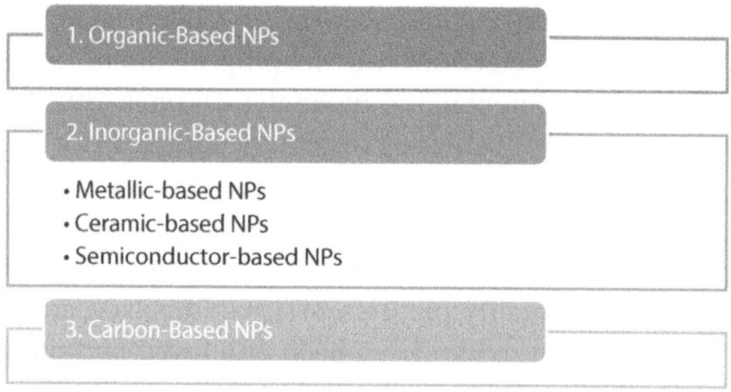

Figure 1.2 Types of NPs.

carbohydrates, proteins, and other organic molecules make up its true composition, including micelles, dendrimers, hollow-core liposomes, and protein complexes like ferritin (Figure 1.3) [36]. Although these NPs are biodegradable and non-carcinogenic, they are thermally sensitive to electromagnetic radiation and light. Moreover, they are typically made up of non-covalent intramolecular connections, which renders them easily extractable from the body and thermally labile. A few significant factors, such as size, shape, morphology, stability, and the ability to carry lipids and proteins, are crucial in determining the many types of organic-based NPs [37, 38].

2. **Inorganic-Based NPs:** These NPs are not made up of carbon or any organic particle. Metallic, ceramic, and semiconductor NPs are the three main subcategories of inorganic NPs.
   A) Metal-Based NPs: They are prepared from metals using chemical and electromechanical processes. Some metal NPs contain a unique thermal and magnetic field

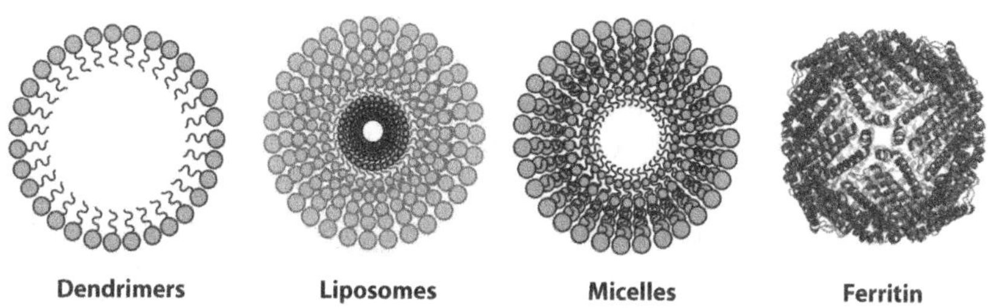

Figure 1.3 Types of organic-based NPs.

that makes them used in various fields like research fields, including the detection and imaging of biomolecules as well as environmental and bioanalytical ones [39, 40]. For instance, gold NPs are used to coat the sample prior to analysis in magnetic NPs such as Fe3O4, gold, and silver NPs for their utilization in these imaging modalities. Typically, this is done to improve the electronic stream so that we can acquire high-quality pictures [41]. Gold, silver, and magnetic NPs (iron oxide) have been employed and modified repeatedly over the years to permit their use as diagnostic and therapeutic agents [42].

B) Ceramic-Based NPs: They are composed of metal or metalloids like titanium and calcium, as well as oxides, carbonates, carbides, and phosphates [43]. They can be found in amorphous, polycrystalline, dense, porous, or hollow forms, and they are often created through heating and subsequent cooling. They exhibit strong heat and chemical resistance. Ceramic NPs demonstrate a good drug delivery agent by adjusting various properties such as size, surface area, porosity, and surface-to-volume ratio. These NPs have been utilized successfully in drug delivery systems for a variety of illnesses, including cancer, glaucoma, and bacterial infections [43].

C) Semiconductor-Based NPs: These NPs resemble numerous metals and non-metals in terms of their characteristics. They are composed of semiconductor materials such as cadmium selenide or cadmium telluride. They are used in water splitting, photo-optics, electronics, imaging, and photocatalysis [44]. However, their potential toxicity and environmental impact are also important considerations in their use and development [45].

3. **Carbon-Based NPs:** These carbon-based NPs are utilized as an alternative to steel for reinforcing constructions. Carbon nanotubes (CNTs), fullerenes, and carbon quantum dots are the three major components of carbon-based NPs (Figure 1.4). A graphene sheet that has been wrapped into a tube makes up the CNT. Because of their 100 times greater strength than steel, these materials are mostly employed for structural reinforcement. CNTs are unique in a way because they are thermally conductive along the length and non-conductive throughout the tube. Carbon

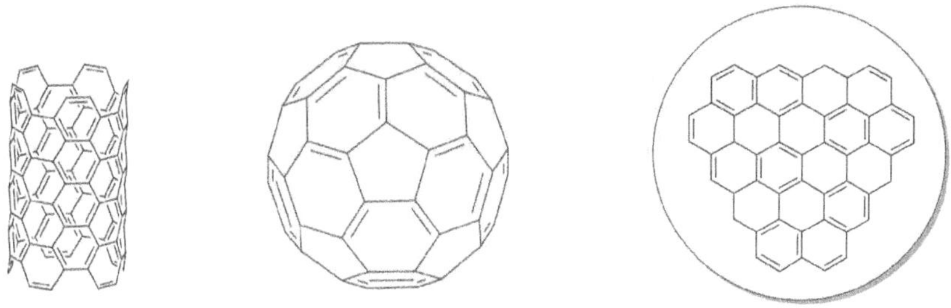

Figure 1.4 Types of carbon-based NPs.

allotropes known as fullerenes feature hollow cage structures made up of at least 60 carbon atoms. The structure of C60 is called buckminsterfullerene, and it looks like a hollow football [46]. However, other forms of fullerenes, such as C70 and C540 fullerenes, have also been made available [47]. They have commercial applications due to their electrical conductivity, structure, high strength, and electron affinity. Carbon quantum dots consist of discrete, quasi-spherical carbon NPs with sizes below 10 nm. The unique characteristics of sp2-hybridized carbon bonds are combined with exceptional physicochemical properties at the nanoscale in carbon-based NPs because of their distinctive electrical conductivity, strength, electron affinity, optical, thermal, and sorption characteristics [48].

1.5 Properties of Nanoparticles

NPs refer to natural, identical, or manufactured material comprising particles either in an unbound state or as aggregate wherein one or more external dimensions is in each size range of 1–1,000 nm [49].

The behavior of NPs is heavily influenced by their unique properties. Their morphological features, particularly their shape and size, can significantly impact their circulation and targeting within the body. Moreover, the size, shape, and distribution of NPs are closely interrelated, and even slight variations in these properties can lead to significant biological effects and impact the overall performance of the NPs [2].

1.5.1 Size

The typical range of NPs is between 1 and 100 nm. There is not a clear-cut boundary to define this cap, though. Because the psychochemical and biological properties of the components do not abruptly change at 100 nm, the largest size that a material can must be designated a nanomaterial is an arbitrary value [49].

NPs ought to circulate till they get to the desired anatomical spot. However, due to NP identification by the reticulum endothelial system (RES), the immune system can aid in their removal RES. Alternatively, NPs can be cleared through mechanical filtration by the lungs, liver, kidneys, or spleen. Regarding particle removal, size is, among other things, a crucial factor [50].

1.5.2 Shape

Particle form (sphere, ring, and disk) may cast a substantial effect on NP circulation, distribution within the body, cellular uptake, and general *in vivo* behavior, according to recent findings.

It is far more likely for particles with irregular geometry to align or tumble in vessel bifurcations or filtering organs. A spherical particle's diameter must be less than 200 nm for it to pass past the spleen.

It can, however, pass through this organ if it has a disk-like shape with a diameter of around 7 m and a height of 150 nm [50].

1.5.3 Surface Area

In comparison to micromaterials or bulk materials, nanomaterials exhibit different surface effects. This is primarily because of three factors: (a) dispersed nanomaterials have a very large surface area and a high particle number per mass unit; (b) the fraction of atoms at the surface is increased; and (c) the atoms situated at the surface in nanomaterials have fewer direct neighbors. The chemical and physical characteristics of nanomaterials differ from those of their larger-dimension counterparts as a result of each of these distinctions [51].

The reactivity of nanomaterials is typically increased by bigger surface areas and surface-to-volume ratios because of the larger reaction surface, which also has a substantial impact on the structure of the materials. The surface effects are significantly influenced by the dispersity of nanomaterials. Strongly attractive interactions between particles can cause nanomaterials to aggregate and agglomerate, which has a detrimental effect on their surface area and nanoscale characteristics.

Optimizing the degree of hydrophilicity/hydrophobicity of the nanomaterial, adjusting the pH and ionic strength of the suspension medium or boosting the zeta potential of nanomaterials can all be used to minimize agglomeration [51].

1.6 Importance of Nanoparticles

NPs have several important applications in healthcare, including the following:

1. **Drug Delivery:** NPs offer several advantages in drug delivery due to their small size, large surface area, and ability to encapsulate or conjugate with drugs [52]. One of the key benefits is their ability to target specific cells or tissues in the body, which allows for more effective therapy with reduced side effects. This is achieved through various mechanisms, such as passive targeting through the enhanced permeability and retention (EPR) effect or active targeting using ligands or antibodies on the NP surface that bind to specific receptors on the target cells [52, 53]. Additionally, NPs can protect drugs from degradation or clearance by the immune system, thereby improving their bioavailability and extending their circulation time [54]. The use of NPs in drug delivery has shown promise in the treatment of various diseases, including cancer [55], inflammatory disorders [42], and infectious diseases [56], among others.
2. **Imaging:** NPs have gained significant attention as contrast agents for various medical imaging modalities due to their unique properties [57]. They can be engineered to have specific sizes, shapes, and surface properties, allowing for targeted delivery and enhanced imaging contrast [52]. Magnetic resonance imaging (MRI) and computed tomography (CT) are two imaging techniques that have greatly benefited from the use of NPs as contrast agents [58, 59]. NPs such as iron oxide- and gadolinium-based particles have been extensively studied for their applications in MRI

[58, 60], while gold and silica NPs are commonly used in CT imaging [59]. These contrast agents offer high sensitivity and specificity, enabling the visualization of specific tissues and organs with high resolution.

3. **Diagnostics and Biomarkers:** NPs have the potential to revolutionize diagnostic testing by providing increased sensitivity, specificity, and accuracy. In diagnostic tests, NPs can be functionalized with biomolecules such as antibodies [61], aptamers [62], and peptides [61] that can recognize and bind to specific disease biomarkers. When the target biomarker is present, the NPs produce a signal that can be easily detected and quantified. This approach enables early detection of diseases and personalized treatment, which can improve patient outcomes and reduce healthcare costs.
4. **Therapeutics:** NPs can be designed to have specific properties that make them effective as therapeutic agents themselves. For example, in cancer [53] therapy, NPs can be used to deliver chemotherapy drugs directly to tumors, improving the efficacy of the treatment and reducing side effects. This is because NPs can selectively accumulate in tumor tissue due to their small size, surface properties, and EPR effect [63]. Additionally, NPs can be functionalized with targeting ligands, such as antibodies or peptides, which can selectively bind to cancer cells, further enhancing their specificity and reducing off-target effects [55]. Furthermore, NPs can be designed to have other therapeutic properties, such as photothermal or photodynamic therapy, where NPs are activated by light to generate heat or reactive oxygen species to destroy cancer cells [64]. The use of NPs as therapeutic agents promises the development of more effective and targeted cancer therapies.
5. **Vaccines:** NPs have gained significant attention as potential carriers for vaccines due to their ability to enhance the immune response and improve vaccine efficacy [65]. By encapsulating vaccine antigens within NPs, the antigens can be protected from degradation and delivered to the desired target cells or tissues, leading to enhanced immune responses. Furthermore, NPs can be designed to target specific cells or tissues, enabling a more targeted and efficient

delivery of the vaccine. This approach has been utilized in the development of NP-based vaccines for a range of infectious diseases, including influenza, human papillomavirus, and hepatitis B [66]. In addition, NP-based vaccine delivery systems can be tailored to specific patient populations, such as elderly individuals or those with compromised immune systems, to improve vaccine effectiveness [67]. Overall, the use of NPs as carriers for vaccines holds great promise for the development of more effective and efficient vaccines.

6. **Wound Healing:** NPs can be used in wound dressings to provide several benefits, including promoting healing and preventing infection [68]. For example, silver NPs have been shown to have antimicrobial properties and can be incorporated into wound dressings to prevent bacterial growth and reduce the risk of infection [69]. Additionally, NPs such as zinc oxide and titanium dioxide have been used in wound dressings to promote wound healing by stimulating cell growth and reducing inflammation [68, 70]. The use of NP-based wound dressings can lead to faster healing times, reduced scarring, and improved overall patient outcomes.
7. **Biomaterials:** NPs have unique properties that make them attractive for use in creating biocompatible materials for implants, prosthetics, and other medical devices [71]. For example, the small size of NPs allows them to penetrate cells and tissues more easily, which can improve the integration of implants and reduce the risk of rejection [72]. Additionally, NPs can be engineered to have specific surface properties that promote cell adhesion, prevent bacterial growth, or improve biocompatibility [73]. By incorporating NPs into materials such as polymers, metals, or ceramics, it is possible to create medical devices that are stronger, more durable, and more compatible with the body. Some examples of medical devices that have been developed using NPs include orthopedic implants [74], dental fillings [72], and cardiovascular stents [75]. The use of NPs in medical devices is an area of active research and holds promise for improving the safety and effectiveness of these devices.

1.7 Conclusion and Future Prospect

The field of NPs has witnessed a remarkable journey, starting from their discovery in the 1850s to the current modern developments. These tiny particles, defined as particles that are smaller than 100 nm in size, have unique properties that differ from their bulk counterparts. The history of NPs can be traced back to Michael Faraday's early works in the 1850s, where he observed the color of gold NPs. Since then, researchers have made significant progress in understanding the properties and applications of NPs. Modern developments in the field have led to the creation of various types of NPs, such as metallic, semiconductor, magnetic, and carbon-based NPs. These developments have led to the creation of innovative materials and devices with improved performance and functionality. The properties of NPs, such as their size, shape, surface area, and surface chemistry, are critical to their applications. Their unique properties, including a high surface-area-to-volume ratio, quantum confinement, and enhanced catalytic activity, make them suitable for various applications. Furthermore, the manipulation and control of these properties have led to the creation of novel materials with unique properties and applications. The importance of NPs cannot be overstated. They have found applications in various fields, including medicine, electronics, energy, and materials science. In medicine, NPs have been used for drug delivery, imaging, and cancer therapy. They have shown great promise in improving the efficacy and safety of drug delivery by targeting specific cells and tissues. The small size of NPs makes them ideal for use in nanotherapeutics. Looking toward the future aspects, the potential applications of NPs are vast, and continued research in this field promises to bring significant advancements in various areas of science and technology. One of the major challenges in the field is the safety and environmental impact of NPs. Therefore, it is essential to continue research in the field of nanotoxicology and environmental impact to minimize any potential risks associated with the use of NPs. In conclusion, NPs have been a subject of fascination for many years, and their unique properties have found applications in various fields. Continued research in this field promises to bring significant advancements in science and technology while addressing any associated risks. The field of NPs is an exciting area of research, with potential applications that are limited only by our imagination. As researchers continue to explore the properties and applications of NPs, we can look forward to a future where these tiny particles have a significant impact on our daily lives.

References

1. Pal, S.L., Jana, U., Manna, P.K., Mohanta, G.P., Manavalan, R., Nanoparticle: An overview of preparation and characterization. *J. Appl. Pharm. Sci.* 01, 228–234, 2011.
2. Barhoum, A., García-Betancourt, M.L., Jeevanandam, J., Hussien, E.A., Mekkawy, S.A., Mostafa, M. *et al.*, Review on natural, incidental, bioinspired, and engineered nanomaterials: History, definitions, classifications, synthesis, properties, market, toxicities, risks, and regulations. *Nanomaterials*, 12, 2, 177, Jan. 6, 2022.
3. Ijaz, I., Gilani, E., Nazir, A., Bukhari, A., Detail review on chemical, physical and green synthesis, classification, characterizations and applications of nanoparticles. *Green Chem. Lett. Rev.*, 13, 3, 223–45, 2020.
4. Hui, Y., Yi, X., Hou, F., Wibowo, D., Zhang, F., Zhao, D. *et al.*, Role of nanoparticle mechanical properties in cancer drug delivery. *ACS Nano*, 13, 7, 7410–24, Jul. 23, 2019.
5. Li, W., Luo, Z., Long, C., Wu, C., Duan, W.H., Shah, S.P., Effects of nanoparticle on the dynamic behaviors of recycled aggregate concrete under impact loading. *Mater. Des.*, 112, 58–66, Dec. 2016.
6. Zhu, M., Wang, R., Nie, G., Applications of nanomaterials as vaccine adjuvants. *Hum. Vaccin. Immunother.*, 10, 9, 2761–74, Sep. 2, 2014.
7. Loretan, C. and Müller, A., Nano, bits, and Feynman's dream: There's plenty of room at the (molecular) bottom. *J. Chem. Educ.*, 100, 3, 1366–70, Mar. 14, 2023.
8. Bayda, S., Adeel, M., Tuccinardi, T., Cordani, M., Rizzolio, F., The history of nanoscience and nanotechnology: From chemical–physical applications to nanomedicine. *Molecules*, 25, 1, 112, Dec. 27, 2019.
9. Freestone, I., Meeks, N., Sax, M., Higgitt, C., The lycurgus cup—A roman nanotechnology. *Gold Bull.*, 40, 4, 270–7, Dec. 2007.
10. Jeevanandam, J., Barhoum, A., Chan, Y.S., Dufresne, A., Danquah, M.K., Review on nanoparticles and nanostructured materials: History, sources, toxicity and regulations. *Beilstein J. Nanotechnol.*, 9, 1050–74, Apr. 3, 2018.
11. Reibold, M., Paufler, P., Levin, A.A., Kochmann, W., Pätzke, N., Meyer, D.C., Carbon nanotubes in an ancient Damascus sabre. *Nature*, 444, 7117, 286–286, Nov. 2006.
12. Pradell, T., Climent-Font, A., Molera, J., Zucchiatti, A., Ynsa, M.D., Roura, P. *et al.*, Metallic and nonmetallic shine in luster: An elastic ion backscattering study. *J. Appl. Phys.*, 101, 10, 103518, May 15, 2007.
13. Thompson, D., Michael Faraday's recognition of ruby gold: The birth of modern nanotechnology: His 1857 lecture to the royal society in London. *Gold Bull.*, 40, 4, 267–9, Dec. 2007.

14. Daniel, M.C. and Astruc, D., Gold nanoparticles: Assembly, supramolecular chemistry, quantum-size-related properties, and applications toward biology, catalysis, and nanotechnology. *Chem. Rev.*, 104, 1, 293–346, Jan. 1, 2004.
15. Baird, D. and Shew, A., Probing the history of scanning tunneling microscopy, *Discovering the Nanoscale*, 5, 37, IOS Press, Amsterdam, 2004.
16. Kinnear, C., Moore, T.L., Rodriguez-Lorenzo, L., Rothen-Rutishauser, B., Petri-Fink, A., Form follows function: Nanoparticle shape and its implications for nanomedicine. *Chem. Rev.*, 117, 17, 11476–521, Sep. 13, 2017.
17. Patra, J.K., Das, G., Fraceto, L.F., Campos, E.V.R., del Pillar Rodriguez-Torres, M., Acosta-Torres, L.S. *et al.*, Nano based drug delivery systems: Recent developments and future prospects. *J. Nanobiotechnol.*, 16, 1, 71, Sep. 19, 2018.
18. Jahangirian, H., Lemraski, E.G., Webster, T.J., Rafiee-Moghaddam, R., Abdollahi, Y., A review of drug delivery systems based on nanotechnology and green chemistry: Green nanomedicine. *Int. J. Nanomed.*, 12, 2957–78, Apr. 2017.
19. Mitchell, M.J., Billingsley, M.M., Haley, R.M., Wechsler, M.E., Peppas, N.A., Langer, R., Engineering precision nanoparticles for drug delivery. *Nat. Rev. Drug Discovery*, 20, 2, 101–24, Feb. 2021.
20. Sanna, V. and Sechi, M., Therapeutic potential of targeted nanoparticles and perspective on nanotherapies. *ACS Med. Chem. Lett.*, 11, 6, 1069–73, Jun. 11, 2020.
21. Guo, B., Huang, Y., Gao, Q., Zhou, Q., Stabilization of microtubules improves cognitive functions and axonal transport of mitochondria in Alzheimer's disease model mice. *Neurobiol. Aging*, 96, 223–32, Dec. 2020.
22. Sun, Z., Zhu, M., Lee, C., Progress in the triboelectric human–machine interfaces (HMIs)-moving from smart gloves to AI/haptic enabled HMI in the 5G/IoT era. *Nanoenergy Adv.*, 1, 81–120, Sep. 19, 2021.
23. Yang, J., Zhang, X., Liu, C., Wang, Z., Deng, L., Feng, C. *et al.*, Biologically modified nanoparticles as theranostic bionanomaterials. *Prog. Mater. Sci.*, 118, 100768, May 2021.
24. Chavda, V.P., Balar, P.C., Patel, S.B., Interventional nanotheranostics in hepatocellular carcinoma. *Nanotheranostics*, 7, 2, 128–41, 2023.
25. Chavda, V.P., Balar, P.C., Patel, S.B., Nanotheranostics-based management of head and neck cancer. *Nanotheranostics*, 7, 2, 202–9, 2023.
26. Anselmo AC, Mitragotri S. Nanoparticles in the clinic: An update. *Bioeng Transl. Med.*, 4: e10143, Sep. 2019.
27. Taratula, O. and Taratula, O.R., Novel nanoparticle-based treatment and imaging modalities. *Pharmaceutics*, 15, 1, 244, Jan. 11, 2023.
28. Afzal, O., Altamimi, A.S.A., Nadeem, M.S., Alzarea, S.I., Almalki, W.H., Tariq, A. *et al.*, Nanoparticles in drug delivery: From history to therapeutic applications. *Nanomaterials*, 12, 24, 4494, Dec. 19, 2022.
29. Abbasi, E., Aval, S.F., Akbarzadeh, A., Milani, M., Nasrabadi, H.T., Joo, S.W. *et al.*, Dendrimers: Synthesis, applications, and properties. *Nanoscale Res. Lett.*, 9, 1, 247, Dec. 2014.

30. Akbarzadeh, A., Rezaei-Sadabady, R., Davaran, S., Joo, S.W., Zarghami, N., Hanifehpour, Y. *et al.*, Liposome: Classification, preparation, and applications. *Nanoscale Res. Lett.*, 8, 1, 102, Dec. 2013.
31. Boisguérin, P., Konate, K., Josse, E., Vivès, E., Deshayes, S., Peptide-based nanoparticles for therapeutic nucleic acid delivery. *Biomedicines*, 9, 5, 583, May 20, 2021.
32. Zhang, C., Wu, L., de Perrot, M., Zhao, X., Carbon nanotubes: A summary of beneficial and dangerous aspects of an increasingly popular group of nanomaterials. *Front. Oncol.*, 11, 693814, Jul. 27, 2021.
33. Zielińska, A., Carreiró, F., Oliveira, A.M., Neves, A., Pires, B., Venkatesh, D.N. *et al.*, Polymeric nanoparticles: Production, characterization, toxicology and ecotoxicology. *Molecules*, 25, 16, 3731, Aug. 15, 2020.
34. Najahi-Missaoui, W., Arnold, R.D., Cummings, B.S., Safe nanoparticles: Are we there yet? *Int. J. Mol. Sci.*, 22, 1, 385, Dec. 31, 2020.
35. Sharma, A., Jain, N., Sareen, R., Nanocarriers for diagnosis and targeting of breast cancer. *BioMed. Res. Int.*, 2013, 1–10, 2013.
36. Pan, K. and Zhong, Q., Organic nanoparticles in foods: Fabrication, characterization, and utilization. *Annu. Rev. Food Sci. Technol.*, 7, 245–66, 2016.
37. Sahlan, M., Irdiani, R., Flamandita, D., Aditama, R., Alfarraj, S., Ansari, M.J. *et al.*, Molecular interaction analysis of Sulawesi propolis compounds with SARS-CoV-2 main protease as preliminary study for COVID-19 drug discovery. *J. King Saud Univ. Sci.*, 33, 1, 101234–101234, 2020/11/17 ed. Jan. 2021.
38. Gujrati, M., Malamas, A., Shin, T., Jin, E., Sun, Y., Lu, Z.R., Multifunctional cationic lipid-based nanoparticles facilitate endosomal escape and reduction-triggered cytosolic siRNA release. *Mol. Pharm.*, 11, 8, 2734–44, Aug. 4, 2014.
39. Khan, I., Saeed, K., Khan, I., Nanoparticles: Properties, applications and tox icities. *Arab. J. Chem.*, 12, 7, 908–31, Nov. 1, 2019.
40. Ealia SA, Saravanakumar MP. A review on the classification, characterisation, synthesis of nanoparticles and their application. In: *IOP Conference Series: Materials Science and Engineering*, 263, 3, 032019,. IOP Publishing, Nov. 1, 2017.
41. Mody, V.V., Siwale, R., Singh, A., Mody, H.R., Introduction to metallic nanoparticles. *J. Pharm. Bioallied Sci.*, 2, 4, 282–9, 2010.
42. Brusini, R., Varna, M., Couvreur, P., Advanced nanomedicines for the treatment of inflammatory diseases. *Adv. Drug Deliv. Rev.*, 157, 161–78, 2020.
43. Thomas, S.C., Harshita, S., Mishra, P.K., Talegaonkar, S., Ceramic nanoparticles: Fabrication methods and applications in drug delivery. *Curr. Pharm. Des.*, 21, 42, 6165–88, 2015.
44. Gupta S.M., and Tripathi M. An overview of commonly used semiconductor nanoparticles in photocatalysis. *High Energy Chem.*, 46, 1–9., Jan. 2012.
45. Sun, S., Murray, C.B., Weller, D., Folks, L., Moser, A., Monodisperse FePt nanoparticles and ferromagnetic FePt nanocrystal superlattices. *Science*, 287, 5460, 1989–92, Mar. 17, 2000.

46. Long, C.M., Nascarella, M.A., Valberg, P.A., Carbon black vs. black carbon and other airborne materials containing elemental carbon: Physical and chemical distinctions. *Environ. Pollut.*, 181, 271–86, Oct. 1, 2013.
47. Ruoff, R.S., Qian, D., Liu, W.K., Mechanical properties of carbon nanotubes: Theoretical predictions and experimental measurements. *C. R. Phys.*, 4, 9, 993–1008, Nov. 1, 2003.
48. Patel, K.D., Singh, R.K., Kim, H.W., Carbon-based nanomaterials as an emerging platform for theranostics. *Mater. Horiz.*, 6, 3, 434–69, Mar. 18, 2019.
49. Soares, S., Sousa, J., Pais, A., Vitorino, C., Nanomedicine: Principles, properties, and regulatory issues. *Front. Chem.*, 6, 360, Aug. 20, 2018.
50. Farjadian, F., Ghasemi, A., Gohari, O., Roointan, A., Karimi, M., Hamblin, M.R., Nanopharmaceuticals and nanomedicines currently on the market: Challenges and opportunities. *Nanomedicine*, 14, 1, 93–126, Jan. 2019.
51. Joudeh, N. and Linke, D., Nanoparticle classification, physicochemical properties, characterization, and applications: A comprehensive review for biologists. *J. Nanobiotechnol.*, 20, 1, 262, Jun. 7, 2022.
52. Rizvi, S.A.A. and Saleh, A.M., Applications of nanoparticle systems in drug delivery technology. *Saudi Pharm. J., Jan*, 26, 1, 64–70, 2018.
53. Bazak, R., Houri, M., El Achy, S., Hussein, W., Refaat, T., Passive targeting of nanoparticles to cancer: A comprehensive review of the literature. *Mol. Clin. Oncol.*, 2, 6, 904–8, Nov. 2014.
54. Chenthamara, D., Subramaniam, S., Ramakrishnan, S.G., Krishnaswamy, S., Essa, M.M., Lin, F.H. *et al.*, Therapeutic efficacy of nanoparticles and routes of administration. *Biomater. Res.*, 23, 20, Nov. 21, 2019.
55. Hosseini, M., Haji-Fatahaliha, M., Jadidi-Niaragh, F., Majidi, J., Yousefi, M., The use of nanoparticles as a promising therapeutic approach in cancer immunotherapy. *Artif. Cells Nanomed. Biotechnol.*, 44, 1–11, Jan. 23, 2015.
56. Zazo, H., Colino, C.I., Lanao, J.M., Current applications of nanoparticles in infectious diseases. *J. Control. Release*, 224, 86–102, Feb. 28, 2016.
57. Han, X., Xu, K., Taratula, O., Farsad, K., Applications of nanoparticles in biomedical imaging. *Nanoscale*, 11, 3, 799–819, Jan. 17, 2019.
58. Estelrich, J., Sánchez-Martín, M.J., Busquets, M.A., Nanoparticles in magnetic resonance imaging: from simple to dual contrast agents. *Int. J. Nanomedicine*, 10, 1727–41, Mar. 6, 2015.
59. Cormode, D.P., Naha, P.C., Fayad, Z.A., Nanoparticle contrast agents for computed tomography: A focus on micelles. *Contrast Media Mol. Imaging*, 9, 1, 37–52, Jan. 2014.
60. Stephen, Z.R., Kievit, F.M., Zhang, M., Magnetite nanoparticles for medical MR imaging. *Mater Today (Kidlington)*, 14, 7–8, 330–8, 2011.
61. Nune, S.K., Gunda, P., Thallapally, P.K., Lin, Y.Y., Forrest, M.L., Berkland, C.J., Nanoparticles for biomedical imaging. *Expert Opin. Drug Deliv.*, 6, 11, 1175–94, Nov. 2009.

62. Shastri, A., McGregor, L.M., Liu, Y., Harris, V., Nan, H., Mujica, M. *et al.*, An aptamer-functionalized chemomechanically modulated biomolecule catch-and-release system. *Nat. Chem.*, 7, 5, 447–54, May 2015.
63. Wu, J., The enhanced permeability and retention (EPR) effect: The significance of the concept and methods to enhance its application. *J. Pers. Med.*, 11, 8, 771, Aug. 6, 2021.
64. Shang, L., Zhou, X., Zhang, J., Shi, Y., Zhong, L., Metal nanoparticles for photodynamic therapy: A potential treatment for breast cancer. *Molecules*, 26, 21, 6532, Oct. 29, 2021.
65. Hess, K.L., Medintz, I.L., Jewell, C.M., Designing inorganic nanomaterials for vaccines and immunotherapies. *Nano Today*, 1, 27, 73–98, Aug. 2019.
66. Diaz-Arévalo D., and Zeng M. Nanoparticle-based vaccines: opportunities and limitations. In: *Nanopharmaceuticals*, 135–150. Elsevier, 2020.
67. Zhao, L., Seth, A., Wibowo, N., Zhao, C.X., Mitter, N., Yu, C. *et al.*, Nanoparticle vaccines. *Vaccine*, 32, 3, 327–37, Jan. 9, 2014.
68. Mihai, M.M., Dima, M.B., Dima, B., Holban, A.M.. Nanomaterials for wound healing and infection control. *Materials*. 12, 13, 2176, Jul. 2019.
69. Paladini F., and Pollini M., Antimicrobial silver nanoparticles for wound healing application: Progress and future trends, *Materials*, 12, 16, 2540, 2019.
70. Kushwaha, A., Goswami, L., Kim, B.S.. Nanomaterial-based therapy for wound healing. *Nanomaterials*, 12, 12, 4, 618, Feb. 2022.
71. Velu, R., Calais, T., Jayakumar, A., Raspall, F.. A comprehensive review on bio-nanomaterials for medical implants and feasibility studies on fabrication of such implants by additive manufacturing, technique. *Materials*, 23, 13, 1, 92, Dec. 2019.
72. Priyadarsini, S., Mukherjee, S., Mishra, M., Nanoparticles used in dentistry: A review. *J. Oral. Biol. Craniofac. Res.*, 8, 1, 58–67, 2018.
73. Wang, L., Hu, C., Shao, L.. The antimicrobial activity of nanoparticles: Present situation and prospects for the future, *Int. J. Nanomed.*, 14:1227–49, Feb. 2017.
74. Smith, W.R., Hudson, P.W., Ponce, B.A., Manoharan, S.R.R., Nanotechnology in orthopedics: A clinically oriented review. *BMC Musculoskelet. Disord.*, 19, 67, Mar. 2, 2018.
75. McDowell, G., Slevin, M., Krupinski, J., Nanotechnology for the treatment of coronary in stent restenosis: A clinical perspective. *Vasc. Cell*, 3, 8, Apr. 18, 2011.

2
Composition of Nanoparticles

Amit Chaudhari[1], Palak Vadodariya[2], Arya Vyas[3], Disha Patel[1] and Divya Teli[1]*

[1]Department of Pharmaceutical Chemistry, L. M. College of Pharmacy, Ahmedabad, Gujarat, India
[2]Department of Pharmacy, Marwadi University, Rajkot, Gujarat, India
[3]Department of Pharmaceutics, L. M. College of Pharmacy, Ahmedabad, Gujarat, India

Abstract
Nanoparticles (NPs) are objects with all three dimensions at the nanoscale (1–100 nm). Nanoparticle compositions can vary in their range from simple material to multiple-layered composite objects. The basic composition comprises the shell or surface layer and the core or bulk layer. The outer surface interfaces and interacts with the surrounding environment, whereas the bulk layer prepares the fundamental center of the NPs. The chemical constitution of the NPs can be analyzed in two ways—surface chemistry and bulk chemistry. Surface chemistry pertains to the chemical composition of the exterior layer of the NP, while bulk chemistry covers all the chemical compounds that make up the NPs. These characteristic methods cover several microscopy and spectroscopy techniques. Nanoparticles transmit their distinct characteristics based on the processes used in their preparation. The basic two synthetic approaches are known: top–down and bottom–up. This chapter focuses on the composition, general synthesis methods, and characterization of NPs.

***Keywords*:** Nanoparticles, composition, characterization, top–down approach, bottom–up approach

2.1 Introduction

Nanoparticles (NPs) have evolved with a wide range of unique physical and chemical properties and found promising novel types of materials

**Corresponding author*: divya.teli@lmcp.ac.in

Vivek P. Chavda and Vasso Apostolopoulos (eds.) Nanocarrier Vaccines: Biopharmaceutics-Based Fast Track Development, (25–62) © 2024 Scrivener Publishing LLC

with diverse utilizations. These materials are outlined as particles with sizes ranging from 1 to 100 nm, a high surface area-to-volume ratio, and quantum confinement lead to disparate characteristics of NPs from their bulk form [1]. The composition of NPs is a pivotal factor that affects properties and performance in various applications. The composition of NPs may be visualized as two main parts: core and shell composition. Core composition refers to the material that makes up the innermost part of the NP, while shell composition refers to the material that surrounds the core. The choice of core and shell materials can affect the size, stability, surface properties, and biocompatibility of the NP, among other properties [2]. Core materials commonly used for NPs include metals, metal oxides, and semiconductors. For example, gold NPs provide low toxicity, unique optical properties, and biocompatibility and are often utilized in the biomedical field, while iron oxide NPs are used for magnetic resonance imaging (MRI) due to their magnetic properties [3]. Shell composition is critical for controlling the surface properties and stability of the NP, as well as for determining its interactions with biological systems. Common shell materials include polymers, lipids, and surfactants. For example, polyethylene glycol (PEG) is often used to coat NPs to improve their stability and biocompatibility, while lipids show application in the targeted delivery system for encapsulating the drug [4]. Moreover, hybrid NPs, such as core–shell structures and alloys, also have a promising class of materials with special properties that can be used in diverse areas [5].

The development of novel synthesis methods has led to the production of NPs with different compositions, which can be tailored at the atomic or molecular level, leading to the emergence of new opportunities for designing and developing advanced materials with desired properties for applications in diverse fields, such as catalysis, electronics, energy, medicine, and environmental science [6]. Additionally, they can also be modified for liberating their cargo upon external factors, like change in temperature or pH that allows use in the controlled release of medicine [7]. Currently, NPs have been found as effective tools for the drug delivery of their special characteristics like a high surface area-to-volume ratio, tunable shape and size, and ability to encapsulate drugs [8]. They can protect drugs from degradation, improve their solubility and bioavailability, and enable targeted delivery to specific cells or tissues [9]. One promising application of NPs in drug delivery is the development of nanocarrier vaccines [10]. Several types of NPs have been explored for use in nanocarrier vaccines, including liposomes, polymeric NPs, and viral-like particles [11].

Within this book chapter, we impart in-depth information on types of NPs and various approaches for synthesizing them with various compositions and properties. We have also discussed the challenges and limitations associated with synthesizing NPs with complex compositions and highlighted recent advances in this field. Finally, we have described the characterization of NPs using microscopic and spectroscopic techniques.

2.2 Types of Nanoparticles

Nanoparticles can be separated into five classes depending on the raw materials used for the synthesis of NPs (Figure 2.1) [12].

Nanoparticles have a complex structure that typically includes two or three layers.

i. The outermost layer is the surface layer. This layer can be modified using various surfactants, metal ions, or polymers to achieve specific properties or functions.
ii. The second layer, the shell layer, is introduced intentionally and is chemically distinct from the core.
iii. The core substance, located in the central region of the NP, is responsible for its unique properties and features.

Together, these layers determine the behavior and performance of the NP, making them valuable tools for a wide range of applications, such as imaging, sensing, and drug delivery.

2.2.1 Polymeric Nanoparticles

Polymeric NPs include polymeric micelles, dendrimer, nanosphere, nanocapsule, polymersome, nanocomplex, and nanogel (Figure 2.2).

2.2.1.1 Polymeric Micelles

Polymeric micelles (PMs) are composed of special core configurations. the hydrophobic blocks within the amphiphilic copolymer show hydrophobic

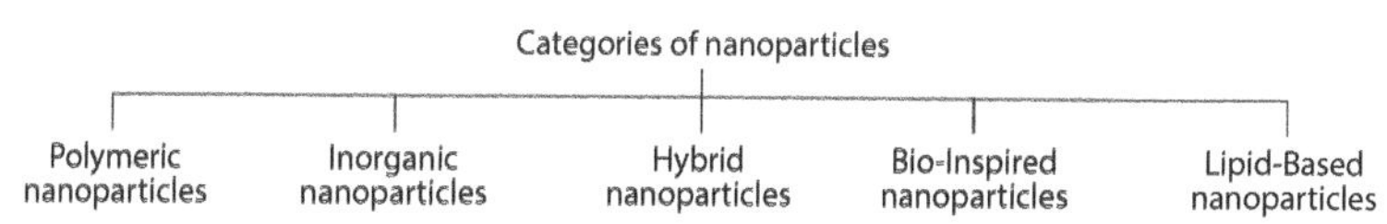

Figure 2.1 Categories of NP.

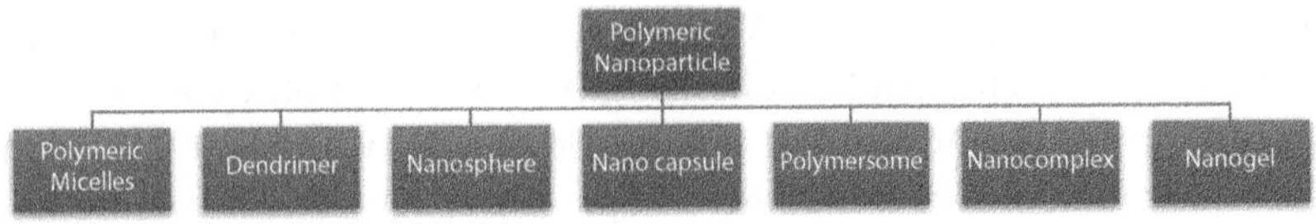

Figure 2.2 Types of polymeric NP.

interactions that drive the self-assembly of PMs, which leads to a hydrophobic core encased in a hydrophilic core [13].

Poly(ethylene oxide)-poly(propylene oxide)-poly(ethylene oxide), poly(ethylene glycol)-poly(lactic acid), poly(lactic-co glycolic acid), and poly(-caprolactone), all of which have been Food and Drug Administration (FDA)-approved for use in biomedical applications, are examples of amphiphilic copolymers frequently used to prepare micelles. PMs have also been created using various hydrophobic regions comprising biodegradable copolymers with covalent linkage with PEG. These hydrophobic segments include hydrophobic poly(l-amino acid) and derivatives, as well as polyanhydrides like poly(sebacic anhydride) (PSA). The hydrophobic drug interacts with a hydrophobic core that results in drug encapsulation into the core of PMs through hydrophobic interactions. Emulsion–solvent evaporation, direct dispersion, solvent evaporation, dialysis, film formation, and nanoprecipitation are a few of the different techniques for making micelles [14].

2.2.1.2 Dendrimer

Unimolecular NPs known as dendrimers have a clearly defined, homogenous, and monodisperse architecture that is divided into three different domains: (i) a central core made up of an atom or molecule with at least two chemical functions that are identical; (ii) branches extending from the core made up of repeating units with at least one branch junction, whose repetition is organized in a geometric progression that produces a series of radially concentric layers known as "generations"; and (iii) a number of terminal functional groups that are situated on the dendritic architecture's surface. The terminology "dendrimer" is raised from the Greek word "dendron," which stands for "tree." For dendrimer synthesis, divergent and convergent techniques are most frequently employed [12–15].

2.2.1.3 Nanosphere

The polymeric NPs known as nanospheres or matrix-type nanodevices are made up of solid spherical polymeric matrices that are extensively focused on applications in drug delivery systems. The polymeric core typically contains an even distribution of biologically active compounds, and the diffusion process releases the loaded active component into the environment [16]. In addition, the surface of nanospheres can absorb medications as their capacity for integrating with hydrophobic medicines at a concentration higher than inherent water solubility. The nanosphere has a high capacity for weak water-soluble drug loading and offers the drug release in a sustained manner. Drug release from nanospheres takes a certain amount of time, depending on the polymer matrix's composition and fluid-absorption capacity [17]. Critical elements that affect the drug-release process from the nanosphere are drug-loading ability and nanosphere size. In release studies, a rapid release rate and the main burst effect are typically seen first at increased drug loading. Additionally, larger nanospheres frequently exhibit a longer sustained release and a lower initial burst effect than smaller ones [18]. Different methods, including emulsification–evaporation and nanoprecipitation, can be used to create nanospheres [17].

2.2.1.4 Nanocapsule

Nanocapsules are referred to as nano-vesicular systems because they have a polymeric membrane around a liquid inside. The surrounding material is a polymer with single layer, with an oily liquid as the core. In building nanocapsules, polyesters like PCL, PLGA, and PLA are utilized frequently. Six traditional processes, namely, emulsion–diffusion, layer-by-layer assembly, double emulsification, emulsion–coacervation, polymer coating, and nanoprecipitation are used to create nanocapsules [19].

The most popular technique for creating polymeric nanocapsules filled with bioactive substances is nanoprecipitation. This technique also popular as interfacial deposition or solvent displacement. To ensure the purity necessary for biomedical applications, the original nanocapsule dispersions can be contaminated by solvents, salts, stabilizers, and cross-linking agents. In order to remove surfactants and solvents, many purification techniques such as diafiltration, dialysis, and cross-flow microfiltration have been extensively researched [19].

2.2.1.5 Polymersome

Due to their intertwined structure, these artificial vesicles have gained popularity in various areas. The polymersome structure contains amphiphilic block polymers with a sphere shape. A hydrophilic core has a tendency to occupy small molecules that are soluble in water. A hydrophobic membrane surrounds this core that may also contain molecules that are lipophilic [20]. By altering the copolymers' hydrophobic to hydrophilic ratios, the thickness of the membranes may be precisely controlled. High-molecular-weight copolymers are typically used in the creation of polymersomes. The macromolecular structure of the polymeric chains causes them to entangle with one another, allowing for further interactions between the copolymers. As a result, they are more stable and durable than liposomes, their natural lipid equivalent [20, 21]. Liposomes are less stable chemically and physically than polymersomes. This is largely because copolymers have greater molecular weights than lipids. Additionally, polymersomes are chemically adaptable because their physicochemical characteristics (vesicle size, membrane thickness, polarity, biodegradation, permeability, stimuli-responsiveness, and targeting capacity) can be altered by choosing the chemical composition of the polymer, the proper molecular weights, and the ratio of hydrophilic and hydrophobic blocks [22, 23].

Two key methods to prepare polymersomes are solvent switching and polymer rehydration. In order to create polymersomes using the solvent-switching method or the phase inversion method, the first step involves the dissolution of amphiphilic copolymers in an organic solvent. Next, the solution is hydrated either by water incorporation within the organic phase or by injection of the organic phase into water. Due to the insoluble state of hydrophobic blocks, the resulting increase in interfacial tension between them and the water, this technique leads copolymers to self-assemble into polymersomes. By using various organic solvents, the polymersomes' size and size distribution can be changed. In order to promote self-assembly, amphiphilic block copolymer films are hydrated in the polymer rehydration method. To create a thin film, the polymer solution is made in organic solvent, which is then evaporated. After hydrating the film with water, subsequent processes of water permeating through polymer layer defects caused by hydration forces, inflating polymer layers, and forming bulges finally produce polymersomes when separated from the surface. Desired polymersomes are created by sequentially extruding through filters with various pore diameters while applying high pressure. The polymersomes synthesized with wide size distribution and are often more stable in the bloodstream than liposomes. The aqueous core of polymersomes can be loaded with hydrophilic, hydrophobic, or amphiphilic agents, which makes them an excellent choice for medication administration, bio-imaging, and diagnostics [24].

2.2.1.6 Nanocomplex

The self-assembly of polymers and biomacromolecular drug payloads, such as protein and nucleic acids, through particular intra- and intermolecular interactions results in functional supramolecular nanocomposites called nanocomplexes. Polyplexes are nanocomplexes created by the condensation of negatively charged nucleic acids by cationic polymers. Typically, polyplex production is kinetically regulated and occurs at high ionic strengths where the interaction of the polycation and polyanion is quick and nearly irreversible, the order in which components are added during the complexation. Both the transfection effectiveness and the size of the resultant polyplex are affected by the process. Proteins can self-assemble into nanocomposites with a variety of materials using electrostatic contacts, hydrogen bonds, hydrophobic interactions, or salt bridges in addition to nucleic acid–based nanocomplexes. Additionally, the fluorination action can be used to synthesize proteins into fluoro-amphiphilic nanocomplexes [12].

2.2.1.7 Nanogel

This hydrogel nanostructure has a submicrometric size range, from 20 to 250 nm with tunable porous structure. The hydrogel can be distinguished from microgels, with the size range of 1–350 m, and *in situ*–forming hydrogels, which facilitate local delivery. The nanogel structure contains different types of polymeric material, to which proteins or molecules are encapsulated. Due to their special characteristics, nanogels can be used for imaging, diagnostics, and medication delivery. Nanogels are a type of hydrogel that continue to exhibit the hydrogel's highly hydrated state and shrinking–swelling characteristics across a variety of environmental settings. They are capable to encapsulate medications with hydrophobic or hydrophilic properties in their intercore due to their three-dimensional (3D) structure, potentially shielding these medications from deterioration while in storage or use (such as degradation by enzymes or the hydrolytic process). Nanogels have effective controlled drug release qualities because, in contrast to ordinary NPs, they have customizable shape and size with sensitivity to external stimuli. Nanogels can also be modified on their surface to increase their circulation time and make them more multifunctional and targeted. Due to the aforementioned benefits, nanogel-based drug delivery systems drug delivery systems (DDSs) have gained popularity and made a big effect recently [25]. Both polymer precursors and heterogeneous polymerization of monomers can be used to create nanogels. The fabrication of nanogels using chemical and physical cross-linking is critical. As the polymer structure includes hydrophilic groups such as -SO3H, -CONH-, -CONH2-, and -OH, nanogels can absorb a huge amount of biological fluids or water without losing their structural integrity. However, because of the presence of cross-links inside the nanogels, they exhibit swelling properties rather than dissolving. Nanogel is a good choice for many applications because of this special characteristic [26].

2.2.2 Inorganic Nanoparticle

Inorganic NPs are divided into five parts:

- Gold NP
- Silica NP
- Magnetic NP
- Quantum dots
- Carbon NP

2.2.2.1 Gold Nanoparticle

Recently, gold NPs (GNPs) have become a desirable alternative for delivering a variety of payloads to their destinations. Small medicinal molecules or huge macromolecules like proteins, DNA, or RNA could be the payloads (vide post). Effective therapy requires that these therapeutic chemicals are released effectively. Internal (such as glutathione (GSH) or pH) or external (such as light) stimuli may cause the release. Significantly, the external stimuli give spatial–temporal control over the release while the internal stimuli function in a biologically controlled manner. For the purpose of loading and unloading drugs, gold NPs take advantage of their special chemical and physical capabilities. The gold core is basically inert and non-toxic, first. The ability to create monodisperse NPs with core diameters ranging from 1 to 150 nm is another benefit. Their ready functionalization, typically through thiol bonds, confers additional versatility (vide post). Additionally, their photophysical capabilities could cause medication release at a remote location [27].

2.2.2.2 Silica Nanoparticle

Mesoporous silica NPs (MSNs) have gained significant attention over the past 10 years as a result of their unique and adaptable physiochemical characteristics. MSNs were first created by Mobil researchers in 1992 and were known as Mobil crystalline materials (MCMs). One of the most diverse and plentiful groups of materials is silica, which may be found both naturally and artificially. Fused quartz, fumed silica, silica gel, and other types of aerogels are a few examples. The Si atom exhibits tetrahedral coordination inside the vast majority of silicates, with four oxygen atoms encircling the core Si atom. The most typical instance can be seen in particular quartzite polymorphs. The size of both types of silica NPs, known as mesoporous and nanoporous, can be adjusted by altering the composition of the surfactants used in their manufacture. Due to their simple and inexpensive large-scale manufacture, hydrophilic nature, superior biocompatibility, large specific surface area, pore volume, and controllable particle size, silica NPs are widely used in a variety of applications. Additionally, silica NPs are considered to be the least hazardous and safest particles for the transport of drugs and DNA [28].

2.2.2.3 *Magnetic Nanoparticle*

A class of NPs known as magnetic NPs (MNPs) are designed particulate materials of 100 nm in size that may be controlled by an external magnetic field. Magnetic elements including iron, nickel, cobalt, and their oxides are frequently found in MNPs. MNPs have been used for MRI, targeted drug and gene delivery, tissue engineering, cell tracking, and bioseparation because of their special capacity to be steered by an external magnetic field. MNPs generate distinctive particle systems that penetrate cell and tissue barriers and provide organ-specific therapeutic and diagnostic modalities when further "functionalized" with medicines and bioactive substances, such as peptides and nucleic acids. Theragnostic is the integration of therapeutic and diagnostic technologies that aims to individualize medicine. It makes use of MNPs, which can be functionalized and simultaneously respond to a magnetic field. MNPs can function as both drug transporters and diagnostic molecular imaging agents because of multilayered functionalization. Cell replacement therapy, organ-specific gene transfer, intraoperative tumor removal surgery, and other applications have already been integrated with MNP-based MRI imaging [29].

2.2.2.4 *Quantum Dots*

Quantum dots (QDs), defined as particles with physical dimensions smaller than the exciton Bohr radius, are nanometer-scale semiconductor crystals made of elements from groups II–VI or III–V. Some of the electrons in such a semiconductor are stimulated into higher energy states when a photon of visible light strikes it. They emit a photon with a frequency unique to that substance when they return to their ground state. Due to their dimensional similarities with biological macromolecules, metal and semiconductor NPs in the size range of 2–6 nm is of great interest [30]. Although different shapes of semiconductor nanocrystals, such as rods and tetra pods, can be made, spherical QDs are the most frequently used for biological applications. The fact that many of the QD attributes, most notably the wavelength of fluorescence emission, are determined by the particle size is one of the most fascinating aspects of QDs. Fluorescence emission may be tuned across a wide wavelength range of 400–2,000 nm by changing the QD size and chemical constitution [31].

2.2.2.5 Nanocarbon

A viable alternative to conventional nanomaterials for a range of biological applications (such as sensing, imaging, and drug transport) has been developed in recent years. This is due to nanocarbon materials' low inherent toxicity, low cost of large-scale synthesis, and adaptable surface functionalization. Carbon is also one of the elements that is found in the largest abundance on earth. Graphene nanoribbons, 0D fullerenes, 1D carbon nanotubes/nano horns, 2D graphene, graphene oxides, and 3D nanodiamonds are some of the carbon allotropes that are present at the nanoscale. Graphene was the most recent allotrope to be found by Andre Geim and Kostya Novoselov in 2004 [32].

2.2.3 Hybrid Nanoparticle

- Cell membrane–coated NP
- Lipid polymer NP
- Organic–inorganic nanocomposite

2.2.3.1 Cell Membrane Coated Nanoparticle

Cell membrane coating has been introduced as a cutting-edge method that uses a straightforward top–down strategy to imbue artificial NPs with the intricate capabilities seen in biological cell membranes. They are gaining popularity due to their unique properties such as improved biocompatibility, reduced absorption by macrophages, prolonged circulation lifetimes, and improved tumor penetration. Various cell types including red blood cells, platelets, white blood cells, cancer cells, stem cells, and microbes have been used to coat synthetic NPs. Applications of these coated NPs include tumor imaging, cancer therapy, the targeting and detection of atherosclerosis, and anti-inflammation. The integrity of the cell must be broken to access the cell membranes for coating the NPs. Various techniques like extrusion and sonication are used for this top-down biomimetic process [33].

2.2.3.2 Lipid Polymer Nanoparticle

In order to capitalize on the unique qualities of polymeric NPs and liposomes that contributed to their early therapeutic effectiveness, as well as to address drawbacks like their structural disintegration, shortened circulation period, and material leakage, a new generation of delivery systems known as polymeric lipid hybrid NPs was developed. This hybrid system has the potential to be a reliable drug delivery system with suitable tissue and molecular targets, high encapsulation efficiency, and unique drug absorption. Polymeric lipid hybrid NPs have developed to address the drawbacks of both lipid and polymeric NPs. The NPs have characteristics of both polymeric and lipid particles, hence the term "hybrid." Drug loading and penetration across the membrane are improved by the lipid, whereas the polymer controls drug release. Polymeric lipid hybrid NPs can make pharmaceuticals more physically stable and biocompatible, opening the door for their potential use in effective drug delivery [34].

Three main elements make up lipid polymer NPs: (i) a core made of a hydrophobic/hydrophilic polymer that efficiently encapsulates both hydrophobic and hydrophilic medicines and produces prolonged release kinetics; (ii) a lipid shell with good biocompatibility, general stability, and decreased drug retention within the polymeric core that surrounds the core; and (iii) an exterior component made of PEG and lipid to improve steric stability, extend circulation duration, and thwart immune detection. For lipid polymer NPs, a one-step and a two-step synthetic method are often two separate options. In the one-step method, the lipid shell, lipid-PEG, drug-loaded polymeric core, and aqueous lipid solution are mixed in a single pot via nanoprecipitation and self-assembly, after dissolving either the free polymeric core or both the polymeric core and hydrophobic drug in an appropriate solvent. The two-step method involves preparing the polymeric core and lipid shell separately and then combining them to create Lipid Polymer NPs. These NPs offer various advantages over traditional drug carriers, including improved physiological properties, favourable release kinetics, and higher biocompatibility [35].

2.2.3.3 Organic–Inorganic Nanocomposite

Hybrid nanomaterials typically comprise inorganic (metal ions, metal clusters or particles, salts, oxides, sulfides, non-metallic elements and their derivatives, etc.) and organic (organic groups or molecules, ligands, biomolecules, pharmaceutical substances, polymers, etc.) components that are brought together by specific interactions, resulting in synergistic enhancements of their functional properties. These interactions range from covalent bonding and complexation to nanoscale binding and self-assembly via various intermolecular interactions such as electrostatic, dispersion, and H-bonding, as well as cooperative interactions in multiple modes. With an endless number of possible combinations, hybrid materials offer vast opportunities for creating distinct, task-specific materials with various components and structural layouts. Organic–inorganic hybrid materials can imitate organic structures, ranging from simple organic molecules to complex molecular structures and organic polymers. The inorganic component of hybrid materials can vary from monometallic species to clusters and nanosized inorganic particles up to extended phases and can be precisely controlled at the atomic level. One practical way to create novel characteristics is through the functionalization of organic polymers with inorganic components [36].

2.2.4 Bioinspired Nanoparticle

- Exosomes
- Protein NPs
- DNA nanostructure

2.2.4.1 Exosomes

Exosomes are naturally occurring nanocarriers of diverse chemicals that function as intercellular messengers to mediate both physiological and pathological processes. They are contained in a phospholipid-bilayer membrane [37]. Most eukaryotic cells secrete exosomes, a subset of extracellular vesicles (EVs), which range in size from 40 to 160 nm in diameter and have an average diameter of 100 nm. Exosomes were first recognized in the late 1980s and were once just thought of as cellular waste. Exosomes, however, constitute a novel route of intercellular communication and contribute to a wide range of biological processes in health and disease, including cancer, as individuals have recently learned with the advancement of research approaches and procedures. Exosomes' ability to convey lipids, metabolites, proteins, and nucleic acids to the target cells allows them to carry out their biological role. The significance of tumor-derived exosomes (TEXs) in cancer is becoming increasingly supported by research. Exosomes and their cargos may act as therapeutic targets, cancer prognostic indicators, or even as carriers for anticancer drugs [38]. Exosomes have been regarded as prospective drug carriers for targeted therapy because of their characteristics of natural generation, nano-scaled size, minimal immunogenicity, and capacity to overcome various biological barriers. However, because of their erratic loading efficiency, quick removal from the systemic circulation, and occasionally subpar targeting capacity, exosomal delivery systems' therapeutic potential is significantly constrained.

2.2.4.2 Protein Nanoparticle

The utilization of natural polymers for the creation of NPs has been very thoroughly reported in the literature. Due to its outstanding qualities, including biodegradability, non-antigenicity, high nutritional value, abundance of renewable sources, and exceptional binding capability of diverse medications, protein nanocarriers are being developed as GRAS (generally regarded as safe) drug delivery methods. In addition to their good functional qualities, such as emulsification, gelation, foaming, and water-binding ability, proteins also have the potential for decreased opsonization by the reticuloendothelial system (RES) through an aqueous steric barrier. Additionally, manufacturing protein NPs is simple and may be scaled up [39]. Additionally, by conjugating additional protein and carbohydrate ligands, surface modification of protein-based NPs is possible. As a result, systemic toxicity is further decreased by enabling tailored distribution to the appropriate tissue and organ. Therefore, using protein NPs for these purposes may show to be a better option for adjusting and enhancing the pharmacokinetic and pharmacodynamic characteristics of the various kinds of drug molecules [40].

2.2.4.3 DNA Nanostructure

DNA nanostructures are artificial, self-assembling nanoscale structures that were rationally created by taking advantage of nucleic acids' exact base-pairing capabilities. Since Seeman's groundbreaking work in the early 1980s, there has been a flurry of activity focused on engineering DNA nanostructures with precise topologies and geometries. The Adenine-Thymine (AT) and Guanine-Cytosine (GC) pairing self-recognition rule in DNA is straightforward and exact, making it simple to hold together several double-helical domains by switching out DNA backbones. Building DNA nanoarchitectures with well-defined sizes and forms, flexible single-stranded DNA (ssDNA), and rigid double-stranded DNA (dsDNA) is based on the structural characteristics of nucleic acids (dsDNA). Additionally, precise control over the flexibility and stress of DNA nanostructures provides the door for the engineering of complicated geometries with twists and bends. Recently, the quickly developing DNA origami technology has opened up new possibilities for building intricate DNA structures. It would be able to create sophisticated nanostructures in one, two, and three dimensions with almost any shape by using DNA origami nanostructures as modular building blocks to create periodic and aperiodic super nanostructures. DNA nanostructures can easily be functionalized with a wide range of chemicals and NPs that are addressable on the nanoscale scale. Thus, artificial functional systems can be created by employing a DNA nanostructured framework with a variety of embellishments [41].

2.2.5 Lipid-Based Nanoparticle

- Liposomes
- Lipoplex
- Solid Lipid NP

2.2.5.1 Liposome

They are small, spherical synthetic vesicles that can be created using cholesterol and safe, natural phospholipids. The hydrophobic and hydrophilic properties of liposomes, in addition to their biocompatibility, make them ideal drug delivery platforms. With respect to lipid composition, surface charge, size, and manufacturing process, liposome properties vary greatly. Additionally, the 'rigidity' or 'fluidity' and the charge of the bilayer are determined by the components used in the bilayer. As spherical vesicles with particle sizes ranging from 30 nm to several micrometers, liposomes are typically identifiable. They are made up of aqueous units surrounded by one or more lipid bilayers, with the polar head groups orientated in the direction of the interior and outer aqueous phases. However, polar lipids can self-assemble into a variety of colloidal particles in addition to the usual bilayer forms that depend on the molecule shape, temperature, and environmental and preparation circumstances. In the pharmaceutical and cosmetic industries, liposomes are widely employed as molecules' carriers. The utilization of liposome encapsulation to develop delivery methods that can entrap unstable substances (such as antimicrobials, antioxidants, flavors, and bioactive components) and safeguard their activity has also been thoroughly researched by the food and farming industries. Both hydrophobic and hydrophilic molecules can be captured by liposomes, which can also prevent the combination's breakdown and release the trapped substances at specific destinations. Liposomes have gained popularity both as an exploratory system and commercially as a drug delivery method due to their biocompatibility, biodegradability, low toxicity, ability to capture both hydrophilic and lipophilic medicines, and capacity to simplify site-specific drug administration to tumor tissues. Many methods are used for the manufacturing of liposomes such as sonication, French pressure cell extrusion, freeze-thawed liposomes, lipid film hydration by hand shaking, non-hand shaking, and freeze drying [42].

2.2.5.2 Lipoplex

Cationic liposomes (CLs) are positively charged, nearly spherical vesicles that are primarily made of cationic lipids. Through electrostatic interactions, CLs effectively condense nucleic acids such as plasmid DNA (pDNA), messenger RNA (mRNA), or small interfering RNA (siRNA) to form complexes known as "lipoplexes" [43].

2.2.5.3 Solid Lipid Nanoparticle

In 1991, solid lipid NPs (SLNs) were developed as a colloidal carrier of nanoscopic size (50–1000 nm) composed of solid lipids (high melting fat matrix) to offer biocompatibility and storage stability while preventing integrated medicine degradation. SLNs are being developed to overcome the drawbacks of conventional colloidal carriers like polymeric NPs and liposomes, such as polymer degradation and cytotoxicity, the lack of a large-scale production method, inferior stability, drug leakage and fusion, phospholipid degradation, high production costs, and sterilization issues. SLNs have a number of unique properties, including reduced toxicity, high surface area, delayed drug release, enhanced cellular absorption, and increased solubility and bioavailability of pharmaceuticals. SLNs made from biodegradable and biocompatible materials can contain both hydrophilic and lipophilic bioactives, making them an effective choice for regulated and targeted drug delivery. The drug is typically distributed or dissolved in the solid core of SLNs, which has a monolayer coating of phospholipids and is hydrophobic, and the type of matrix and position of the medication in the formulation affects drug release from SLNs [44].

2.3 Composition of Nanoparticles

Various polymers are used in preparation of NPs. Polymers can be natural polymers or synthetic polymers. The polymers ought to be biodegradable, biocompatible, and adaptable to the body (i.e., non-toxic, non-antigenic, and non-toxic). Natural polymers include chitosan, albumin, sodium alginate, etc., and synthetic polymers include polylactides (PLAs), polyglycolides (PGAs), poly(lactide co-glycolides) (PLGAs), poly(vinyl alcohol), poly(acrylic acid), poly(ethylene glycol), poly(ethylene glycol), poly(methyl methacrylate), and poly(methacrylic acid) [45].

2.3.1 Chitosan

Chitosan is a molecule having a cellulose-like carbohydrate backbone structure made up of two repeating units, *N*-acetyl-D-glucosamine and D-glucosamine, connected by (1–4) glycosidic linkages. Obtained by alkaline deacetylation of chitin, chitosan has numerous amino groups along its chain, giving it distinctive chemical properties. While chitin is relatively inert, chitosan is highly soluble in diluted acetic, citric, and tartaric acid solutions but insoluble in phosphoric or sulfuric acid at pH levels below 6.5. Chitosan is available in low and high molecular weights ranging from 3,800 to 20,000 Da, with varying degrees of deacetylation. Its properties, especially when creating micro- and NPs, are significantly influenced by its molecular weight and degree of deacetylation. Chitosan is widely used as a polymeric drug carrier due to its biocompatibility, biodegradability, and low toxicity and has also shown some therapeutic activities such as reducing cholesterol, promoting wound healing, and exhibiting anti-ulcer and antibacterial properties. Pharmaceutical companies have successfully employed chitosan-based drug carriers to deliver a range of pharmaceuticals, including anticancer medicines, proteins/peptides, growth factors, antibiotics, anti-inflammatory drugs, and as a technique in vaccine delivery and gene therapy. Chitosan's mucoadhesive nature and ease of chemical modification make it a suitable option for medicinal formulations via a variety of administration routes [46].

Aside from being non-toxic, mucoadhesive, hemocompatible, and biodegradable, chitosan also exhibits good biological qualities such as anticancer, antioxidant, and antibacterial capabilities. These characteristics make chitosan a very appealing biomaterial for various biological applications. Due to the presence of active hydroxyl and amino groups, chitosan can undergo various chemical processes such as alkylation, carboxylation, hydroxylation, and esterification. These processes introduce dangling groups into the chitosan, causing it to lose its crystal structure and thus become more soluble. These chitosan derivatives are better suitable for usage as carriers in biomedical industry thanks to their increased physicochemical and biological properties [47].

Ohya and colleagues published the first description of chitosan NPs (ChNP) in 1994. They utilized ChNP made by emulsifying and cross-linking to administer the antitumor medication 5-fluorouracil intravenously [48]. Large amounts of lone-pair electrons are present in chitosan-based NPs, which also show a strong affinity for materials with vacant orbitals. They are utilized in biosensors, fractionated pictures, medication delivery, and gene delivery. The consistency and particle size of the produced microspheres serve as the foundation for this function of ChNPs. The quantity of antigen adsorption and distribution is influenced by particle size, and this impacts the immune response. The function of microspheres is influenced by their structural characteristics, the size of their surface micropores, and the rate at which they release antigen. Chitosan NPs were produced by chitosan solution coating, cross-linking, solvent evaporation, spray drying, or precipitation [48].

2.3.2 Albumin

It is the most abundant plasma protein, comprising 35–50 g/L of human serum and having a molecular weight of 66.5 kDa. With low immunogenicity, biocompatibility, and biodegradability, albumin is thought to be harmless. The most crucial function is that it transports different substances in the blood plasma. Aside from supplying nutrients to the perivascular tissue, albumin breakdown can release amino acids. These characteristics make human serum albumin (HSA) and bovine serum albumin (BSA), the two albumin species that have been the subject of the vast majority of research, an excellent material for creating NPs for drug delivery [49]. Different proteins, including albumin and gelatin, can be used to create NPs. Among these, albumin shows promise as a material for making NPs, and numerous experiments have been done to determine how well albumin works as a carrier. The FDA authorized Abraxane1, the first human serum albumin-based NPs containing paclitaxel, in 2005. According to reports, the new formulation increases the drug's solubility and provides a number of advantages over the traditional paclitaxel therapy [50].

A considerable amount of medicine can be absorbed into the particle matrix, thanks to the various drug binding sites found in albumin, making albumin-based NP carrier systems an appealing method. Albumin-based

NPs could enable the electrostatic adsorption of positively (such as ganciclovir) or negatively (such as oligonucleotide) charged molecules without the need for additional compounds because of the defined albumin primary structure and high content of charged amino acids (such as lysine). Additionally, coacervation, controlled desolvation, or emulsion creation are simple methods for producing albumin NPs in soft circumstances. They are smaller (50–300 nm) than microparticles and, in general, have more controlled release characteristics than liposomes, which may increase patient compliance and acceptance. Commercially, albumins can be generated in large amounts from soy, milk, and cereals as well as from egg white (ovalbumin), bovine serum (bovine serum albumin, BSA), and human serum (human serum albumin, HSA) [51]. Numerous albumin-based NPs have been created in the lab over the years. The production procedure, chemical make-up, and characteristics of the various albumin NPs varied greatly. Depending on whether the medicine is covalently bonded to the albumin and if it has been cross-linked into a matrix or gel, these albumin NPs can be divided into various classes. Albumin NPs are typically formed via cross-linking with glutaraldehyde or other agents, resulting in a cross-linked albumin matrix that can either chemically or physically entrap the drug payload [52].

2.3.3 Polylactic Acid

It has shown to be a remarkably adaptable material with intriguing qualities including biocompatibility and biodegradability. Polylactic acid (PLA) has therefore been researched for a variety of therapeutic uses, such as a nanoparticulate antigen and a drug delivery vehicle [53]. It is an aliphatic polyester connected by ester linkages between its monomer units and has gained significant importance in the biomedical sector due to its diverse applications such as suture threads, bone fixation screws, and drug delivery devices. It has various intriguing qualities that make it a great choice for biological applications. The degradation of PLA occurs through hydrolysis, wherein water molecules break the ester linkages forming the polymer's backbone, resulting in the natural breakdown of PLA. This reduces the need for additional procedures to remove the device, enhancing patient rehabilitation and optimizing healthcare expenses. Currently, the primary mechanisms of degradation and the key variables affecting the rate of hydrolysis are well understood. Thus, by carefully adjusting a few polymer parameters, degradation rates and mechanical qualities can be customized, leading to the creation of biomedical devices that are best suited for each individual application. Because lactic acid and its short oligomers are recognized and digested by the body, degradation products from PLA have an innate biocompatibility that inhibits the development of important immune responses. Additionally, PLA can be processed using tried-and-true techniques like injection molding and extrusion. These characteristics make PLA a viable alternative for the emerging paradigm of nanomedicine, which contends that nanomaterials—whose size is comparable to biologically interesting substances like proteins or viruses—are dispersed throughout the body and can be ingested by cells. The creation of NPs, which may be used for the targeted delivery of medications, vaccines, and genes, has opened up hitherto unimaginable possibilities for nanomedicine [54].

For the preparation of PLA-based NPs, numerous approaches are helpful. Four categories are used to group the techniques. Traditional emulsion-based techniques falling under category 1 include single, double, and multiple emulsions. Category 2 includes procedures that rely on precipitation, such as nanoprecipitation, fast supercritical fluid expansion into liquid, salting out, and dialysis. Category 3 includes direct compositing techniques like spray drying, melting, and *in situ* microparticle formation. New techniques in category 4 include template/mold-based methods and microfluidic methods. The medication encapsulation method affects additional criteria. The medicine is either disseminated in polymer matrices or is confined inside "capsule" particles [55]. They can be prepared using different ways, with the most popular being the emulsion solvent evaporation method. In preclinical cancer models, PLA NPs have been engaged to deliver a variety of antitumor hydrophobic medications. Temozolomide was delivered to rodent glioma cells *in vitro* using PLA NPs. The outcomes show that temozolomide's antitumor effect is maintained when given using NPs. The use of PLA NPs in the treatment of leukemia has been researched. It has use in intranasal and transdermal drug delivery [56].

2.3.4 Polylactide-co-glycolide (PLGA)

Polylactide-co-glycolide is one of the typical polymers used to create NPs. Typically, biodegradable polymers break down into oligomers and individual monomers, which the body metabolizes and excretes through regular processes. The physical characteristics of the polymer can be used to precisely control drug release kinetics and degradation. One of the most widely used degradable polymers is PLGA. Due to PLGA's exceptional biocompatibility, biodegradability, and mechanical strength, it has attracted a lot of interest. Both a polycondensation process and the ring-opening polymerization of cyclic diesters can be used to create PLGA. In order to synthesize PLGA and polylactic acid, ring-opening polymerization is now favored due to its quicker reaction times and greater rates of monomer conversion. Since then, a variety of pharmaceutical and biological applications have made use of these polymers. These polymers have been used to create a variety of polymeric devices, including microspheres, microcapsules, NPs, pellets, implants, and films. Additionally, they are simple to construct into different delivery systems for a variety of active agents, including vaccines, peptides, proteins, and micromolecules, some of which are currently approved by the Food and Drug Administration for use in drug administration. Methods for preparation of PLGA NPs include emulsification solvent evaporation, emulsification solvent diffusion, emulsification reverse salting-out, and nanoprecipitation. Various anticancer drugs are incorporated with PLGA NPs like paclitaxel, docetaxel, cisplatin, and doxorubicin [57].

2.3.5 Polyacrylate

Since 1843, when acrylic acid was first produced, its esters have been known to polymerize into polyacrylates. Depending on the cross-linking agent used in the manufacture, they can have varying degrees of hydrophilicity and are bio-inert, non-degradable synthetic polymers. Acrylates have a fundamental chemical structure, and a variety of polymers can be created by combining various monomers. By combining two or more polyacrylates to meet the needs of a particular application, different chemical moieties can be substituted to create a variety of polyacrylate polymers with tailored physical properties. Examples are polyacrylic acid, polymethacrylic acid, poly methyl acrylate, and poly methyl methacrylate. The dispersion of prepared polymer and the polymerization of monomers are two general approaches that can be used to prepare or manufacture polyacrylate NPs. In the first category, there are processes like nanoprecipitation, emulsification, dialysis, solvent diffusion, salting out, and phase inversion temperature, while in the second category, there are processes like conventional, micro/mini-emulsion, interfacial polymerization, and controlled radical polymerization. Additionally, various cutting-edge techniques, like nanospray drying and supercritical fluid technologies, have been applied to create polyacrylate NPs. The application of polyacrylate NP is in lung cancer, colorectal cancer, stomach (gastric) cancer, liver cancer (hepatocellular carcinoma), and breast Cancer. Various drug like erlotinib, curcumin, cisplatin, paclitaxel, doxorubicin, and idarubicin are used with this polymer. Polyacrylates are homopolymers or copolymers with USFDA GRAS status. Examples include sodium polyacrylates, basic methacrylate, methyl methacrylate- (MMA-), and others. Their transformation into different copolymeric forms and NPs, however, might be hazardous and restrict their usage in medication administration. It has been claimed that polyacrylate aerosols with 30-nm-sized NPs produced fibrosis, non-specific inflammation, and foreign body granulomas in the pleura. Although there is conflicting evidence in the literature on polyacrylate NP toxicity, these particles are being used to transport a number of therapeutic medicines with promising efficacy, which undoubtedly offers enormous promise for the treatment of cancer [58].

2.4 Synthesis of Nanoparticles

Nanotechnology arises with novel applications, and simultaneously several affords are made for a synthesizing variety of NPs with variable sizes and shapes. There are two types of approaches utilized in NP synthesis from ancient times. One is the breakdown (top–down) approach where an external force is applied to starting material that results in fragmentation and finally smaller particle formation. Another is the buildup (bottom–up) approach that utilizes atomic-level liquid or gas as starting material, and after the process, is transformed into NPs. A schematic representation is shown in Figure 2.3.

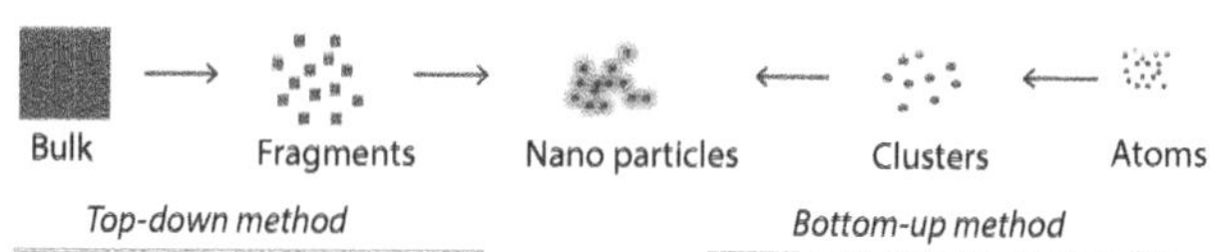

Figure 2.3 Approaches for NP synthesis.

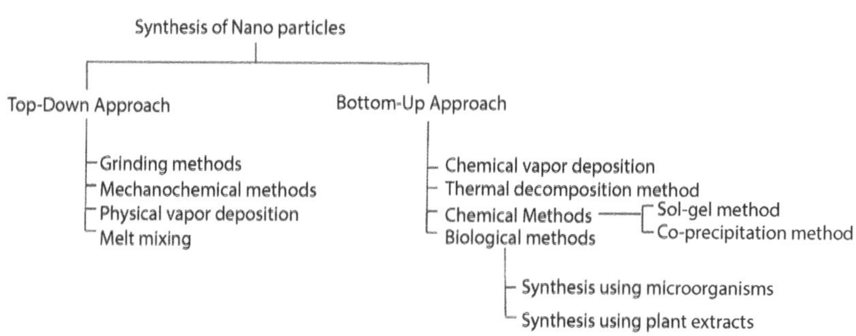

Figure 2.4 Methods for the synthesis of NPs.

NPs are utilized in devised areas such as biotechnology, healthcare, electronics, and optics. The need for effective, durable, and safe non-material is again in focus in synthetic techniques. Breakdown and buildup approaches are further classified based on their process techniques where widely applied methods are summarized in Figure 2.4.

2.4.1 Top–Down Approach

The method begins with larger in size material that is reduced to NPs through operational sequences performed over them. The method is a destructive type of synthesis with high consistency and well control in the NP size and shape, suitable for industrial large-scale production [59]. Demerits include required expensive types of equipment and uncontrolled particle growth, and the impurity may enter during abrasion with milling elements. Thus, this method has a limit where a high degree of purity is required.

2.4.1.1 Ball Milling

Different varieties of mills such as tumbler, vibratory, planetary, and rod mills and widely utilized for large-scale production. The ball mill consists of a blank cylindrical container that occupies dancer metallic balls made up of steel or carbide. The principle behind this method is attrition and impact (Figure 2.5). This method is based on the grinding of bulk; it is not preferred for soft starting materials. Non-crystalline particles of cobalt, gold, chromium, and iron are synthesized. The method operated with the 2:1 ratio of balls to the material. The container is loaded with material and

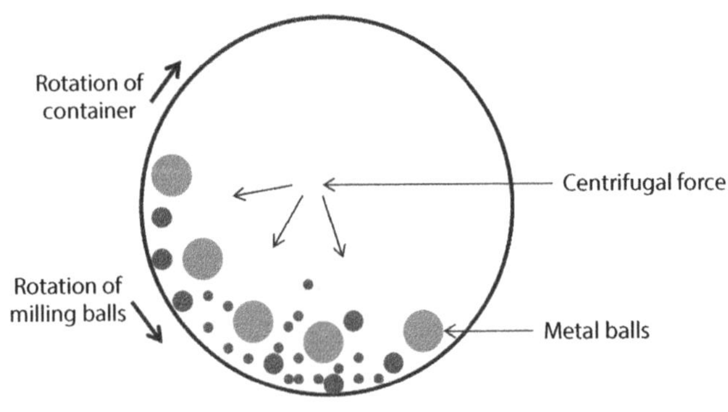

Figure 2.5 Schematic representation of the ball milling process.

metal balls and inert gas that rotated around the central axis. During this process, materials are pressed between the container wall and the balls, which reduces the size of the materials. The frequency of rotation and duration of the process influence synthesizing NPs of optimum size. Nano berberine powder with 570.7 nm average size uniformity was prepared by the ball milling method [60]. Planetary ball milling was utilized to enhance maximum plasma concentration apigenin NP preparation [61].

2.4.1.2 *Physical Vapor Deposition (PVD)*

This method comprises conversion of the solid or liquid state material into a vapor state through evaporation; this vapor is then converted to fine NPs by rapid cooling. The arc discharge method is also utilized to facilitate the evaporation of starting material. This method has popularity in the synthesis of thin film polymer [62]. PVD is applied to modify the surface properties by vapor deposition without change in the physicochemical change in the substrate. Coating materials like TiN, CrN, and TiAlCN are efficiently deposited with PVD to prepare a nano-coating on medical devices, which improves hardness and resistance toward corrosion [63].

2.4.1.3 *Melt Mixing*

The melt mixing method includes mixing molten materials with solvent from pre-emulsion and finally, on cooling, converting them into NPs. graphene oxide–silica nanohybrids with different grafting rates were prepared using this method [64]. Raj Kumar *et al.* enhanced the bioavailability of pharmaceutical compounds that are poorly water-soluble through lipid NPs synthesized by melt mixing [65]. Similarly, lipid NPs loaded with fenofibrate were also synthesized using this method [66].

2.4.1.4 *Pulse Laser Ablation*

Synthesis of NPs facilitated by laser. The high-pulsed laser is targeted toward the material placed inside liquid media or vacuum chamber that modifies the material into a plasma state and finally into an NP mixture [67]. The second-harmonic laser and multiple-pulse laser are widely utilized for NP production. Through the alteration of the sub-pulse number, the average size of NPs can be manipulated in wider ranges from 120 to 4 nm [68]. Antonino *et al.* have applied this method for the synthesis of diverse NPs of platinum, palladium, and copper hydroxide [69].

2.4.2 Bottom-Up Approach

This approach is the constructive method that starts with the atomic level and, through specific steps, converts it into nanostructures [70]. Since procedures involve a solvent system, these methods are also denoted as "wet" methods. The approach is principally based on self-assembly where the physical forces acted on the nanoscale that unit atomic particles into stable nanostructures. This approach is widely applied at the laboratory scale, and compared to the top–down approach, it is an energy-efficient, cost-effective, and rapid method.

2.4.2.1 Chemical Vapor Deposition (CVD)

This method comprises the deposition of a thin film of vaporized coating on the substrate at around 300–1200°C. Vapor will chemically interact with the heated substrate, which results in solid phase coating on the surface [71]. Modified chemical vapor deposition methods are thermally activated CVD, plasma-enhanced CVD, and photoinduced CVD [72]; these synthetic methods are used for the production of carbon nanotubes (CNTs) like carbon sources, catalysts, and substrates with more emphasis on CNT catalysts [73]. The requirement of special apparatus and the formation of highly toxic gaseous as by-products are disadvantages of the CVD method [74]. CVD coating with strong chemical bonding provides more durability under a wide range of conditions. The growth of high-quality transition metal dichalcogenide monolayers of molybdenum chemicals is applied on the substrate using CVD [75].

2.4.2.2 Thermal Decomposition Method

This method is exerted with an endothermic phenomenon where chemical decomposition is produced on heating. The chemical bonding within the material is broken at high temperatures. This is the simple method preferred in the production of metal oxides like iron-oxide magnetic NPs and other types of inorganic NPs like cobalt NPs. This method is applied for the synthesis of mono-dispersed metal NPs that are widely used to reduce the side effects of anticancer drugs [76]. This method synthesizes uniform-shaped and -sized NPs with diverse magnetic, optical, electronic, and catalytic properties [77]. Patsula *et al.* utilized the same method for the production of superparamagnetic Fe_3O_4 NPs coated with PMG-P that are used in MRI [78].

2.4.2.3 Chemical Methods

a) Sol–gel method

The *sol–gel* method comprising the hydrolysis of material that forms the stable stage of the mixture is produced and referred to as sol. Further processes carried out are condensation, drying processes, and heating of the metal substrate in a liquid medium that resulted in high-quality NPs as well as mixed oxide composites (Figure 2.6). Silica has emerged as an excellent competitor as a nanocarrier in nanomedicine and is synthesized using the sol–gel method [79]. A diverse variety of metal NPs such as TiO_2, ZnO, SnO_2, and WO_3 that have photocatalysis and biomedical applications are efficiently synthesized using this method [80].

b) Co-precipitation method

The co-precipitation method or the solvent displacement method comprises a wet chemical operation, where the synthesis of nanomaterial is conducted through the rapid flow of solvent containing polymer through a pure phase of polymer and lastly the mixing of the whole solution. Varieties in the shape and size of nanomaterial are obtained by manipulating interfacial tension between them [81]. Nanoparticles of iron oxide are widely utilized in cancer diagnosis and treatment therapy, synthesized using an aqueous mixture of ferrous and ferric salts, and then treated with the base at an environmental or higher temperature that gives iron hydroxide intermediates that, on forming a precipitate, produce iron oxide NPs [82].

2.4.2.4 Biological Methods

a) Synthesis using microorganisms

In recent years, the synthesis of NPs using microorganisms has gained more attention due to its cost-effectiveness and eco-friendliness. It can be classified into intracellular and extracellular synthesis according to the location where NPs are formed. Bacteria are easy to genetically mold

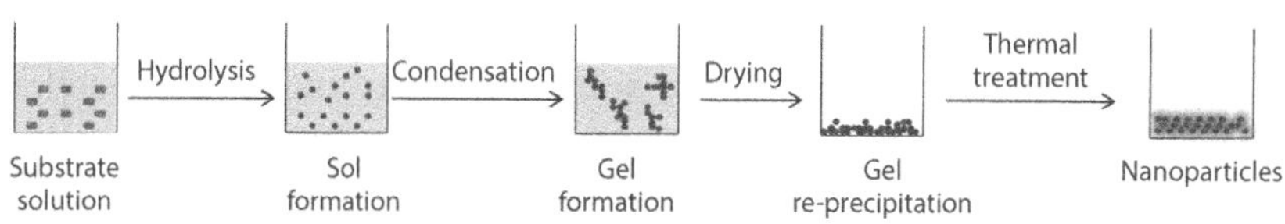

Figure 2.6 Step sequence of the sol–gel method for the NP synthesis.

and manipulate for metal ion biomineralization as compared to other microbes. Due to environmental heavy metal pollution, bacteria have developed some natural defense mechanisms such as extracellular precipitation, intracellular sequestration, and change in metal ion concentration; these mechanisms have been fruitfully utilized by bacteria for NP synthesis [83]. Inorganic NPs including metallic NPs, oxide NPs, sulfide NPs, and other typical NPs are biosynthesized using this method [84].

b) Synthesis using plant extracts

The utilization of plants is a green and cost-effective way of NP synthesis. This type of technique in nanotechnology explores a new way of environment sustainability development. Plant extracts provide a vital character for NP synthesis. Different plant organs are used for NP syntheses such as the leaf, stem, root, and seed; however, exact mechanisms that play a role in synthesis remain to be elucidated. Various plant materials such as terpenoids, alkaloids, flavonoids, plant proteins, and vitamins provide a crucial role in the synthesis of NPs. The method is typically one-pot synthesis, also recognized as green synthesis or a green process of manufacture of NPs [85]. Silver NPs are synthesized with wide application in medical science as antimicrobial and antibacterial properties [86]. Other NPs include gold, silver, copper, palladium, platinum, zinc oxide, and titanium dioxide produced with plant-mediated synthesis [87].

2.5 Nanoparticle Characterization by Various Instrumental Techniques

Characterization is an important technique that helps to explore the material's structure and composition and determine whether the method was successful. These techniques include both qualitative as well as quantitative methods. They have been discussed in the sections below.

2.5.1 Dynamic Light Scattering (DLS)

This technique has wider utilization for the characterization of NP size in colloidal solutions in nanometer to submicrometer ranges. These colloidal solutions are in Brownian motion where DLS measures the scattered light by the particles using Stokes–Einstein assumptions to determine the hydrodynamic diameter of NPs. It is quick, easy to operate, highly sensitive, and reproducible for homogenous, monodisperse samples. The drawback is that the particle has to be in suspension and undergo continuous Brownian motion [88, 89].

2.5.2 Zeta Potential

NPs will have a charge on their surface and upon the application of electrical field, the particles start to migrate as the result of the charge and the electrical field interaction. Zeta potential is raised due to the difference between oppositely charged ions' potential within the fluid layer and the fluid with the suspended particles. Zeta potential is related to the stability of the particles. The higher zeta potential represents more electrostatic repulsion leading to more stabilization as well as minimal aggregation [90, 91].

2.5.3 Microscopic Techniques to Characterize Nanoparticles

2.5.3.1 Scanning Electron Microscopy (SEM)

In this technique, the high-energy electron beam is bombarded on sample surface where electron beam interacts with the sample atoms to generate a signal in the form of secondary electrons or back-scattered electrons. These electrons create X-rays that reveal the morphology and composition of the sample. The 3D detailed image can be obtained with high resolution up to ×300,000, and these images are black and white as they are produced without light waves. In SEM, the NP-containing solution must be converted to dry powder and the non-conductive materials are coated with conductive materials before it is mounted on a sample holder [92, 93].

2.5.3.2 Transmission Electron Microscopy (TEM)

In this microscopic method, the electron beam is imparted through the sample and a 2D image is formed using these transmitted electrons. For characterization, the ideal thickness of the specimen is <100 nm. Also, local heating of the sample by absorbed kinetic energy can cause structural damage, which can be overcome by cryo-TEM as it analyzes the sample in the frozen-hydrated state leading to minimal modification of the sample. The information about the dispersion and distribution of NPs in matrices of fiber, nanocomposites, and nanocoating can be obtained by TEM images [94, 95].

2.5.4 Spectroscopic Techniques to Characterize Nanoparticles

2.5.4.1 Ultraviolet-Visible Spectroscopy (UV-Vis)

This type of spectroscopy carried out using UV-visible spectrophotometer mainly consists of a light source, sample and reference beams, a monochromator, and a detector. The beam that travels from the light source to the detector with no interaction with the sample is the reference beam. The beam that interacts with the sample in presence of ultraviolet light of continuously changing wavelength is the sample beam. The energy is absorbed by the sample when the emitted wavelength is similar to the energy level that promotes electrons to a higher-energy molecular orbital. The difference in the intensities of the sample and reference is recorded by the detector. The characterization of NPs by UV-Vis can give information about optical properties, concentration, size, agglomeration, and hints on the shape of NPs [96, 97].

2.5.4.2 Raman Spectroscopy

This spectroscopy is carried out with Raman or inelastic scattering of monochromatic light and is used to study rotational, vibrational, and other low-frequency modes in a system. In Raman spectroscopy, a laser beam illuminates the sample in which molecules get excited from the ground state to the excited state and are relaxed into a vibrational excited state and creates Stokes Raman scattering. If the molecule is already in the vibrational state, then it is known as anti-Stokes–Raman scattering. Raman spectroscopy helps to study the chemical bonding of drugs and polymers as the vibrational data are unique to every bonding within the molecule. Stimulated Raman spectroscopy, surface-enhanced Raman spectroscopy, and resonance Raman spectroscopy are some advanced versions of Raman spectroscopy [98, 99].

2.5.4.3 Fourier Transform Infrared Spectroscopy (FTIR)

This method also termed "vibrational spectroscopy" as it causes the vibrational transition in the molecule by the absorption of infrared (IR) radiations. The IR region extends from 800 to 1,000 μm, where most studies are confined to the middle IR region (2–15 μm) due to the high absorption of organic molecules in this region. When the applied IR frequency is equal to the natural frequency, IR radiation is absorbed and recorded as a peak in the spectra. Every functional group absorbs specific frequencies of IR giving characteristic peak values. FTIR is applied to determine the structural properties of NPs and to investigate the molecular interaction between drugs and the polymer. This method also utilized for the identification of the existing functional groups and molecules, to prove the loading and coating of the drug on the nanocarriers as well as to ensure the formation of bindings [100, 101].

2.5.5 X-Ray Diffraction Method (XRD)

X-rays are electromagnetic radiation having a shorter wavelength and are produced when electrically charged particles with sufficient energy are decelerated. These X-rays are collimated and directed to the sample where the diffracted X-rays are detected by the detector and a spectrograph is obtained. Most of the scattering is eliminated as destructive interference. Constructive interference is seen when the waves are moving in phase with each other. The angle of diffraction is measured by Bragg's equation. XRD data provide information about crystallinity, the orientation of the crystallites, crystallite size, and the phase composition in semi-crystalline polymers. PNPs and polymer-layered silicate composites can also be characterized by XRD with appropriate accessories [102, 103].

2.6 Understanding Nanotoxicity: Potential Risks and Implications

Because of their smaller size, NPs have unique characteristics that make them appealing for a variety of uses, including drug delivery, electronics, and environmental remediation. The expanded use and production of NPs, on the other hand, raise concerns about their possible toxicity to human health and the environment. Understanding the potential risks and implications of NP toxicity is essential to ensure the safety of human health and the environment. The toxicity profile of NPs can be determined by several factors, including their size, shape, surface charge, surface coating, and solubility, which can influence their ability to penetrate cells and tissues and cause damage [104]. Depending on their properties and interactions with cells and tissues, different types of NPs can have variable toxicities. Metal NPs such as silver, gold, and metal oxide have been shown to cause toxicity by damaging DNA, producing oxidative stress, and causing inflammation [105]. Carbon-based NPs, such as carbon nanotubes and fullerenes, have been observed to cause lung damage by inducing inflammation and fibrosis [106]. Similarly, polymeric NPs used for drug delivery and imaging applications were also found to induce inflammation, especially when they accumulate in organs and tissues. Susceptible groups, such as the diseased, neonates, pregnant, and aged populations, may be more easily affected by nanomaterial exposure [107]. The mechanism of nanotoxicity involves the interaction of NPs with cells and their components. Once NPs enter the body, they can penetrate cells and reach organelles, including the nucleus. Nanoparticles can induce oxidative stress, which can lead to damage to cellular components such as DNA, lipids, and proteins. They can also interfere with cellular signaling pathways, disrupting normal cellular functions. In addition, NPs can elicit an immune response, which can cause inflammation and tissue damage [108]. Therefore, it is essential to evaluate the toxicity of NPs before their wide-scale utilization in various fields, including medicine, energy, and consumer goods. Additionally, preventive strategies, such as using protective equipment and adopting safe handling practices, should be implemented to minimize the risk of exposure to NPs in occupational and environmental settings.

2.7 Conclusion

NPs are a class of nanomaterials that have attracted significant attentiveness in various fields, including biomedical engineering, materials science, and environmental science, among others. These materials can be synthesized with a spacious range of compositions and properties, including size, shape, and surface charge, and can be engineered to carry drugs, genes, and other biomolecules. In this book chapter, we have provided an extensive overview of the different types of NPs and approaches for synthesizing them with various compositions and properties. We have also discussed recent advances in this field and described the challenges and limitations associated with synthesizing nanostructures with complex compositions. Furthermore, we have highlighted the importance of characterizing nanoscale particles using microscopic and spectroscopic techniques. As research in this field continues to advance, NPs are poised to become an increasingly important tool in various applications, particularly in drug delivery and vaccine development. Although the toxicity of NPs should not be ignored, it is crucial to continue the research and implement safety measures to minimize the hazardous effects associated with NPs.

References

1. Khan, I., Saeed, K., Khan, I., Nanoparticles: Properties, applications and toxicities. *Arab. J. Chem.*, 12, 908–931, 2019.

2. Pelaz, B., Del Pino, P., Maffre, P., Hartmann, R., Gallego, M., Rivera-Fernández, S., De La Fuente, J.M., Nienhaus, G.U., Parak, W.J., Surface functionalization of nanoparticles with polyethylene glycol: Effects on protein adsorption and cellular uptake. *ACS Nano*, 9, 6996–7008, 2015. doi:10.1021/acsnano.5b01326.
3. Dreaden, E.C., Alkilany, A.M., Huang, X., Murphy, C.J., El-Sayed, M.A., The golden age: Gold nanoparticles for biomedicine. *Chem. Soc. Rev.*, 41, 2740–2779, 2012. doi:10.1039/c1cs15237h.
4. Tamburro, D., Fredolini, C., Espina, V., Douglas, T.A., Ranganathan, A., Ilag, L., Zhou, W., Russo, P., Espina, B.H., Muto, G. *et al.*, Multifunctional core-shell nanoparticles: Discovery of previously invisible biomarkers. *J. Am. Chem. Soc.*, 133, 19178–19188, 2011. doi:10.1021/ja207515j.
5. Sailor, M.J. and Park, J.H., Hybrid nanoparticles for detection and treatment of cancer. *Adv. Mater.*, 24, 3779–3802, 2012. doi:10.1002/adma.201200653.
6. Salari, N., Faraji, F., Torghabeh, F.M., Faraji, F., Mansouri, K., Abam, F., Shohaimi, S., Akbari, H., Mohammadi, M., Polymer-based drug delivery systems for anticancer drugs: A systematic review. *Cancer Treat. Res. Commun.*, 32, 100605, 2022. doi:10.1016/j.ctarc.2022.100605.
7. Palanikumar, L., Al-Hosani, S., Kalmouni, M., Nguyen, V.P., Ali, L., Pasricha, R., Barrera, F.N., Magzoub, M., PH-responsive high stability polymeric nanoparticles for targeted delivery of anticancer therapeutics. *Commun. Biol.*, 3, 95, 2020. doi:10.1038/s42003-020-0817-4.
8. Fraceto, L.F., Campos, E.V.R., Habtemariam, S., del Pillar Rodriguez-Torres, M., Patra, J.K., Grillo, R., Acosta-Torres, L.S., Sharma, S., Shin, H.-S., Diaz-Torres, L.A. *et al.*, Nano based drug delivery systems: Recent developments and future prospects. *J. Nanobiotechnology*, 16, 1–33, 2018.
9. Mitchell, M.J., Billingsley, M.M., Haley, R.M., Wechsler, M.E., Peppas, N.A., Langer, R., Engineering precision nanoparticles for drug delivery. *Nat. Rev. Drug Discovery*, 20, 101–124, 2021. doi:10.1038/s41573-020-0090-8.
10. Petkar, K.C., Patil, S.M., Chavhan, S.S., Kaneko, K., Sawant, K.K., Kunda, N.K., Saleem, I.Y., An overview of nanocarrier-based adjuvants for vaccine delivery. *Pharmaceutics*, 13, 455, 2021. doi:10.3390/pharmaceutics13040455.
11. Guerrini, G., Magrì, D., Gioria, S., Medaglini, D., Calzolai, L., Characterization of nanoparticles-based vaccines for COVID-19. *Nat. Nanotechnol.*, 17, 570–576, 2022. doi:10.1038/s41565-022-01129-w.
12. Yin, L. and Zhong, Z., Nanoparticles, in: *Biomaterials Science*, pp. 453–483, 2020.
13. Kataoka, K., Harada, A., Nagasaki, Y., Block copolymer micelles for drug delivery: Design, characterization and biological significance. *Adv. Drug Deliv. Rev.*, 47, 113–131, 2001. doi:10.1016/S0169-409X(00)00124-1.
14. Cagel, M., Tesan, F.C., Bernabeu, E., Salgueiro, M.J., Zubillaga, M.B., Moretton, M.A., Chiappetta, D.A., Polymeric mixed micelles as nanomedicines: Achievements and perspectives. *Eur. J. Pharm. Biopharm.*, 113, 211–228, 2017. doi:10.1016/j.ejpb.2016.12.019.
15. Boas, U. and Heegaard, P.M.H., Dendrimers in drug research. *Chem. Soc. Rev.*, 33, 43, 2004. doi:10.1039/b309043b.

16. Steichen, S.D., Caldorera-Moore, M., Peppas, N.A., A review of current nanoparticle and targeting moieties for the delivery of cancer therapeutics. *Eur. J. Pharm. Sci.*, 48, 416–427, 2013. doi:10.1016/j.ejps.2012.12.006.
17. García, M.C., Nano-and microparticles as drug carriers, in: *Engineering Drug Delivery Systems*, pp. 71–110, Elsevier, Woodhead Publishing, Inc., Sawston, Cambridge, 2020.
18. Nicolas, J., Mura, S., Brambilla, D., Mackiewicz, N., Couvreur, P., Design, functionalization strategies and biomedical applications of targeted biodegradable/biocompatible polymer-based nanocarriers for drug delivery. *Chem. Soc. Rev.*, 42, 1147–1235, 2013. doi:10.1039/C2CS35265F.
19. Mora-Huertas, C.E., Fessi, H., Elaissari, A., Polymer-based nanocapsules for drug delivery. *Int. J. Pharm.*, 385, 113–142, 2010. doi:10.1016/j.ijpharm.2009.10.018.
20. Bleul, R., Thiermann, R., Maskos, M., Techniques to control polymersome size. *Macromolecules*, 48, 7396–7409, 2015. doi:10.1021/acs.macromol.5b01500.
21. Thambi, T., Park, J.H., Lee, D.S., Stimuli-responsive polymersomes for cancer therapy. *Biomater. Sci.*, 4, 55–69, 2016. doi:10.1039/C5BM00268K.
22. Palivan, C.G., Goers, R., Najer, A., Zhang, X., Car, A., Meier, W., Bioinspired polymer vesicles and membranes for biological and medical applications. *Chem. Soc. Rev.*, 45, 377–411, 2016. doi:10.1039/C5CS00569H.
23. Guan, L., Rizzello, L., Battaglia, G., Polymersomes and their applications in cancer delivery and therapy. *Nanomedicine*, 10, 2757–2780, 2015. doi:10.2217/nnm.15.110.
24. Lee, J.S. and Feijen, J., Polymersomes for drug delivery: Design, formation and characterization. *J. Control. Release*, 161, 473–483, 2012. doi:10.1016/j.jconrel.2011.10.005.
25. Yin, Y., Hu, B., Yuan, X., Cai, L., Gao, H., Yang, Q., Nanogel: A versatile nano-delivery system for biomedical applications. *Pharmaceutics*, 12, 290, 2020. doi:10.3390/pharmaceutics12030290.
26. Zhang, H., Zhai, Y., Wang, J., Zhai, G., New progress and prospects: The application of nanogel in drug delivery. *Mater. Sci. Eng. C*, 60, 560–568, 2016. doi:10.1016/j.msec.2015.11.041.
27. Ghosh, P., Han, G., De, M., Kim, C., Rotello, V., Gold nanoparticles in delivery applications☆. *Adv. Drug Deliv. Rev.*, 60, 1307–1315, 2008. doi:10.1016/j.addr.2008.03.016.
28. Jeelani, P.G., Mulay, P., Venkat, R., Ramalingam, C., Multifaceted application of silica nanoparticles. A review. *Silicon*, 12, 1337–1354, 2020. doi:10.1007/s12633-019-00229-y.
29. Shubayev, V.I., Pisanic, T.R., Jin, S., Magnetic nanoparticles for theragnostics. *Adv. Drug Deliv. Rev.*, 61, 467–477, 2009. doi:10.1016/j.addr.2009.03.007.
30. Jamieson, T., Bakhshi, R., Petrova, D., Pocock, R., Imani, M., Seifalian, A.M., Biological applications of quantum dots. *Biomaterials*, 28, 4717–4732, 2007. doi:10.1016/j.biomaterials.2007.07.014.

31. Bailey, R.E., Smith, A.M., Nie, S., Quantum dots in biology and medicine. *Phys. E Low-Dimens. Syst. Nanostructures*, 25, 1–12, 2004. doi:10.1016/j.physe.2004.07.013.
32. Panwar, N., Soehartono, A.M., Chan, K.K., Zeng, S., Xu, G., Qu, J., Coquet, P., Yong, K.-T., Chen, X., Nanocarbons for biology and medicine: Sensing, imaging, and drug delivery. *Chem. Rev.*, 119, 9559–9656, 2019. doi:10.1021/acs.chemrev.9b00099.
33. Liu, L., Bai, X., Martikainen, M.-V., Kårlund, A., Roponen, M., Xu, W., Hu, G., Tasciotti, E., Lehto, V.-P., Cell membrane coating integrity affects the internalization mechanism of biomimetic nanoparticles. *Nat. Commun.*, 12, 5726, 2021. doi:10.1038/s41467-021-26052-x.
34. Sivadasan, D., Sultan, M.H., Madkhali, O., Almoshari, Y., Thangavel, N., Polymeric lipid hybrid nanoparticles (PLNs) as emerging drug delivery platform—A comprehensive review of their properties, preparation methods, and therapeutic applications. *Pharmaceutics*, 13, 1291, 2021. doi:10.3390/pharmaceutics13081291.
35. Mohanty, A., Uthaman, S., Park, I.-K., Utilization of polymer-lipid hybrid nanoparticles for targeted anti-cancer therapy. *Molecules*, 25, 4377, 2020. doi:10.3390/molecules25194377.
36. Ananikov, V.P., Organic–inorganic hybrid nanomaterials. *Nanomaterials*, 9, 1197, 2019. doi:10.3390/nano9091197.
37. Chen, H., Wang, L., Zeng, X., Schwarz, H., Nanda, H.S., Peng, X., Zhou, Y., Exosomes, a new star for targeted delivery. *Front. Cell Dev. Biol.*, 9, 751079, 2021. doi:10.3389/fcell.2021.751079.
38. Dai, J., Su, Y., Zhong, S., Cong, L., Liu, B., Yang, J., Tao, Y., He, Z., Chen, C., Jiang, Y., Exosomes: Key players in cancer and potential therapeutic strategy. *Signal Transduction Targeting Ther.*, 5, 145, 2020. doi:10.1038/s41392-020-00261-0.
39. Elzoghby, A.O., Samy, W.M., Elgindy, N.A., Protein-based nanocarriers as promising drug and gene delivery systems. *J. Control. Release*, 161, 38–49, 2012. doi:10.1016/j.jconrel.2012.04.036.
40. Jain, A., Singh, S.K., Arya, S.K., Kundu, S.C., Kapoor, S., Protein nanoparticles: Promising platforms for drug delivery applications. *ACS Biomater. Sci. Eng.*, 4, 3939–3961, 2018. doi:10.1021/acsbiomaterials.8b01098.
41. Pei, H., Zuo, X., Zhu, D., Huang, Q., Fan, C., Functional DNA nanostructures for theranostic applications. *Acc. Chem. Res.*, 47, 550–559, 2014. doi:10.1021/ar400195t.
42. Akbarzadeh, A., Rezaei-Sadabady, R., Davaran, S., Joo, S.W., Zarghami, N., Hanifehpour, Y., Samiei, M., Kouhi, M., Nejati-Koshki, K., Liposome: Classification, preparation, and applications. *Nanoscale Res. Lett.*, 8, 102, 2013. doi:10.1186/1556-276X-8-102.
43. Liu, C., Zhang, L., Zhu, W., Guo, R., Sun, H., Chen, X., Deng, N., Barriers and strategies of cationic liposomes for cancer gene therapy. *Mol. Ther. Methods Clin. Dev.*, 18, 751–764, 2020. doi:10.1016/j.omtm.2020.07.015.

44. Mishra, V., Bansal, K., Verma, A., Yadav, N., Thakur, S., Sudhakar, K., Rosenholm, J., Solid lipid nanoparticles: Emerging colloidal nano drug delivery systems. *Pharmaceutics*, 10, 191, 2018. doi:10.3390/pharmaceutics10040191.
45. Nagavarma, B.V.N., Yadav, H.K.S., Ayaz, A., Vasudha, L.S., Shivakumar, H.G., Different techniques for preparation of polymeric nanoparticles-A review. *Asian J. Pharm. Clin. Res.*, 5, 16–23, 2012.
46. Ahmed, T. and Aljaeid, B., Preparation, characterization, and potential application of chitosan, chitosan derivatives, and chitosan metal nanoparticles in pharmaceutical drug delivery. *Drug Des. Devel. Ther.*, 10, 483, 2016. doi:10.2147/DDDT.S99651.
47. Zhao, D., Yu, S., Sun, B., Gao, S., Guo, S., Zhao, K., Biomedical applications of chitosan and its derivative nanoparticles. *Polymers*, 10, 462, 2018. doi:10.3390/polym10040462.
48. Divya, K. and Jisha, M.S., Chitosan nanoparticles preparation and applications. *Environ. Chem. Lett.*, 16, 101–112, 2018. doi:10.1007/s10311-017-0670-y.
49. Jiang, Y. and Stenzel, M., Drug delivery vehicles based on albumin-polymer conjugates. *Macromol. Biosci.*, 16, 791–802, 2016. doi:10.1002/mabi.201500453.
50. Wilson, B., Lavanya, Y., Priyadarshini, S.R.B., Ramasamy, M., Jenita, J.L., Albumin nanoparticles for the delivery of gabapentin: Preparation, characterization and pharmacodynamic studies. *Int. J. Pharm.*, 473, 73–79, 2014. doi:10.1016/j.ijpharm.2014.05.056.
51. Elzoghby, A.O., Samy, W.M., Elgindy, N.A., Albumin-based nanoparticles as potential controlled release drug delivery systems. *J. Control. Release*, 157, 168–182, 2012. doi:10.1016/j.jconrel.2011.07.031.
52. Desai, N., Albumin-drug nanoparticles, in: *Drug Delivery Oncology*, pp. 1133–1161, 2011.
53. Da Silva, J., Jesus, S., Bernardi, N., Colaço, M., Borges, O., Poly (D,L-lactic acid) nanoparticle size reduction increases its immunotoxicity. *Front. Bioeng. Biotechnol.*, 7, 137, 2019. doi:10.3389/fbioe.2019.00137.
54. Casalini, T., Rossi, F., Castrovinci, A., Perale, G., A perspective on polylactic acid-based polymers use for nanoparticles synthesis and applications. *Front. Bioeng. Biotechnol.*, 7, 259, 2019. doi:10.3389/fbioe.2019.00259.
55. Lee, B.K., Yun, Y., Park, K., PLA micro-and nano-particles. *Adv. Drug Deliv. Rev.*, 107, 176–191, 2016. doi:10.1016/j.addr.2016.05.020.
56. Tyler, B., Gullotti, D., Mangraviti, A., Utsuki, T., Brem, H., Polylactic acid (PLA) controlled delivery carriers for biomedical applications. *Adv. Drug Deliv. Rev.*, 107, 163–175, 2016. doi:10.1016/j.addr.2016.06.018.
57. Rafiei, P. and Haddadi, A., A robust systematic design: Optimization and preparation of polymeric nanoparticles of PLGA for docetaxel intravenous delivery. *Mater. Sci. Eng. C*, 104, 109950, 2019. doi:10.1016/j.msec.2019.109950.

58. Petkar, K.C., Polyacrylate nanoparticles as a promising tool for anticancer therapeutics, in: *Polymeric Nanoparticles as a Promising Tool for Anticancer Therapeutics*, pp. 35–56, Elsevier, Academic Press, Inc., Cambridge, Massachusetts, 2019.
59. Fu, X., Cai, J., Zhang, X., Li, W.D., Ge, H., Hu, Y., Top-down fabrication of shape-controlled, monodisperse nanoparticles for biomedical applications. *Adv. Drug Deliv. Rev.*, 132, 169–187, 2018. doi:10.1016/j.addr.2018.07.006.
60. Tran, T.V.A., Dao, V.H., Pham, Q.T., Ha, T.H., Nguyen, T.M.D., Nguyen, V.T., Ngo, T.S., Trinh, T.D., Nguyen, T.H., Nguyen, V.K., Preparation of berberine chloride nanoparticles by ball milling method. *VNU J. Sci. Med. Pharm. Sci.*, 38, 44–53, 2022. doi:10.25073/2588-1132/vnumps.4437.
61. Alshehri, A., Ibrahim, M., Alshehri, S., Alshora, D., Elzayat, E., Meanazel, O., Alsaadi, B., Sherbiny, G., Osman, S., Optimization of apigenin nanoparticles prepared by planetary ball milling: *In vitro* and *in vivo* studies. *Green Process. Synth.*, 12, 20228107, 2023. doi:10.1515/gps-2022-8107.
62. Wang, Y., Jeong, H., Chowdhury, M., Arnold, C. B., Priestley, R. D., Exploiting physical vapor deposition for morphological control in semi-crystalline polymer films. *Polym. Cryst.*, 1, e10021, 2018. doi.org/10.1002/pcr2.10021.
63. Geyao, L., Yang, D., Wanglin, C., Chengyong, W., Development and application of physical vapor deposited coatings for medical devices: A review. *Procedia CIRP*, 89, 250–262, 2020. doi:10.1016/j.procir.2020.05.149.
64. Jiang, W., Sun, C., Zhang, Y., Xie, Z., Zhou, J., Kang, J., Cao, Y., Xiang, M., Preparation of well-dispersed graphene oxide-silica nanohybrids/poly (lactic acid) composites by melt mixing. *Polym. Test.*, 118, 107912, 2023. doi:10.1016/j.polymertesting.2022.107912.
65. Kumar, R., Singh, A., Garg, N., Acoustic cavitation assisted hot melt mixing technique for solid lipid nanoparticles formulation, characterization, and controlled delivery of poorly water soluble drugs. *J. Drug Deliv. Sci. Technol.*, 54, 101277, 2019. doi:10.1016/j.jddst.2019.101277.
66. Kumar, R., Singh, A., Sharma, K., Dhasmana, D., Garg, N., Siril, P.F., Preparation, characterization and *in vitro* cytotoxicity of fenofibrate and nabumetone loaded solid lipid nanoparticles. *Mater. Sci. Eng. C*, 106, 110184, 2020. doi:10.1016/j.msec.2019.110184.
67. Zhang, D., Gökce, B., Barcikowski, S., Laser synthesis and processing of colloids: Fundamentals and applications. *Chem. Rev.*, 117, 3990–4103, 2017. doi:10.1021/acs.chemrev.6b00468.
68. Yu, J., Nan, J., Zeng, H., Size control of nanoparticles by multiple-pulse laser ablation. *Appl. Surf. Sci.*, 402, 330–335, 2017. doi:10.1016/j.apsusc.2017.01.094.
69. Scandurra, A., Iacono, V., Boscarino, S., Scalese, S., Grimaldi, M.G., Ruffino, F., Alkaline electro-sorption of hydrogen onto nanoparticles of Pt, Pd, Pt 80 Pd 20 and Cu (OH) 2 obtained by pulsed laser ablation. *Nanomaterials*, 13, 561, 2023.

70. De Oliveira, P.F.M., Torresi, R.M., Emmerling, F., Camargo, P.H.C., Challenges and opportunities in the bottom-up mechanochemical synthesis of noble metal nanoparticles. *J. Mater. Chem. A*, 8, 16114–16141, 2020. doi:10.1039/d0ta05183g.
71. Bhaviripudi, S., Mile, E., Steiner, S.A., Zare, A.T., Dresselhaus, M.S., Belcher, A.M., Kong, J., CVD synthesis of single-walled carbon nanotubes from gold nanoparticle catalysts. *J. Am. Chem. Soc.*, 129, 1516–1517, 2007. doi:10.1021/ja0673332.
72. Zhao, X., Wei, C., Gai, Z., Yu, S., Ren, X., Chemical vapor deposition and its application in surface modification of nanoparticles. *Chem. Pap.*, 74, 767–778, 2020. doi:10.1007/s11696-019-00963-y.
73. Shah, K.A. and Tali, B.A., Synthesis of carbon nanotubes by catalytic chemical vapour deposition: A review on carbon sources, catalysts and substrates. *Mater. Sci. Semicond. Process.*, 41, 67–82, 2016. doi:10.1016/j.mssp.2015.08.013.
74. Adachi, M., Tsukui, S., Okuyama, K., Nanoparticle synthesis by ionizing source gas in chemical vapor deposition. *Jpn. J. Appl. Phys.*, 42, L77, 2003. doi:10.1143/JJAP.42.L77.
75. Aras, F.G., Yilmaz, A., Tasdelen, H.G., Ozden, A., Ay, F., Perkgoz, N.K., Yeltik, A., A review on recent advances of chemical vapor deposition technique for monolayer transition metal dichalcogenides (MX2: Mo, W; S, Se, Te). *Mater. Sci. Semicond. Process.*, *148*, 106829, 2022. doi:10.1016/j.mssp.2022.106829.
76. Odularu, A.T., Metal nanoparticles: Thermal decomposition, biomedicinal applications to cancer treatment, and future perspectives. *Bioinorg. Chem. Appl.*, 2018, 9354708, 2018. doi:10.1155/2018/9354708.
77. Varanda, L.C., Souza, C.G.S., Moraes, D.A., Neves, H.R., Souza Junior, J.B., Silva, M.F., Bini, R.A., Albers, R.F., Silva, T.L., Beck, W., Size and shape controlled nanomaterials based on modified polyol and thermal decomposition approaches. A brief review. *An. Acad. Bras. Cienc.*, 91, 1–32 2019, 10.1590/0001-3765201920181180
78. Patsula, V., Kosinová, L., Lovrić, M., Ferhatovic Hamzić, L., Rabyk, M., Konefal, R., Paruzel, A., Šlouf, M., Herynek, V., Gajović, S. *et al.*, Superparamagnetic Fe3O4 nanoparticles: Synthesis by thermal decomposition of iron (III) glucuronate and application in magnetic resonance imaging. *ACS Appl. Mater. Interfaces*, 8, 7238–7247, 2016. doi:10.1021/acsami.5b12720.
79. Gonçalves, M.C., Sol-gel silica nanoparticles in medicine: A natural choice. Design, synthesis and products. *Molecules*, 23, 1–26, 2018. doi:10.3390/molecules23082021.
80. Parashar, M., Shukla, V.K., Singh, R., Metal oxides nanoparticles via sol–gel method: A review on synthesis, characterization and applications. *J. Mater. Sci. Mater. Electron.*, 31, 3729–3749, 2020. doi:10.1007/s10854-020-02994-8.
81. Sun, Z., Wu, B., Ren, Y., Wang, Z., Zhao, C.X., Hai, M., Weitz, D.A., Chen, D., Diverse particle carriers prepared by co-precipitation and phase separation: Formation and applications. *ChemPlusChem*, 86, 49–58, 2021. doi:10.1002/cplu.202000497.

82. Hernández-Hernández, A.A., Aguirre-Álvarez, G., Cariño-Cortés, R., Mendoza-Huizar, L.H., Jiménez-Alvarado, R., Iron oxide nanoparticles: synthesis, functionalization, and applications in diagnosis and treatment of Cancer. *Chem. Pap.*, 74, 3809–3824, 2020. doi:10.1007/s11696-020-01229-8.
83. Jadoun, S., Chauhan, N.P.S., Zarrintaj, P., Barani, M., Varma, R.S., Chinnam, S., Rahdar, A., Synthesis of nanoparticles using microorganisms and their applications: A review, *Environ. Chem. Lett.*, 5, 3153–3197, 2022.
84. Li, X., Xu, H., Chen, Z.S., Chen, G., Biosynthesis of nanoparticles by microorganisms and their applications. *J. Nanomater.*, 1–16, 2011, doi:10.1155/2011/270974.
85. Sajjad, S., Leghari, S.A.K., Ryma, N.U.A., Farooqi, S.A., *Green Synthesis of Metal-Based Nanoparticles and Their Applications*, ISBN 9781119418900, Scrivener Publishing, Inc.: Massachusetts, United States, 2018.
86. Bano, S., Agrawal, M., Pal, D., A review on green synthesis of silver nanoparticle through plant extract and its medicinal applications. *J. Indian Chem. Soc.*, 97, 983–988, 2020.
87. Jadoun, S., Arif, R., Jangid, N.K., Meena, R.K., Green synthesis of nanoparticles using plant extracts: A review. *Environ. Chem. Lett.*, 19, 355–374, 2021. doi:10.1007/s10311-020-01074-x.
88. Fischer, K. and Schmidt, M., Pitfalls and novel applications of particle sizing by dynamic light scattering. *Biomaterials*, 98, 79–91, 2016. doi:10.1016/j.biomaterials.2016.05.003.
89. Ramos, A.P., Dynamic light scattering applied to nanoparticle characterization, in: *Nanocharacterization Techniques*, pp. 99–110, Elsevier, William Andrew Publishing, Inc. Norwich, New York, 2017.
90. Honary, S. and Zahir, F., Effect of zeta potential on the properties of nanodrug delivery systems-a review (part 1). *Trop. J. Pharm. Res.*, 12, 255–264, 2013. doi:10.4314/tjpr.v12i2.19.
91. Xu, R., Progress in nanoparticles characterization: Sizing and zeta potential measurement. *Particuology*, 6, 112–115, 2008. doi:10.1016/j.partic.2007.12.002.
92. Mohan, A.C. and Renjanadevi, B., Preparation of zinc oxide nanoparticles and its characterization using scanning electron microscopy (SEM) and x-ray diffraction(XRD). *Procedia Technol.*, 24, 761–766, 2016. doi:10.1016/j.protcy.2016.05.078.
93. Pal, S.L., Jana, U., Manna, P.K., Mohanta, G.P., Manavalan, R., Nanoparticle: An overview of preparation and characterization. *J. Appl. Pharm. Sci.*, 01, 228–234, 2011.
94. Anjum, D.H., Characterization of nanomaterials with transmission electron microscopy. *IOP Conf. Ser. Mater. Sci. Eng.*, 146, 1–29, 2016. doi:10.1088/1757-899X/146/1/012001.
95. Walther, T., Transmission electron microscopy of nanostructures, in: *Microscopy Methods in Nanomaterials Characterization*, pp. 105–134, 2017, doi:10.1016/B978-0-323-46141-2.00004-3.

96. Begum, R., Farooqi, Z.H., Naseem, K., Ali, F., Batool, M., Xiao, J., Irfan, A., Applications of UV/Vis spectroscopy in characterization and catalytic activity of noble metal nanoparticles fabricated in responsive polymer microgels: A review. *Crit. Rev. Anal. Chem.*, 48, 503–516, 2018. doi:10.1080/10408347.2018.1451299.
97. Quevedo, A.C., Guggenheim, E., Briffa, S.M., Adams, J., Lofts, S., Kwak, M., Lee, T.G., Johnston, C., Wagner, S., Holbrook, T.R. *et al.*, UV-Vis spectroscopic characterization of nanomaterials in aqueous media. *J. Vis. Exp.*, 176, 1–11, 2021. doi:10.3791/61764.
98. Gouadec, G. and Colomban, P., Raman spectroscopy of nanomaterials: How spectra relate to disorder, particle size and mechanical properties. *Prog. Cryst. Growth Charact. Mater.*, 53, 1–56, 2007. doi:10.1016/j.pcrysgrow.2007.01.001.
99. Kumar, D., Malleswara Rao, D., AppaRao, D., Rama Mohan, J., Ramakrishna Rao, P., Sundar Singh, C., Ram Narayana, D., Raman spectroscopy in nanotechnology applications. *Int. J. Acad. Res.*, 4, 2348–7666, 2017.
100. Faghihzadeh, F., Anaya, N.M., Schifman, L.A., Oyanedel-Craver, V., Fourier transform infrared spectroscopy to assess molecular-level changes in microorganisms exposed to nanoparticles. *Nanotechnol. Environ. Eng.*, 1, 1–16, 2016. doi:10.1007/s41204-016-0001-8.
101. Baudot, C., Tan, C.M., Kong, J.C., FTIR spectroscopy as a tool for nanomaterial characterization. *Infrared Phys. Technol.*, 53, 434–438, 2010. doi:10.1016/j.infrared.2010.09.002.
102. Sharma, R., Bisen, D.P., Shukla, U., Sharma, B.G., X-ray diffraction: A powerful method of characterizing nanomaterials. *Recent Res. Sci. Technol.*, 4, 77–79, 2012.
103. Holder, C.F. and Schaak, R.E., Tutorial on powder x-ray diffraction for characterizing nanoscale materials. *ACS Nano*, 13, 7359–7365, 2019. doi:10.1021/acsnano.9b05157.
104. Khanna, P., Ong, C., Bay, B.H., Baeg, G.H., Nanotoxicity: An interplay of oxidative stress, inflammation and cell death. *Nanomaterials*, 5, 1163–1180, 2015. doi:10.3390/nano5031163.
105. Sengul, A.B. and Asmatulu, E., Toxicity of metal and metal oxide nanoparticles: A review. *Environ. Chem. Lett.*, 18, 1659–1683, 2020. doi:10.1007/s10311-020-01033-6.
106. Yuan, X., Zhang, X., Sun, L., Wei, Y., Wei, X., Cellular toxicity and immunological effects of carbon-based nanomaterials. *Part. Fibre Toxicol.*, 16, 18, 2019. doi:10.1186/s12989-019-0299-z.
107. Voigt, N., Henrich-Noack, P., Kockentiedt, S., Hintz, W., Tomas, J., Sabel, B.A., Toxicity of polymeric nanoparticles *in vivo* and *in vitro*. *J. Nanopart. Res.*, 16, 2379, 2014. doi:10.1007/s11051-014-2379-1.
108. Fu, P.P., Xia, Q., Hwang, H.-M., Ray, P.C., Yu, H., Mechanisms of nanotoxicity: Generation of reactive oxygen species. *J. Food Drug Anal.*, 22, 64–75, 2014. doi:10.1016/j.jfda.2014.01.005.

3

Nanotechnology and Vaccine Development

Keshava L. Jetha[1,2*], Praful D. Bharadia[1] and Manish P. Patel[1]

[1]Department of Pharmaceutics and Pharmaceutical Technology, L. M. College of Pharmacy, Navrangpura, Ahmedabad, Gujarat, India
[2]Gujarat Technological University, Chandkheda, Ahmedabad, Gujarat, India

Abstract
Nanotechnology-based vaccines are a promising area of research that offers numerous advantages over traditional vaccines. They have the potential to provide enhanced immunogenicity, reduced side effects, improved stability, and multi-functional capabilities. Nanoparticle vaccines can be administered through various routes, including intramuscular injection, subcutaneous injection, intradermal injection, oral delivery, nasal delivery, and topical application. However, the large-scale manufacturing and uniformity in regulatory guidelines are major challenges associated with nanoparticle-based vaccines that needs to be addressed. Despite these challenges, the development of nanotechnology-based vaccines offers an exciting opportunity to improve vaccine efficacy and accessibility.

Keywords: Nanovaccine development, preclinical studies, nanocarriers, vaccine delivery

3.1 Introduction

'Nano', when used as a prefix to anything, is 'one billionth' fraction, or 10^{-9} of something, which was first described by Nobel laureate Richard Feynman at the American Physical Society on December 1959. The lecture was entitled "There's plenty of room at the bottom." Over the years, there have been many revolutionary ideas across various fields that have supported Feynman's ideas of modifying matter at the atomic scale [1].

**Corresponding author*: keshavjetha@gmail.com

Vivek P. Chavda and Vasso Apostolopoulos (eds.) Nanocarrier Vaccines: Biopharmaceutics-Based Fast Track Development, (63–94) © 2024 Scrivener Publishing LLC

Nanotechnology is the engineering of materials and devices at the nanometer scale, which is 1–100 nm in size. In medicine and healthcare, nanotechnology has shown tremendous potential in enhancing the diagnosis, treatment, and prevention of diseases [2, 3].

At the nanoscale, materials exhibit unique physical and chemical properties that can be harnessed for medical applications [4]. For example, nanoparticles can be engineered to target specific cells or tissues in the body, enabling more precise drug delivery [5]. Nanoscale materials also have a larger surface area-to-volume ratio, which can improve their reactivity and effectiveness [6].

Nanotechnology has shown tremendous potential in the field of vaccine development. In recent years, there has been a surge in research exploring the use of nanotechnology for vaccine delivery and formulation. This has resulted in the development of several novel approaches that have the potential to enhance the efficacy and safety of vaccines [7, 8].

3.2 Overview of Vaccine Development

Vaccines are one of the most effective public health interventions in human history, saving millions of lives every year. Vaccine development involves a complex process that typically takes many years and involves several stages of research and clinical testing [9].

The main stages of vaccine development [10] include

- **Discovery and preclinical testing:** Researchers identify potential vaccine targets and test them in laboratory and animal studies to assess their safety and efficacy.
- **Clinical testing:** If the preclinical studies are successful, the vaccine candidate moves into clinical testing in humans. This involves three phases of clinical trials, with each phase designed to assess different aspects of the vaccine's safety and efficacy.
- **Regulatory review and approval:** If the clinical trials are successful, the vaccine is submitted to regulatory agencies for approval. The regulatory agencies evaluate the safety and efficacy data and decide on whether to approve the vaccine for use.
- **Manufacturing, distribution, and post-marketing surveillance (PMS):** Once the vaccine is approved, it needs to

be manufactured and distributed on a large scale to make it available to the public. Postmarketing surveillance is a critical aspect of vaccine development and involves the monitoring of vaccine safety and effectiveness after the vaccine has been approved and is available to the public. Do refer to Figure 3.1.

Vaccine development faces several challenges [11, 12], including

- **Complexity of vaccine targets:** Developing vaccines for some diseases, such as HIV and malaria, has proven to be very challenging due to the complexity of the pathogens and the immune responses they elicit.
- **Cost and time:** Developing vaccines is a time-consuming and expensive process, with estimates ranging from $200 million to $1 billion and taking up to 15 years.
- **Regulatory requirements:** Vaccines are highly regulated products, and the regulatory requirements for approval can be lengthy and complex.
- **Vaccine hesitancy:** Vaccine hesitancy, or the reluctance to receive vaccines, has become a major challenge in recent years, fueled by misinformation and a lack of trust in vaccines and the pharmaceutical industry.

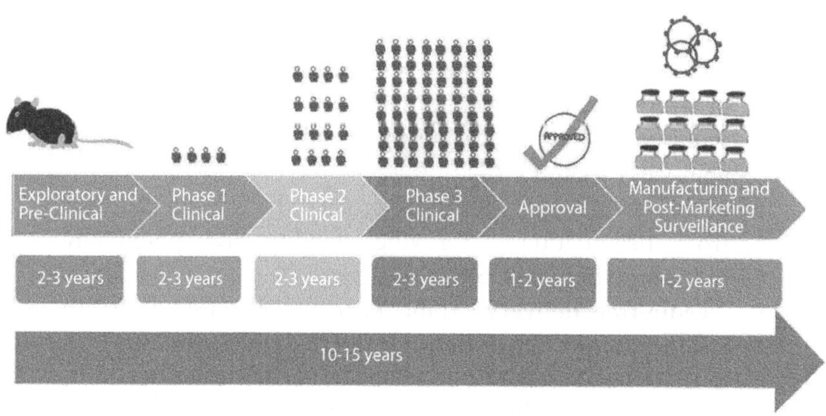

Figure 3.1 Process of vaccine development adopter under CC BY 4 FROM (11).

3.3 Advantages of Nanoparticles in Vaccine Delivery

Nanoparticles have gained increasing attention as potential carriers for vaccines due to their unique properties such as high surface area, small size, and the ability to encapsulate or adsorb antigens [13]. We will study about properties of nanoparticles in detail in Chapter 5. Based on the advanced properties of nanoparticles, they offer numerous advantages such as

1. **Targeted delivery:** Nanoparticles can be engineered to target specific cells or tissues in the body, improving vaccine efficacy and reducing the number of doses needed. By selectively delivering the vaccine antigen to the cells or tissues that need it, targeted delivery can enhance the immune response and reduce the potential for adverse effects on healthy tissues [14].
2. **Controlled release:** Nanoparticles can be designed to release the vaccine antigen over an extended period of time, leading to a sustained immune response and potentially reducing the number of doses needed [15].
3. **Improved immune response:** Nanoparticles can enhance the immune response by facilitating the uptake of the vaccine antigen by immune cells and promoting antigen presentation to the immune system. Nanoparticles enhance the immune response by improving the delivery of the vaccine antigen to immune cells [15].
4. **Adjuvant properties:** Some nanoparticles serve as adjuvants, enhancing the immune response to the vaccine antigen. Nanoscale adjuvants can be designed to mimic the structure and function of pathogens, triggering a stronger and longer-lasting immune response. Furthermore, the use of nanoscale adjuvants can reduce the amount of vaccine required, potentially increasing the accessibility and affordability of vaccines [16].
5. **Stability:** Nanoparticles can protect the vaccine antigen from degradation, improving stability and potentially extending shelf life [15].
6. **Needle-free delivery:** Some nanoparticle-based vaccines can be administered through alternative routes, such as oral or nasal delivery, reducing the need for injections. Microneedle patches, which are made up of thousands of tiny needles, can be used to deliver vaccines painlessly and efficiently [17].

3.4 Types of Nanoparticles as Vaccine Carriers

Nanoparticles can be engineered from a wide range of materials and can be used as carriers for vaccines. Here are some of the most commonly used types of nanoparticles as vaccine carriers:

3.4.1 Liposomes

Liposomes are spherical nanoparticles composed of a phospholipid bilayer that can encapsulate vaccine antigens. They can be used to deliver both hydrophilic and hydrophobic antigens and can be designed to release the antigen in a controlled manner. Liposomes are biocompatible, biodegradable, and non-toxic, making them a promising candidate for vaccine delivery [18–20].

3.4.2 Polymer-Based Nanoparticles

Polymer-based nanoparticles can be engineered from a variety of materials, including polylactic acid, poly(lactic-co-glycolic acid) (PLGA), and polyethylene glycol. They can be designed to encapsulate vaccine antigens, and the release rate of the antigen can be controlled by altering the properties of the polymer [21–23].

3.4.3 Virus-Like Particles (VLPs)

VLPs are self-assembling nanoparticles that mimic the structure of viruses but do not contain genetic material. They can be designed to display the vaccine antigen on their surface or encapsulate it inside the particle. VLPs are highly immunogenic, making them a promising candidate for vaccine development [24–26].

3.4.4 Nanogels

Nanogels are cross-linked polymer networks that can encapsulate vaccine antigens. They can be designed to release the antigen in a controlled manner and can be engineered to target specific cells or tissues in the body [19, 27, 28].

3.4.5 Inorganic Nanoparticles

Inorganic nanoparticles, such as gold, silica, and iron oxide, can also be used as carriers for vaccines. They can be engineered to target specific cells

Table 3.1 Types of nanoparticles along with their advantage and available marketed formulation.

Type of nanoparticle	Advantage	Example	Reference
Liposomes	• Biocompatible • Suitable for both hydrophilic and hydrophobic antigens • Ligand targeting	• Fendrix for Hepatitis B • Epaxal for Hepatitis A	[31, 32]
Polymeric	• Possible for different sizes • Can have a surface charge • Sustained release of vaccine	• Gardasil and Cervarix vaccines for human papillomavirus • Fluad and FluMist Quadrivalent vaccines for influenza	[33, 34]
Inorganic	• For imaging and cell targeting due to their unique optical and magnetic property	• Adju-Phos vaccine for foot-and-mouth disease	[35]
Virus-like particles	• Mimic structure of the virus • Highly immunogenic • Very strong antibody and cellular immune response	• Recombivax HB and Engerix-B vaccines for hepatitis B virus	[36]

or tissues and can be designed to release the vaccine antigen in a controlled manner [15, 29, 30].

Each type of nanoparticle has its unique properties and advantages for vaccine delivery. The choice of nanoparticle will depend on the specific vaccine antigen, the target cells or tissues, and the desired release profile. Some examples of the same have been mentioned in Table 3.1.

Overall, the application of nanotechnology in vaccine development has shown great promise in enhancing the efficacy and safety of vaccines, as well as improving their accessibility and affordability. While there are still challenges to be overcome, such as scaling up production and ensuring safety, the future of nanotechnology in vaccine development is bright.

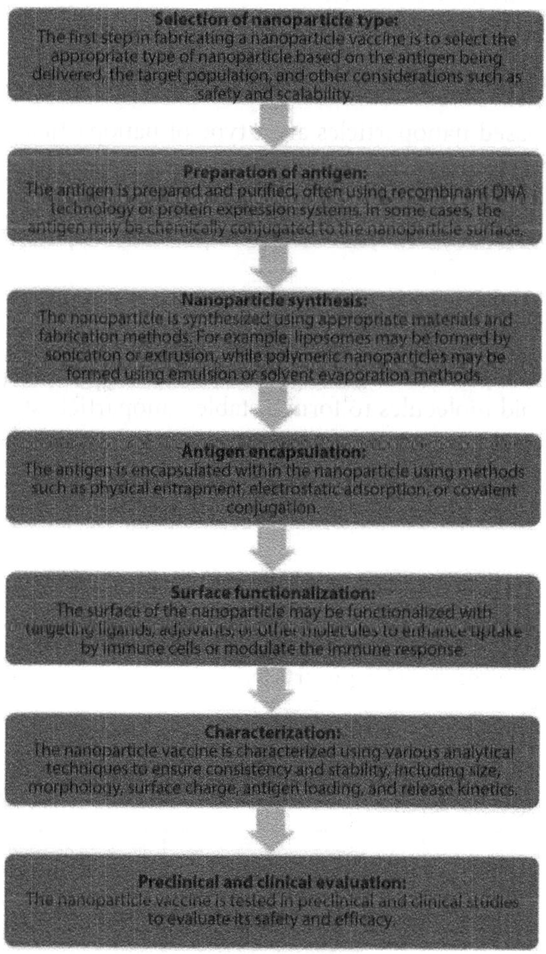

Figure 3.2 Development of a nanoparticle-based vaccine.

3.5 Development of Nanoparticle-Based Vaccine

The fabrication of nanoparticle vaccines involves several steps, which may vary depending on the type of nanoparticle being used. However, some general steps involved in the fabrication of nanoparticle vaccines [37–40] have been explained in Figure 3.2.

Nanoparticle-based carriers are emerging as a promising approach for delivering vaccines and therapeutic agents due to their unique physicochemical properties, such as a high surface area-to-volume ratio, tunable size and surface chemistry, and the ability to protect and deliver payloads to target cells and tissues. Each carrier has its advantages and challenges [41]. Some of this includes carriers such as the following.

3.5.1 Viral Vector–Based Nanoparticle

Viral vector–based nanoparticles are a type of nanoparticle used for vaccine delivery that utilizes a viral vector as the carrier. The viral vector is genetically modified to remove its pathogenic properties and to express the antigen of interest [42]. They are shown in the Table 3.2 below.

3.5.2 Lipid-Based Nanoparticles

Lipid-based nanoparticle vaccines are a type of nanoparticle-based carrier that utilizes lipid molecules to form a stable nanoparticle structure. These nanoparticles can encapsulate a wide range of vaccine payloads, including proteins, nucleic acids, and small molecules, and can be designed to target specific cells and tissues [47]. Table 3.3 summarizes various lipid-based vaccines.

Table 3.2 Viral vector–based nanoparticle.

Viral-vector carrier	Vaccine	Disease	Reference
Adenovirus-based nanoparticle	Astrazeneca, Johnsons	SARS-COV-2	[43, 44]
Lentivirus-based nanoparticle	LV-305	Cancer	[45]
Retrovirus-based nanoparticle	Moloney MLV-VLP	Hepatitis C and HIV	[46]

Table 3.3 Lipid-based nanoparticle vaccines.

Vaccine	Disease	Loading	Reference
Pfizer-BioNTech	COVID-19	mRNA	[48]
Moderna	COVID-19	mRNA	[48]
Novavax	COVID-19	Protein antigens	[49]
Ionis Pharmaceuticals	Huntington's disease	Antisense oligonucleotides	[50]

3.5.3 DNA-Based Nanoparticles

DNA-based nanoparticle vaccines are a type of vaccine that utilizes DNA molecules to encode the antigenic proteins of a pathogen. The DNA is encapsulated within a nanoparticle carrier, which protects it from degradation and facilitates its delivery to cells. Once inside cells, the DNA is translated into the antigenic protein, which triggers an immune response and leads to the production of antibodies and T cells that can recognize and neutralize the pathogen. Table 3.4 summarizes various DNA-based vaccines.

3.5.4 mRNA-Based Nanoparticles

mRNA-based vaccine nanoparticles are a type of vaccine that utilizes messenger RNA (mRNA) molecules to encode the antigenic proteins of a pathogen. The mRNA is encapsulated within a nanoparticle carrier, which protects it from degradation and facilitates its delivery to cells. Once inside cells, the mRNA is translated into the antigenic protein, which triggers an immune response and leads to the production of antibodies and T cells that can recognize and neutralize the pathogen [54, 55]. Refer to Table 3.5.

3.5.5 Protein-Based Nanoparticles

Protein delivery nanoparticle vaccines are a type of vaccine that utilize nanoparticles to deliver proteins or protein fragments from a pathogen. These vaccines work by stimulating an immune response against the pathogen's proteins, which can then protect the individual from infection by the pathogen [58]. Refer to Table 3.6.

Each of these carriers has been discussed in detail in Part 2: Chapters 6–10 of this book.

Table 3.4 DNA-based nanoparticle vaccines.

Vaccine	Disease	Loading	Reference
VGX-3100 by Inovio Pharmaceuticals	Cervical dysplasia	HPV16 and HPV18 E6 and E7 antigens	[51]
Vical's Vaxfectin-adjuvanted DNA vaccines	Influenza, cytomegalovirus, and herpes simplex virus	-	[52]
OncoVAX by Vaxil Bio	Cancer	MUC1	[53]

Table 3.5 mRNA-based nanoparticle vaccines.

Vaccine	Disease	Nanoparticles	Reference
Pfizer-BioNTech	COVID-19	Lipid-based	[48]
Moderna	COVID-19	Lipid-based	[48]
Moderna mRNA-1647 vaccine	Cytomegalovirus infection	Lipid-based	[56]
BioNTech mRNA cancer vaccine	Cancer	Lipid-based	[57]

3.6 Adjuvants and their Role in Vaccine Development

Adjuvants are substances that are added to vaccines to enhance the body's immune response to the antigen (the part of the vaccine that stimulates the immune system) [64]. They are included in vaccine formulations to improve the magnitude, duration, and quality of the immune response to the vaccine [35, 64].

Adjuvants work by activating various immune cells, including antigen-presenting cells (APCs) such as dendritic cells, which help to initiate and direct the immune response [65]. They can also increase the number of immune cells that respond to the vaccine and improve the ability of these cells to recognize and respond to the antigen [64].

Table 3.6 Protein-based nanoparticle vaccines.

Vaccine	Disease	Protein	Reference
Recombivax HB by Merck and Engerix-B by GSK	Hepatitis B	Hepatitis B surface antigen (HBsAg) protein	[59, 60]
Gardasil 9 by Merck	Human papillomavirus vaccine	L1 protein	[61]
Menactra by Sanofi Pasteur and Menveo by GlaxoSmithKline	Meningococcal disease	Recombinant protein	[62]
Prevnar 13 by Pfizer and Pneumovax 23 by Merck	Pneumococcal	Polysaccharide or protein subunits	[63]

Some adjuvants include

6.1 **Alum**: This is the only adjuvant that is currently licensed for use in human vaccines. It consists of aluminum salts that are used to enhance the immune response to certain antigens [66]. Examples: Engerix-B [59], Recombivax HB [60], and Gardasil 9 [61]

6.2 **Oil-in-water emulsions**: These adjuvants are based on a mixture of oil and water and are designed to enhance the uptake and presentation of the antigen by immune cells [67]. Examples: Fluad [68] and Pandemrix [69]

6.3 **Particulate adjuvants:** These adjuvants are made up of small particles that can enhance the immune response by targeting APCs and improving the uptake and presentation of the antigen [70]. Examples: Gardasil 9 [61] and Cervarix [71]

6.4 **Toll-like receptor agonists:** These adjuvants activate immune cells by binding to Toll-like receptors, which are involved in detecting the presence of pathogens [72]. Examples: Gardasil 9 [61] and Heplisav-B [73]

6.5 Cytokines: These adjuvants are proteins that are involved in regulating the immune response. They can be used to enhance the immune response to certain antigens [74]. Example: Engerix-B [59]

Adjuvants play a critical role in the development of vaccines, as they can enhance the immune response to the vaccine and improve its efficacy. By including adjuvants in vaccine formulations, it may be possible to use smaller amounts of antigen, which can reduce costs and minimize potential side effects associated with the vaccine [75]. Adjuvants are also important for developing vaccines against certain diseases that may not elicit a strong immune response on their own [64, 75].

3.7 Nanoscale Adjuvants

Nanoscale adjuvants are in the range of 10–100 nm in size and can be made from a variety of materials, including lipids, proteins, and nucleic acids [76]. Nanoscale adjuvants can be fabricated using a variety of techniques, such as emulsion-based methods, nanoprecipitation, and self-assembly, among others [77].

There are several types of nanoscale adjuvants, including

7.1 Lipid-based adjuvants: Lipid-based adjuvants are a type of vaccine adjuvant that can enhance the immune response to vaccine antigens. These adjuvants are composed of lipids, such as phospholipids and cholesterol, and can form nanoparticles through self-assembly. The resulting nanoparticles can encapsulate the vaccine antigen and enhance its uptake by APCs, which can lead to a more robust immune response [78].
Examples of lipid-based adjuvants include MF59, which is composed of squalene, a naturally occurring oil, and other lipids [68, 79], and AS03, which is composed of squalene, tocopherol, and a synthetic surfactant [79].

7.2 Particulate adjuvants: Particulate adjuvants are a type of vaccine adjuvant that can enhance the immune response to vaccine antigens by delivering antigens directly to APCs. These adjuvants are composed of particles, such as aluminum salts or virus-like particles, that can improve the stability and immunogenicity of the vaccine antigen [70]. Examples of particulate adjuvants include alum [80], which is aluminum salt, and AS04, which is composed of a combination of aluminum hydroxide and monophosphoryl lipid A [81].

7.3 **Polymer-based adjuvants:** These are composed of polymers, such as PLGA, and can be fabricated into nanoparticles using methods such as nanoprecipitation or emulsion-based methods [21–23]. Examples of polymer-based adjuvants include CpG oligodeoxynucleotides and poly-ICLC (Poly-ICLC is a synthetic double stranded RNA comprising of polyriboinosinic-poly ribocytidylic acid (Poly IC) stabilized with l-lysine (L) and carboxymethylcellulose (C)) [82].

3.8 Advantages

Nanoadjuvants are a promising class of materials that have emerged in recent years as a powerful tool for enhancing the immune response to vaccines. Nanoadjuvants offer several advantages over traditional adjuvants [16, 64, 66, 67, 75, 76, 78], including

8.1 **Enhanced immunogenicity:** One of the main advantages of using nanoadjuvants in vaccine development is their ability to enhance the immunogenicity of the vaccine. Traditional adjuvants often work by non-specifically activating the immune system, which can lead to side effects and a less targeted immune response. Nanoadjuvants, on the other hand, can be designed to target specific immune cells and activate APCs, leading to a more robust and longer-lasting immune response. They can also increase antigen uptake, which can further enhance the immune response.

8.2 **Reduced side effects:** Traditional adjuvants can cause local and systemic side effects, such as pain, redness, and fever. These side effects are often due to the non-specific activation of the immune system by the adjuvant. Nanoadjuvants can reduce these side effects by using targeted delivery mechanisms and requiring lower doses. By delivering the adjuvant only to the specific cells or tissues that need it, the rest of the body can be spared from any unwanted immune response.

8.3 **Improved stability:** Nanoadjuvants can improve the stability and shelf life of vaccines. By protecting the antigen from degradation and increasing its solubility, the vaccine can remain effective for a longer period of time. This is particularly important for vaccines that need to be transported and stored under difficult conditions, such as in remote or resource-poor areas.

8.4 **Multifunctional:** Nanoadjuvants can be designed to perform multiple functions, such as targeting specific cells, enhancing antigen uptake, and providing a sustained release of the antigen. This can allow for the development of vaccines that are tailored to specific diseases or populations. For example, a vaccine for cancer may require the sustained release of the antigen to generate a strong and prolonged immune response.

8.5 **Improved safety:** Nanoadjuvants are generally considered safe and biocompatible, with minimal toxicity and immunogenicity. This is because they are often made from natural or biodegradable materials and can be designed to be easily cleared from the body. In addition, their targeted delivery mechanisms can reduce the risk of unwanted immune responses or systemic toxicity, making them a safer alternative to traditional adjuvants.

3.9 Techniques for Nanoscale Adjuvants

Due to their small size and specific surface characteristics, nanoadjuvants can influence the interaction of nanoparticles with biological systems. Therefore, characterizing the physicochemical properties of nanoscale adjuvants is critical for developing effective vaccine formulations [83]. Several techniques are available to characterize nanoscale adjuvants.

9.1 **Dynamic light scattering (DLS)** [84] is a commonly used technique that measures the size distribution of nanoparticles in solution. DLS can also provide information about the stability and aggregation state of the particles.

9.2 **Transmission electron microscopy (TEM)** [85] is a powerful imaging technique that can visualize the morphology, size, and shape of nanoparticles at high resolution. TEM can also be used to examine the internal structure of nanoparticles and confirm their identity.

9.3 **Fourier transform infrared spectroscopy (FTIR)** [86] is a technique that measures the absorption and transmission of infrared radiation by the sample. FTIR can provide information about the chemical composition and molecular structure of nanoparticles, as well as the nature of their interactions with other molecules.

9.4 Other techniques that can be used to characterize nanoscale adjuvants include **X-ray diffraction** [85], **thermogravimetric analysis** [87], **and surface plasmon resonance spectroscopy** [88], among others.

By understanding the physicochemical properties of nanoscale adjuvants, researchers can optimize their design for improved efficacy and safety. For example, optimizing particle size, surface charge, and antigen loading can enhance their uptake by APCs and promote a more robust immune response. Therefore, the characterization of nanoscale adjuvants is a crucial step in developing effective vaccine formulations.

3.10 Route of Administration for Vaccines

Nanoparticle vaccines offer versatility in terms of administration routes, providing flexibility in selecting the most appropriate route for a particular disease. The route of administration can affect the type and magnitude of the immune response induced by the vaccine, as well as its safety and efficacy. Some examples of routes of administration for nanoparticle vaccines and their associated diseases are as follows:

10.1 **Intramuscular injection:** Intramuscular injection is the most common route of administration for nanoparticle vaccines [89]. In this method, the vaccine is injected directly into a muscle, usually, the deltoid muscle in the upper arm or the vastus lateralis muscle in the thigh. The muscle has a rich blood supply and contains immune cells, which allows for efficient uptake of the vaccine and activation of the immune response [90]. Examples of intramuscularly administered nanoparticle vaccines include Gardasil [61] and Cervarix [71] (human papillomavirus (HPV) vaccine).

10.2 **Subcutaneous injection:** Subcutaneous administration is another route of vaccine delivery for nanoparticle vaccines. In this method, the vaccine is injected into the subcutaneous tissue, which is the layer of tissue located between the skin and the underlying muscle. Subcutaneous injection is often used for vaccines that require a slower release and a longer duration of action, as the vaccine is absorbed more slowly from the subcutaneous tissue than from the muscle [91]. This route is commonly used for vaccines against infectious diseases, such as Comirnaty and Moderna COVID-19 vaccines [92–95].

10.3 Intradermal injection: Intradermal administration is a route in which the vaccine is injected into the dermis layer, which is just below the epidermis. This method requires a shorter needle and a smaller volume of vaccine compared to intramuscular injection [96]. Examples of nanoparticle vaccines that can be administered intradermally include Intanza, the seasonal influenza vaccine [97].

10.4 Oral administration: Oral administration is a convenient and non-invasive route for delivering nanoparticle vaccines. In this method, the vaccine is delivered orally, either as a pill or a liquid. The vaccine must survive the harsh acidic environment of the stomach and the enzymes in the gastrointestinal tract to reach the target cells and induce an immune response [98]. Examples of nanoparticle vaccines that can be administered orally include Rotarix, the rotavirus vaccine [99].

10.5 Mucosal administration: Mucosal vaccine delivery is a method of administering vaccines to the mucosal surfaces of the body, such as the respiratory tract, gastrointestinal tract, and genitourinary tract, rather than through traditional injection routes [100]. Examples of nanoparticle vaccines that can be administered through mucosa include OPV Trivalent, which is the oral poliovirus vaccine [101].

10.6 Nasal administration: Intranasal vaccine delivery is a method of administering vaccines through the nasal mucosa. This route of administration offers several advantages over traditional injection routes. The nasal mucosa is a highly permeable and immunologically active surface that can generate both systemic and mucosal immune responses [102]. Examples of nanoparticle vaccines that can be administered nasally include FluMist, an influenza vaccine [103].

10.7 Topical administration: Topical administration is a route of vaccine delivery that involves the application of the vaccine directly to the skin or mucous membranes. This method of delivery is particularly useful for vaccines designed to protect against pathogens that enter the body through the skin or mucosal surfaces, such as herpes simplex virus or HPV. Topical vaccines can be administered using a variety of formulations, including creams, gels, patches, and sprays [100]. Examples of nanoparticle vaccines that can be administered topically include Veregen, the HPV vaccine [104].

The route of administration for vaccines is an important consideration when designing effective vaccination strategies. The choice of administration route can affect the immune response elicited, as well as the safety and tolerability of the vaccine. This will be discussed in Part-3: Chapters 11 to 14 of this book.

3.11 Recent Advances in Nanotechnology-Based Vaccines

Nanotechnology-based vaccine development has seen several recent advances that have the potential to transform vaccine design and delivery. Here are some examples:

11.1 **mRNA vaccines**: mRNA-based vaccines, such as the Pfizer-BioNTech and Moderna COVID-19 vaccines, use nanotechnology to encapsulate the mRNA in lipid nanoparticles. This allows for targeted delivery and protection of the mRNA from degradation [20, 105, 106].

11.2 **Virus-like particles (VLPs):** VLPs are nanoparticles that mimic the structure of viruses but lack the genetic material required for replication. They can be used as vaccine antigens and adjuvants, and several VLP-based vaccines are currently in clinical trials for diseases such as influenza, HPV, and COVID-19 [24].

11.3 **Self-assembling peptides:** Self-assembling peptides can form nanofibers that can be used as vaccine carriers and adjuvants. They have the potential to improve vaccine stability, increase antigen uptake, and stimulate a stronger immune response [107, 108].

11.4 **Metal-based nanoparticles:** Metal-based nanoparticles, such as gold and iron oxide nanoparticles, have been explored as potential vaccine carriers and adjuvants. They can enhance antigen delivery and stimulate the immune system through their unique physical and chemical properties [109, 110].

11.5 **Nanoparticle-based adjuvants:** Several new nanoscale adjuvants, such as CpG oligonucleotides and Toll-like receptor agonists, have been developed that can enhance the immune response and reduce the required vaccine dose [83, 111].

3.12 The Regulatory Perspective of Nanoparticle-Based Vaccine Development

Nanotechnology-based vaccines have the potential to revolutionize the field of vaccinology by offering several advantages over traditional vaccines, such as improved efficacy, stability, and targeted delivery. However, the development of these vaccines also presents several challenges, particularly in terms of regulatory approval and safety considerations.

One of the key regulatory challenges for nanotechnology-based vaccines is the lack of specific guidelines and regulations governing their development and approval [112]. Currently, regulatory agencies such as the US Food and Drug Administration (FDA) [113] and the European Medicines Agency [114] do not have specific guidelines for the development of nanotechnology-based vaccines. This can create uncertainty and delays in the approval process, as regulatory agencies must assess the safety and efficacy of these vaccines on a case-by-case basis.

Another important consideration is the potential toxicity of nanoparticles, which are a key component of many nanotechnology-based vaccines. Nanoparticles may be engineered from a variety of materials, including metals, polymers, and lipids, and their physical and chemical properties can influence their toxicity [115, 116]. Therefore, it is important to evaluate the toxicity of nanoparticles, as well as their distribution and metabolism in the body, to ensure their safety for use in vaccines.

In addition to their potential toxicity, there are concerns about the long-term effects of nanoparticles on human health, particularly if they accumulate in the body over time. More research is needed to understand the long-term effects of exposure to nanoparticles, as well as their potential to induce inflammatory or immune responses in the body [117].

Furthermore, there may be concerns about the environmental impact of nanoparticles, particularly if they are released into the environment during vaccine production or disposal. Regulatory agencies must consider the potential environmental impact of nanoparticles and ensure that appropriate measures are in place to minimize any potential harm [118].

Overall, regulatory agencies must balance the potential benefits of nanotechnology-based vaccines with the need to ensure their safety for human use and the environment. This requires a thorough understanding of the toxicity and potential environmental impact of nanoparticles, as well as the development of specific guidelines and regulations governing their use in vaccines.

3.13 Future Prospects

The future prospects of nanotechnology-based vaccine development are promising. Advances in manufacturing, formulation, and characterization techniques are expected to enable the production of safer, more effective, and more accessible vaccines. Nanovaccines may also provide newer opportunities in the following.

13.1 Customization of vaccines: Nanotechnology-based vaccines hold great promise for the future of immunization. One of their most exciting features is the ability to personalize vaccines for individual patients based on their genetic makeup and immune response. By tailoring vaccines to each patient's unique characteristics, nanovaccines have the potential to increase efficacy and reduce adverse effects [119]. For example, a personalized nanovaccine could be designed to target specific cancer cells or pathogens that are resistant to traditional treatments [28]. Furthermore, nanovaccines can be engineered to target specific cells or tissues, which can increase the potency of the immune response and reduce off-target effects [28]. As research continues in this area, nanovaccines are poised to revolutionize the field of immunology and improve public health worldwide.

13.2 Targeted delivery: Nanoparticles have unique properties that enable their customization to target specific cells or tissues, which has significant implications for vaccine design. By modifying the surface properties of nanoparticles, they can be designed to target specific cells or tissues, such as cancer cells or infected cells, while avoiding healthy cells. This targeted approach can increase the efficacy of vaccines while reducing the potential for side effects [120, 121]. This targeted approach can also reduce the potential for off-target effects, which can occur with traditional chemotherapy [28]. Similarly, in infectious disease vaccines, nanoparticles can be designed to target specific cells or tissues that are infected with the pathogen. This targeted approach can enhance the immune response against the pathogen while reducing potential side effects associated with traditional vaccines [120].

13.3 Combining multiple antigens: Nanoparticles have the potential to carry multiple antigens, which is a major advantage over traditional vaccine approaches. This allows for the development of combination vaccines that can protect against multiple diseases simultaneously [39, 122]. By incorporating multiple antigens into a single nanoparticle-based vaccine, the immune system can be stimulated to produce a broader range of protective responses. This could lead to more effective vaccines that offer protection against multiple diseases in a single dose.

13.4 Needle-free vaccination: Nanotechnology-based vaccines have the potential to be administered using non-invasive methods, such as oral, nasal, or transdermal delivery, which can be more convenient and less painful than traditional injection-based methods [17, 123]. Non-invasive delivery methods may also reduce the risk of needle-stick injuries and eliminate the need for trained healthcare professionals to administer vaccines, making them more accessible to people in resource-limited settings.

13.5 Development of advanced manufacturing processes: Nanoparticle-based vaccines have gained attention due to their potential to improve vaccine efficacy, but their manufacturing process is complex and challenging. Precise control over size, shape, and composition is required to ensure consistent and reproducible batches of nanoparticles [39,

121]. Additionally, scaling up production for mass vaccination campaigns can be difficult, as the processes used to produce small quantities of nanoparticles may not be feasible for large-scale manufacturing.

13.6 **Regulatory uniformity:** The regulatory approval of nanoparticle-based vaccines is crucial to ensure their safety and efficacy for use in humans. However, the complex nature of these vaccines presents unique regulatory challenges that may require new testing protocols. For instance, the size, shape, and chemical composition of nanoparticles can affect their behavior in the body, including their toxicity and immunogenicity. Therefore, regulatory agencies must establish uniform standards for nanoparticle-based vaccines to ensure their safety and efficacy across different countries and regions [40, 124]. Additionally, there is a need for clear guidelines on the manufacture, quality control, and distribution of nanoparticle-based vaccines to ensure their consistent performance and availability.

13.7 **Cost-effectiveness:** Nanoparticle-based vaccines have the potential to be more expensive to manufacture than traditional vaccines, due to the complexity of their production and the cost of the raw materials used. This could limit their accessibility to patients in low-resource settings, particularly in developing countries where vaccine-preventable diseases are more prevalent [40, 125]. However, advancements in nanotechnology and manufacturing processes are expected to reduce the cost of nanoparticle-based vaccines in the future, making them more affordable and accessible.

13.8 **Public perception:** Nanotechnology-based vaccines are relatively new and can raise concerns among the public about their safety and effectiveness. The public's perception of these vaccines can affect their acceptance and uptake, and education is essential to address these concerns. It is important to have clear and transparent communication about the benefits and potential risks of these vaccines to build trust and confidence in their use [126]. It is also crucial to involve patients and the public in discussions about the development and deployment of these vaccines to ensure that their needs and concerns are heard and addressed. A collaborative and inclusive approach to vaccine development and deployment can help ensure the success of these innovative technologies.

13.9 **Intellectual property:** The development of nanotechnology-based vaccines requires substantial investment in research and development to optimize their design and production methods. The high cost associated with this research and development can make these vaccines more expensive than traditional vaccines. Moreover, there may be intellectual property issues associated with the development of nanoparticle-based vaccines that need to be addressed to ensure equitable access for all individuals who need them [3, 127]. The issue of intellectual property rights can also impact the development of generic or biosimilar versions of these vaccines, which may limit their availability and affordability. Policymakers need to consider these issues and work to ensure that these vaccines are accessible to all who need them, regardless of their socioeconomic status or geographic location.

A detailed discussion of applications and advances in nanoparticles has been discussed in Part 4 and Part 5: Chapters 15–20.

3.14 Conclusion

Nanoparticle-based vaccines offer many promising advantages over traditional vaccines, such as enhanced immunogenicity, reduced side effects, improved stability, and multifunctionality. They can also be administered through various routes, including non-invasive methods, and can be personalized for individual patients. However, their complex manufacturing process and regulatory challenges may limit their accessibility and affordability, and public concerns about their safety and efficacy need to be addressed through education. Nevertheless, the potential benefits of nanoparticle-based vaccines make them an exciting area of research and development that could revolutionize the way we prevent and treat diseases.

References

1. Benelmekki M. An introduction to nanoparticles and nanotechnology. In: *Designing Hybrid Nanoparticles*, Morgan & Claypool Publishers, 2015 Apr 1.
2. Anjum, S., Ishaque, S., Fatima, H., Farooq, W., Hano, C., Abbasi, B.H. *et al.*, Emerging applications of nanotechnology in healthcare systems: Grand challenges and perspectives. *Pharmaceuticals*, 14, 8, 707, Jul. 21, 2021.
3. Singh, L., Kruger, H.G., Maguire, G.E.M., Govender, T., Parboosing, R., The role of nanotechnology in the treatment of viral infections. *Ther. Adv. Infect. Dis.*, 4, 4, 105–31, Jul. 2017.
4. Zhang Y, Poon K, Masonsong GS, Ramaswamy Y, Singh G. Sustainable nanomaterials for biomedical applications. *Pharmaceutics*, 12, 15, 3, 922, 2023 Mar.
5. Carissimi, G., Montalbán, M.G., Fuster, M.G., Víllora, G., Carissimi, G., Montalbán, M.G. *et al.*, Nanoparticles as drug delivery systems. *21st Century Nanostructured Materials-Physics, Chemistry, Classification, and Emerging Applications in Industry, Biomedicine, and Agriculture*, IntechOpen, 2021, [Internet] [cited 2023 Mar 29] Available from: https://www.intechopen.com/chapters/78717.
6. Joudeh N, Linke D. Nanoparticle classification, physicochemical properties, characterization, and applications: A comprehensive review for biologists. *J. Nanobiotechnology*, 7, 20, 1, 262, 2022 Jun.
7. Panda, A.K., Nanotechnology in vaccine development. *Proceedings of the National Academy of Sciences, Section B: Biological Sciences*. Oct;82:13–27, India, 2012.
8. Shin, M.D., Shukla, S., Chung, Y.H., Beiss, V., Chan, S.K., Ortega-Rivera, O.A., Wirth, D.M., Chen, A., Sack, M., Pokorski, J.K., Steinmetz, N.F.. COVID-19 vaccine development and a potential nanomaterial path forward. *Nat. Nanotechnol.*, 15, 8, 646–55, 2020 Aug.
9. Rodrigues, C.M., Plotkin, S.A.. Impact of vaccines; health, economic and social perspectives. *Front. Microbiol.* 11, 1526, 2020 Jul. 14 .
10. Strategies I of M (US) C on the CVIPA, Mitchell, V.S., Philipose, N.M., Sanford, J.P., Stages of vaccine development, in: *The Children's Vaccine Initiative: Achieving the Vision*, National Academies Press, US, 1993, [Internet] [cited 2023 Mar 30]. Available from: https://www.ncbi.nlm.nih.gov/books/NBK236428/.
11. Sharma, O., Sultan, A.A., Ding, H., Triggle, C.R.. A review of the progress and challenges of developing a vaccine for COVID-19. *Front. Immunol.*, 11, 585354, 2020 Oct 14.

12. Kennedy, R.B., Ovsyannikova, I.G., Palese, P., Poland, G.A., Current challenges in vaccinology. *Front. Immunol.*. 11, 1181, 2020 Jun 25.
13. Chattopadhyay, S., Chen, J.Y., Chen, H.W., Hu, C.M.. Nanoparticle vaccines adopting virus-like features for enhanced immune potentiation. *Nanotheranostics*. 1, 3, 244, 2017.
14. Hans, M.L. and Lowman, A.M., Biodegradable nanoparticles for drug delivery and targeting. *Curr. Opin. Solid State Mater. Sci.*, 6, 4, 319–27, Aug. 1, 2002.
15. Pati, R., Shevtsov, M., Sonawane, A.. Nanoparticle vaccines against infectious diseases. *Front. Immunol.*, 9: 2224, 2018 Oct 4.
16. Zhu, M., Wang, R., Nie, G.. Applications of nanomaterials as vaccine adjuvants. *Hum. Vaccin. Immunother.*, 10, 9, 2761–74, 2014 Sep 2.
17. Kim, Y.C., Park, J.H., Prausnitz, M.R., Microneedles for drug and vaccine delivery. *Adv. Drug Deliv. Rev.*, 64, 14, 1547–68, Nov. 2012.
18. Nakhaei, P., Margiana, R., Bokov, D.O., Abdelbasset, W.K., Kouhbanani, M.A.J., Varma, R.S. *et al.*, Liposomes: Structure, biomedical applications, and stability parameters with emphasis on cholesterol. *Front. Bioeng. Biotechnol.*, 9, 705886, Sep. 9, 2021.
19. Sriwidodo, Umar, A.K., Wathoni, N., Zothantluanga, J.H., Das, S., Luckanagul, J.A., Liposome-polymer complex for drug delivery system and vaccine stabilization. *Heliyon*, 8, 2, e08934, Feb. 12, 2022.
20. Tenchov, R., Bird, R., Curtze, A.E., Zhou, Q.. Lipid nanoparticles— From liposomes to mRNA vaccine delivery, a landscape of research diversity and advancement. *ACS Nano*, 15, 1, 16982–7015, 2021 Jun 28.
21. Kim, K.T., Lee, J.Y., Kim, D.D., Yoon, I.S., Cho, H.J., Recent progress in the development of poly(lactic-co-glycolic acid)-based nanostructures for cancer imaging and therapy. *Pharmaceutics*, 11, 6, 280, Jun. 14, 2019.
22. Makadia, H.K. and Siegel, S.J., Poly lactic-co-glycolic acid (PLGA) as biodegradable controlled drug delivery carrier. *Polymers*, 3, 3, 1377–97, Sep. 1, 2011.
23. Elmowafy, E.M., Tiboni, M., Soliman, M.E.. Biocompatibility, biodegradation and biomedical applications of poly (lactic acid)/poly (lactic-co-glycolic acid) micro and nanoparticles. *J. Pharm. Investig.*, 49, 347–80, 2019 Jul 1.
24. Nooraei, S., Bahrulolum, H., Hoseini, Z.S., Katalani, C., Hajizade, A., Easton, A.J. *et al.*, Virus-like particles: Preparation, immunogenicity and their roles as nanovaccines and drug nanocarriers. *J. Nanobiotechnology*, 19, 1, 59, Feb. 25, 2021.
25. Yan, D., Wei, Y.Q., Guo, H.C., Sun, S.Q., The application of virus-like particles as vaccines and biological vehicles. *Appl. Microbiol. Biotechnol.*, 99, 24, 10415–32, 2015.
26. Donaldson, B., Al-Barwani, F., Young, V., Scullion, S., Ward, V., Young, S., Virus-like particles, a versatile subunit vaccine platform, in: *Subunit Vaccine Delivery*, pp. 159–80, Aug. 28, 2014.

27. Soni, K.S., Desale, S.S., Bronich, T.K., Nanogels: An overview of properties, biomedical applications and obstacles to clinical translation. *J. Control. Release*, 240, 109–26, Oct 28, 2016.
28. Yu, M., Yang, W., Yue, W., Chen, Y., Targeted cancer immunotherapy: Nanoformulation engineering and clinical translation. *Adv. Sci.*, 9, 35, 2204335, 2022.
29. Hess, K.L., Medintz, I.L., Jewell, C.M., Designing inorganic nanomaterials for vaccines and immunotherapies. *Nano Today*, 27, 73–98, Aug. 2019.
30. Poon, C. and Patel, A.A., Organic and inorganic nanoparticle vaccines for prevention of infectious diseases. *Nano Express*, 1, 1, 012001, Apr. 2020.
31. Tejada-Pérez, J.J., Vázquez-Vicente, J.J., Herrera-Burgos, M.R., Martín-Martín, F.G., Parrón-Carreño, T., Alarcón-Rodríguez, R., Fendrix® vaccine effectiveness in healthcare workers who are non-responsive to engerix B® vaccination. *Vaccines*, 9, 3, 279, Mar. 19, 2021.
32. Bovier PA. Epaxal®: A virosomal vaccine to prevent hepatitis A infection. *Expert Rev.* Vaccine, 7, 8, 1141–50, 2008 Oct 1.
33. Wong, S.S. and Webby, R.J., Traditional and new influenza vaccines. *Clin. Microbiol. Rev.*, 26, 3, 476–92, Jul. 2013.
34. Cheng, L., Wang, Y., Du, J., Human papillomavirus vaccines: An updated review. *Vaccines*, 8, 3, 391, Jul. 16, 2020.
35. Bidart, J., Kornuta, C., Gammella, M., Gnazzo, V., Soria, I., Langellotti, C. *et al.*, A new cage-like particle adjuvant enhances protection of foot-and-mouth disease vaccine. *Front. Vet. Sci.*, 7, 396, Jul. 31, 2020.
36. Lacson, E., Teng, M., Ong, J., Vienneau, L., Ofsthun, N,, Lazarus, J.M., Antibody response to Engerix-B® and Recombivax-HB® hepatitis B vaccination in endstage renal disease. *Hemodial. Int.*,9, 4, 367–75, 2005 Oct.
37. Zhao, L., Seth, A., Wibowo, N., Zhao, C.X., Mitter, N., Yu, C. *et al.*, Nanoparticle vaccines. *Vaccine*, 32, 3, 327–37, Jan. 9, 2014.
38. Khan, I., Saeed, K., Khan, I.. Nanoparticles: Properties, applications and toxicities. *Arab. J. Chem.*, 12, 7, 908–31, 2019 Nov 1.
39. Curley, S.M., and Putnam, D.. Biological nanoparticles in vaccine development. *Front. Bioeng. Biotechnol*, 10, 867119, 2022 Mar 23.
40. Diaz-Arévalo, D., and Zeng, M.. Nanoparticle-based vaccines: opportunities and limitations. In: *Nanopharmaceuticals*, 135–150, Elsevier, 2020 Jan 1.
41. Chavda, V.P., Balar, P.C., Patel, S.B., Nanotheranostics-based management of head and neck cancer. *Nanotheranostics*, 7, 2, 202–9, 2023.
42. Chavda, V.P., Valu, D.D., Parikh, P.K., Tiwari, N., Chhipa, A.S., Shukla, S. *et al.*, Conventional and novel diagnostic tools for the diagnosis of emerging SARS-CoV-2 variants. *Vaccines*, 11, 2, 374, Feb. 6, 2023.
43. Chavda, V.P., Bezbaruah, R., Valu, D., Patel, B., Kumar, A,, Prasad, S., Kakoti, B.B., Kaushik, A., Jesawadawala, M.. Adenoviral vector-based vaccine platform for COVID-19: Current status. *Vaccines*. 11, 2, 432, 2023 Feb 13.

44. Chavda, V.P., Bezbaruah, R., Valu, D., Patel, B., Kumar, A., Prasad, S. *et al.*, Adenoviral vector-based vaccine platform for COVID-19: Current status. *Vaccines*, 11, 2, 432, Feb. 13, 2023.
45. Somaiah, N., Block, M.S., Kim, J.W., Shapiro, G.I., Do, K.T., Hwu, P. *et al.*, First-in-class, first-in-human study evaluating LV305, a dendritic-cell tropic lentiviral vector, in sarcoma and other solid tumors expressing NY-ESO-1. *Clin. Cancer Res.*, 25, 19, 5808–17, Oct. 1, 2019.
46. Pitoiset, F., Vazquez, T., Levacher, B., Nehar-Belaid, D., Dérian, N., Vigneron, J. *et al.*, Retrovirus-based virus-like particle immunogenicity and its modulation by toll-like receptor activation. *J. Virol.*, 91, 21, e01230–17, Oct. 13, 2017.
47. Chavda, V.P., Balar, P.C., Patel, S.B., Interventional nanotheranostics in hepatocellular carcinoma. *Nanotheranostics*, 7, 2, 128–41, 2023.
48. Heinz, F.X., and Stiasny, K.. Distinguishing features of current COVID-19 vaccines: Knowns and unknowns of antigen presentation and modes of action, *NPJ Vaccines*, 6, 1, 104, 2021 Aug 16.
49. Chung, Y.H., Beiss, V., Fiering, S.N., Steinmetz, N.F.. COVID-19 vaccine frontrunners and their nanotechnology design. *ACS Nano*, 14, 10, 12522–37, 2020 Oct 9.
50. Rook, M.E. and Southwell, A.L., Antisense oligonucleotide therapy: From design to the huntington disease clinic. *Biodrugs*, 36, 2, 105–19, 2022.
51. Akhatova, A., Chan, C.K., Azizan, A., Aimagambetova, G.. The efficacy of therapeutic DNA vaccines expressing the human papillomavirus E6 and E7 oncoproteins for treatment of cervical cancer: systematic review. *Vaccines*, 10, 1,:53, 2021 Dec. 31
52. Saade, F. and Petrovsky, N., Technologies for enhanced efficacy of DNA vaccines. *Expert Rev. Vaccines*, 11, 2, 189–209, Feb. 2012.
53. Yang, B., Jeang, J., Yang, A., Wu, T.C., Hung, C.F., DNA vaccine for cancer immunotherapy. *Hum. Vaccin. Immunother.*, 10, 11, 3153–64, Jan. 27, 2015.
54. Chavda, V.P., Jogi, G., Dave, S., Patel, B.M., Nalla, L.V., Koradia, K., mRNA-based vaccine for COVID-19: They are new but not unknown! *Vaccines*, 11, 3, 507, Feb. 22, 2023.
55. Chavda, V.P., Soni, S., Vora, L.K., Soni, S., Khadela, A., Ajabiya, J., mRNA-based vaccines and therapeutics for COVID-19 and future pandemics. *Vaccines*, 10, 12, 2150, Dec. 15, 2022.
56. Scarpini, S., Morigi, F., Betti, L., Dondi, A., Biagi, C., Lanari, M.. Development of a vaccine against human cytomegalovirus: advances, barriers, and implications for the clinical practice. *Vaccines*. 9, 6, 551, 2021 May 25.
57. Miao, L., Zhang, Y., Huang, L.. mRNA vaccine for cancer immunotherapy, *Mol. Cancer*, 20, 1, 1–23, 2021 Dec.
58. Chavda, V.P., Nalla, L.V., Balar, P., Bezbaruah, R., Apostolopoulos, V., Singla, R.K. *et al.*, Advanced phytochemical-based nanocarrier systems for the treatment of breast cancer. *Cancers*, 15, 4, 1023, Feb. 6, 2023.
59. Keating, G.M. and Noble, S.. Recombinant Hepatitis B vaccine (Engerix-B®) A review of its immunogenicity and protective efficacy against Hepatitis B. *Drugs*. 63, 10, 1021-51, 2003 May

60. Lacson, E., Teng, M., Ong, J., Vienneau, L., Ofsthun, N., Lazarus, J.M., Antibody response to engerix-B and recombivax-HB hepatitis B vaccination in end-stage renal disease. *International Symposium on Home Hemodialysis*, vol. 9, pp. 367–75, Oct. 2005.
61. Lexchin, J., Arya, N., Singh, S., Gardasil®–the new HPV vaccine: The right product, the right time? A commentary. *Healthc. Policy*, 5, 4, 26–36, May 2010.
62. Pizza, M., Bekkat-Berkani, R., Rappuoli, R., Vaccines against meningococcal diseases. *Microorganisms*, 8, 10, 1521, Oct. 3, 2020.
63. Buchwald, U.K., Andrews, C.P., Ervin, J., Peterson, J.T., Tamms, G.M., Krupa, D. *et al.*, Sequential administration of Prevnar 13™ and PNEUMOVAX™ 23 in healthy participants 50 years of age and older. *Hum. Vaccin. Immunother.*, 17, 8, 2678–90, 2021.
64. Coffman, R.L., Sher, A., Seder, R.A., Vaccine adjuvants: Putting innate immunity to work. *Immunity*, 33, 4, 492–503, Oct. 29, 2010.
65. Chavda, V.P., Patel, A.B., Vora, L.K., Apostolopoulos, V., Uhal, B.D., Dendritic cell-based vaccine: The state-of-the-art vaccine platform for COVID-19 management. *Expert Rev. Vaccines*, 21, 10, 1395–403, Oct. 3, 2022.
66. Facciolà, A., Visalli, G., Laganà, A., Di Pietro, A.. An overview of vaccine adjuvants: Current evidence and future perspectives. *Vaccines*, 10, 5, 819, 2022 May 22.
67. Awate, S., Babiuk, L.A., Mutwiri, G.. Mechanisms of action of adjuvants. Frontiers in immunology. 2013 May 16;4:114
68. Cruz-Valdez, A., Valdez-Zapata, G., Patel, S.S., Castelli, F.V., Garcia, M.G., Jansen, W.T. *et al.*, MF59-adjuvanted influenza vaccine (FLUAD®) elicits higher immune responses than a non-adjuvanted influenza vaccine (Fluzone®): A randomized, multicenter, Phase III pediatric trial in Mexico. *Hum. Vaccin. Immunother.*, 14, 2, 386–95, Jan. 3, 2018.
69. Hallberg, P., Smedje, H., Eriksson, N., Kohnke, H., Daniilidou, M., Öhman, I., Yue, Q.Y.,Cavalli, M., Wadelius, C., Magnusson, P.K., Landtblom, A.M.. Pandemrix-induced narcolepsy is associated with genes related to immunity and neuronal survival, *EBioMedicine*,. 40, 595–604, 2019 Feb 1.
70. Awate, S., Babiuk, L.A., Mutwiri, G., Mechanisms of action of adjuvants. *Front. Immunol.*, 4, 114, May 16, 2013.
71. Monie, A., Hung, C.F., Roden, R., Wu, T.C., Cervarix™: A vaccine for the prevention of HPV 16, 18-associated cervical cancer. *Biol. Targets Ther.*, 2, 1, 107–13, Mar. 2008.
72. Janssens, S. and Beyaert., R., Role of toll-like receptors in pathogen recognition. *Clin. Microbiol Rev.*, 16, 4, 637–46, Oct. 2003.
73. Champion, C.R., Heplisav-B: A hepatitis B vaccine with a novel adjuvant. *Ann. Pharmacother.*, 55, 6, 783–91, Jun. 2021.
74. Chaplin, D.D.. Overview of the immune response. *J. Allergy Clin. Immunol.*. 125, 2, S3–23, 2010 Feb 1.

75. Mohan, T., Verma, P., Rao, D.N., Novel adjuvants & delivery vehicles for vaccines development: A road ahead. *Indian J. Med. Res.*, 138, 5, 779–95, Nov. 2013.
76. Jin, S., Li, S., Wang, C., Liu, J., Yang, X., Wang, P.C. *et al.*, Biosafe nanoscale pharmaceutical adjuvant materials. *J. Biomed. Nanotechnol.*, 10, 9, 2393–419, Sep. 2014.
77. Pingale, P., Kendre, P., Pardeshi, K., Rajput, A., An emerging era in manufacturing of drug delivery systems: Nanofabrication techniques. *Heliyon*, 9, 3, e14247, Mar. 1, 2023.
78. Chatzikleanthous, D., O'Hagan, D.T., Adamo, R., Lipid-based nanoparticles for delivery of vaccine adjuvants and antigens: Toward multicomponent vaccines. *Mol. Pharm.*, 18, 8, 2867–88, Aug. 2, 2021.
79. Wilkins, A.L., Kazmin, D., Napolitani, G., Clutterbuck, E.A., Pulendran, B., Siegrist, C.A. *et al.*, AS03-and MF59-adjuvanted influenza vaccines in children. *Front. Immunol.*, 8, 1760, Dec. 13, 2017.
80. Oleszycka, E. and Lavelle, E.C., Immunomodulatory properties of the vaccine adjuvant alum. *Curr. Opin. Immunol.*, 28, 1–5, Jun. 1, 2014.
81. Didierlaurent, A.M., Morel, S., Lockman, L., Giannini, S.L., Bisteau, M., Carlsen, H. *et al.*, AS04, an aluminum salt- and TLR4 agonist-based adjuvant system, induces a transient localized innate immune response leading to enhanced adaptive immunity. *J. Immunol. (Baltimore Md 1950)*, 183, 10, 6186–97, Nov. 15, 2009.
82. Wong, J.P., Christopher, M.E., Viswanathan, S., Karpoff, N., Dai, X., Das, D. *et al.*, Activation of toll-like receptor signaling pathway for protection against influenza virus infection. *Vaccine*, 27, 25, 3481–3, May 26, 2009.
83. Alfagih, I.M., Aldosari, B., AlQuadeib, B., Almurshedi, A., Alfagih, M.M.,.Nanoparticles as adjuvants and nanodelivery systems for mRNA-based vaccines. *Pharmaceutics*. 13, 1, 45, 2020 Dec 30.
84. Ashizawa K. Nanosize particle analysis by dynamic light scattering (DLS). Yakugaku Zasshi (ed.) *Journal of the Pharmaceutical Society of Japan*, 139, 2, 237–48, 2019 Jan. 1
85. Guerrini, G., Vivi, A., Gioria, S., Ponti, J., Magrì, D., Hoeveler, A. *et al.*, Physicochemical characterization cascade of nanoadjuvant–antigen systems for improving vaccines. *Vaccines*, 9, 6, 544, May 21, 2021.
86. Kamnev, A.A., Dyatlova, Y.A., Kenzhegulov, O.A., Vladimirova, A.A., Mamchenkova, P.V., Tugarova, A.V., Fourier transform infrared (FTIR) spectroscopic analyses of microbiological samples and biogenic selenium nanoparticles of microbial origin: Sample preparation effects. *Molecules (Basel Switz.)*, 26, 4, 1146, Feb. 21, 2021.
87. Dongargaonkar, A.A. and Clogston, J.D., Quantitation of surface coating on nanoparticles using thermogravimetric analysis. *Methods Mol. Biol. Clifton NJ*, 1682, 57–63, 2018.
88. Eddin, F.B.K. and Fen, Y.W., The principle of nanomaterials based surface plasmon resonance biosensors and its potential for dopamine detection. *Molecules*, 25, 12, 2769, Jan. 2020.

89. Anderluzzi, G., Lou, G., Woods, S., Schmidt, S.T., Gallorini, S., Brazzoli, M. *et al.*, The role of nanoparticle format and route of administration on self-amplifying mRNA vaccine potency. *J. Control. Release*, 342, 388–99, Feb. 2022.
90. Zuckerman, J.N., The importance of injecting vaccines into muscle. *BMJ*, 321, 1237–8, 7271, Nov. 18, 2000.
91. Hansen, S. and Lehr, C., Nanoparticles for transcutaneous vaccination. *Microb. Biotechnol.*, 5, 2, 156–67, Mar. 2012.
92. Papini, F., Grassi, N., Guglielmi, G., Gattini, V., Rago, L., Bisordi, C. *et al.*, Covid-19 vaccine management (Comirnaty and mrna-1273 Moderna) in a teaching hospital in Italy: A short report on the vaccination campaign. *Environ. Health Prev. Med.*, 26, 99, 2021.
93. Chavda, V.P., Patel, A.B., Pandya, A., Vora, L.K., Patravale, V., Tambuwala, Z.M. *et al.*, Co-infection associated with SARS-CoV-2 and their management. *Future Sci. OA*, 8, 9, FSO819, Oct. 2022.
94. Chavda, V.P. and Apostolopoulos, V., COVID-19 vaccine design and vaccination strategy for emerging variants. *Expert. Rev. Vaccines*, 21, 10, 1359–61, Oct. 3, 2022.
95. Chavda, V.P., Vora, L.K., Vihol, D.R., COVAX-19&xxx24C7; vaccine: Completely blocks virus transmission to non-immune individuals. *Clin. Complement. Med. Pharmacol.*, 1, 1, 100004, Dec. 1, 2021.
96. Kim, Y.C., Jarrahian, C., Zehrung, D., Mitragotri, S., Prausnitz, M.R., Delivery systems for intradermal vaccination. *Intradermal Immunization*, 351, 77–112, Apr 7, 2011.
97. Atmar, R.L., Patel, S.M., Keitel, W.A., Intanza(®): A new intradermal vaccine for seasonal influenza. *Expert Rev. Vaccines*, 9, 12, 1399–409, Dec. 2010.
98. Zhu, Q., and Berzofsky, J.A.. Oral vaccines: Directed safe passage to the front line of defense. *Gut Microbes*. 4, 3, 246–5, 2013 May 1.
99. Buyse, H., Vinals, C., Karkada, N., Han, H.H., The human rotavirus vaccine Rotarix™ in infants: An integrated analysis of safety and reactogenicity. *Hum. Vaccin. Immunother.*, 10, 1, 19–24, 2014.
100. Shakya, A.K., Chowdhury, M.Y.E., Tao, W., Gill, H.S., Mucosal vaccine delivery: Current state and a pediatric perspective. *J. Control. Release*, 240, 394–413, Oct. 28, 2016.
101. Tevi-Benissan, C., Okeibunor, J., du Châtellier, G.M., Assefa, A., Biey, J.N.M., Cheikh, D. *et al.*, Introduction of inactivated poliovirus vaccine and trivalent oral polio vaccine/bivalent oral polio vaccine switch in the African region. *J. Infect. Dis.*, 216, suppl_1, S66–75, Jul. 1, 2017.
102. Ramvikas, M., Arumugam, M., Chakrabarti, S.R., Jaganathan, K.S., Nasal vaccine delivery, in: *Micro and Nanotechnology in Vaccine Development*, pp. 279–301, 2017.
103. Al-Halifa, S., Gauthier, L., Arpin, D., Bourgault, S., Archambault, D.. Nanoparticlebased vaccines against respiratory viruses. *Front. Immunol.*. 10, 22, 2019 Jan 24.

104. Muñoz-Santos, C., Pigem, R., Alsina, M., New treatments for human papillomavirus infection. *Actas Dermosifiliogr. Engl. Ed.*, 104, 10, 883–9, Dec. 1, 2013.
105. Wilson, B. and Geetha, K.M., Lipid nanoparticles in the development of mRNA vaccines for COVID-19. *J. Drug Deliv. Sci. Technol.*, 74, 103553, Aug. 2022.
106. Hou, X., Zaks, T., Langer, R., Dong, Y.. Lipid nanoparticles for mRNA delivery. *Nat. Rev. Mater.*, 6, 12, 1078–94, 2021 Dec.
107. Abdullah, T., Bhatt, K., Eggermont, L.J., O'Hare, N., Memic, A., Bencherif, S.A., Supramolecular self-assembled peptide-based vaccines: Current state and future perspectives. *Front. Chem.*, 8, 598160, Oct. 30, 2020.
108. O'Neill, C.L., Shrimali, P.C., Clapacs, Z.P., Files, M.A., Rudra, J.S., Peptide-based supramolecular vaccine systems. *Acta Biomaterialia*. 133, 153–67, 2021 Oct 1.
109. Luo, Y.H., Chang, L.W., Lin, P., Metal-based nanoparticles and the immune system: Activation, inflammation, and potential applications. *BioMed. Res. Int.*, 2015, 143720, 2015.
110. Marques Neto, L.M., Kipnis, A., Junqueira-Kipnis, A.P., Role of metallic nanoparticles in vaccinology: implications for infectious disease vaccine development. *Front. Immunol.*, 8, 239, 2017 Mar 8.
111. Petkar, K.C., Patil, S.M., Chavhan, S.S., Kaneko, K., Sawant, K.K., Kunda, N.K. *et al.*, An overview of nanocarrier-based adjuvants for vaccine delivery. *Pharmaceutics*, 13, 4, 455, Mar. 27, 2021.
112. Desai, N., Challenges in development of nanoparticle-based therapeutics. *AAPS J.*, 14, 2, 282–95, Mar. 10, 2012.
113. Farjadian, F., Ghasemi, A., Gohari, O., Roointan, A., Karimi, M., Hamblin, M.R..Nanopharmaceuticals and nanomedicines currently on the market: challenges and opportunities. *Nanomedicine*, 14, 1, 93–126, 2019 Jan.
114. Oualikene-Gonin, W., Sautou, V., Ezan, E., Bastos, H., Bellissant, E., Belgodère, L., Maison, P., Ankri, J., Barbot, J., Barouki, R., Bellissant, É., Regulatory assessment of nano-enabled health products in public health interest. Position of the scientific advisory board of the French National Agency for the Safety of Medicines and Health Products. *Frontiers in Public Health*, 11, p. 1125577, 2023.
115. De Jong, W.H. and Borm, P.J., Drug delivery and nanoparticles: Applications and hazards. *Int. J. Nanomedicine*, 3, 2, 133–49, Jun 2008.
116. Sharma, S., Parveen, R., Chatterji, B.P., Toxicology of nanoparticles in drug delivery. *Curr. Pathobiol. Rep.*, 9, 4, 133–44, 2021.
117. Hoet, P.H., Brüske-Hohlfeld, I., Salata, O.V., Nanoparticles–known and unknown health risks. *J. Nanobiotechnology*, 2, 12, Dec. 8, 2004.
118. Ray, P.C., Yu, H., Fu, P.P., Toxicity and environmental risks of nanomaterials: Challenges and future needs. *J. Environ. Sci. Health C Environ. Carcinog. Ecotoxicol. Rev.*, 27, 1, 1–35, Jan. 2009.
119. Zhou, J., Kroll, A.V., Holay, M., Fang, R.H., Zhang, L., Biomimetic nanotechnology towards personalized vaccines. *Adv. Mater.*, 32, 13, e1901255, Apr. 2020.

120. Feng C, Li Y, Ferdows BE, Patel DN, Ouyang J, Tang Z, Kong N, Chen E, Tao W. Emerging vaccine nanotechnology: From defense against infection to sniping cancer. *Acta Pharmaceutica Sinica B*, 12, 5, 2206–23, 2022 May 1.
121. Bo, Y., and Wang H.. Materials-based vaccines for infectious diseases. Wiley Interdiscip. Rev. Nanomed. *Nanobiotechnol.*, 14, 5, e1824, 2022 Sep.
122. Gregory, A.E., Titball, R., Williamson, D.. Vaccine delivery using nanoparticles. *Front. Cell. Infect. Microbiol.* 3, 13, 2013 Mar 25.
123. Mangla, B., Javed, S., Sultan, M.H., Ahsan, W., Aggarwal, G., Kohli, K., Nanocarriers-assisted needle-free vaccine delivery through oral and intranasal transmucosal routes: A novel therapeutic conduit. *Front. Pharmacol.*, 12, 757761, Jan. 11, 2022.
124. Guerrini, G., Magrì, D., Gioria, S., Medaglini, D., Calzolai, L., Characterization of nanoparticles-based vaccines for COVID-19. *Nat. Nanotechnol.*, 17, 6, 570–6, Jun. 2022.
125. Plotkin, S., Robinson, J.M., Cunningham, G., Iqbal, R., Larsen, S., The complexity and cost of vaccine manufacturing–An overview. *Vaccine*, 35, 33, 4064–71, Jul. 24, 2017.
126. Cordina, M., Lauri, M.A., Lauri, J., Attitudes towards COVID-19 vaccination, vaccine hesitancy and intention to take the vaccine. *Pharm. Pract.*, 19, 1, 2317, 2021.
127. Patra, J.K., Das, G., Fraceto, L.F., Campos, E.V., Rodriguez-Torres, M.D., Acosta-Torres, L.S., Diaz-Torres, L.A., Grillo, R., Swamy, M.K., Sharma, S., Habtemariam, S., Nano based drug delivery systems: recent developments and future prospects. *J. Nanobiotechnology*, 16, 1:1–33, 2018 Dec.

4

Nanoparticle Formulations: A Sustainable Approach to Biodegradable and Non-Biodegradable Products

Amandeep Singh[1], Shreni Parikh[2], Nutan Sethi[2], Sachin Patel[3], Nrupal Modi[4] and Kaushika Patel[2*]

[1]Department of Pharmaceutics, ISF College of Pharmacy, Moga, Punjab, India
[2]Department of Pharmaceutical Technology, L. J. Institute of Pharmacy, L J University, Ahmedabad, Gujarat, India
[3]F&D - AGM, Stallion Laboratories PVT Ltd., Ahmadabad, Gujarat, India
[4]F & D, Catalent Pharma Solutions, Somerset, NJ, United States

Abstract

Nanoparticle-based formulations are one of the most sought-after dosage forms in the current scenario due to the potential benefits and versatile roles in the alleviation of diseases, especially cancer. Nanoparticles bag a specialized application in targeted and tailor-made formulations due to the excellent *in vivo* performance apart from contributing to desired *in vitro* characteristics. In the present chapter, the emphasis is given to the formulation aspects involved in the nanoparticulate dosage forms. The various types of nanoparticles (NPs) prevalent are discussed along with the techniques available for the manufacturing of NPs. Nanoparticles may be synthesized by matrixing or encapsulation with different types of polymers. The role of polymers, namely, biodegradable, non-biodegradable, and natural polymers, in NP designing is discussed in depth along with the description of the research work executed utilizing these polymers. The hurdles and challenges faced in the scalability of the NP-based formulations are also described. These challenges hold back the pharmaceutical industries to undergo the scale-up and approval process for the NP-based formulations. The fabrication of such products for oral application is again a distinguished area of research. Apart from the synthetic molecules, recently, extensive research is being undertaken for the NP synthesis of natural products having potential therapeutic benefits limited by the

**Corresponding author*: kaushikaspatel@gmail.com

Vivek P. Chavda and Vasso Apostolopoulos (eds.) Nanocarrier Vaccines: Biopharmaceutics-Based Fast Track Development, (95–152) © 2024 Scrivener Publishing LLC

stability and the bioavailability parameters. The present work summarizes all such products comprising NPs.

Keywords: Nanoparticle, biodegradable, natural, non-biodegradable

4.1 Introduction

One cannot help but note numerous improvements in medical practice when compared with that of the previous century. Many new medications have been created to successfully treat complex conditions, but some of them have severe side effects, so the benefit does not always justify the cost [1].

On the other hand, several treatments are highly effective *in vitro* but, if orally administered, are unable to resist the endogenous enzymes present in the gastrointestinal (GI) tract, leaving them effectively useless *in vivo* [2]. There remains a need improvement in drug delivery systems and targeting, despite significant advancements in recognizing therapeutic targets and engineering and producing better drug molecules. Over the recent decades, a diverse variety of scientific fields have devoted significant attention to nanotechnology, especially the development of NPs [3]. Because they can be customized for specific drug delivery, increased bioavailability, and controlled drug release of medication from a single dose, NPs made of synthetic and natural polymers (both biodegradable and non-biodegradable) have drawn the most attention. Through all the adaptation, the system could indeed stop endogenous enzymes from degrading the drug [4].

Secondly, the development of novel drug delivery techniques provides a unique chance for commercial sales to diversify. Because of their small size and large surface area, as well as their capacity to cross the blood–brain barrier (BBB), enter the pulmonary system, and be absorbed through the close connections of skin endothelial cells, the nanomaterials have enhanced solubility and thus improved bioavailability [4, 5].

Even though there are major drawbacks with NPs all through the formulation, drug loading, and scale-up procedures, they have a great chance of being granted Food and Drug Administration (FDA) and the European Medicines Agency (EMA) approval. FDA-approved NPs peaked between 2001 and 2005; after that, they started to decline, which the 2008 global financial crisis may have impacted. Considering this, it features the largest portion of all nano-carriers [6].

4.2 Types of Nanoparticles

Table 4.1 discusses the various kinds of NPs depicted in Figure 4.1.

Table 4.1 Types of nanoparticles.

Sr. no.	Type of nanoparticle	Significance	Reference
1	Carbon-based nanoparticles	These are composed of carbon atoms arranged in pentagonal or hexagonal arrays, and their allotropic form, i.e., fullerene, produces a hollow football-like structure composed of C60 and named Buckminster fullerenes. Another form is carbon nanotubes, which are subdivided into single-walled carbon nanotubes and multi-walled carbon nanotubes. They are employed economically because of their electron affinity, structure, and high strength, among other factors.	[7]
2	Ceramic nanoparticles	Due to their porosity, surface-to-volume ratio, surface area, and size, these inorganic solids provide a higher applicability in drug delivery systems. The drug is used to treat cancer, glaucoma, and bacterial infections.	[7]
3	Metal nanoparticles	These NPs have several applications in the domains of detection, bioanalysis, biomolecule imaging, and the environment. They are produced by chemical, photochemical, and electrochemical processes from metal precursors such as gold, copper, silver, and aluminum. They may ultimately absorb tiny molecules and possess high surface energy.	[7]

(Continued)

Table 4.1 Types of nanoparticles. (*Continued*)

Sr. no.	Type of nanoparticle	Significance	Reference
4	Lipid nanoparticles	Lipid NPs are composed of lipids and have a spherical form; their inner surface is composed of soluble lipophilic molecules. Consequently, they are frequently employed in RNA release systems and cancer treatment. Liposomes and beta-cyclodextrin are examples.	[7]
5	Polymeric nanoparticles	These are organic Nanoparticles (NPs). Due to their biodegradability and biocompatibility, they are widely employed for medication delivery and diagnostic reasons. The merits of this NPS include sustained release, selective targeting, and internal and external drug protection. They consist of different shapes and structures, such as nanocapsules and nanospheres.	[7]
5(a)	Nanocapsules	It contains the medication in its core, which is surrounded by a polymeric membrane.	[8]
5(b)	Nanospheres	It encapsulated the medicament inside the polymeric matrix and dispersed it uniformly.	[8]
6	Polymeric micelles	They are the micellar form of a block copolymer composed of lipophilic and lipophobic monomer units. They enhance the water solubility of weakly soluble medicines, hence increasing their transpermeability across physiological barriers. Owing to its lipophilicity and tiny size, they are not up taken by the RES (reticular endothelium system) for considerable time.	[8]

(*Continued*)

Table 4.1 Types of nanoparticles. (*Continued*)

Sr. no.	Type of nanoparticle	Significance	Reference
7	Semiconductor nanoparticles	They are frequently used in DNA hybridization, cell labeling, gene therapy, medical and electrical devices, and cancer treatment as well. Ex. Quantum dots.	[8]
8	Paramagnetic nanoparticles	These NPs have a size less than 100 nm and are controlled by a magnetic field. These NPs target certain organs for their identification and are used for both diagnostic and therapeutic reasons.	[8]
9	Solid lipid nanoparticles (SLNs)	These NPs are composed of liposomes, emulsions, and polymeric NPs and are used for the sustained release of different medications. SLNs have greater tolerability, enhanced bioavailability, biodegradability, and a compact size.	[8]
10	Nano-emulsions	They are commonly employed in drug delivery systems because to their low water solubility, high bioavailability, and heterogeneous structure. Due to their diminutive size and low surface tension between the aqueous and oil phases, they exhibit a rapid rate of drug distribution inside the body.	[8]

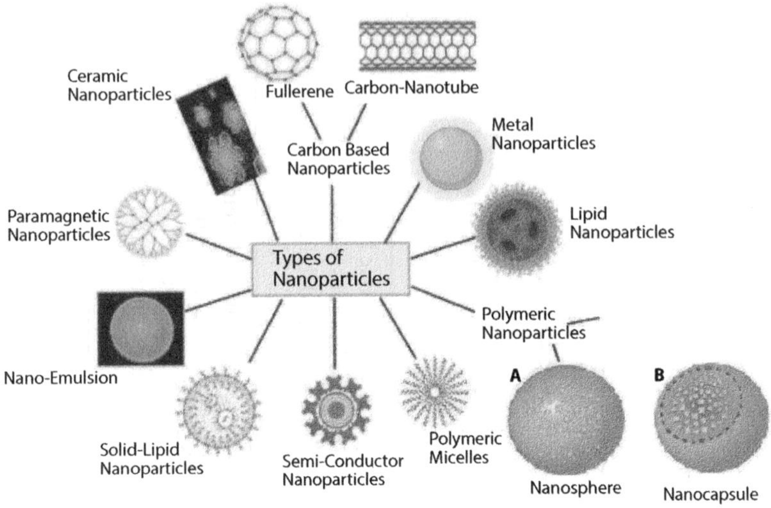

Figure 4.1 Different types of nanoparticles.

4.3 Preparation of Nanoparticles

Depending on their physicochemical characteristics, such as size, drug molecules, and loading needs, nano-formulation may be prepared using a variety of techniques as discussed in Table 4.2. Drug solubility, the amount of time that aqueous and organic solvents is mixed, the kinds of surfactants used, and the solvent of choice are just a few of the factors that come into play while creating polymeric NPs [9]. The general steps of manufacturing are explained in Figure 4.2 below.

4.4 Factors Affecting Selection of Method

A number of variables, including pressure, pH, reaction temperature, light intensity, leaf extract, drug concentration, $AgNO_3$ concentration, and reaction duration, all have a significant impact on the manufacturing of NPs. Modification of these factors is necessary to manufacture NPs in the correct size and to study their unique activity [18].

Temperature: Nanoparticle production occurs between 30 and 100 °C. The majority of the reaction is conducted at room temperature, which

Table 4.2 Different methods to formulate nanoparticles.

Sr no.	Method	Description
Performed Polymer		
1	**Solvent evaporation method**	It was the first technique for creating polymeric NPs, and the pharmaceutical sector may use it extensively. It primarily entails two processes: 1) Formation of single emulsions, such as oil in water (o/w), and 2) double emulsions, such as water in oil in water (w/o/w). Solvents including chloroform, ethyl acetate, and chloromethane are volatile oils that are used in these processes. An emulsifying agent is used to create the first emulsion, which is subsequently evaporated by vigorous continuous stirring. The hardened NPs are then removed by ultracentrifugation, washed in distilled water, and then lyophilized [9, 10].
2	**Nanoprecipitation/ solvent displacement method**	The foundation of this technique is the creation of a lyophilic solution through the interfacial deposition of a polymer, which is an emulsification/solidification technique following the displacement of a partially polar solvent that is miscible with water, which is why it is also named the solvent displacement method. Three phases make up this preparation: nucleation, growth, and aggregation. Using a magnetic stirrer, the organic phase is gradually introduced to the aqueous phase while swirling moderately in order to produce a specific range of NPs by fast diffusion of polymeric solution in a non-solvent phase. This approach uses a variety of organic solvents, including acetone and ethanol. It is a quick and simple way of making nanospheres and nanocapsules [9, 11].

(*Continued*)

Table 4.2 Different methods to formulate nanoparticles. (*Continued*)

Sr no.	Method	Description
3	**Salting-out method**	With the aid of high concentrations of salt and sucrose for a powerful salting-out effect, along with the addition of a stabilizer and a salting agent, the organic layer is emulsified in the aqueous layer using this method without the use of any externally applied forces, and after extensive washing, we obtain drug–polymer droplets. Both lyophilic medicines and heat-labile chemicals may use it extensively [9, 12].
4	**Ionotropic gelation**	This process is often used to create hydrogel NP formulation from polymers like chitosan and alginate. When the drug and chitosan combination is dissolved in an alkaline tripolyphosphate solution and stirred continuously, aggregates are formed at a pH that promotes cross-linking [9, 13].
5	**Supercritical fluid technology**	It is a method for producing NPs without the use of a solvent, and it may be broadly divided into two types: fast expansion of critical solution and supercritical anti-solvent (RESS). Typically, CO_2 is utilized in this procedure as a supercritical fluid to treat temperatures over its critical temperature and pressure and produce pure NPs [9, 14].
Monomer Polymerization		
1	**Emulsion polymerization**	In this phase, an organic continuous phase and surfactant mixture is treated using the ultrasonication method to create monomer droplets in the aqueous continuous phase. After the polymerization process, the result is a suspension of NPs, and the surfactant, along with other protective soluble polymers, prevents the initial aggregation of NPs [9].

(*Continued*)

Table 4.2 Different methods to formulate nanoparticles. (*Continued*)

Sr no.	Method	Description
2	**Mini-emulsion polymerization**	To establish the equilibrium state for the creation of core shell NPs using this approach, a low-molecular-mass component, such as a costabilizer, is needed. High-tension- generating equipment, such as ultrasound, may also be used. Water, a surfactant, a monomer combination, an initiator, and a costabilizer are used in this procedure [9, 15].
3	**Micro-emulsion polymerization**	There are much fewer chains of particles and smaller particles when kinetically compared to mini-emulsion polymerization. A water-soluble initiator is combined with the water phase to create a microemulsion that is thermodynamically stable and includes swelling micelles. This approach employs a large quantity of surfactant, which reduces the interfacial tension at the o/w contact to zero and completely coats the particles. Polymeric chains are first recognized as tiny droplets, to which osmotic and elastic management of chains leads to destabilization, increasing the average size of the chain and forming the final product [9, 16].
4	**Interfacial polymerization**	When radical polymerization, polycondensation, and polyaddition are used as cross-linking processes at the interface of two liquids, two reactive monomers that are dissolved in continuous and dispersed phases are polymerized to produce hollow particles that are nanoscale in size. By utilizing aprotic solvents like acetone and acetonitrile, interfacial polymerization of the monomer at the w/o interface in w/o micro-emulsion precipitates to generate nanocapsules [9, 17].

(*Continued*)

Table 4.2 Different methods to formulate nanoparticles. (*Continued*)

Sr no.	Method	Description
5	**Controlled/ living radical polymerization**	Aqueous dispersed systems, which are crucial for industrial applications, can acquire living radical polymerization through three different processes: atom transfer, radical polymerization (ATRP), nitroxide-mediated polymerization (NMP), and reversible addition and fragmentation transfer chain polymerization (RAFT) [9].

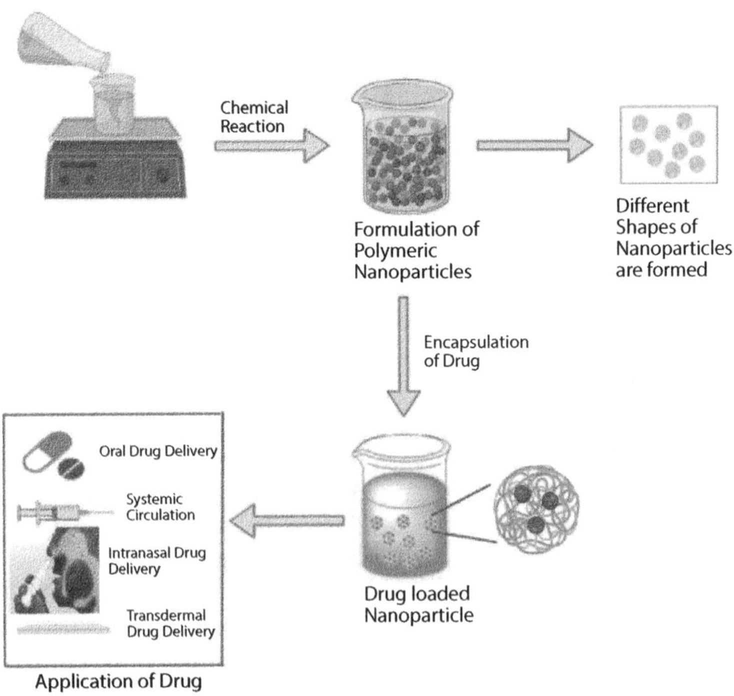

Figure 4.2 Preparation of nanoparticles.

prolongs the reaction's duration. However, by raising the temperature, we can improve it. The creation of NPs reduces and stability increases as the reaction mixture's temperature rises [18].

The size, shape, and pH of the NPs are all significantly influenced by pH. At lower pH levels, NPs form erratic aggregates and are shown to be bigger in size. At different pH levels, the creation of NPs with the required size and shape is accomplished. To avoid aggregation at alkaline pH, NPs are dispersed at the colloidal stage as clusters. The produced gold NPs display hexagonal, irregular, icosahedral multi-twinned, tetrahedral, rod-shaped, and decahedral structures at pH 2, 3, and 4 [18–20].

4.4.1 Pressure

Pressure is a key factor in modifying the size and shape of NPs to achieve uniform shape and desired size during synthesis. Utilizing biological agents speeds up the reduction of metal ions under conditions of substantial pressure [20].

4.4.2 Particle Shape and Size

When estimating the characteristics of NPs, particle shape and size are crucial factors. This component aids in identifying the physical characteristics of NPs, such as melting point, which falls as nanometer-scale particle size increases. When two NPs have comparable energies but distinct atomic configurations, the form of the particles may be readily changed. Chemical characteristics have an impact on the dynamic behavior and form of the NPs during production [20].

4.4.3 Environment

Environmental conditions influence the physicochemical characteristics of NPs during their production as well as the nature of the particles. A single NP absorbs or interacts with additional components from the environment and creates core–shell NPs as a result of oxidation or corrosion. The coating in a biological system results in the formation of thicker and bigger NPs. Therefore, the biological environment has a significant impact on the creation of NPs and on their chemical and physical makeup. Zinc-sulfide NPs shift from being crystalline to being non-crystalline as the environment changes from being wet to being dry. Cerium nitrate NPs undergo a chemical change when suspended in a peroxide-containing solution [21–23].

4.4.4 Pore Size

Porosity, which is a factor in terms of the generated NPs' pore size, has an impact on the effectiveness and use of NPs in the pharmaceutical sector. This component is crucial for the immobilization of biomolecules onto NPs, a process that is often used in the biomedical industry for drug delivery [18, 24].

4.4.5 Particular Method or Technique

There are several ways to create NPs, including physical, chemical, and biological approaches, each of which has advantages and disadvantages of its own. The most important aspect influencing the quality and quantity of the produced NPs is the choice of a specific chemical process [18].

4.4.6 Cost of Preparation

The expense associated with creating the NPs must be managed and adjusted. The primary element influencing the synthesis of NPs is their profitable manufacture. The synthesis of NPs utilizing a chemical process result in a larger yield of product in a shorter length of time and is more cost-effective [18].

4.4.7 Proximity

When an isolated NP interacts with other NPs, its characteristics are changed, and new modified NPs are created as a result of the altered NPs' changing behavior. These variables have a substantial impact on the NPs in many ways, including substrate interactions, particle charging, and magnetic characteristics [18, 25].

4.4.8 Time

Nanoparticles may aggregate, contract, or grow in size and quantity as a result of being stored for a prolonged length of time. For instance, tetra chloroauric acid and trisodium citrate dihydrate interact during the creation of gold NPs, which is a time-dependent process [18, 26, 27].

4.4.9 Other Variables

The quality and amount of the produced NPs may fluctuate depending on the procedure used for purification. Chromatography methods are used to separate the NPs based on variations in the coefficients of the mobile phase and stationary phase. The separation of the NPs based on their solubility, which is dissolved in two separate miscible liquids, typically water and organic solvents, may be accomplished using the electrophoresis or chromatography approach. The use of NPs in the pharmaceutical and biomedical sectors depends heavily on this separation of the particles [18, 28–30].

4.5 Polymers Used in NP Formulation

The bio-engineered material known as polymers has undergone substantial investigation in the past because of its adaptability and capability to meet the most stringent standards for drug delivery systems. For a particular targeted medication administration, polymeric NPs are crucial in reducing drug toxicity and improving drug bioavailability. Modern drug delivery systems modify the polymers' structural characteristics to provide them observable biological activity [31]. In the drug delivery system, polymers are often employed because of their superior surface and bulk characteristics. The ability of polymers to load and release medicinal molecules is crucial. Certain polymeric NP systems may pass the BBB. The polymeric structure protects the medication against physical and chemical deterioration, increasing its durability. It is possible to identify the precise molecular target for treating cancer when polymers are coupled with certain biomarkers or antibiotics [32]. It is conjugated with a polymer that acts as an inert carrier to improve pharmacokinetic (PK) and pharmacodynamic (PD) properties such as plasma $t_{1/2}$, increases the solubility of low-molecular-weight drugs, promotes the stability of biopharmaceuticals, reduces immunogenicity, and has potential for targeted delivery of biopharmaceuticals [33]. Three different kinds of polymers used in the creation of NPs as named below are discussed in Table 4.3, and their mechanisms are demonstrated in Figure 4.3.

1. Biodegradable polymer
2. Non-biodegradable polymer
3. Natural polymer

4.6 Nanoparticle Formulations Based on Biodegradable Polymers

Modern research indicates that the production of bioactive macromolecules loaded with NPs for optimal pharmacological effects *in vivo* in the biopharmaceuticals business serves as a carrier for drug delivery systems. Poly (D, L-lactic acid) (PLA) is a biocompatible, biodegradable, and

Table 4.3 Types of polymers used in nanoparticles.

Sr no.	Types of polymers	Details	Examples
1.	**Biodegradable polymer**	Through enzymatic or non-enzymatic mechanisms inside the body, biodegradation converts the biodegradable polymer into a biocompatible or harmless by-product [34]. Numerous systems, including tissue engineering, controlled drug delivery systems (CDDS), regenerative medicine, sustained release drug delivery systems (SRDDSs), nucleic acid, anticancer medication delivery, protein, and vaccines, depend on biodegradable polymers [35]. The ideal characteristics of a biodegradable polymer are 1) It should not excrete toxic products that are harmful to the body or the environment, 2) It should have fine permeability and therapeutic properties, 3) It should be abundantly available in nature and inexpensive, and 4) It should not cause persistent inflammatory effects. 5) When the particles are circulated throughout the body, the relationship between them should be readily breakable. 6) Drugs released gradually rather than all at once [36].	• **Polyesters:** ➤ Poly (lactic acid) ➤ Poly (ε- caprolactone) ➤ Poly (dioxanomes) ➤ Poly dianones ➤ Polyhydroxybutyrate ➤ Poly (glycolic acid) ➤ Poly (B-maleic acid) ➤ Poly (hydroxybutyrate) • **Polyanhydrides:** ➤ Poly sebacic acid ➤ Poly adipic acid ➤ Poly terephthalic acid • **Phosphorus Based:** ➤ Polyphosphates ➤ Polyphosphazenes ➤ Polyphosphonates

(*Continued*)

Table 4.3 Types of polymers used in nanoparticles. (*Continued*)

Sr no.	Types of polymers	Details	Examples
		Ester-containing polymers, which in nature might be aliphatic ester, aromatic ester, or copolymers of both, make up the majority of biodegradable polymers. The ester bond of this polymer is broken as a result of an enzymatic process [37]. Biodegradable polymers degrade at different rates depending on variables such as pH, bond type, water content, enzyme concentration, solubility, temperature, and the presence of microorganisms. Metal catalysts, condensation reaction, and ring opening polymerization may all be used to create this polymer [37]. Gases, salts, and biomass are produced as byproducts of the degradation of biodegradable polymers [38]. There are two ways by which biodegradable polymers are broken down: 1) physical disintegration and 2) biological processes. In which biological processes take place via aerobic (presence of oxygen) and anaerobic (absence of oxygen) processes, physical breakdown includes photodegradation and hydrolysis that results in partial and total destruction. The leftover or uninterested parts of physical processes are taken up by microorganisms, which then simplify them. When there are no oligomers or monomers remaining, the biodegradation process is complete. Enzymatic conversion of adenosine triphosphates and polymeric byproducts such nitrogen gas, water, biomass, carbon dioxide, methane, and minerals yield all of the oligomers and monomers [39].	• **Polyamide:** ➤ Poly(imino carbonates) ➤ Polyamonoacids • **Others:** ➤ Polyurethanes ➤ Polyorthoesters ➤ Polycyanoacrylate ➤ Poly (trimethyl carbonates)

(*Continued*)

Table 4.3 Types of polymers used in nanoparticles. (*Continued*)

Sr no.	Types of polymers	Details	Examples
2.	**Non-biodegradable polymer**	When used as a drug carrier, non-biodegradable polymers do not degrade and stay stable until the drug has reached the intended target organ [38]. Non-biodegradable polymers provide advantages such as the capacity to resolve microbial degradation, improve resource efficiency, and lower carbon footprint during manufacture. The non-biodegradable polymers have important qualities such as resistance to biodegradation, water resistance, cost effectiveness, and ease of production [40].	• **Acrylic Polymers:** ➤ Polymethacrylate ➤ Polymethyl methacrylate ➤ Polyhydro ethyl methacrylate • **Silicones:** ➤ Colloidal silica ➤ Poly dimethyl siloxanes ➤ Cellulose Derivatives: ➤ Ethyl cellulose ➤ Cellulose acetate ➤ Carboxymethyl cellulose • **Others:** ➤ Poloxamine ➤ Ethyl vinyl acetate ➤ Polaxamers ➤ Polyvinyl pyrrolidone

(*Continued*)

Table 4.3 Types of polymers used in nanoparticles. (*Continued*)

Sr no.	Types of polymers	Details	Examples
3.	**Natural polymer**	When acting as a natural polymer-based drug carrier, polysaccharides and proteins, such as chitosan and hyaluronic acid, exhibit the greatest similarity and the least amount of interfering qualities. The drug delivery system is made up of these natural polymers. Wound healing, tissue engineering, and anticancer treatment often make use of low-molecular-weight medicinal compounds, DNA, and proteins loaded with natural polymers. Drug delivery systems are made up of natural polymers, their derivatives, and other materials that go through chemical and morphological changes that cause them to bind to the targeted location, which increases drug delivery and yields therapeutic benefit [41]. Natural polymer-based scaffolds are very mechanically adaptable, have a high water-holding capacity, and closely mimic an extracellular matrix. Natural polymers go through many reactions such as ligand conjugation and cross-linking when they serve as a drug carrier for the synthesis of therapeutically active proteins [42].	• **Polysaccharide Based:** ➤ Chitosan ➤ Hyaluronic acid ➤ Dextran ➤ Cyclodextrin ➤ Agarose • **Protein Based:** ➤ Elastin ➤ Albumin ➤ Gelatin ➤ Collagen

(*Continued*)

Table 4.3 Types of polymers used in nanoparticles. (*Continued*)

Sr no.	Types of polymers	Details	Examples
		Natural materials are employed extensively in contrast to artificial materials because of their beneficial qualities, such as dominance in biocompatibility, sourcing from natural resources, ease of application, affordability, and accessibility [42, 43]. When a natural polymer's functional group is changed, new materials are created that may react with other materials to create new hybrid materials with modified physicochemical characteristics [41]. Due to their greater affinity for cell receptors, some natural polymers control cellular processes including proliferation, migration, and adhesion.	

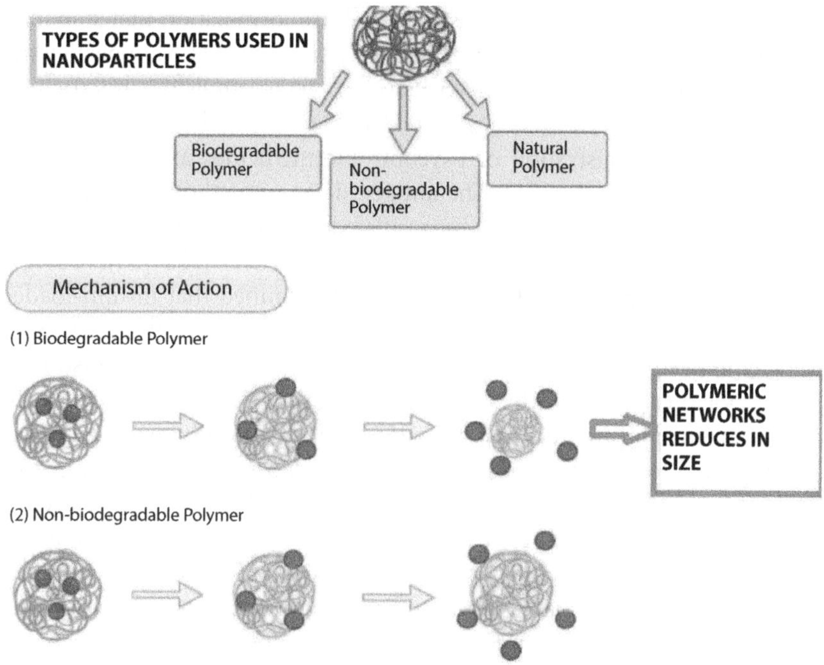

Figure 4.3 Mechanism of nanoparticle polymers.

non-toxic polymer that is extensively employed for the administration of a variety of medicinal medicines [44]. Inhaling *in vitro* transcribed mRNA via the nasal cavity does not improve its efficacy; thus, it is coupled with poly (beta amino esters) (hPBAEs) to improve medication delivery to the epithelium of the lung [45]. As a method for extending the shelf life of the formulation, polyethylene glycol (PEG) water molecules are removed [46]. Lysosomes, Fab, and IgG are utilized to treat posterior eye problems when they are suspended in pentablock (PB) copolymer based on PEG, polylactic acid, and polycaprolactone) by double emulsion solvent evaporation [47]. Nile Red (NR) and Doxorubicin (DOX) are encapsulated in dextran-g-poly (o-nitrobenzyl acrylate) copolymers (Dexg-PNBA), a photophobia nanocarrier that provides prolonged release for the nanomedicines during drug administration [48]. Rebecca A. Chmielowski produced ferulic acid-based poly (anhydride ester) NPs that are efficacious at the onset of the athero-inflammatory cascade in cardiovascular disease [49]. The food delivery system uses oligo-hyaluronic acid curcumin (OHC) biodegradable amphiphilic polymer for the delivery of curcumin (Cur) and resveratrol (Res) [50]. Polymersomes are biodegradable polymeric vesicles encasing hydrophilic molecules in the aqueous compartment and hydrophobic molecules in the vesicular bilayer, resulting in a sustained release formulation that stabilized the particle's system to boost drug delivery efficiency [51] Nanoparticles of biodegradable polymers, nanosuspensions, nano-emulsions/microemulsions, SLNs, polymeric micelles, and dendrimers are the carriers for the efficient oral administration of psychotropic medicines and to maximize therapeutic effectiveness [52]. The antigen delivery method including protein antigen ovalbumin (OVA) encapsulated in hydrophilic polyester poly (lactide-co-hydroxymethyl glycolic acid) (pLHMGA) NPs [53]. Using a Box–Behnken experimental design, it was determined that insulin-loaded poly (hydroxybutyrate-co-hydroxyvalerate) (PHBV) NPs (INS-PHBV-NPs) were employed for sustained release formulation insulin administration [54]. A polymer grafted with NPs of cerium oxide was created to prevent the coating from splitting during drying [55]. Therapeutically, several carriers are employed to transport anticancer medicines. Nanoprecipitation is used to create cytarabine-loaded poly (-caprolactone) (PCL), which decreases dose-related toxicity and so has anticancer efficacy [56]. By the combination of Celastrol (Cela) and ferrocene (Fc) with pH-sensitive NPs (PP/H NPs) prepared by poly (lactic-co-glycolic acid) modified with -cyclodextrin (PLGA–CD), polyethyleneimine grafted with benzimidazole (PEI-BM), and low-molecular-weight heparin (LMWH), the tumor-inhibiting rate is significantly decreased [57]. Fucoidan is a polysaccharide (marine polymer) derived from a variety of brown algae, and it exhibits binary-targeted systems used in the treatment of pancreatic cancer [58]. Doxorubicin-loaded redox-sensitive NPs (DOX-SS-NPs) are a bioresponsive nano-drug used in cancer treatment that can be produced by two techniques, including the oil-in-water emulsion approach and the cosolvent removal method [59, 60]. In order to promote the release of siRNA in malignant cells, solid poly (disulfide amide) (PDSA)/cationic lipid core and (lipid–PEG) based on biodegradable and redox-responsive nanoparticles are encapsulated with biodegradable and redox-responsive properties [61]. Noscapine (NOS) is encapsulated in PEG-co-poly (-caprolactone) block copolymer to boost its colloidal stability in the blood, hence increasing its half-life and its availability in the body, so advantageous against the tumor [62]. As2O3 is mixed with nanodrugs such as liposomes, albumin-based nanodrugs, polymer micelles, and silica-based nanodrugs in order to minimize its toxicity and create antimitotic action [63]. Poly (lactic-co-glycolic acid) (PLGA) NPs encapsulated with paclitaxel and hyaluronic acid (HA-PTX-PLGA) are effective against triple-negative breast cancer cells. They are manufactured via a modified oil-in-water emulsion process [64, 65]. The Arg-Gly-Asp peptide sequence (RGD peptide) that is a generic cell-targeting peptide modified, redox-sensitive, prodrug-based lipid polymer NPs used for the administration of paclitaxel (PTX) and cisplatin (CDDP) and generate synergistic impact on malignant cell in lungs [66]. Recent research by Xuanyu Lia and Xingyu Jiang on microfluidic chips demonstrates that when its inner core is combined with PLGA NPs, it serves as an effective vehicle for medication delivery in the treatment of cancer [67]. Docetaxel-loaded SLNs offer antimitotic action [68]. PLA serves as a carrier for the natural flavonoid Fisetin, which has anticancer action, and is synthesized via the emulsification solvent diffusion technique [69].

PLGA is a biodegradable polymer that is recommended by several scientists owing to its low toxicity, excellent biocompatibility, and high bioavailability, as well as its US FDA approval. When loaded with therapeutic macromolecules, the use of PLGA NPs in the design, formulation, and production processes is for the controlled distribution of pharmaceuticals throughout the body [70]. Preparation of hybrid lipid–polymer NPs adjuvant by the oil-in-water single emulsion method by association of cationic surfactant dimethyl dioctadecyl ammonium (DDA) bromide and immunopotentiator trehalose-6,6'-dibehenate (TDB) with biodegradable poly (DL-lactic-co-glycolic acid) (PLGA) NPs [71] Flufenamic acid (NSAID) transdermal distribution is enhanced by iontophoresis when loaded with PLGA [72]. Due to its stable structure, encapsulating Saquinavir (SQV) with PLGA enhances its bioavailability [73] Duloxetine (DLX) is encapsulated with PLGA and Polycaprolactone (PCL) for the efficient delivery of CNS drugs to the specific target in the brain in order to overcome first pass metabolism, unpredictable bioavailability, and degradation of its acid-sensitive molecule [74] Retinoic acid is transported through the skin via PLGA and nanostructured lipid carriers [75] Quercetin is a possible therapeutic medicine used to treat Alzheimer's disease; however, it cannot cross the BBB. Thus, it is encapsulated with PLGA to boost its therapeutic efficacy and reduce its negative effects [76]. Fluazinam (Flu) is encapsulated in PLGA NPs (NPs) using polyvinyl alcohol (PVA) and sodium dodecyl sulfate (SDS) surfactants using the Shirasu Porous Glass (SPG) membrane emulsification process [77]. Chrysin (Chr) is surrounded by PLGA and has anti-inflammatory and damaged cell regeneration properties in the human body [78].

4.7 Nanoparticle Formulations Based on Non-Biodegradable Polymers

Drug delivery methods that are not biodegradable are effective in treating chronic illnesses that call for long-term medicine usage. They feature an interatomic form of bonding based on their structural activity relationship (SAR), which makes it difficult for microbes to break the link and digest them, providing exceptional durability and strength. As a result, it delays the bacterial breakdown of the system. Utilizing polyurethanes, silicones, polyacrylates, or copolymers like poly(ethylene vinyl acetate), non-biodegradable implantable systems that are delivered by a simple method have been created. Vancomycin entrapped in polymethyl methacrylate (PMMA) is used to treat post-traumatic and post-operative osteomyelitis by eradicating infection [79]. To treat bacterial infections and produce tissues restorations, 2-hydroxyethyl methacrylate (HEMA) and the cross-linking monomer trimethylolpropane trimethacrylate (TMPT) are combined with fibroblast growth factor-2 (FGF-2) or cetylpyridinium chloride (CPC), which is hydrophilic in nature and absorbs protein and water-soluble antimicrobial [80]. When oleic acid–reinforced PEGylated polymethacrylate (OLA-PEG-E-RLPO) is used to encapsulate atorvastatin calcium (ATC), the antidyslipidemic action is amplified and bioavailability is improved [81]. The use of polymethylmethacrylate beads boosts gentamicin's antibacterial effectiveness against the staphylococcal strain [82]. Ethyl vinyl acetate is entrapped in a nanocomposite of montmorillonite (organo-MMT) nanofiller that exhibits great toughness, thermostability, and biostability and is used as an insulating material for implanted devices [83]. When lornoxicam (LX), a hydrophobic model drug, is loaded with poly (ethylene oxide)-poly (propylene oxide), which is combined with highly hydrophilic poloxamers, Synperonic PE/P84, and Synperonic PE/F127, as well as the hydrophobic poloxamine counterpart (Tetronic T701) by the comicellization process, it increases the corneal penetration power of the ophthalmic drug [84]. Prednisone, a medication that dissolves only sporadically, becomes more soluble when it is encased in povidone (polyvinylpyrrolidone, PVP) nanofibers [85]. Griseofulvin (GF) increases the rate of solubility to create antifungal action when combined with hydroxypropyl methyl cellulose (HPMC) [86]. In order to create electrorheological fluids using the freeze-drying method, silicon oil suspension is confined inside extremely porous chitosan particles to increase the fluid's sedimentation stability [87].

Dendrimer is a non-biodegradable cytotoxic drug carrier that is used for targeted medication therapy to treat cancer. There are other cytotoxic drug carriers as well. These are the vehicles that transport antineoplastic drugs, which prolong the cytotoxic agent's half-life, payload, and solubility while reducing renal clearance [88]. Poly[2-(dimethylamino) ethyl methacrylate] and poly (-amino ester) (PDMAEMA-b-PAE-b-PDMAEMA), a cationic based tri-block copolymer, is employed as a gene carrier to treat cervical tumors [89]. A non-biodegradable glucose polymer called hydroxyethylcellulose serves as a vehicle for the administration of methotrexate (MTX), a drug used to treat tumors [90]. To counteract the carcinogenic impact, doxorubicin (DOX-FA-MSNPs or DOX-NAG-MSNPs) is encapsulated with mesoporous silica NPs and transported via GLUT [91].

Due to its high solubility in stomach PH and high thermoplastic nature, Eudragit is an appropriate acrylic material for film coating of controlled release of drugs. Additionally, it serves as a non-medicinal component in formulation, is non-toxic, and non-irritating [92]. Aspart-insulin, a short-acting insulin analogue, is delivered by insulin loaded with a combination of biodegradable polyester (poly(e-caprolactone)) and a polycationic non-biodegradable acrylic polymer (Eudragit 1 RS) [93]. For the effective drug delivery system, low-molecular-weight heparin is loaded with Eudragit polymer mixed with [poly-"-caprolactone, PCL, and poly (D,L-lactic-co-glycolic acid) 50/50, PLGA] that is manufactured utilizing solvent evaporation and ultrasonic probe [94]. Amphotericin-B that is positively charged is enclosed in positively charged polymer. Eudragit RL 100 is created using solvent displacement or nanoprecipitation methods, and it treats eye irritation and functions as an antimycotic [95]. The solvent evaporation process was used to create the glipizide drug-loaded Eudragit RS100, which results in prolonged drug releases, maximum entrapment efficiency with minimal particle size, fewer side effects, reduced dosage frequency, and improved patient compliance [96]. Oseltamivir phosphate (OSE), which is biocompatible and possesses catatonic qualities that trigger apoptosis utilizing the intrinsic route, is combined with pegylated Eudragit-based nanoparticles. Lung cancer may be treated using effective medication delivery systems [97]. When coupled with the biodegradable polymer PLGA, two negatively charged, non-biodegradable polymers called Eudragit RS100 and RL100 work as antibiotics against *Staphylococcus aureus* and *Pseudomonas aeruginosa* [98].

Potential inorganic–organic drug carrier mesoporous silica polymer has a sophisticated pore structure that binds huge amounts of drug to boost the system's binding capacity [99]. Additionally, when it is coated with Eudragit polymers, it prevents early drug disintegration in gastric PH and reduces meropenem efflux by acting as an antagonist against p-glycoprotein protein efflux pumps. Meropenem-loaded porous silica (MCM-48) polymer provides a better loading capacity of the drug [100]. When di-acrylated Pluronic F127 (F127-DA) is covalently attached to polydimethylsiloxane (PDMS) chips, the result is a non-swelling property that is useful for treating cells and tissues [101]. Since colchicine has low bioavailability, gastrointestinal side effects, and a prolonged first pass effect, it is loaded with mesoporous silica NPs (MSNs) and combined with a self-healing hydrogel to create transdermal patches that allow the medicine to be absorbed easily via the skin [102]. Tacrolimus enhances the topical administration of medicines that are poorly soluble when combined with hydrogel and MSNs to treat atopic dermatitis (AD) [103]. The co-spray dried mesoporous silica (SBA-15) that is combined with artemisinin (ART) and mefloquine (MFQ) to create a single dosage form boosts the solubility of the partly soluble antimalarial medication [104].

4.8 Nanoparticle Formulations Based on Natural Polymers

Several natural polymers have been utilized till date for the synthesis of NPs [105]. Natural polymers consist of massive organic long chains derived from natural sources. Additional polymerization or condensation polymerization is used to synthesize them. They are primarily composed of three structures: polynucleotides, polyamides, and nucleic acid blocks. They were significantly less hazardous than synthetic polymers. Double hydrophilic thermo-responsive pullulan-g-poly(N-isopropylacrylamide) (P-g-pNIPAM) copolymers are employed in the creation of indomethacin-loaded NPs, and they are fabricated by dialysis or nanoprecipitation [106]. Bovine serum albumin is combined with NPs for pulmonary administration. Albumin serves as a vehicle for the sustained release of antigen following injection and plays a crucial role in the development of novel vaccines used to treat infectious diseases or cancer [107]. Rutin encapsulated in the polymeric system of zein-based nanosystems has antioxidant action [108]. The NP dispersion within the hydrogel serves as the platform for the pharmaceutical medication delivery system [109]. Nanoformulations of LipoPOx (lipopolyoxazolines) including polyoxazoline blocks and lipid derivatives are explored for cutaneous and intravenous drug delivery [110]. Phthalate angico gum is a semicrystalline, hydrophobic polymer that is employed in drug delivery systems [111]. By enhancing the creation of spherical nanosystems with a limited size distribution, super-refined polyoxyethylene [2] oleyl ether achieves the greatest results. Polyoxyethylene [2] oleyl ether-stabilized gliadin-based nanosystems serve as carriers for the delivery of diverse active chemicals [112]. Trans-Resveratrol delivery when loaded with Polyoxyethylene 40 Stearate-based nanostructured lipid carrier confers the highest Tyrosinase inhibitory action over Glyceryl Behenate [113], antileishmanial activity of Eudragit-coated Amphotericin B (AmB), and piperine (Pip)-loaded guar gum NPs [114]. Albendazole (ABZ) is only moderately water soluble; hence, it is encapsulated with cyclodextrin (CD) or hydroxypropyl–cyclodextrin [HPCD] to create anthelmintic action [115] Curcumin-loaded pluronic micelles have extensive use as antioxidants, anticancer agents, and painkillers [116]. When dexamethasone (Dex) is encapsulated in liposomes, there are no adverse effects on human bone marrow–derived mesenchymal stem cells (hBM-SCs) and the safety and efficacy of regenerative medicine is enhanced [117]. Propolis, which is primarily composed of wax and resins coated with SLNs, serves as a vehicle for the local administration of diclofenac and flavonoids for the treatment of allergic rhinitis [118].

Chitosan is the commercial natural linear amino polysaccharide polymer because it satisfies all regulatory requirements and serves as the carrier for nano drug delivery systems. Chitosan has been utilized extensively as a natural biopolymer for many years due to its polycationic nature, biocompatibility, non-toxicity, biodegradability, and non-allergenicity. Chitosan-coated ZnO NPs facilitate the distribution of curcumin and improve the therapeutic effect on numerous incurable diseases, such as Hepatitis-C and Alzheimer's disease [119]. Chitosan that has been cross-linked with cinnamaldehyde exhibits antibacterial action [120]. As an oral drug delivery strategy for diabetes, insulin-loaded NPs are created through self-gelation using chitosan and naturally produced water-soluble snail mucin [121]. When chitosan is combined with NaF and tripolyphosphate, stable NPs are formed that display continuous delivery of fluorine and offers prevention against tooth caries development [122]. When water-insoluble pesticides were encapsulated with Paraquat-loaded porous carbon NPs (PCNs) and then coated with chitosan, they exhibited enhanced fungicidal activity and decreased cytotoxicity to human normal cells [123] Chitosan serves as a nanocarrier for lupulone and xanthohumol and has broad antibacterial action [124]. Clove essential oil–loaded chitosan NPs enhance antifungal effectiveness of clove essential oil [125]. Chitosan NPs serve as carriers for *S. pneumoniae* infection–treating cpl-1 [126]. 5-fluorouracil [5FU] and doxorubicin [Dox], when combined with a chitosan–gold aptamer, improve glioblastoma treatment [127]. Protein-encapsulated BSA-loaded chitosan/oligonucleotide (Dz13Scr) NPs are employed for illness management [128]. When synthesized with sericin-loaded hydrogel, silver NPs coated with chitosan demonstrate wound-healing and antibacterial capabilities [129]. Cytarabine-loaded chitosan NPs (CCNPs) demonstrate anticancer efficacy *in vitro* [130]. N, N, N-trimethyl chitosan (TMC), which is the hydrophilic derivative of chitosan (CHT, a linear cationic polysaccharide), is utilized in the manufacture of medicines [131]. Protein carriers for pulmonary and nasal transmucosal administration include chitosan, carrageenan, and tripolyphosphate [132]. Rutin NPs coated in chitosan are designed to cure cerebral ischemia by targeting the brain [133]. By using the nanoemulsion process, paclitaxel-containing chitosan NPs that are effective against a variety of malignancies can be produced [134]. Preparation of liraglutide-loaded chitosan-coated calcium-alginate nanocapsules is used for oral sustained delivery of liraglutide to treat diabetes [135] through the inclusion of a chitosan-based platform with dextran-grafted maghemite nanoarchitectures (DM) and L-arginine (L-Arg) amino acid as a bioactive agent in tissue engineering and regenerative medicine [136]. Chitosan encased in naringenin (NAR) NPs stabilized by tripolyphosphate protects human cells from free radicals and demonstrates anticancer action [137]. Lecithin functions as an anionic surfactant capped with chitosan and its inclusion with Ibuprofen for the treatment of dry eye condition [138]. Gelatin is a polymer that is naturally water-soluble. To synthesize diclofenac sodium, NPs of gelatin were utilized [139]. *Aspergillus niger* was inhibited by the gelatin-based nanocomposite films that contained chitin NPs [140] Gelatin NPs containing moxifloxacin are employed for its potent ocular administration and to control its subsequent release in the cornea of the eye [141].

Highly suggested for the delivery of oncological medications is the field of nanotechnology. Doxorubicin-loaded mesoporous silica NPs (MSNs)-poly-L-ornithine (PLO) and carboxymethyl lentinan (LC) demonstrate exceptional antiproliferative activity against human cervical cancer cell lines (Hela) [142]. Gemcitabine (2',2'-difluoro-2'-deoxycytidine; dFdC) is a strong anticancer agent that is utilized in the *in vivo* delivery of macromolecular prodrugs and nanomedicines [143]. Tamoxifen (TMX)-loaded papain- grafted S-protected hyaluronic acid-lithocholic acid co-block (PAP-HA-ss-LCA) included the self-nanoemulsifying drug delivery system (SNEDDS) utilized for breast cancer treatment [144]. Melatonin-loaded forsterite NPs enhance hydrogels' biological and mechanical characteristics. Teriflunomide (TFM)-loaded nano lipid-based (TNLC) carbopol–gellan gum *in situ* gel [TNLCGHG] is therapeutically suggested for glioma, brain, and spine tumors [145]. Bortezomib-loaded polypropyleneimine and polyamidoamine dendrimers are employed in the treatment of multiple myeloma and mantle cell lymphoma [146]. Metallic NPs are widely employed in the field of biomedicine due to their numerous benefits, which include the ability to increase the therapeutic index of medications and to serve as a vehicle for sustained drug release [147]. The mixture of tyrosine and starch is utilized in the synthesis of carbohydrate polymer-inspired silver NPs that exhibit antifilaricidal and antimosquitocidal action [148] As a vaccination, gold glycoNPs (GNPs) are loaded with a listeriolysin O peptide LLO91–99 (GNP-LLO) for the treatment of listeria infection [149].

Gold NPs (AuNPs) coated with l-cysteine methyl ester hydrochloride and poly(ethylene glycol) (PEG) are used for siRNA delivery [150].

4.9 Challenges in NPs from Laboratory to Industrial Scale-Up

The continued development of NP manufacturing necessitates market and economic efficiency. From a commercial standpoint, the nanomedicine business is estimated to increase at a rate of USD 350.8 billion by 2025. Despite these advantages, the fundamental issues for NPs s in nanomedicine research have shown to be a significant impediment to getting NPs to the clinical stage [151].

Two examples of NP-based pharmaceuticals that have acquired US approval from The U.S. Food and Drug Administration are Doxil® (Janssen Products, LP, PA, USA) and Abraxane® (Celgene Corporation, NJ, USA) for the treatment of advanced breast cancer and ovarian cancer, correspondingly.

However, researchers are still confronted with multiple challenges in their efforts to enhance patient benefits throughout diagnosis and treatment related to NPs [152].

The scale-up step comprises the incorporation of methods and the transfer of technologies for the large-scale industrial production of nanomaterials. Any preparative methodology may fail to translate from a laboratory level to an industrial level owing to technical obstacles in small-scale preparation. A well-planned scale-up approach will also ensure the accuracy of the nanomedicine, cost-effectiveness, and timely product launch [153].

Another critical element is the operational limits, which include the durability of the materials employed in the processes and the use of harmful solvents (e.g., the usage of dichloromethane or chloroform as an organic phase). Due to the hazardous potential of organic solvents, pharmaceutical businesses are increasingly avoiding their use, even for coating tablets and granules using the aqueous coating method. As a result, pharmaceutical companies must discover novel methods for producing nanomedicines employing aqueous solvents or low-toxicity solvents [154]. Before selecting ingredients, solvent, NP development technique, cost, and acceptability of the finished product, the research team must proceed cautiously. When a laboratory process is scaled up, the desired properties of NPs can be lost [155]. For example, in research on NP scale-up utilizing the emulsion method, it was discovered that when impeller speed and agitation time increased, particle size fell while entrapment effectiveness stayed unchanged [156]. Other problems include reproducibility and cost-effectiveness, delineation of dangerous potentials related to distinct nano-surface properties and reactivity [157], consideration of stability, shelf-life, and deterioration at the nanostructures. Composition-specific technological stumbling blocks specific considerations such as short shelf-life, temperature sensitivity, and radioactive nanomedicine may necessitate a thoughtful process design [158, 159]. Practical experimental designs could provide efficient solutions to the issues of nanomedicine scale-up. Once scaled up, nanomedicines will be more widely available on the market for the treatment of a wide range of diseases.

4.10 Nanoparticle-Based Approved & Marketed Formulations

There have been more newly developed drugs in recent years with significant lipophilicity, high molecular weight, and low water solubility of approximately 40% [159]. Poor water solubility characterizes nearly 90% of molecules in the discovery pipeline and of drugs with market approval. The inability of recently developed medicines to dissolve well in water is a major factor in their failure [160]. It is well known that poor dissolution and low solubility can result in insufficient drug delivery and decreased bioavailability. Recently, various research groups have explored NP-based therapeutic agents for the management of various ailments (Table 4.2). To speed up the dissolution of these molecules, many different approaches are used, including formulations based on NPs, lipid-based drug delivery systems, prodrugs, amorphous solid dispersions, salt formation, cocrystals, and cyclodextrin complexes after huge research, and continuous efforts for the optimization and quality aspects many of the NP-based formulations are approved for the market as given in Table 4.4. While Table 4.5 summarizes recent NP-based approaches for the treatment of various diseases.

4.11 Future Aspects & Conclusion

In recent tenure, various research groups are focusing on the development and optimization of NPs for the management of various ailments. Only a few studies have been conducted in preclinical and clinical settings. There are undoubtedly several reasons for this reluctance, ranging from

Table 4.4 Recently global market-approved NPs-based formulations (From the year 2000 onwards).

Formulation type	Drug	Name of company	Application	Year of approval	References
Liposomal vincristine	Vincristine	Marqibo (Acrotech)	Acute lymphocytic blood clot	2012	[161]
Inorganic/metallic NPs	Ferric carboxymaltose	Injectafer/(Am Regent)	Chronic kidney disease and iron deficiency anemia	2013	[162]
Inorganic/metallic NPs	Ferric carboxymaltose	Injectafer/(Am Regent)	Chronic kidney disease and iron deficiency anemia	2013	[162]
Polymer–protein conjugate (PEGylated IFNbeta-1a)	Peginterferon beta-1A	Plegridy (Biogen)	Multiple sclerosis	2014	[163]
Nanocrystal	Dantrolene sodium	Ryanodex (Eagle Pharmaceutical)	Malignant benign hypothermia	2014	[164]
Liposomal irinotecan	Irinotecan	Onivyde (Ipsen Inc)	Pancreatic cancer	2015	[165]

(Continued)

Table 4.4 Recently global market-approved NPs-based formulations (From the year 2000 onwards). (*Continued*)

Formulation type	Drug	Name of company	Application	Year of approval	References
PEGylated factor VIII	Polymer–protein conjugate (PEGylated factor VIII)	Adynovate (Baxalta)	Hemophilia	2015	[163]
Liposomes	Cytarabine: daunorubicin	VYXEOS (Celator Pharms)	Acute myeloid leukemia	2017	[152]
PLGA hydrogel	Triamconolone acetonide	Zilretta (Flexion Theraps Inc)	Osteoarthritis	2017	[166]
Glycopegylated coagulation factor IX	Factor IX	Rebinyn (Novo Nordisk Inc)	Hemophilia	FDA 2017	[166]
Lipid NP	Patisiran sodium	Onpattro	Transthyretin (TTR)-mediated amyloidosis	2018	[152]
Liposomal NP	Pfizer-BioNTech	Pfizer-BioNTech	COVID-19	2020 (EUAs)	[167]
Liposomal NP	Moderna	Moderna	COVID-19	2020 (EUAs)	[168]

Table 4.5 Lists of recent NP-based approaches for the treatment of various diseases [169–193].

Nanoparticle type	Drug	Polymer	Method	Outcome	References
Hybrid NPs	Budesonide + siRNA	Poly (DL-lactic-co-glycolic acid)	Double emulsion solvent evaporation method	The observed results indicated that hybrid nanoparticles could be a therapy for chronic obstructive pulmonary disease management.	[169]
Alginate modified-PLGA NPs	Amikaci and moxifloxacin	Poly [DL-lactic-co-glycolic acid]	Double emulsion solvent evaporation method	The amikacin-moxifloxacin alginate encapsulated PLGA NPs showed promise for additional *in vivo* testing studies.	[170]
Solid lipid NPs	Ketoconazole	Lecithin E-80 and phosphatidylcholine	Hot homogenization in conjunction with ultrasonication	Solid lipid NP–loaded formulations could be investigated as an alternate platform for highly lipophilic medicines.	[171]

(*Continued*)

Table 4.5 Lists of recent NP-based approaches for the treatment of various diseases [169–193]. (*Continued*)

Nanoparticle type	Drug	Polymer	Method	Outcome	References
Stealth lipid Nanoparticles	Astaxanthin	Stearic acid, Poloxamer 188	Solvent-diffusion technique	Preliminary experiments prove that stealth lipid nanoparticles could be useful and be considered promising carriers for systemic delivery of Astaxanthin to treat Alzheimer's illness.	[172]
Chitosan NPs	Curcumin	Chitosan	Ionic gelation method	It was feasible to achieve pH-stimulated and sustained transport of curcumin to the colon.	[173]
Solid lipid NPs	Levofloxacin + doxycycline	Pluronic F1–27	Hot emulsification method	This combined tactic will consist of tremendous promise in the treatment of bacterial meningitis.	[174]

(*Continued*)

Table 4.5 Lists of recent NP-based approaches for the treatment of various diseases [169–193]. (*Continued*)

Nanoparticle type	Drug	Polymer	Method	Outcome	References
Chitosan NPs	S-adenosyl-L-methionine	Chitosan	Ionotropic gelation method	Chitosan NPs containing S-adenosyl-L-methionine were successfully manufactured for colonic delivery.	[175]
pH-responsive NP	Flurbiprofen	Polyvinyl alcohol	Nanoprecipitation technique	The flurbiprofen-loaded pH-sensitive NP-based transdermal patch has the potential to successfully control rheumatoid arthritis.	[176]
Chitosan NPs	siRNA	Chitosan	Ionotropic gelation	The observed preclinical outcomes pave the way for clinical trials in Huntington's disease patients.	[177]

(*Continued*)

Table 4.5 Lists of recent NP-based approaches for the treatment of various diseases [169–193]. (*Continued*)

Nanoparticle type	Drug	Polymer	Method	Outcome	References
Chitosan Nanoparticle	Levofloxacin	Chitosan	Ionic gelation method	The levofloxacin-loaded chitosan NP *in situ* gel method was discovered to be an excellent carrier for levofloxacin ocular administration.	[178]
Coated NPs	dexamethasone	Poly(lactic-*co*-glycolic acid)	Single emulsion method	This research takes advantage of the unique properties of biological membrane coatings to engineer extra targeting specificities using naturally occurring target–ligand interactions.	[179]

(*Continued*)

Table 4.5 Lists of recent NP-based approaches for the treatment of various diseases [169–193]. (*Continued*)

Nanoparticle type	Drug	Polymer	Method	Outcome	References
Hybrid nanoparticles	Galantamine	Poly(l-lactic acid) and poly(lactide-co-glycolide)	Modified double emulsification method	Administered Galantamine encapsulated in PLGA-coated carbon nanoforms was successfully delivered *in vivo* for Alzheimer's disease management.	[180]
Functionalized NPs	Curcumin and resveratrol	Chitosan	Ionic-gelation method	The findings indicated that HA-functionalized coloaded NPs may have considerable therapeutic potential for the management of chronic diabetic wounds.	[181]

(*Continued*)

Table 4.5 Lists of recent NP-based approaches for the treatment of various diseases [169–193]. (*Continued*)

Nanoparticle type	Drug	Polymer	Method	Outcome	References
Polymeric NPs	Betamethasone	Low-molecular-weight chitosan, hyaluronic acid	High-pressure homogenization–evaporation method	This could be a potential nanodelivery method for efficient betamethasone skin targeting.	[182]
Lipid nanoparticles	Treprostinil	Propanol, butanol, pentanol	Esterification	These findings suggest further research into Treprostinil-based lipid nanoparticles as a potential therapeutic for pulmonary issues.	[183]
Solid lipid NPs	Eugenia caryophyllata essential oil	Tween 80, Poloxamer 188	High-shear homogenization and ultrasound methods	The findings of this investigation demonstrated the efficacy of these formulations against human infections.	[184]

(*Continued*)

Table 4.5 Lists of recent NP-based approaches for the treatment of various diseases [169–193]. (*Continued*)

Nanoparticle type	Drug	Polymer	Method	Outcome	References
Lipid nanoparticles	Ropinirole	Dynasan-114, Poloxamer 188	Hot melt emulsification coupled with ultra-wave sonication method	The findings revealed that lipid NPs can be used as an alternate delivery method for better ropinirole distribution.	[185]
Solid lipid NPs	Carvedilol	Glyceryl palmitostearate, lauroyl macrogolglycerides	Ultra-sonication technique	The results showed that solid lipid NPs loaded with carvedilol increased oral bioavailability by more than twofold.	[186]
Chitosan NPs	Metformin	Chitosan	Ionic gelation	Chitosan NPs (CS-NPs) were studied as a potentially effective medication delivery platform for oral administration in chronic kidney disease.	[187]

(*Continued*)

Table 4.5 Lists of recent NP-based approaches for the treatment of various diseases [169–193]. (*Continued*)

Nanoparticle type	Drug	Polymer	Method	Outcome	References
Polymeric NPs	Quercetin	Eudragit -S100	Nanoprecipitation method	The findings suggest a viable nanomedicine for colon-targeted quercetin administration in colon cancer.	[188]
Chitosan NPs	Pramipexole dihydrochloride	Chitosan	Ionic gelation method	Pramipexole-loaded chitosan NPs can be used for successful brain targeting via a non-invasive nasal route for Parkinson's disease treatment.	[189]
Solid lipid Nanoparticles	Dimethyl fumarate	Soy lecithin, tween 80,	Hot emulsion and ultrasonication method	The current formulation has superior properties and a longer half-life for the better management of multiple sclerosis, according to research.	[190]

(*Continued*)

Table 4.5 Lists of recent NP-based approaches for the treatment of various diseases [169–193]. (*Continued*)

Nanoparticle type	Drug	Polymer	Method	Outcome	References
Chitosan NPs	Doxorubicin	Chitosan,	Coprecipitation method	This novel biocompatible formulation has the potential to be exploited as a promising theranostic agent for glioblastoma diagnosis and treatment.	[191]
Solid lipid NPs	Olmesartan medoxomil	Poloxamer 188 and Poloxamer 407	Hot homogenization method	The developed Molmesartan medoxomil-loaded solid lipid NPs provide better and more effective drug delivery	[192]

(*Continued*)

Table 4.5 Lists of recent NP-based approaches for the treatment of various diseases [169–193]. (*Continued*)

Nanoparticle type	Drug	Polymer	Method	Outcome	References
Solid lipid NPs	Ferulic acid	Chitosan, Compritol 888 ATO,	Hot homogenization technique	The preclinical data showed increased anti-efficacy, Alzheimer's superior nasal mucoadhesion and penetration, and prolonged drug release.	[193]

regulatory, quality, and technical concerns to the need for a paradigm and cultural shift to enable the pharmaceutical industry to accept these novel technologies. The researchers may also conduct an experiment combining probiotics with the drug. Nowadays, probiotics-based research is conducted by various research groups. Probiotics are lyophilized microorganisms when administered to the host give beneficial effects. In our previous research, we have encapsulated *Saccharomyces boulardii* as a probiotic along with the mesalamine in the coated microparticles. The obtained results were excellent against ulcerative colitis [194, 195].

However, there are still a few potential drawbacks that need to be addressed first, such as the lack of sophisticated equipment for precise and scalable nanomaterial synthesis, the difficulty in determining its efficacy and safety, and some limitations of particular materials. Subsequent generations could discover it simple to produce NPs because of the following critical components: cost-effective and extremely effective, biocompatible, toxin-free, and without any side effects on the pathophysiological system, such thrombosis or inflammation. Moreover, they should be able to traverse several biological barriers and have proper targeting in addition to being stable in blood and resistant to RES clearance, loaded with molecules that could be released freely, and stable in blood and resistant to RES clearance.

Based on all the discussions undertaken in this work, it may be summarized that nanoparticle-based formulations offer several challenges along with the applications in the pharmaceutical area. The possible use of different methods and different types of polymers in the manufacturing allow the feasibility of novelty and applicability in the relevant field. There is still a scope to improvise such products in terms of stability, cost-effectiveness, and scalability. Since last few decades, NPs have been a favored area of research and still, it will remain in the upcoming decades due to a vast scope of exploration possible. Emphasis should be given by the pharmaceutical industries to include such novel products in the pipeline of the approval process to enhance the number of NP-based marketed formulations in the upcoming time.

References

1. Rizvi, S.A.A. and Saleh, A.M., Applications of nanoparticle systems in drug delivery technology. *Saudi Pharm. J.*, Elsevier B.V., 26, 64–70, 2018.
2. Mulder, D.J., Noble, A.J., Justinich, C.J., Duffin, J.M., A tale of two diseases: The history of inflammatory bowel disease. *J. Crohn's Colitis* Elsevier, 8, 341–8, 2014.

3. Hu, Q., Lu, Y., Luo, Y., Recent advances in dextran-based drug delivery systems: From fabrication strategies to applications. *Carbohydr. Polym.* Elsevier Ltd, 264, 117999, 2021.
4. Cui, M., Pan, H., Su, Y., Fang, D., Qiao, S., Ding, P. *et al.*, Opportunities and challenges of three-dimensional printing technology in pharmaceutical formulation development. *Acta Pharm. Sin. B Chin. Acad. Med. Sci.*, 11, 2488–504, 2021.
5. Manzari, M.T., Shamay, Y., Kiguchi, H., Rosen, N., Scaltriti, M., Heller, D.A., Targeted drug delivery strategies for precision medicines. *Nat. Rev. Mater. Nat. Res.*, 6, 351–70, 2021.
6. Patel, J.K. and Pathakk, Y.V. (Eds.), *Emerging Technologies for Nanoparticle Manufacturing*. Springer, Cham, 23 Jun 2021. https://doi.org/10.1007/978-3-030-50703-9
7. Khan, I., Saeed, K., Khan, I., Nanoparticles: Properties, applications and toxicities. *Arab. J. Chem.* Elsevier B.V., 12, 908–31, 2019.
8. Mazayen, Z.M., Ghoneim, A.M., Elbatanony, R.S., Basalious, E.B., Bendas, E.R., Pharmaceutical nanotechnology: From the bench to the market. *Future J. Pharm. Sci.*, 8, 1, 12, Jan. 15, 2022.
9. George, A., Shah, P.A., Shrivastav, P.S., Natural biodegradable polymers based nano-formulations for drug delivery: A review. *Int. J. Pharm.* Elsevier B.V., 561, 244–64, 2019.
10. Zohri, M., Gazori, T., Mirdamadi, S.S., Asadi, A., Haririan, I., Polymeric nanoparticles: Production, applications and advantage. *Internet J. Nanotechnol.*, 3, 1, 2009.
11. Mishra, B., Patel, B.B., Tiwari, S., Colloidal nanocarriers: A review on formulation technology, types and applications toward targeted drug delivery. *Nanomed. Nanotechnol. Biol. Med.* Elsevier Inc., 6, 9–24, 2010.
12. Ganachaud, F. and Katz, J.L., Nanoparticles and nanocapsules created using the ouzo effect: Spontaneous emulsification as an alternative to ultrasonic and high-shear devices. *ChemPhysChem*, Wiley-VCH Verlag, 6, 209–16, 2005.
13. Kaloti, M. and Bohidar, H.B., Kinetics of coacervation transition versus nanoparticle formation in chitosan-sodium tripolyphosphate solutions. *Colloids Surf. B Biointerfaces*, 81, 1, 165–73, Nov. 2010.
14. Mohanraj, V.J. and Chen, Y., Nanoparticles-a review. *Trop. J. Pharm. Res.*, 5, 561–573, 2006. [Internet] Available from: http://www.tjpr.freehosting.net.
15. Wu, M., Dellacherie, E., Durand, A., Marie, E., Poly(n-butyl cyanoacrylate) nanoparticles via miniemulsion polymerization (1): Dextran-based surfactants. *Colloids Surf. B Biointerfaces*, 69, 1, 141–6, Feb. 15, 2009.
16. Rao, J.P. and Geckeler, K.E., Polymer nanoparticles: Preparation techniques and size-control parameters. *Prog. Polym. Sci. (Oxford)*, Elsevier Ltd, 36, 887–913, 2011.

17. Landfester, K., Musyanovych, A., Mailänder, V., From polymeric particles to multifunctional nanocapsules for biomedical applications using the miniemulsion process. *J. Polym. Sci. A Polym. Chem.*, 48, 3, 493–515, Feb. 1, 2010.
18. Lade, B.D. and Shanware, A.S., Phytonanofabrication: Methodology and factors affecting biosynthesis of nanoparticles, in: *Smart Nanosystems for Biomedicine, Optoelectronics and Catalysis*, IntechOpen, 2020.
19. Patil, S. and Chandrasekaran, R., Biogenic nanoparticles: A comprehensive perspective in synthesis, characterization, application and its challenges. *J. Genet. Eng. Biotechnol.* Springer Sci. Business Media Deutschland GmbH, 18, 5, 2020.
20. Patra, J.K. and Baek, K.H., Green nanobiotechnology: Factors affecting synthesis and characterization techniques. *J. Nanomater.* Hindawi Limited, 3, 2014.
21. Kuchibhatla, S.V.N.T., Karakoti, A.S., Baer, D.R., Samudrala, S., Engelhard, M.H., Amonette, J.E. *et al.*, Influence of aging and environment on nanoparticle chemistry: Implication to confinement effects in nanoceria. *J. Phys. Chem. C*, 116, 26, 14108–14, Jul. 5, 2012.
22. Sarathy, V., Tratnyek, P.G., Nurmi, J.T., Baer, D.R., Amonette, J.E., Chun, C.L. *et al.*, Aging of iron nanoparticles in aqueous solution: Effects on structure and reactivity. *J. Phys. Chem. C*, 112, 7, 2286–93, Feb. 21, 2008.
23. Lynch, I., Cedervall, T., Lundqvist, M., Cabaleiro-Lago, C., Linse, S., Dawson, K.A., The nanoparticle-protein complex as a biological entity; A complex fluids and surface science challenge for the 21st century. *Adv. Colloid Interface Sci.*, 134–135, 167–74, 2007.
24. Ruckenstein, E. and Kong, X.Z., Control of pore generation and pore size in nanoparticles of poly (styrene-methyl methacrylate-acrylic acid), *Journal of Applied Polymer Science*. 72, 419-426, 1999.
25. Baer, D.R., Amonette, J.E., Engelhard, M.H., Gaspar, D.J., Karakoti, A.S., Kuchibhatla, S. *et al.*, Characterization challenges for nanomaterials, in: *Surface and Interface Analysis*, pp. 529–37, 2008.
26. Baer, D.R., Surface characterization of nanoparticles: Critical needs and significant challenges, *Journal of Surface Analysis*. 17(3), 163-169, Jan 2011.
27. Yazdani, S., Daneshkhah, A., Diwate, A., Patel, H., Smith, J., Reul, O. *et al.*, Model for gold nanoparticle synthesis: Effect of pH and reaction time. *ACS Omega*, 6, 26, 16847–53, Jul. 6, 2021.
28. Jimenez, V.L., Leopold, M.C., Mazzitelli, C., Jorgenson, J.W., Murray, R.W., HPLC of monolayer-protected gold nanoclusters. *Anal. Chem.*, 75, 2, 199–206, Jan. 15, 2003.
29. Hanauer, M., Pierrat, S., Zins, I., Lotz, A., Sönnichsen, C., Separation of nanoparticles by gel electrophoresis according to size and shape. *Nano Lett.*, 7, 9, 2881–5, Sep. 2007.
30. Kowalczyk, B., Lagzi, I., Grzybowski, B.A., Nanoseparations: Strategies for size and/or shape-selective purification of nanoparticles. *Curr. Opin. Colloid Interface Sci.*, 16, 135–48, 2011.

31. Patra, J.K., Das, G., Fraceto, L.F., Campos, E.V.R., Rodriguez-Torres, M.D.P., Acosta-Torres, L.S. *et al.*, Nano based drug delivery systems: Recent developments and future prospects. *J. Nanobiotechnology*, BioMed. Cent. Ltd., 16, 3, 2018.
32. Srivastava, A., Yadav, T., Sharma, S., Nayak, A., Kumari, A.A., Mishra, N., Polymers in drug delivery. *J. Biosci. Med. (Irvine)*, 04, 01, 69–84, 2016. [Internet] Available from: http://www.scirp.org/journal/doi.aspx?DOI=10.4236/jbm.2016.41009.
33. Priya, V.S.V., Roy., H.K., Jyothi, N., Prasanthi, N.L., Polymers in drug delivery technology, types of polymers and applications. *Sch. Acad. J. Pharm.*, 5, 7, 305–8, Jul. 2016.
34. Doppalapudi, S., Jain, A., Khan, W., Domb, A.J., Biodegradable polymers-An overview. *Polym. Adv. Technol.* John Wiley Sons Ltd, 25, 427–35, 2014.
35. Rahman, M. and Hasan, M.R., Synthetic biopolymers, Common Synthetic Biopolymers and their Applications, Springer, pp. 1–43, 2019. Polymers and Polymeric Composites: A Reference Series book series (POPOC) https://link.springer.com/bookseries/15068
36. Prajapati, S.K., Jain, A., Jain, A., Jain, S., Biodegradable polymers and constructs: A novel approach in drug delivery. *Eur. Polym. J.* Elsevier Ltd, 120, 6–12, 2019.
37. Nath, K., Bhattacharyya, S.K., Das, N.C., Biodegradable polymeric materials for EMI shielding, in: *Materials for Potential EMI Shielding Applications*, pp. 165–78, Elsevier, 2020. https://doi.org/10.1016/B978-0-12-817590-3.00010-5
38. Imazato, S., Kitagawa, H., Tsuboi, R., Kitagawa, R., Thongthai, P., Sasaki, J., II, Non-biodegradable polymer particles for drug delivery: A new technology for "bio-active" restorative materials. *Dent. Mater. J.* Japanese Soc. Dental Mater. Devices, 36, 524–32, 2017.
39. Zeng, S.H., Duan, P.P., Shen, M.X., Xue, Y.J., Wang, Z.Y., Preparation and degradation mechanisms of biodegradable polymer: A review, in: *IOP Conference Series: Materials Science and Engineering*, Institute of Physics Publishing, 2016.
40. Andreeßen, C. and Steinbüchel, A., Recent developments in non-biodegradable biopolymers: Precursors, production processes, and future perspectives. *Appl. Microbiol. Biotechnol.* Springer Verlag, 103, 143–57, 2019.
41. Tong, X., Pan, W., Su, T., Zhang, M., Dong, W., Qi, X., Recent advances in natural polymer-based drug delivery systems. *React. Funct. Polym.* Elsevier B.V., 148, 1–2, 2020.
42. Akash, M.S.H., Rehman, K., Chen, S., Natural and synthetic polymers as drug carriers for delivery of therapeutic proteins. *Polym. Rev.* Taylor Francis Inc., 55, 371–406, 2015.
43. Qi, X., Wei, W., Li, J., Liu, Y., Hu, X., Zhang, J. *et al.*, Fabrication and characterization of a novel anticancer drug delivery system: Salecan/poly(methacrylic acid) semi-interpenetrating polymer network hydrogel. *ACS Biomater. Sci. Eng.*, 1, 12, 1287–99, Dec. 14, 2015.

44. Lee, B.K., Yun, Y., Park, K., PLA micro- and nano-particles. *Adv. Drug Deliv. Rev.* Elsevier B.V., 107, 176–91, 2016.
45. Patel, A.K., Kaczmarek, J.C., Bose, S., Kauffman, K.J., Mir, F., Heartlein, M.W. *et al.*, Inhaled nanoformulated mRNA polyplexes for protein production in lung epithelium. *Adv. Mater.*, 31, 8, Feb. 22, 2019.
46. Maraldi, M., Ferrari, R., Auriemma, R., Sponchioni, M., Moscatelli, D., Concentration of polymer nanoparticles through dialysis: Efficacy and comparison with lyophilization for PEGylated and zwitterionic systems. *J. Pharm. Sci.*, 109, 8, 2607–14, Aug. 1, 2020.
47. Agrahari, V., Patel, S.P., Dhall, N., Aulgur, Z., Thukral, S., Yang, X. *et al.*, Nanoparticles in thermosensitive gel based composite nanosystem for ocular diseases. *Drug Deliv. Transl. Res.*, 8, 2, 422–35, Apr. 1, 2018.
48. Soliman, S.M.A., el Founi, M., Vanderesse, R., Acherar, S., Ferji, K., Babin, J. *et al.*, Light-sensitive dextran-covered PNBA nanoparticles to continuously or discontinuously improve the drug release. *Colloids Surf. B Biointerfaces*, 182, Oct. 1, 2019.
49. Chmielowski, R.A., Abdelhamid, D.S., Faig, J.J., Petersen, L.K., Gardner, C.R., Uhrich, K.E. *et al.*, Athero-inflammatory nanotherapeutics: Ferulic acid-based poly (anhydride-ester) nanoparticles attenuate foam cell formation by regulating macrophage lipogenesis and reactive oxygen species generation. *Acta Biomater.*, 57, 85–94, Jul. 15, 2017.
50. Guo, C., Yin, J., Chen, D., Co-encapsulation of curcumin and resveratrol into novel nutraceutical hyalurosomes nano-food delivery system based on oligo-hyaluronic acid-curcumin polymer. *Carbohydr. Polym.*, 181, 1033–7, Feb. 1, 2018.
51. Mohammadi, M., Ramezani, M., Abnous, K., Alibolandi, M., Biocompatible polymersomes-based cancer theranostics: Towards multifunctional nanomedicine. *Int. J. Pharm.* Elsevier B.V., 519, 287–303, 2017.
52. Dening, T.J., Rao, S., Thomas, N., Prestidge, C.A., Oral nanomedicine approaches for the treatment of psychiatric illnesses. *J. Control. Release* Elsevier B.V., 223, 137–56, 2016.
53. Rahimian, S., Kleinovink, J.W., Fransen, M.F., Mezzanotte, L., Gold, H., Wisse, P. *et al.*, Near infrared labeled, ovalbumin loaded polymeric nanoparticles based on a hydrophilic polyester as model vaccine: *In vivo* tracking and evaluation of antigen-specific CD8+ T cell immune response. *Biomaterials*, 37, 469–77, Jan. 1, 2015.
54. Bayrami, S., Esmaili, Z., SeyedAlinaghi, S.A., Moghadam, S.R.J., Bayrami, S., Javar, H.A. *et al.*, Fabrication of long-acting insulin formulation based on poly (3-hydroxybutyrate-co-3-hydroxyvalerate) (PHBV) nanoparticles: preparation, optimization, characterization, and *in vitro* evaluation. *Pharm. Dev. Technol.*, 24, 2, 176–88, Feb. 7, 2019.
55. Bignon, C., Amigoni, S., Devers, T., Guittard, F., Barrier cream based on CeO2 nanoparticles grafted polymer as an active compound against the penetration of organophosphates. *Chem. Biol. Interact.*, 267, 17–24, Apr. 1, 2017.

56. Jan, N., Madni, A., Rahim, M.A., Khan, N.U., Jamshaid, T., Khan, A. *et al.*, *In vitro* anti-leukemic assessment and sustained release behaviour of cytarabine loaded biodegradable polymer-based nanoparticles. *Life Sci.*, 267, Feb. 15, 1–2, 2021.
57. Qian, Y., Zhang, J., Xu, R., Li, Q., Shen, Q., Zhu, G., Nanoparticles based on polymers modified with pH-sensitive molecular switch and low molecular weight heparin carrying Celastrol and ferrocene for breast cancer treatment. *Int. J. Biol. Macromol.*, 183, 2215–26, Jul. 31, 2021.
58. Etman, S.M., Abdallah, O.Y., Elnaggar, Y.S.R., Novel fucoidan based bioactive targeted nanoparticles from Undaria Pinnatifida for treatment of pancreatic cancer. *Int. J. Biol. Macromol.*, 145, 390–401, Feb. 15, 2020.
59. Gonzalez-Fajardo, L., Ndaya, D., Kasi, R.M., Lu, X., Influence of the method of preparation on the characteristics and performance of cholesterol-based polymeric nanoparticles for redox-triggered release of doxorubicin in tumor cells. *Int. J. Pharm.*, 571, Nov. 25, 118701, 2019.
60. Dong, Y., Du, P., Liu, P., pH-responsive intramolecular FRET-based self-tracking polymer prodrug nanoparticles for real-time tumor intracellular drug release monitoring and imaging. *Int. J. Pharm.*, 588, 1–4, Oct. 15, 2020.
61. Xu, X., Wu, J., Liu, S., Saw, P.E., Tao, W., Li, Y. *et al.*, Redox-responsive nanoparticle-mediated systemic RNAi for effective cancer therapy. *Small*, 14, 41, Oct. 11, 2018.
62. Shalaby, K.S., Soliman, M.E., Bonacucina, G., Cespi, M., Palmieri, G.F., Sammour, O.A. *et al.*, Nanoparticles based on linear and star-shaped poly (ethylene glycol)-poly(ε-caprolactone) copolymers for the delivery of antitubulin drug. *Pharm. Res.*, 33, 8, 2010–24, Aug. 1, 2016.
63. Fei, W., Li, C., Tao, J., Cai, X., Yao, W., Ye, Y. *et al.*, Construction of arsenic-metal complexes loaded nanodrugs for solid tumor therapy: A mini review. *Int. J. Pharm.* Elsevier B.V., 583, 2020.
64. Cerqueira, B.B.S., Lasham, A., Shelling, A.N., Al-Kassas, R., Development of biodegradable PLGA nanoparticles surface engineered with hyaluronic acid for targeted delivery of paclitaxel to triple negative breast cancer cells. *Mater. Sci. Eng. C*, 76, 593–600, Jul. 1, 2017.
65. Mittal, P., Vardhan, H., Ajmal, G., Bonde, G.V., Kapoor, R., Mittal, A. *et al.*, Formulation, optimization, hemocompatibility and pharmacokinetic evaluation of PLGA nanoparticles containing paclitaxel. *Drug Dev. Ind. Pharm.*, 45, 3, 365–78, Mar. 4, 2019.
66. Wang, G., Wang, Z., Li, C., Duan, G., Wang, K., Li, Q. *et al.*, RGD peptide-modified, paclitaxel prodrug-based, dual-drugs loaded, and redox-sensitive lipid-polymer nanoparticles for the enhanced lung cancer therapy. *Biomed. Pharmacother.*, 106, 275–84, Oct. 1, 2018.
67. Li, X. and Jiang, X., Microfluidics for producing poly (lactic-co-glycolic acid)-based pharmaceutical nanoparticles. *Adv. Drug Deliv. Rev.* Elsevier B.V., 128, 101–14, 2018.

68. Albano, J.M.R., de Morais Ribeiro, L.N., Couto, V.M., Messias, M.B., da Silva, G.H.R., Breitkreitz, M.C. *et al.*, Rational design of polymer-lipid nanoparticles for docetaxel delivery. *Colloids Surf. B Biointerfaces*, 175, 56–64, Mar. 1, 2019.
69. Feng, C., Yuan, X., Chu, K., Zhang, H., Ji, W., Rui, M., Preparation and optimization of poly (lactic acid) nanoparticles loaded with fisetin to improve anti-cancer therapy. *Int. J. Biol. Macromol.*, 125, 700–10, Mar. 15, 2019.
70. Ding, D. and Zhu, Q., Recent advances of PLGA micro/nanoparticles for the delivery of biomacromolecular therapeutics. *Mater. Sci. Eng. C*, Elsevier Ltd, 92, 1041–60, 2018.
71. Rose, F., Wern, J.E., Ingvarsson, P.T., van de Weert, M., Andersen, P., Follmann, F. *et al.*, Engineering of a novel adjuvant based on lipid-polymer hybrid nanoparticles: A quality-by-design approach. *J. Control. Release*, 210, 48–57, Jul. 28, 2015.
72. Malinovskaja-Gomez, K., Labouta, H.I., Schneider, M., Hirvonen, J., Laaksonen, T., Transdermal iontophoresis of flufenamic acid loaded PLGA nanoparticles. *Eur. J. Pharm. Sci.*, 89, 154–62, Jun. 30, 2016.
73. Li, Q., Xia, D., Tao, J., Shen, A., He, Y., Gan, Y. *et al.*, Self-assembled core-shell-type lipid-polymer hybrid nanoparticles: Intracellular trafficking and relevance for oral absorption. *J. Pharm. Sci.*, 106, 10, 3120–30, Oct. 1, 2017.
74. Singh, G., Sarwal, A., Sharma, S., Prasad, P., Kuhad, A., Ali, W., Polymer-based prolonged-release nanoformulation of duloxetine: Fabrication, characterization and neuropharmacological assessments. *Drug Dev. Ind. Pharm.*, 47, 1, 12–21, 2021.
75. Hung, C.F., Chen, W.Y., Hsu, C.Y., Aljuffali, I.A., Shih, H.C., Fang, J.Y., Cutaneous penetration of soft nanoparticles via photodamaged skin: Lipid-based and polymer-based nanocarriers for drug delivery. *Eur. J. Pharm. Biopharm.*, 94, 94–105, Aug. 1, 2015.
76. Sun, D., Li, N., Zhang, W., Zhao, Z., Mou, Z., Huang, D. *et al.*, Design of PLGA-functionalized quercetin nanoparticles for potential use in Alzheimer's disease. *Colloids Surf. B Biointerfaces*, 148, 116–29, Dec. 1, 2016.
77. Wang, Q., Shen, M., Li, W., Li, W., Zhang, F., Controlled-release of fluazinam from biodegradable PLGA-based microspheres. *J. Environ. Sci. Health B*, 54, 10, 810–6, Oct. 3, 2019.
78. Firouzi-Amandi, A., Dadashpour, M., Nouri, M., Zarghami, N., Serati-Nouri, H., Jafari-Gharabaghlou, D. *et al.*, Chrysin-nanoencapsulated PLGA-PEG for macrophage repolarization: Possible application in tissue regeneration. *Biomed. Pharmacother.*, 105, 773–80, Sep. 1, 2018.
79. Luo, S., Jiang, T., Yang, Y., Yang, X., Zhao, J., Combination therapy with vancomycin-loaded calcium sulfate and vancomycin-loaded PMMA in the treatment of chronic osteomyelitis. *BMC Musculoskelet. Disord.*, 17, 1, Dec. 22, 2016.
80. Imazato, S., Kitagawa, H., Tsuboi, R., Kitagawa, R., Thongthai, P., Sasaki, J., II, Non-biodegradable polymer particles for drug delivery: A new technology

for "bio-active" restorative materials. *Dent. Mater. J. Japanese Soc. Dental Mater. Devices*, 36, 524–32, 2017.

81. El-Say, K.M., Ahmed, T.A., Aljefri, A.H., El-Sawy, H.S., Fassihi, R., Abou-Gharbia, M., Oleic acid–reinforced PEGylated polymethacrylate transdermal film with enhanced antidyslipidemic activity and bioavailability of atorvastatin: A mechanistic *ex-vivo/in-vivo* analysis. *Int. J. Pharm.*, 608, 1, Oct. 25, 2021.
82. Neut, D., van de Belt, H., van Horn, J.R., van der Mei, H.C., Busscher, H.J., Residual gentamicin-release from antibiotic-loaded polymethylmethacrylate beads after 5 years of implantation. *Biomaterials*, 24, 10, 1829–31, 2003.
83. Osman, A.F., Tuty, T.F., Rakibuddin, M., Hashim, F., Johari, S.A.T.T., Ananthakrishnan, R. *et al.*, Pre-dispersed organo-montmorillonite (organo-MMT) nanofiller: Morphology, cytocompatibility and impact on flexibility, toughness and biostability of biomedical ethyl vinyl acetate (EVA) copolymer. *Mater. Sci. Eng. C*, 74, 194–206, May 1, 2017.
84. Salama, A.H. and Shamma, R.N., Tri/tetra-block co-polymeric nanocarriers as a potential ocular delivery system of lornoxicam: *In-vitro* characterization, and *in-vivo* estimation of corneal permeation. *Int. J. Pharm.*, 492, 1–2, 28–39, Jul. 17, 2015.
85. Poller, B., Strachan, C., Broadbent, R., Walker, G.F., A minitablet formulation made from electrospun nanofibers. *Eur. J. Pharm. Biopharm.*, 114, 213–20, May 1, 2017.
86. Thakkar, M., Islam, M.S., Railkar, A., Mitra, S., Antisolvent precipitative immobilization of micro and nanostructured griseofulvin on laboratory cultured diatom frustules for enhanced aqueous dissolution. *Colloids Surf. B Biointerfaces*, 196, Dec. 1, 2020.
87. Kuznetsov, N.M., Zagoskin, Y.D., Vdovichenko, A.Y., Bakirov, A.V., Kamyshinsky, R.A., Istomina, A.P. *et al.*, Enhanced electrorheological activity of porous chitosan particles. *Carbohydr. Polym.*, 256, 1–2, Mar. 15, 2021.
88. Pérez-Herrero, E. and Fernández-Medarde, A., Advanced targeted therapies in cancer: Drug nanocarriers, the future of chemotherapy. *Eur. J. Pharm. Biopharm.* Elsevier B.V., 93, 52–79, 2015.
89. Cordeiro, R.A., Santo, D., Farinha, D., Serra, A., Faneca, H., Coelho, J.F.J., High transfection efficiency promoted by tailor-made cationic tri-block copolymer-based nanoparticles. *Acta Biomater.*, 47, 113–23, Jan. 1, 2017.
90. Ciekot, J., Psurski, M., Jurec, K., Boratyński, J., Hydroxyethylcellulose as a methotrexate carrier in anticancer therapy. *Invest. New Drugs*, 39, 1, 15–23, Feb. 1, 2021.
91. Kumar, P., Tambe, P., Paknikar, K.M., Gajbhiye, V., Folate/N-acetyl glucosamine conjugated mesoporous silica nanoparticles for targeting breast cancer cells: A comparative study. *Colloids Surf. B Biointerfaces*, 156, 203–12, Aug. 1, 2017.
92. Patra, C.N., Priya, R., Swain, S., Jena, G.K., Panigrahi, K.C., Ghose, D., Pharmaceutical significance of Eudragit: A review. *Future J. Pharm. Sci.*, 3, 1, 33–45, Jun. 2017.

93. Damgé, C., Socha, M., Ubrich, N., Maincent, P., Poly (ε-caprolactone)/eudragit nanoparticles for oral delivery of aspart-insulin in the treatment of diabetes. *J. Pharm. Sci.*, 99, 2, 879–89, 2010.
94. Hoffart, V., Ubrich, N., Simonin, C., Babak, V., Vigneron, C., Hoffman, M. *et al.*, Low molecular weight heparin-loaded polymeric nanoparticles: Formulation, characterization, and release characteristics. *Drug Dev. Ind. Pharm.*, 28, 9, 1091–9, 2002.
95. Das, S., Suresh, P.K., Desmukh, R., Design of Eudragit RL 100 nanoparticles by nanoprecipitation method for ocular drug delivery. *Nanomedicine*, 6, 2, 318–23, Apr. 2010.
96. Saharan, P., Bahmani, K., Saharan, S.P., Preparation, optimization and *in vitro* evaluation of glipizide nanoparticles integrated with Eudragit RS-100. *Pharm. Nanotechnol.*, 7, 1, 72–85, Mar. 20, 2019.
97. Yurtdaş-Kırımlıoğlu, G., Güleç, K., Görgülü, Ş., Kıyan, H.T., Oseltamivir phosphate loaded pegylated-Eudragit nanoparticles for lung cancer therapy: Characterization, prolonged release, cytotoxicity profile, apoptosis pathways and *in vivo* anti-angiogenic effect by using CAM assay. *Microvasc. Res.*, 139, 104251, Jan. 1, 2022.
98. Dillen, K., Vandervoort, J., van den Mooter, G., Ludwig, A., Evaluation of ciprofloxacin-loaded Eudragit® RS100 or RL100/PLGA nanoparticles. *Int. J. Pharm.*, 314, 1, 72–82, 109881, May 11, 2006.
99. Kierys, A., Zaleski, R., Grochowicz, M., Gorgol, M., Sienkiewicz, A., Polymer–mesoporous silica composites for drug release systems. *Microporous Mesoporous Mater.*, 294, Mar. 1, 2020.
100. Raza, A., Alavi, S.E., Sime, F.B., Han, F.Y., Roberts, J.A., Popat, A. *et al.*, Microfluidic assembly of pomegranate-like hierarchical microspheres for efflux regulation in oral drug delivery. *Acta Biomater.*, 126, 277–90, May 1, 2021.
101. Shen, C., Li, Y., Wang, Y., Meng, Q., Non-swelling hydrogel-based microfluidic chips. *Lab. Chip*, 19, 23, 3962–73, Dec. 7, 2019.
102. Mohamed, A.L., Elmotasem, H., Salama, A.A.A., Colchicine mesoporous silica nanoparticles/hydrogel composite loaded cotton patches as a new encapsulator system for transdermal osteoarthritis management. *Int. J. Biol. Macromol.*, 164, 1149–63, Dec. 1, 2020.
103. Parekh, K., Hariharan, K., Qu, Z., Rewatkar, P., Cao, Y., Moniruzzaman, M. *et al.*, Tacrolimus encapsulated mesoporous silica nanoparticles embedded hydrogel for the treatment of atopic dermatitis. *Int. J. Pharm.*, 608, 121079, Oct. 25, 2021.
104. Letchmanan, K., Shen, S.C., Ng, W.K., Tan, R.B.H., Dissolution and physicochemical stability enhancement of artemisinin and mefloquine co-formulation via nano-confinement with mesoporous SBA-15. *Colloids Surf. B Biointerfaces*, 155, 560–8, Jul. 1, 2017.

105. Falsafi, S.R., Wang, Y., Ashaolu, T.J., Sharma, M., Rawal, S., Patel, K. *et al.*, Biopolymer nanovehicles for oral delivery of natural anticancer agents. *Adv. Funct. Mater.* John Wiley Sons Inc., 33, 2209419, 1–29, 2022.
106. Constantin, M., Bucătariu, S., Stoica, I., Fundueanu, G., Smart nanoparticles based on pullulan-g-poly (N-isopropylacrylamide) for controlled delivery of indomethacin: Paper dedicated to the 150th anniversary of the Romanian Academy. *Int. J. Biol. Macromol.*, 94, 698–708, Jan. 1, 2017.
107. da Silva, N.I.O., Salvador, E.A., Franco, I.R., de Souza, G.A.P., de Souza Morais, S.M., Rocha, R.P. *et al.*, Bovine serum albumin nanoparticles induce histopathological changes and inflammatory cell recruitment in the skin of treated mice. *Biomed. Pharmacother.*, 107, 1311–7, Nov. 1, 2018.
108. Gagliardi, A., Paolino, D., Costa, N., Fresta, M., Cosco, D., Zein- vs PLGA-based nanoparticles containing rutin: A comparative investigation. *Mater. Sci. Eng. C*, 118, 111538, Jan 1, 2021.
109. Pitorre, M., Gazaille, C., Pham, L.T.T., Frankova, K., Béjaud, J., Lautram, N. *et al.*, Polymer-free hydrogel made of lipid nanocapsules, as a local drug delivery platform. *Mater. Sci. Eng. C*, 126, 112188, Jul. 1, 2021.
110. Simon, L., Marcotte, N., Devoisselle, J.M., Begu, S., Lapinte, V., Recent advances and prospects in nano drug delivery systems using lipopolyoxazolines. *Int. J. Pharm.*, 585, 119536, Jul. 30, 2020.
111. de Oliveira, T.C., de Jesus Oliveira, A.C., Patriota, Y.B.G., Chaves, L.L., de Oliveira Silva Ribeiro, F., de Paula, R.C.M. *et al.*, Eco-friendly synthesis of phthalate angico gum towards nanoparticles engineering using Quality by Design (QbD) approach. *Int. J. Biol. Macromol.*, 190, 801–9, Nov. 1, 2021.
112. Voci, S., Gagliardi, A., Salvatici, M.C., Fresta, M., Cosco, D., Development of polyoxyethylene (2) oleyl ether-gliadin nanoparticles: Characterization and *in vitro* cytotoxicity. *Eur. J. Pharm. Sci.*, 162, 105849, Jul. 1, 2021.
113. Fachinetti, N., Rigon, R.B., Eloy, J.O., Sato, M.R., dos Santos, K.C., Chorilli, M., Comparative study of glyceryl behenate or polyoxyethylene 40 stearate-based lipid carriers for trans-resveratrol delivery: Development, characterization and evaluation of the *in vitro* tyrosinase inhibition. *AAPS PharmSciTech*, 19, 3, 1401–9, Apr. 1, 2018.
114. Ray, L., Karthik, R., Srivastava, V., Singh, S.P., Pant, A.B., Goyal, N. *et al.*, Efficient antileishmanial activity of amphotericin B and piperine entrapped in enteric coated guar gum nanoparticles. *Drug Deliv. Transl. Res.*, 11, 1, 118–30, Feb. 1, 2021.
115. Pacheco, P.A., Rodrigues, L.N.C., Ferreira, J.F.S., Gomes, A.C.P., Veríssimo, C.J., Louvandini, H. *et al.*, Inclusion complex and nanoclusters of cyclodextrin to increase the solubility and efficacy of albendazole. *Parasitol. Res.*, 117, 3, 705–12, Mar. 1, 2018.
116. Vaidya, F.U., Sharma, R., Shaikh, S., Ray, D., Aswal, V.K., Pathak, C., Pluronic micelles encapsulated curcumin manifests apoptotic cell death and inhibits pro-inflammatory cytokines in human breast adenocarcinoma cells. *Cancer Rep.*, 2, 1, e1133, Feb 1, 2019.

117. Monteiro, N., Martins, A., Ribeiro, D., Faria, S., Fonseca, N.A., Moreira, J.N. *et al.*, On the use of dexamethasone-loaded liposomes to induce the osteogenic differentiation of human mesenchymal stem cells. *J. Tissue Eng. Regen. Med.*, 9, 9, 1056–66, Sep. 1, 2015.
118. Rassu, G., Cossu, M., Langasco, R., Carta, A., Cavalli, R., Giunchedi, P. *et al.*, Propolis as lipid bioactive nano-carrier for topical nasal drug delivery. *Colloids Surf. B Biointerfaces*, 136, 908–17, Dec. 1, 2015.
119. Hu, Q. and Luo, Y., Chitosan-based nanocarriers for encapsulation and delivery of curcumin: A review. *Int. J. Biol. Macromol.* Elsevier B.V., 179, 125–35, 2021.
120. Gadkari, R.R., Suwalka, S., Yogi, M.R., Ali, W., Das, A., Alagirusamy, R., Green synthesis of chitosan-cinnamaldehyde cross-linked nanoparticles: Characterization and antibacterial activity. *Carbohydr. Polym.*, 226, Dec. 15, 2019.
121. Mumuni, M.A., Kenechukwu, F.C., Ofokansi, K.C., Attama, A.A., Díaz, D.D., Insulin-loaded mucoadhesive nanoparticles based on mucin-chitosan complexes for oral delivery and diabetes treatment. *Carbohydr. Polym.*, 229, 115506, Feb. 1, 2020.
122. Nguyen, S., Escudero, C., Sediqi, N., Smistad, G., Hiorth, M., Fluoride loaded polymeric nanoparticles for dental delivery. *Eur. J. Pharm. Sci.*, 104, 326–34, Jun. 15, 2017.
123. Dong, J., Liu, X., Chen, Y., Yang, W., Du, X., User-safe and efficient chitosan-gated porous carbon nanopesticides and nanoherbicides. *J. Colloid Interface Sci.*, 594, 20–34, Jul. 15, 2021.
124. Leonida, M.D., Belbekhouche, S., Benzecry, A., Peddineni, M., Suria, A., Carbonnier, B., Antibacterial hop extracts encapsulated in nanochitosan matrices. *Int. J. Biol. Macromol.*, 120, 1335–43, Dec. 1, 2018.
125. Hasheminejad, N., Khodaiyan, F., Safari, M., Improving the antifungal activity of clove essential oil encapsulated by chitosan nanoparticles. *Food Chem.*, 275, 113–22, Mar. 1, 2019.
126. Gondil, V.S., Dube, T., Panda, J.J., Yennamalli, R.M., Harjai, K., Chhibber, S., Comprehensive evaluation of chitosan nanoparticle-based phage lysin delivery system; A novel approach to counter *S. pneumoniae* infections. *Int. J. Pharm.*, 573, 118850, Jan. 5, 2020.
127. Sathiyaseelan, A., Saravanakumar, K., Mariadoss, A.V.A., Wang, M.H., pH-controlled nucleolin targeted release of dual drug from chitosan-gold based aptamer functionalized nano drug delivery system for improved glioblastoma treatment. *Carbohydr. Polym.*, 262, 117907, Jun. 15, 2021.
128. Wong, C.Y., Martinez, J., Al-Salami, H., Dass, C.R., Quantification of BSA-loaded chitosan/oligonucleotide nanoparticles using reverse-phase high-performance liquid chromatography. *Anal. Bioanal. Chem.*, 410, 27, 6991–7006, Nov. 1, 2018.

129. Verma, J., Kanoujia, J., Parashar, P., Tripathi, C.B., Saraf, S.A., Wound healing applications of sericin/chitosan-capped silver nanoparticles incorporated hydrogel. *Drug Deliv. Transl. Res.*, 7, 1, 77–88, Feb. 1, 2017.
130. Deepa, G., Sivakumar, K.C., Sajeevan, T.P., Molecular simulation and *in vitro* evaluation of chitosan nanoparticles as drug delivery systems for the controlled release of anticancer drug cytarabine against solid tumours. *3 Biotech.*, 8, 12, 1, Dec. 1, 2018.
131. Kulkarni, A.D., Vanjari, Y.H., Sancheti, K.H., Patel, H.M., Belgamwar, V.S., Surana, S.J. *et al.*, New nasal nanocomplex self-assembled from charged biomacromolecules: N, N, N-Trimethyl chitosan and dextran sulfate. *Int. J. Biol. Macromol.*, 88, 476–90, Jul. 1, 2016.
132. Rodrigues, S., Cordeiro, C., Seijo, B., Remuñán-López, C., Grenha, A., Hybrid nanosystems based on natural polymers as protein carriers for respiratory delivery: Stability and toxicological evaluation. *Carbohydr. Polym.*, 123, 369–80, Jun. 5, 2015.
133. Ahmad, N., Ahmad, R., Naqvi, A.A., Alam, M.A., Ashafaq, M., Samim, M. *et al.*, Rutin-encapsulated chitosan nanoparticles targeted to the brain in the treatment of Cerebral Ischemia. *Int. J. Biol. Macromol.*, 91, 640–55, Oct. 1, 2016.
134. Gupta, U., Sharma, S., Khan, I., Gothwal, A., Sharma, A.K., Singh, Y. *et al.*, Enhanced apoptotic and anticancer potential of paclitaxel loaded biodegradable nanoparticles based on chitosan. *Int. J. Biol. Macromol.*, 98, 810–9, May 1, 2017.
135. Shamekhi, F., Tamjid, E., Khajeh, K., Development of chitosan coated calcium-alginate nanocapsules for oral delivery of liraglutide to diabetic patients. *Int. J. Biol. Macromol.*, 120, 460–7, Dec. 1, 2018.
136. Scialla, S., Barca, A., Palazzo, B., D'Amora, U., Russo, T., Gloria, A. *et al.*, Bioactive chitosan-based scaffolds with improved properties induced by dextran-grafted nano-maghemite and l-arginine amino acid. *J. Biomed. Mater. Res. A*, 107, 6, 1244–52, Jun. 1, 2019.
137. Kumar, S.P., Birundha, K., Kaveri, K., Devi, K.T.R., Antioxidant studies of chitosan nanoparticles containing naringenin and their cytotoxicity effects in lung cancer cells. *Int. J. Biol. Macromol.*, 78, 87–95, Jul. 1, 2015.
138. Dukovski, B.J., Juretić, M., Bračko, D., Randjelović, D., Savić, S., Moral, M.C. *et al.*, Functional ibuprofen-loaded cationic nanoemulsion: Development and optimization for dry eye disease treatment. *Int. J. Pharm.*, 576, 87-95, Feb. 25, 2020.
139. Koletti, A.E., Tsarouchi, E., Kapourani, A., Kontogiannopoulos, K.N., Assimopoulou, A.N., Barmpalexis, P., Gelatin nanoparticles for NSAID systemic administration: Quality by design and artificial neural networks implementation. *Int. J. Pharm.*, 578, 119118, Mar. 30, 2020.
140. Sahraee, S., Milani, J.M., Ghanbarzadeh, B., Hamishehkar, H., Physicochemical and antifungal properties of bio-nanocomposite film based on gelatin-chitin nanoparticles. *Int. J. Biol. Macromol.*, 97, 373–81, Apr. 1, 2017.

141. Mahor, A., Prajapati, S.K., Verma, A., Gupta, R., Iyer, A.K., Kesharwani, P., Moxifloxacin loaded gelatin nanoparticles for ocular delivery: Formulation and *in-vitro, in-vivo* evaluation. *J. Colloid Interface Sci.*, 483, 132–8, Dec. 1, 2016.
142. Cai, D., Shi, L., Long, R., Ren, G., Wang, S., Liu, Y., A carboxymethyl lentinan layer by layer self-assembly system as a promising drug chemotherapeutic platform. *Carbohydr. Polym.*, 261, 117847, Jun. 1, 2021.
143. Dubey, R.D., Saneja, A., Gupta, P.K., Gupta, P.N., Recent advances in drug delivery strategies for improved therapeutic efficacy of gemcitabine. *Eur. J. Pharm. Sci.* Elsevier B.V., 93, 147–62, 2016.
144. Batool, A., Arshad, R., Razzaq, S., Nousheen, K., Kiani, M.H., Shahnaz, G., Formulation and evaluation of hyaluronic acid-based mucoadhesive self-nanoemulsifying drug delivery system (SNEDDS) of tamoxifen for targeting breast cancer. *Int. J. Biol. Macromol.*, 152, 503–15, Jun. 1, 2020.
145. Gadhave, D., Rasal, N., Sonawane, R., Sekar, M., Kokare, C., Nose-to-brain delivery of teriflunomide-loaded lipid-based carbopol-gellan gum nanogel for glioma: Pharmacological and *in vitro* cytotoxicity studies. *Int. J. Biol. Macromol.*, 167, 906–20, Jan. 15, 2021.
146. Chaudhary, S., Gothwal, A., Khan, I., Srivastava, S., Malik, R., Gupta, U., Polypropyleneimine and polyamidoamine dendrimer mediated enhanced solubilization of bortezomib: Comparison and evaluation of mechanistic aspects by thermodynamics and molecular simulations. *Mater. Sci. Eng. C*, 72, 611–9, Mar. 1, 2017.
147. Chandrakala, V., Aruna, V., Angajala, G., Review on metal nanoparticles as nanocarriers: current challenges and perspectives in drug delivery systems. *Emergent Mater.* Springer Nat., 6, 1597, 2022.
148. Saha, S.K., Roy, P., Saini, P., Mondal, M.K., Chowdhury, P., Babu, S.P.S., Carbohydrate polymer inspired silver nanoparticles for filaricidal and mosquitocidal activities: A comprehensive view. *Carbohydr. Polym.*, 137, 390–401, Feb. 10, 2016.
149. Rodriguez-Del Rio, E., Marradi, M., Calderon-Gonzalez, R., Frande-Cabanes, E., Penadés, S., Petrovsky, N. *et al.*, A gold glyco-nanoparticle carrying a listeriolysin O peptide and formulated with AdvaxTM delta inulin adjuvant induces robust T-cell protection against listeria infection. *Vaccine*, 33, 12, 1465–73, Mar. 17, 2015.
150. Rahme, K., Guo, J., Holmes, J.D., O'Driscoll, C.M., Evaluation of the physicochemical properties and the biocompatibility of polyethylene glycol-conjugated gold nanoparticles: A formulation strategy for siRNA delivery. *Colloids Surf B Biointerfaces.*, 135, 604–12, 2015.
151. Ragelle, H., Danhier, F., Préat, V., Langer, R., Anderson., D.G., Nanoparticle-based drug delivery systems: A commercial and regulatory outlook as the field matures. *Expert Opin. Drug Deliv.* Taylor Francis Ltd, 14, 851–64, 2017.
152. Anselmo, A.C. and Mitragotri, S., Nanoparticles in the clinic: An update. *Bioeng. Transl. Med.*, 4, 3, e10143, Sep. 2019.

153. Shah, S., Dhawan, V., Holm, R., Nagarsenker, M.S., Perrie, Y., Liposomes: Advancements and innovation in the manufacturing process. *Adv. Drug Deliv. Rev.* Elsevier B.V., 154–155, 102–22, 2020.
154. Zheng, C., Li, M., Ding, J., Challenges and opportunities of nanomedicines in clinical translation. *Bio Integration*, 2, 2, 57–60, 2021.
155. Wu, L.P., Wang, D., Li, Z., Grand challenges in nanomedicine. *Mater. Sci. Eng. C* Elsevier Ltd, 106, 110302, 110302, 2020.
156. Colombo, A.P., Briançon, S., Lieto, J., Fessi, H., Project, design, and use of a pilot plant for nanocapsule production. *Drug development and industrial pharmacy*, 27, 10, 1063–1072, Jan 2011. [Internet] Available from: www.dekker.com.
157. Agrahari, V. and Agrahari, V., Facilitating the translation of nanomedicines to a clinical product: Challenges and opportunities. *Drug Discovery Today* Elsevier Ltd, 23, 974–91, 2018.
158. Colby, A.H., Liu, R., Doyle, R.P., Merting, A., Zhang, H., Savage, N. *et al.*, Pilot-scale production of expansile nanoparticles: Practical methods for clinical scale-up. *J. Control. Release*, 337, 144–54, Sep. 10, 2021.
159. Kumar, R., Dalvi, V., Siril, P.F., Dalvi, S.V., Siril, F., Nanoparticle-based drugs and formulations: Current status and emerging applications. *ACS Applied Nano Materials*, 3, 6, 4944–4961, May 2020.
160. Kalyane, D., Raval, N., Maheshwari, R., Tambe, V., Kalia, K., Tekade, R.K., Employment of enhanced permeability and retention effect (EPR): Nanoparticle-based precision tools for targeting of therapeutic and diagnostic agent in cancer. *Mater. Sci. Eng. C* Elsevier Ltd, 98, 1252–76, 2019.
161. Silverman, J.A. and Deitcher, S.R., Marqibo® (vincristine sulfate liposome injection) improves the pharmacokinetics and pharmacodynamics of vincristine. *Cancer Chemother. Pharmacol.*, 71, 555–64, 2013.
162. Lyseng-Williamson, K.A., Keating, G.M., Cavill, I., Koutroubakis, I.E., Littlewood, T.J., Qunibi, W.Y., Ferric carboxymaltose a review of its use in iron-deficiency anaemia. *Drugs*, 69, 739-756, Apr 2009.
163. Choi, Y.H. and Han, H.K., Nanomedicines: Current status and future perspectives in aspect of drug delivery and pharmacokinetics. *J. Pharm. Investig.* Springer Netherlands, 48, 43–60, 2018.
164. Patra, J.K., Das, G., Fraceto, L.F., Campos, E.V.R., Rodriguez-Torres, M.D.P., Acosta-Torres, L.S. *et al.*, Nano based drug delivery systems: Recent developments and future prospects. *J. Nanobiotechnology* BioMed. Cent. Ltd., 16, 71, 2018.
165. Zhang, H., Onivyde for the therapy of multiple solid tumors. *Onco Targets Ther.* Dove Med. Press Ltd., 9, 3001–7, 2016.
166. Farjadian, F., Ghasemi, A., Gohari, O., Roointan, A., Karimi, M., Hamblin, M.R., Nanopharmaceuticals and nanomedicines currently on the market: Challenges and opportunities. *Nanomedicine* Future Med. Ltd., 14, 93–126, 2019.

167. Bouazzaoui, A., Abdellatif, A.A.H., Al-Allaf, F.A., Bogari, N.M., Al-Dehlawi, S., Qari, S.H., Strategies for vaccination: Conventional vaccine approaches versus new-generation strategies in combination with adjuvants. *Pharmaceutics*, MDPI AG, 13, 1–20, 2021.
168. Abdellatif, A.A.H., Tawfeek, H.M., Abdelfattah, A., El-Saber Batiha, G., Hetta, H.F., Recent updates in COVID-19 with emphasis on inhalation therapeutics: Nanostructured and targeting systems. *J. Drug Deliv. Sci. Technol.* Editions Sante, 63, 102435, 2021.
169. Leng, D., Thanki, K., Fattal, E., Foged, C., Yang, M., Engineering of budesonide-loaded lipid-polymer hybrid nanoparticles using a quality-by-design approach. *Int. J. Pharm.*, 548, 2, 740–6, Sep. 15, 2018.
170. Abdelghany, S., Parumasivam, T., Pang, A., Roediger, B., Tang, P., Jahn, K. *et al.*, Alginate modified-PLGA nanoparticles entrapping amikacin and moxifloxacin as a novel host-directed therapy for multidrug-resistant tuberculosis. *J. Drug Deliv. Sci. Technol.*, 52, 642–51, Aug. 1, 2019.
171. Dudhipala, N. and Ay, A.A., Amelioration of ketoconazole in lipid nanoparticles for enhanced antifungal activity and bioavailability through oral administration for management of fungal infections. *Chem. Phys. Lipids*, 232, 104953, Oct. 1, 2020.
172. Santonocito, D., Raciti, G., Campisi, A., Sposito, G., Panico, A., Siciliano, E.A. *et al.*, Astaxanthin-loaded stealth lipid nanoparticles (AST-SSLN) as potential carriers for the treatment of alzheimer's disease: Formulation development and optimization, *Nanomaterials*, 11, 2, 391, 2021, Available from: https://doi.org/10.3390/nano11020391.
173. Raj, P.M., Raj, R., Kaul, A., Mishra, A.K., Ram, A., Biodistribution and targeting potential assessment of mucoadhesive chitosan nanoparticles designed for ulcerative colitis: Via scintigraphy. *RSC Adv.*, 8, 37, 20809–21, 2018.
174. Hady, M.A., Sayed, O.M., Akl, M.A., Brain uptake and accumulation of new levofloxacin-doxycycline combination through the use of solid lipid nanoparticles: Formulation; Optimization and *in-vivo* evaluation. *Colloids Surf. B Biointerfaces*, 193, 111076, Sep. 1, 2020.
175. Ergin, A.D., Bayindir, Z.S., Yüksel, N., Characterization and optimization of colon targeted S adenosyl L methionine loaded chitosan nanoparticles. *J. Res. Pharm.*, 23, 5, 914–26, 2019.
176. Khan, D., Qindeel, M., Ahmed, N., Khan, A.U., Khan, S., Rehman, A.U., Development of novel pH-sensitive nanoparticle-based transdermal patch for management of rheumatoid arthritis. *Nanomedicine*, 15, 6, 603–24, Mar. 1, 2020.
177. Sava, V., Fihurka, O., Khvorova, A., Sanchez-Ramos, J., Enriched chitosan nanoparticles loaded with siRNA are effective in lowering Huntington's disease gene expression following intranasal administration. *Nanomedicine*, 24, 102119, Feb. 1, 2020.
178. Zafar, A., Imam, S.S., Bukhari, S.N.A., Ahmad, J., Ali, A., Formulation and optimization of levofloxacin loaded chitosan nanoparticle for ocular

delivery: *In-vitro* characterization, ocular tolerance and antibacterial activity. *Int. J. Biol. Macromol.*, 108, 650–9, Mar. 1, 2018.

179. Park, J.H., Jiang, Y., Zhou, J., Gong, H., Mohapatra, A., Heo, J. *et al.*, Genetically engineered cell membrane-coated nanoparticles for targeted delivery of dexamethasone to inflamed lungs. *Appl. Sci. Eng. Sci. Adv.*, 7, eabf7820, 2021.
180. Nanaki, S.G., Spyrou, K., Bekiari, C., Veneti, P., Baroud, T.N., Karouta, N. *et al.*, Hierarchical porous Carbon—PLLA and PLGA hybrid nanoparticles for intranasal delivery of galantamine for alzheimer's disease therapy. *Pharmaceutics*, 12, 3, 227, Mar. 1, 2020.
181. Hussain, Z., Pandey, M., Choudhury, H., Ying, P.C., Xian, T.M., Kaur, T. *et al.*, Hyaluronic acid functionalized nanoparticles for simultaneous delivery of curcumin and resveratrol for management of chronic diabetic wounds: Fabrication, characterization, stability and *in vitro* release kinetics. *J. Drug Deliv. Sci. Technol.*, 57, 101747, Jun. 1, 2020.
182. Pandey, M., Choudhury, H., Gunasegaran, T.A.P., Nathan, S.S., Md, S., Gorain, B. *et al.*, Hyaluronic acid-modified betamethasone encapsulated polymeric nanoparticles: Fabrication, characterisation, *in vitro* release kinetics, and dermal targeting. *Drug Deliv. Transl. Res.*, 9, 2, 520–33, Apr. 15, 2019.
183. Leifer, F.G., Konicek, D.M., Chen, K.J., Plaunt, A.J., Salvail, D., Laurent, C.E. *et al.*, Inhaled treprostinil-prodrug lipid nanoparticle formulations provide long-acting pulmonary vasodilation. *Drug Res.*, 68, 11, 605–14, May 23, 2018.
184. Bazzaz, B.S.F., Khameneh, B., Namazi, N., Iranshahi, M., Davoodi, D., Golmohammadzadeh, S., Solid lipid nanoparticles carrying Eugenia caryophyllata essential oil: The novel nanoparticulate systems with broad-spectrum antimicrobial activity. *Lett. Appl. Microbiol.*, 66, 6, 506–13, Jun. 1, 2018.
185. Dudhipala, N. and Gorre, T., Neuroprotective effect of ropinirole lipid nanoparticles enriched hydrogel for parkinson's disease: *In vitro, ex vivo*, pharmacokinetic and pharmacodynamic evaluation. *Pharmaceutics*, 12, 5, 448, May 1, 2020.
186. El-Say, K.M. and Hosny, K.M., Optimization of carvedilol solid lipid nanoparticles: An approach to control the release and enhance the oral bioavailability on rabbits. *Plos One*, 13, 8, e0203405, Aug. 1, 2018.
187. Wang, J., Chin, D., Poon, C., Mancino, V., Pham, J., Li, H. *et al.*, Oral delivery of metformin by chitosan nanoparticles for polycystic kidney disease. *J. Control. Release*, 329, 1198–209, Jan. 10, 2021.
188. Sunoqrot, S. and Abujamous, L., pH-sensitive polymeric nanoparticles of quercetin as a potential colon cancer-targeted nanomedicine. *J. Drug Deliv. Sci. Technol.*, 52, 670–6, Aug. 1, 2019.
189. Raj, R., Wairkar, S., Sridhar, V., Gaud, R., Pramipexole dihydrochloride loaded chitosan nanoparticles for nose to brain delivery: Development,

characterization and *in vivo* anti-Parkinson activity. *Int. J. Biol. Macromol.*, 109, 27–35, Apr. 1, 2018.

190. Ojha, S. and Kumar, B., Preparation and statistical modeling of solid lipid nanoparticles of dimethyl fumarate for better management of multiple sclerosis. *Adv. Pharm. Bull.*, 8, 2, 225–33, 2018.
191. Gholami, L., Tafaghodi, M., Abbasi, B., Daroudi, M., Oskuee, R.K., Preparation of superparamagnetic iron oxide/doxorubicin loaded chitosan nanoparticles as a promising glioblastoma theranostic tool. *J. Cell. Physiol.*, 234, 2, 1547–59, Feb. 1, 2019.
192. Pandya, N.T., Jani, P., Vanza, J., Tandel, H., Solid lipid nanoparticles as an efficient drug delivery system of olmesartan medoxomil for the treatment of hypertension. *Colloids Surf. B Biointerfaces*, 165, 37–44, May 1, 2018.
193. Saini, S., Sharma, T., Jain, A., Kaur, H., Katare, O.P., Singh, B., Systematically designed chitosan-coated solid lipid nanoparticles of ferulic acid for effective management of Alzheimer's disease: A preclinical evidence. *Colloids Surf. B Biointerfaces*, 205, 111838, Sep. 1, 2021.
194. Singh, A., Mandal, U.K., Narang, R.K., Development and characterization of enteric coated pectin pellets containing mesalamine and Saccharomyces boulardii for specific inflamed colon: *In vitro* and *in vivo* evaluation. *J. Drug Deliv. Sci. Technol.*, 62, 22–34, Apr. 1, 2021.
195. Sahoo, S. and Singh, A., Letter to the editor: 3D printing has an imperative role in colitis management. *Assay Drug Dev. Technol. NLM (Medline)*, 20, 295–7, 2022.

5

Nanoparticle Properties: Size, Shape, Charge, Inertness, Efficacy, Morphology

Kajal P. Baviskar[1]*, Brijesh M. Shah[2], Anjali P. Bedse[3], Shilpa S. Raut[3], Suchita P. Dhamane[4] and Dhara J. Dave[5]

[1]*Department of Pharmaceutical Chemistry, K K Wagh College of Pharmacy, Nashik, Affiliated to Dr Babasaheb Ambedkar Technological University, Lonere, Raigad, Maharashtra, India*
[2]*Department of Formulation and Development, Anodyne Nanotech, Inc., Boston, MA, USA*
[3]*Department of Pharmaceutics, K K Wagh College of Pharmacy, Nashik, Affiliated to Dr Babasaheb Ambedkar Technological University, Lonere, Raigad, Maharashtra, India*
[4]*Department of Pharmaceutics JSPM's Jayawantrao Sawant College of Pharmacy & Research, Hadapsar, Pune, Affiliated to Savitribai Phule Pune University, Pune, Maharashtra, India*
[5]*Guardian Burnaby Square Pharmacy, Burnaby, British Columbia, Canada*

Abstract
As a constantly expanding class of materials with several applications, NPs have drawn enormous interest. They are being used at an accelerated rate in diverse fields like cosmetics, food industry, and electronics. In the sphere of modern medicine too, NPs have become an inevitable paradigm. Due to the extremely high ratio of atoms on their surface to those inside the particle, NPs have what are known as quantum characteristics. As properties like shape, size, and morphology of NPs diverge from those of bulk materials, their catalytic characteristics improve and so is their applicability. Cellular interactions, behavior of NPs, and their effects are influenced by various properties of NPs, size, shape, and charge being the most prominent. However, while opening the new horizons, unique properties of NPs can also account for toxicity. Thus, it is crucial to understand properties of NPs. The chapter focuses on important properties of NPs. Cellular interactions and

**Corresponding author*: kajal29191@gmail.com

Vivek P. Chavda and Vasso Apostolopoulos (eds.) Nanocarrier Vaccines: Biopharmaceutics-Based Fast Track Development, (153–192) © 2024 Scrivener Publishing LLC

toxicity of NPs have been discussed in brief. In addition, the characterization techniques for determining surface properties have been tabulated.

Keywords: Nanoparticles, properties, size, shape, charge, toxicity

5.1 Introduction

Over the past 10 years, the nanotechnology field has experienced tremendous growth, giving rise to a plethora of nanoparticles (NPs). Nanoparticles have made their way into a variety of industries, such as food, medicine, and cosmetics. Particles with one dimension between 1 and 100 nm are known as NPs [1]. As compared to larger particles, NPs have unique characteristics. These characteristics, such as their compact size and comparatively large surface area, make them attractive in a range of fields, such as electronics, cosmetics, and medicine [2]. NPs are employed as pharmaceutical drug carriers having uses in both therapy and diagnostics. Their therapeutic relevance is influenced by a variety of factors, including their physical and chemical characteristics, drug loading effectiveness, drug release, and, most significantly, minimum or no carrier toxicity [3]. NPs have become potential players in medical fields in recent years, with applications ranging from contrast agents in medical imaging to carriers for the transfer of genes to specific cells. Tremendous advancements in the nanotechnology field have led to myriads of NPs. Polymeric NPs, especially PLGA and chitosan, are being extensively used for sustained and controlled release [4]. Solid NPs include gold, silver, and iron. The use of gold NPs in radiation, treatment, and diagnostics of cancer has been suggested based on their size, shape, and surface characteristics. Numerous medicinal applications, including drug administration, gene therapy, and imaging, are increasingly using carbon-based NPs [1]. One of the most popular drug delivery technologies is lipid-based NPs, with many US Food and Drug Administration (FDA)-approved formulations currently being used and many more in clinical or preclinical development [5]. For the delivery of vaccinations and anticancer medications, nanoemulsions (NEs) have been widely used. NEs are colloidal dispersions that can be utilized to transport drugs made up of molecules that are only partially soluble in water [6].

Although NPs have a lot of potential for different uses, their toxicity cannot be completely disregarded. As evident from literature, NPs can accumulate in cells, thereby inducing organ-specific toxicity [1]. Toxicity of NPs can be attributed to their different properties. To explore more applications of NPs while avoiding toxicity, it becomes essential to understand their different properties. This chapter aims to focus on different properties of NPs, mainly their size, shape, charge, inertness efficacy, and morphology. Again, interaction of NPs with cells is crucial as it can alter the cellular responses, and interaction is dependent on properties of NPs. Brief discussion of interaction of NPs with cells is also covered. In the end, interconnection between toxicity of NPs and properties is described.

5.2 Applications of Nanoparticle Formulations

Nanoparticulate carriers have shown remarkable progress as a biomedical, theranostic, engineering, and targeting tool in the last few years [7, 8]. A recent global example wherein NPs being used as a delivery platform to deliver mRNA vaccine against coronavirus disease is the most prominent approach to prevent COVID-19 infection [9–12]. Applications of NPs are widely scattered among nanomedicine and nanoengineering fields, but, considering the theme of the current section, revealing different forms of NPs with their applications would be suitable. In general, NPs are broadly categorized into polymeric (dendrimers, micelles, and macromolecular NPs), lipidic (liposomes, solid lipid NPs, and nanostructured lipid carriers), hybrid (exosomes and niosomes), and non-polymeric (nanotubes, silica NPs, and metallic NPs) systems [13–16]. Although polymeric and lipidic NPs are widely used novel carriers in drug delivery, metallic NPs being non-polymeric hold great potential for theranostic applications [17].

Polymeric NPs comprise either natural or synthetic polymer, wherein therapeutic agents are enclosed or dispersed within the polymer matrix. Polymeric micelles with a size <100 nm, containing a hydrophobic core and hydrophobic cell, is a widely used approach for the delivery of poorly water-soluble chemotherapeutic agents. Additionally, surface PEGylation improves the passive transport of drugs toward tumors [18]. Micellar formulations of paclitaxel (PTX) [19–21] and docetaxel (DTX) [22, 23] have been reported by several groups wherein Genexol-PM and Nanoxel-PM, respectively, are the modern micellar formulations under clinical investigations [24]. Several comprehensive reviews on the status and applications of polymeric micelles have been published elsewhere [25, 26]. Protein- and polysaccharide-based NPs are widely studied for ocular delivery, anticancer drug delivery, and inflammatory bowel disease and imaging studies. Abraxane®, an albumin-based NP formulation of paclitaxel, is approved for lung and pancreatic cancer [2]. Pilocarpine polymeric nanohydrogels for glaucoma [27] and latanoprost-loaded chitosan–hyaluronic acid NPs for reducing intraocular pressure [28] are recent updates in the ocular segment. Plant-derived polymers zein and pectin have been used to design colon-targeted nanoformulations of indomethacin and budesonide, respectively [29]. Dendrimers sized <10 nm contain several surface branches for the binding/conjugation of drugs, imaging agents, amino acids, and surface modifiers. Magnetic NPs modified with a polyamidoamine (G5) dendrimer have been reported to improve gene delivery into tumor cells and inhibit tumor growth [30]. As an imaging tool, binding of the radiolabeled agent to the dendrimer (G5) showed high uptake and accumulation of ^{99m}Tc-dendrimer complex into tumor cells under SPECT imaging [31].

Lipid NPs involve the similar lipid composition of cell membranes that enable enhanced lipid–lipid exchange and speed up the flux of actives from NPs into the cell membrane [8]. Liposomes are vesicular systems containing phospholipids and cholesterol bilayer membranes. PEGylated and positively charged liposomes provide longer biological circulation. Doxil®, a doxorubicin containing PEGylated liposomes, is approved for the treatment of breast cancer [14]. PEGylation improves targeted delivery and enhances biological circulation. Delivery of mRNA and DNA macromolecules [32] through lipid NPs has progressed rapidly in the last few years. COVID-19 vaccine is one such example whereby mRNA is encapsulated within positively charged lipids, phospholipids, and cholesterol, thus protecting mRNA from destructive enzymes and releasing into cell cytoplasm [33–35]. A journey of lipid NPs in last few years has launched several epic anticancer products, *viz.*, Marqibo® (Vincristine sulfate), Onivyde® (Irinotecan), Vyxeox® (Daunorubicin), and Lipodox® (Doxorubicin) [36] [22]. Solid lipid NPs and nanostructured lipid carriers are other lipid carriers that are studied to achieve sustained release of drugs in pulmonary, oral, and parenteral administrations [8, 9, 13, 14].

Non-polymeric NPs including gold NPs, nanotubes and quantum dots have been widely used as diagnostic tools and biomedical imaging. Doxorubicin-laden PEGylated gold nanorods have been developed to provide photothermal and chemical effects in cancer treatment by inducing cell apoptosis superiorly against free drugs [37]. Campothecin conjugated with multiwalled carbon nanotubes showed improved solubility and antitumor activity against marketed campothecin injection while significantly higher circulation time (3.6 h) was observed under SPECT and gamma imaging [38].

5.3 Interaction with Cells

Nano carriers, owing to their nanometer size (1–100 nm) that matches with cellular organelles and transport pathways, allows them to interact with the cells [39]. This interaction influences cell physiology and may produce positive or negative intracellular responses to NPs. Therefore, it is crucial to understand how NPs interact and what parameters impact or improve cellular uptake while avoiding adverse events like cell death, cellular stress, and damage to the biomolecules. After NPs are administered, cellular interaction occurs and their localization into cellular compartments takes place via different uptake mechanisms as shown in Figure 5.1 [40]. The size, shape, and surface charge of the NPs are the most common physicochemical parameters that widely influence cellular interactions, while rigidity, coating, and topography have a slight influence.

NP–cellular interaction is not only NP dependent but also cell type and cell size dependent. There is no optimum size of NPs reported for cellular internalization due to differences in the size of individual cell types, while literature supports particles between 50 and 100 nm in size are adequate [39–41]. NPs with smaller sizes have weak interaction but higher diffusion,

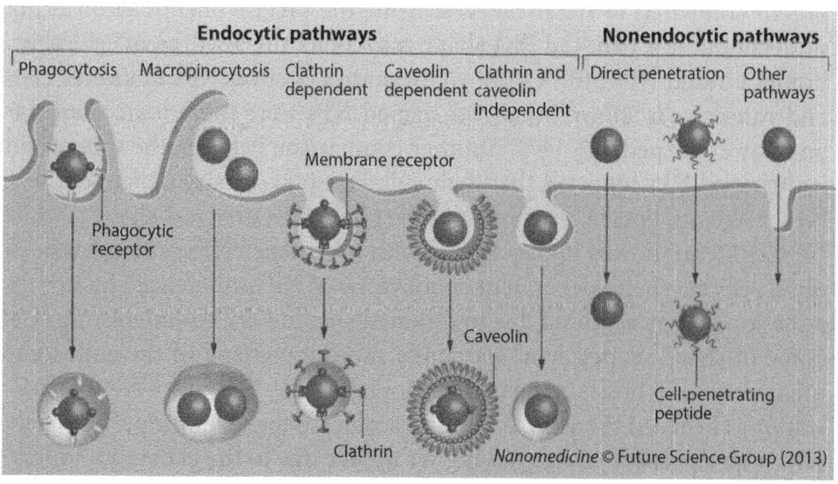

Figure 5.1 Cellular uptake pathways (endocytic and non-endocytic) of nanomaterials. Clathrin- and caveolae-independent endocytosis, clathrin-dependent endocytosis, macropinocytosis, phagocytosis, and caveolae-dependent endocytosis are endocytic pathways whereas non-endocytic cellular uptake occurs via direct penetration and other pathways. (Reprinted with permission from [40]).

while larger NPs have strong interaction and poor diffusion [42]. Shang *et al.* have studied the phenomenon of size-dependent accumulation and internalization with small- and larger-sized NPs. They found that small-sized silver NPs (≈10 nm) could not endocytose on their own without prior accumulation on the plasma membrane, while 100 nm NPs were self-endocytosed without accumulation [43]. The authors concluded that the cellular uptake of small NPs requires local interaction to trigger cellular internalization. Describing individual cellular uptake mechanisms is, however, beyond the scope of this chapter and has been covered elsewhere [40, 44]. In general and as reported in the literature, NPs with size in the range of 50–100 nm are internalized through clathrin-mediated endocytosis, while larger-sized NPs (>100 nm) follow phagocytosis and the smallest NPs (<5 nm) are diffused through cell membranes causing cytotoxicity and damages DNA. Shi *et al.* studied cellular internalization of different-sized (20 nm, 80 nm, and 12 μm) hydroxyapatite NPs on human endothelial cells and concluded that NPs were uptaken via clathrin- and caveolin-mediated endocytosis and distributed into the cytoplasm, while only a small amount of micron-sized particles was observed in cells by micropinocytosis [45].

Studies have shown that the shape of NPs has an influence on cellular internalization, efficacy, and cytotoxicity. Gold NPs with flower shapes showed higher uptake and antibacterial activity against *Staphylococcus aureus* compared to spherical NPs; while the precise phenomenon is not understood, it is believed that shape may affect the bending of cell membranes around NPs [46]. Cytotoxicity studies of silver NPs in kidney cells and cancer cells showed that rod-shaped NPs were more toxic than star and flower-shaped NPs [47]. Another comparison between the sphere and rod-shaped NPs reported that their cellular uptake is size dependent. For NPs <100 nm, uptake was higher with sphere NPs over rod-shaped, while NPs >100 nm showed higher uptake with rod-shaped NPs [48]. It has been observed that shape-dependent endocytosis of NP is influenced by surface adhesion energy acquired due to different shapes [39]. An interesting outcome of sharp-shaped nanodiamonds showed piercing of the endosome membrane and thus escaped to the cytoplasm [49]. The fate of NP internalization is believed to be the shape and surface-dependent property. Sphere NPs with a positive charge has higher uptake due to the generation of disordered membrane regions; thus, increasing entropy and sphere shape also favors wrapping by a cell membrane, while negatively charged NPs induce a more ordered membrane region causing a drop in entropy and making them poorly internalized [39]. In contrast, NPs coated with cationic ligands (polyamidoamine) to promote cellular entry did not seem promising since membrane thinning and hole formation effects were observed [50].

5.4 Properties of Nanoparticles

Due to their ability to connect bulk material to their molecular or atomic structure, NPs are of profound scientific interest. Although it is frequently alluring to think of NPs as simple molecules, they are rather complicated mixtures. Even if they may seem to dissolve like other small molecules and absorb light-like dyes, their real behavior is frequently subtle and is typically the result of the varied components of the material [51]. An extraordinarily high surface area-to-volume ratio is one of the causes behind some of their unique properties. Thus, even the most basic NP will have a surface chemistry that differs significantly from that of the basic components. Regardless of size, a substance's physical and chemical properties are constant in the bulk; however, at the nanoscale, these properties are reliant on more or less distinct atomic or molecular phenomena [52]. They have substantially lower melting points and more reactivity due to their special chemical and physical characteristics [53].

5.4.1 Classification of Nanoparticle Properties

5.4.1.1 Physicochemical Properties

i. Size and size distribution

For NPs, size includes one of the following:

a. Total physical dimension(s) determined by the atomic structure; b. Diffusion or sedimentation behavior (due to adsorption of matrix components onto the NP surface and agglomeration/aggregation of particles in the matrix) influences the effective size of the particle in a particular matrix. c. Effective size of the NP, influenced by mass or electron distribution. More insights on size have been described in the next section of the chapter [54].

ii. Shape

NPs are available not only in spherical shape but also in geometric as well as irregular shapes [55]. Owing to the shape, even NPs having similar dimensions and same compositions can demonstrate different behaviors. More details about shape have been discussed in the next section [55].

iii. Surface charge
The dangling bonds, the adsorption of charged molecules, or the grafting of charged molecules are the factors accounting for a net charge on the NP surface. The charge can have a profound effect on the aggregation of NPs, their interactions with biological samples and fluids, and their uptake by cells and tissues. This property has been discussed in the next section of this chapter [56].

iv. Porosity
Porous NPs have garnered a lot of interest in the medical and pharmaceutical areas during the past 20 years for their use in the distribution and targeting of drugs as well as for diagnosing diseases. Due to porosity, which can be many orders of magnitude greater than that of solid particles of the same size, NPs have a significantly increased surface-to-volume ratio. Additionally, toxicity is affected by porosity [57].

5.4.1.2 Optical Properties

Emission and adsorption are optical properties that are attributed to electron transition. This transition occurs between the highest occupied and lowest unoccupied molecular orbital. The surface plasmon resonance effect (i.e., resonance of the valence electron bands of the particles with light radiations) is attributed to the color of NPs [58]. Optical properties of NPs encompass absorption, reflectance and transmittance, elastic light scattering second harmonic generation, third-order non-linear optical properties, surface-enhanced Raman scattering, etc. [59]. Optical characterization of NPs can be accomplished by spectroscopic techniques like UV-visible, Raman, photoluminescence, and Brillouin [59].

5.4.1.3 Magnetic Properties

Nanomaterials acquire magnetic characteristics as a result of the interaction between electron charges and magnetic spins of the constituents. The constituent atoms of nanostructures experience distinct magnetic interaction with nearby atoms due to enhanced surface-to-volume ratio, which results in different magnetic characteristics. Super-paramagnetism is observed in magnetic NPs as thermal variations in the magnetization vector occur due to the low energy of magnetic anisotropy [60, 61]. In superparamagnetic materials, neither remanence norcoercivity exists. The core and coating are the two major components of MNPs. The coating stabilizes and shields the core from the chemical effects of the medium, whereas the core exhibits a dominant quantum effect and frequently contains magnetic elements like Fe, Ni, or Co as well as their related oxides [62]. Bulk materials of gold and platinum are non-magnetic in nature; on the other hand, nano-sized materials of gold and platinum exhibit magnetic properties. The magnetic NPs exhibit promising applications like cell sorting/targeting, sensing, drug delivery, bioseparation, and magnetic resonance imaging where contrast agents can be tagged with a recognition moiety, bioprocessing, enzyme immobilization, immunoassays, gene transfection/detection systems, and radiotherapy [62]. Iron oxide NPs are extensively used for drug delivery due to better magnetic characteristics, more possibility for surface modification, and colloidal and chemical stability. Physical properties like pH, morphology, and size affect the stability of magnetic NPs [63].

5.4.1.4 *Catalytic Properties*

Correlations exist between the surface chemistry of NPs and its catalytic properties. Catalytic activity for two different formulations of ceric oxide NPs (Nanoceria) under biologically relevant conditions (aqueous buffered near neutrality) was done by X-ray photoelectron spectroscopy (XPS) analysis [64]. This study reported that synthetic methods, size, and surface chemical properties significantly influence catalytic properties. Nanoceria are catalytic antioxidants with broad-spectrum activity in multiple types of ocular diseases Nanoceria are catalytic antioxidants that imitate the enzymatic activities of superoxide dismutase (SOD) and catalase [65].

5.4.1.5 *Mechanical Properties*

In comparison to bulk materials, NPs exhibit different mechanical properties due to the surface and quantum effects [66]. Diverse interaction forces between NPs or with the surface (Keesom force, Debye force, and London forces) are attributed to different mechanical properties of NPs. Other relevant interaction forces are structural, solvation, hydration, and electrostatic and electrical double-layer forces [67]. The biological performance of NPs depends on the mechanical property of NPs. One or more of the significant biological processes that are influenced by NP mechanical characteristics include blood circulation, tumor accumulation/penetration, and cancer cell internalization. The physicochemical properties like the size, shape, and charge of NPs must be taken into consideration in order to be employed efficiently. The section below elaborates on some of the important properties of NPs [67].

5.4.2 Different Properties

5.4.2.1 Size

The size of the NPs ranges from 1 to 100 nm [68, 69]. An average molecule can have around 1–25 atoms, giving it a radius of less than 1 to about 10 nm. An NP contains 50–200,000 atoms; dimensions can range from a few nanometers to several hundred nanometers [2, 69]. Based on the size, NPs may be classified as ultradispersion systems (1–50 nm), high-dispersion systems (50–100 nm), or sedimentable particles (100–1,000 nm). Sedimentable particles comprise either large particles or fractal-type particles [52]. A substance's properties change when it approaches the atomic scale. This is due to the material's performance being dominated by its surface atoms due to the high surface area-to-volume ratio [69, 70]. As compared to bulk materials like powders, plates, and sheets, NPs have a higher surface area-to-volume ratio due to their extremely small size. Chemistry of solids occurs at these surfaces; hence, increased surfaces mean increased chemical reactivity [69]. Particle size and size distribution can be regarded as a crucial property of NP systems. The biological fate of NPs, cellular uptake, toxicity, and targeting ability of NP systems is determined by the size as well as the size distribution of NPs [71]. Chithrani *et al.* reported that spheres in the range of 50 nm were taken up more effectively by the cells compared to both smaller and larger spheres in the range of 10–100 nm [72]. Again, the size of NPs may prevent their efficient clearance from the body. Potential toxicity is amplified in the absence of such clearance or their biodegradation into biologically benign components [73]. One of the main challenges to ensure that NPs can attain the necessary concentration in the target tissue is the fast clearance of NPs from the bloodstream through the mononuclear phagocyte system (MPS) and reticuloendothelial system (RES) [68, 74, 75]. Measurements of blood circulation half-life and analyses of the biodistribution of NPs in the blood, RES organs, and targeted tissue have successfully shown temporary evasion of the RES post-injection [74, 75]. Changes in pharmacokinetic parameters of NPs to prevent rapid clearance from blood has been accomplished by optimizing their sizes [74, 76]. Additionally, they can affect drug release, drug loading, and NP stability [71]. NPs having a diameter of less than 100 nm can cross blood vessels [53]. Particle size also influences drug release. In smaller particles, the majority of the drug is at or near the surface of the particle, which leads to rapid drug release as compared to larger particles, wherein the drug is encapsulated in larger cores and diffuses out slowly [77]. Particle size also affects stability. For instance, when storing and transporting NP dispersion, smaller particles are also more likely to aggregate [78]. Betancor *et al.*'s *in vitro* research showed that the rate of breakdown of poly (lactic-co-glycolic acid) increases with particle size [76]. Different applications of NPs based on their size are depicted in Table 5.1.

The most common methods to determine NP size are scanning electron microscopy (SEM), transmission electron microscopy (TEM), and atomic force microscopy (AFM). These are based on direct visualization via microscopy. Size and shape can be directly observed via SEM and TEM. It is possible to measure the thickness of the particles and combine the analysis of the size distribution with statistics. Furthermore, after slicing the samples, the microstructure of the NPs can be examined.

5.4.2.2 Shape

Nanomaterials can be found in many different shapes, such as fibers, rings, tubes, spheres, and planes. Spherical NOs are considered as zero dimensional, whereas, nanowires, nanotubes, and nanobelts are one dimensional. Thin filaments like nanodiscs, nanosheets, nanomembranes, and nanoplates are categorized as two dimensional [53]. As of the current scenario, spherical shape dominates [88]. Most of research to date has been on isotropic particles, which are spherical or pseudospherical, and various synthetic techniques have been devised for producing them with moderate-to-good control on their sizes and size distributions [89].

Most of research has been done on isotropic particles due to their ease of synthesis and existing lack of knowledge regarding the links between structure and activity. Developments in self-assembly techniques have also resulted in a process for synthesizing anisotropic polymeric NPs. Nonetheless, over the last two decades, non-spherical NPs like nanorods, nanobelts, nanocubes, and discs have been synthesized [88]. Owing to the decreased symmetry, anisotropic particles exhibit novel and unique chemical and physical behavior and hence have grabbed considerable attention. Also, shape affects the behavior of NPs [90].

Recent research has revealed that the efficiency of cancer therapy delivery in terms of targeting, biodistribution, and circulation duration is significantly influenced by the shape of the NPs. Compared to similarly sized

Table 5.1 Applications of NPs based on size.

Nanoparticle Type	Diameter Size Range	Application	Ref
Liposomes	50–450 nm	To impart stability to the drug, improved biodistribution of the drug, useful for delivery of hydrophilic and hydrophobic drugs; biocompatible and biodegradable	[79, 80]
Polymeric micelles	1–100 nm	Enhancement of stability and bioavailability; to deliver anticancer drugs and for ocular delivery of drugs; diagnostic imaging and drug and gene delivery	[81, 82] [83] [84]
Dendrimers	5–20 nm	Anticancer Drug Carriers Magnetic resonance imaging contrast agents Tissue Regenerators Cell Repair Vaccine Development	[85]
Nanocrystals	within the 1,000 nm range	Increased solubility and bioavailability, better patient compliance, and fast onset of action	[86]
Quantum dots	2–10 nm	Sustained and/or controlled release of anticancer drugs	[87]

nanospheres, NPs shaped into non-spherical forms may have distinctly advantageous features [90]. According to Qi *et al.*, shape might influence the melting point of NPs, and the effect of particle form on the melting temperature increases as the particle size decreases [31]. Owing to the biocompatibility, nanosized hydroxyapatite has been considered as a drug delivery vehicle. As demonstrated by Zhao *et al.*, nanosized hydroxyapatite demonstrated shape and cell dependent cytotoxicity. It was observed that needle-shaped and plate-shaped NPs were more cytotoxic as compared to spherical and rod-shaped NPs [91]. The shape of nanomaterials influences how they are distributed in the body, interactions with blood vessels, their transportation across endothelial cells, cellular uptake, and their elimination from the human body [92]. Gold nanorods have more potent cancer cell–killing capacity as compared to gold nanospheres. When compared to their spherical counterparts, non-spherical particles are phagocytized by cells differently. For instance, disk/oblate-shaped particles are more prone to phagocytosis while ellipsoid particles are uptaken less actively by macrophages [14]. Additionally, compared to spherical micelles, filamentous micelles possess a circulation duration that is about 10 times longer [92]. Nonetheless, stability of non-spherical NPs should also be considered. For instance, as reported by Phan *et al.*, nanocubes with surface facets showed worsened catalytic activity as atomic rearrangement led to the formation of nanospheres having lower energy and less reactive surface facets [93].

Table 5.2 depicts different shapes of metallic NPs and their preparation methods and some of the shapes of NPs are discussed below:

i. Nano cubes

Young *et al.* synthesized silver nanocubes having a diameter of 30–50 nm. Silver nitrate was reduced employing ethylene glycol as a reducing agent, while polyvinylpyrrolidone (PVP) was used as a capping agent [94]. Huang *et al.* synthesized gold nanocubes employing an electrochemical method. Acetone was used as solvent along with cationic surfactant solution. Hexadecyltrimethylammonium bromide (CTAB) and tetradodecylammonium bromide (TTAB) were used as primary and cosurfactants, respectively [95].

ii. Nanorods

For synthesizing nanorods, thermal, photochemical, and electrochemistry-based template methods can be employed [89]. Apart from serving as tumor-targeting agents and gene- and drug-delivery carriers, nanorods absorb near-infrared radiation and convert this energy into heat energy, thereby,

Table 5.2 Different shapes of NPs and their preparation.

Shape	Metal	Method	Reducing agent/solvent	Stabilizer/surfactant	Reference
Nanospheres	Silver	Chemical	$NaBH_4$	Dodecanoic acid (DDA)	[111]
	Silver	Chemical	Paraffin	Oleylamine	[112]
Nanocubes	Silver	Chemical	Ethylene glycol	PVP	[94]
	Gold	Photochemical	Carboxymethylated chitosan	Carboxymethylated 2-8 nm cubic (Huang *et al.*, 2008) chitosan (CMCTS)	[95]
	Gold	Electrochemical	Acetone	Hexadecyltrimethylammonium bromide CTAB Tetradodecylammonium bromide (ttab)	[113]
Nanorods	Gold	Electrochemical	Ascorbic acid	CTAB	[46]
	Gold	Photochemical	4-(2- hydroxyethoxy) phenyl-(2-hydroxy-2-propyl)ketone	CTAB	[114]
	Gold	Seed mediated	Ascorbic acid	CTAB	[115]
Nanostars	Silver	Chemical	L- ascorbic acid	-	[116]
	Silver	Seed mediated	Ascorbic acid	PVP, citrate	[117]
	Gold	Seed growth	Ascorbic acid	CTAB	[118]

killing the diseases tissue by photothermal therapy [89]. In addition to having a greater surface area and ERP impact than other nanocarriers, gold nanorods also have peculiar optical features [96]. The electron localization on a gold nanorod surface imparts unique size and shape-dependent characteristics that are not found in other colloidal particles. Gold nanorods are being increasingly used in nuclear medicine. For instance, Zhang *et al.* produced 131I-labeled gold nanorods conjugated to the cyclic RGD (a peptide) for targeting integrin $\alpha v\beta 3$-positive B16F10 cells implanted in mice [97]. Cu-integrated gold NPs with the RGD peptide showed enhanced tumor affinity in a glioblastoma xenograft model of mice; it was also effectively used for PET image–assisted photothermal therapy [98].

iii. Triangular Nanoprism

These are prismatic plate-like nanostructures that are also referred to as nanoprisms, nanotriangles, nanoplates, or nanodisks [89] (millstone), Three pointed vertices or "points" make up triangle nanoprisms, which greatly affect their optical and electrical characteristics [99, 100]. To get a huge aspect ratio, flat, triangular, hexagonal, or circular plates have been produced as nanoprisms [101]. Different photochemical routes like employing visible and ultraviolet radiations, and radiolysis have been used to develop nanoprisms. Thermal synthesis of nanoprisms have also been demonstrated by various researchers [99].

iv. Nanostars

Over the last few years, there has been rising interest in development of branched NPs, often referred to as nanostars. Colloid chemistry methods have been used to generate branching metallic NPs very well. The size of nanostars can range from 45 to 160 nm. Due to localized surface plasmon resonances, branched NPs exhibit intriguing optical characteristics. They are excellent substrates for surface enhanced Raman scattering (SERS), and they can be candidates for excellent catalytic activity [102].

Preparation Methods of Non-Spherical Nanoparticles

Mechanical as well as chemical methods are available for synthesis of NPs. Mechanical methods include procedures like particle replication in lithography, film stretching, and template assembly. The seed growth method, seedless growth method, and hydrothermal growth are some of the chemical methods. Lithographic techniques can be further divided into non-wetting templates and microfluidic systems [92]. In non-wetting templates, fluorinated material's non-wetting properties are utilized to drive raw materials into a mold that may be customized, where they are cross-linked photochemically. A variety of materials, including lipid complexes like tripropylenglycol diacrylate, polyethylene-glycol (PEG), and polylactic-co-glycolic acid (PLGA) or their combinations can be used to create particles. These particles can also be loaded with drugs by simply incorporating the drugs into the raw material [103]. Microfluidic devices are potential methods for creating non-spherical particles of various sizes, shapes, porosities, and materials [43, 104].

In the film-stretching technology, during the production process, spherical particles are inserted in a polymer film and stretched in one or two dimensions. Hydrogen bonding in the liquid state plays a significant role in the process of stretching. Temperature as well as viscosity of the liquefied particles also influence tensile strength [105]. In template assembly methods, spheres are arranged in precise arrays and connected by crosslinking chemicals or heat [106].

To create non-spherical colloidal structures, self-assembly methods and microfluidics are frequently coupled [107]. Seed growth method comprises two stages. In the initial stage, spherical NPs are produced by a specific reducing agent. In the subsequent stage, these are treated with growth solutions of same or different metal ions and ligands, to be reduced at the seed surface through heterogeneous nucleation [108].

Sol–gel synthesis, chemical bath deposition techniques, polyol reduction, electrodeposition, and the oleylamine-mediated technique are further approaches to chemical synthesis [109, 110]. Refer to Table 5.2.

5.4.2.3 Charge

Charge of the NPs can influence their behavior. The surface charge plays a key role in predicting agglomeration and intercellular uptake [119]. The loading efficiencies and release kinetics of polymeric NPs can be adjusted precisely by modifying attributes like composition, stability, responsiveness, and surface charge [120, 121]. Surface charge is also a significant factor in the initial adsorption of NPs onto the cell membrane [122]. Cell uptake of anionic (negative surface charge) NPs is by caveolin-mediated or clathrin-mediated endocytosis due to the specific interaction mechanism of these NPs with receptors [123]. Arviso *et al.* proposed perturbation of the membrane potential by positively charged gold particles with flipping of membrane areas as the route of uptake. Anionic and cationic NPs are both taken up by cells in the same way, but biological effects are typically larger when uptake rates are high. The varied uptake preferences of phagocytic and non-phagocytic cells for cationic and anionic NPs may have an impact on the efficiency and selectivity of NPs for imaging and drug delivery [56]. During a recent screening of inorganic NPs for the oral administration of protein medicines, smaller, negatively charged silica NPs improved intestinal penetration by opening tight junctions, eliminating the need for cellular absorption for transport through the epithelial barrier [124]. For improved delivery of anticancer drugs, a two-faced, zwitterionic polymer-based NP with sensitivity to the acid environment of tumor is designed. The NPs have no charge under physiological conditions and have a long circulation time. However, once these enter the acidic extracellular tumor environment, which surrounds tumors, the NPs acquire a positive charge and get activated. The positively charged NPs are quickly absorbed by the tumor cells, enhancing the therapeutic effects in the treatment of cancer [125]. The positively charged NPs strongly adhere to the negatively charged tumor ECM, reducing permeation. As intracellular and muco-penetrating applications that need endosomal escape, like in many cancer and therapeutic applications, targeting surface markers takes precedence over NP charge [125].

For NP absorption and penetration into cells, NP charge must be taken into consideration. For a variety of diagnostic and therapeutic purposes, the charge is crucial for internalization, and uptake, within cells as well as preventing cell uptake [126]. Negatively charged NPs have a lower phagocytic uptake, which prolongs their period of blood circulation. In contrast, owing to interaction of positively charged particles with anionic cell membrane, they are more prone to phagocytosis [127].

However, other researchers have shown that negatively charged NPs can bind to accessible cationic sites on the surface of macrophage and be recognized by scavenger receptors, facilitating uptake by RES, while neutral and cationic NPs can decrease the uptake of particles by RES and are cleared more quickly than negatively charged ones [127].

The surface potential of the NPs should be negative in order to increase their half-life in circulation. In comparison to anionic NPs, cationic NPs more significantly disturb plasma membrane integrity, produce more mitochondrial and lysosomal damage, and result in a greater number of autophagosomes. While charge density and hydrophobicity are equally significant, non-phagocytic cells typically take up cationic NPs to a greater extent; phagocytic cells, on the other hand, prefer to take up anionic NPs. For the majority of NPs, cationic surface charge is correlated with better cellular absorption and increased cytotoxicity in non-phagocytic cells. Anionic NPs appear to induce apoptosis, but cationic NPs appear to damage the plasma membrane to a larger extent. In phagocytic cells, anionic NPs are more effectively absorbed and have a greater cytotoxic effect [128]. Due to repulsion forces, anionic NPs may have difficulty adhering to the surface of cells, but cationic NPs, if too positively charged, may disrupt cell membranes and potentially induce cytotoxicity. Cationic NPs cause membrane damage either directly or by detaching adsorbed polymers (e.g., polyethylenimine). Anionic particles cause intracellular damage [129]. As per studies on the impact of charge density and type (positive and negative) in non-phagocytic cells, charged polystyrene and iron oxide particles are more readily absorbed as compared to uncharged counterparts [130–132].

Studies have revealed that when charged groups are present on the surface, positively charged particles are taken up more effectively than negatively charged ones. It was demonstrated that cells ingest positively charged gold and silver particles, hydroxylapatite, superparamagnetic iron oxide particles, silicon dioxide, poly (lactic acid), chitosan, and polymeric, lipidic and polystyrene particles to a greater extent than their respective anionic counterparts [56]. Carboxylated polystyrene particles bind more preferentially with serum than positively charged whereas positively charged CeO particles bind BSA (bovine serum albumin) better when compared to negatively charged particles [133, 134].

Surface charge can be determined by measuring the zeta potential. Zeta potential is important to predict the stability and toxicity of NPs. It also determines the specificity with cellular targets and the efficacy. A larger value of zeta potential indicates high repulsive force between particles, the system being more stable. While a smaller value indicates aggregation tendency of NPs. Zeta potential can be influenced by pH, the type of electrolyte, and concentration. Zeta potential can be measured by electrophoretic light scattering (ELS), acoustic, and electroacoustic, ELS being the most widely used method. Zeta potential has more recently been measured using commercial devices. For instance, Zetasizer Nano series (Malvern, UK) can efficiently eliminate electroosmosis and analyze samples in non-aqueous dispersions [135].

5.4.2.4 *Inertness*

Considering the size in nanometers, NPs have great potential for breaking through the body's numerous biological barriers. These in particular may be able to pass the blood–brain barrier, which could lead to new options for drug delivery to the brain. Additionally, nanosize permits entry into the cell and its many compartments, along with the nucleus.

NPs are especially employed as drug delivery systems to diminish toxicity and side effects of drugs. However, up to recently, it was not realized that carrier systems themselves may present risks to the patient. In comparison to the dangers caused by chemicals in traditional drug delivery methods, employing NPs for drug administration poses more serious risks. The composition or nature of materials used to formulate NPs is mainly responsible for the significant interaction with cells and tissues and hence, toxicity. Nowadays, different types of non-toxic nanomaterials are being studied for drug delivery. Biological substances like albumin, gelatin, and phospholipids as well as various polymers and solid metals such as gold and silver are preferred to produce inert or non-toxic NPs [136].

The drugs are entrapped in the NPs for increased delivery or uptake by target cells and/or decreased toxicity of the free drug to non-target organs. For reduced toxicity, an increase in the therapeutic index is must. This can be accomplished by developing long-lived and target-specific NPs. Use of biodegradable non-toxic material and surface modification with polyethylene glycol (PEG) can be viewed as potential solution [137–139].

Thermosensitive NPs could be employed for selective release of the drug after specific localization. For instance, it was observed that Doxorubicin-loaded alternating multiblock copolymers (MBCs) presented enhanced cytotoxicity to tumor cells *in vitro* at 42°C compared to 37°C [140]. The unique reactive characteristics may result in the toxicity of NPs. The tests and procedures available may detect different risks associated with the use of these NPs but sometimes may not ascertain all potential risks. Thus, additional assays (based on different types of material like biological versus non-biological origin) may be required. NPs have different physical and chemical properties when compared with micron-sized particles, which may lead to altered body distribution, passage through the blood–brain barrier, and induction of blood coagulation pathways. Thus, pharmacokinetics and distribution studies of NPs become essential. Currently, basic understanding of *in vivo* distribution of NPs both at the organ and cellular levels is lacking. In populations exposed to the environment, the effects of combustion-derived NPs primarily affect people who are already ill. As healthy animals and volunteers are nearly often used in typical preclinical screening, particle hazards could only be discovered much later [141].

5.4.2.5 Efficacy

NPs facilitate the specific targeting of various organs leading to the accumulation at the site of application. The biocompatible and biodegradable NPs reduce the systemic toxicities associated with drug molecules with enhanced therapeutic activity. NPs are usually small nanospheres made up of materials engineered at very molecular and atomic levels. As a result, NPs can easily penetrate and biodistribute in the biological system of the human body as compared to larger materials. The physical–chemical, electrochemical, magnetic, and biological characteristics of NPs are all distinct [142]. NPs have a small particle size and surface characteristic; it is helpful to manipulate the particles to achieve active and passive targeting when the drug is given by a non-oral route of administration [143]. A nanoparticulate drug delivery system is highly efficient for the treatment of cancer, autoimmune disease, tuberculosis, cardiovascular disease, and neurodegenerative diseases [144, 145]. NPs are potential carrier systems to sustain and control drug release owing to smaller particle size and surface area. NPs can reduce the side effects by proper drug transportation of the drug molecule, drug distribution in the body, and subsequent clearance [146]. The drug can be incorporated into the matrix using various excipients with no chemical reaction and due to the fact that this drug loading is relatively high as compared to other forms of particles that preserve the activity of the drug [147].

NPs are said to aid the efficient targeting and delivery of poorly aqueous soluble molecules at targeted sites and avoid drug vanishing in the gastrointestinal region. Because of having the typical uptake mechanism of absorptive endocytosis, nano drugs have higher oral bioavailability. Nanostructures can circulate in the bloodstream for an extended period and can release the drug as specified; hence, chances of plasma fluctuations are less and ultimately reduce the adverse reactions [120]. Many nanoformulations, such as polymeric NPs, liposomes, lipid NPs, nano-emulsion, and proliposomes have significant benefits for herbal medicine, including improved bioavailability, solubility, pharmacological action, protection from toxic effects, higher stability enhancement, sustained delivery, improved macrophage distribution, and resistance to chemical and physical deterioration [148]. In the oral drug delivery system, poor aqueous solubility and metabolic stability are the major concern for successful oral drug delivery. Hence, to improve these problems for hydrophobic drugs, there are different approaches like the production of NPs in nanotechnology to overcome the challenges that are associated with the oral administration of the drugs [149]. NPs also contributed to the targeted drug delivery systems and improved efficiency and reduced side effects. Nowadays, most NPs are developed for only one type of disease, but many researchers have developed a combination of drugs and NPs. Hence therapeutic NPs are directed toward multitherapeutic NPs that can be applied to treat more than one disease. The NP's drug delivery system is beneficial to reduce the drug dose, enhance bioavailability, increase penetration, and achieve therapeutic efficacy. It lowers the cost and reduces side effects [150].

5.4.2.6 Morphology

The surface morphology of NPs has a great impact on the physicochemical properties and biological applications [151]. The morphological studies deal with the size, shape, and structure of the NPs. Nowadays, various nanomaterials like dendrimers, liposomes, nanocapsules, nanorods, nanotubes, and polymeric micelles with different morphological characteristics are available as novel drug delivery systems [152]. The control of the morphology of the NPs is quite essential to achieve the desired therapeutic, physical, and chemical outcomes. The morphology of the NPs has a very great significance that determines the physicochemical properties of the NPs [153]. Apart from this, various properties like solubility, permeation, tissue distribution, circulation time, cell interactions, and flow behavior are strongly related to the size and shape of the NPs [154]. Considering all the potential importance of the size, shape, and structure of the NPs, the study of the surface morphology of the NPs is quite important in drug delivery and biomedical applications. The surface morphology studies of the NPs can be possible with advanced instrumental characterization techniques. These techniques include scanning electron microscopy (SEM), transmission electron microscopy (TEM), atomic force microscopy (AFM), nuclear magnetic resonance (NMR), and thermal analysis [155]. These advanced techniques can be used efficiently to determine the surface as well as internal morphology of the NPs. Similarly, the size distribution pattern of the NPs in suspension can be determined using a particle size analyzer. Based on the surface morphology of the NPs, they are classified as nanospheres (circular or spherical form), nanorods (rod-shaped), nanotubes (tubular-shaped), nano boxes (box-shaped), nanoclusters (clusters of the NPs), and nano films (thin film–shaped NPs) [155, 156]. These different morphologies of the NPs are due to the different manufacturing procedures, solvent effect, and excipient characteristics. Nowadays, the surface morphology of the NPs is considered a crucial aspect during the development of drug delivery technologies due to its tremendous impact on various parameters including drug release patterns, cellular uptakes, and diffusion [156]. The computational modeling along with morphologically tailor-made NPs would help to better understand the role of size and shape in biological membranes and cell lines [157]. The morphology of the NPs needs to be controlled considering the site of application of the NPs. The NPs intended for application in the ocular region need to be spherical rather than any other shape, while for topical application nano films or spherical NPs are widely acceptable to cover the maximum skin surface area [154]. In the case of cancer treatment, to facilitate enhanced permeation and retention (EPR) morphology adaptive nanomaterials having a size lesser than 600 nm are highly recommended that can extravasate from the leaky tumor vasculature [158]. The morphology of adaptive nanomaterials is widely used in the case of cancer nanomedicines, and these materials have tremendous potential to solve various biological barriers. The morphology of the NPs can be altered by using biodegradable block copolymer amphiphilic, which is considered a potential tool to engineer the morphology of the NPs ranging from spherical to tubular to worm-like micelles [158]. In the current scenario, research related to NPs is mainly directed toward understanding the relationship between the magnetic, electrical, and optical properties of the NPs with morphology (size, shape, and surface characteristics). This advancement in the study of NP morphology provided a great platform for the engineering of the NPs, especially in biomedical and pharmaceutical fields [145, 159].

As evident from the above section, properties of NPs influence their applicability; thus, it is crucial to characterize them. Table 5.3 depicts some of the important characterization techniques for NPs.

5.5 Role of Physicochemical Properties in Nanoparticle Toxicity

Nanomaterials being nanosized have gained considerable interest in the last two decades mainly for their use in the drug delivery and diagnosis. Novel carries like polymeric NPs, lipidic NPs, and metallic NPs have shown promising data in terms of efficacy, stability, and compatibility, while few of them are marketed and few have made their way up to the clinical level. However, their toxicological challenges remain a huge concern. This is because, in addition to synthesis materials, over the time, key properties of NPs like, size, surface-to-volume ratio, charge, and reactivity might alter, leading to uncontrolled cellular uptake, toxic effects, and damage to the biomolecules [162]. In general, compared to other NPs, metal NPs are reported to have higher toxicity concerns, owing to their smaller size making them enter organs via ingestion or inhalation and exert toxic effects [163, 164]. NPs for inhaled delivery have been known to cause respiratory toxicity since the inhalation system receives the total cardiac output. NPs interact with protein molecules and organize around them; depending on several physicochemical properties of NPs, this interaction may lead to protein unfolding, fibrillation, thiol crosslinking, and loss of

Table 5.3 Characterization techniques for determining various properties.

Characterization techniques/tools	**Information available**	**Ref**
Electron spectroscopies		[160]
Auger electron spectroscopy (AES)	Surface composition Surface enrichment or depletion of elements Information on layering or coatings	
X-ray photoelectron spectroscopy (XPS)	Size of NPs Collections of particles in a single layer or in powder form Electrical properties of NPs and coatings Presence of a functional group on the surface along with its nature Surface composition and chemical state	
Incidention methods		[160]
Time-of-flight secondary-ion mass spectrometry (TOF-SIMS)	Basic composition Presence of surface coatings/ contaminants Functional groups on surface	
Low-energy ion scattering (LEIS)	Surface enrichment measurements Effects of size	
Scanning probe microscopies		
Scanning tunneling microscopy (STM)	Scanning tunneling microscopy (STM size distribution of particles deposited or grown on a surface	
Atomic-force microscopy (AFM)	3D visualization of NPs distributed on a flat surface Qualitative and quantitative data on physical properties like size, morphology, surface texture, and roughness	

(*Continued*)

Table 5.3 Characterization techniques for determining various properties. (*Continued*)

Characterization techniques/tools	Information available	Ref
X ray diffraction	Crystalline structure and atomic arrangements of nanomaterials Phase transitions and solid-state reactions	[53]
Dynamic light scattering	Hydrodynamic particle size Size distribution Surface functionalization Characterizes stability in suspensions	[161]
Zeta potential	Charge on NPs Stability of NPs in suspension	[161]

enzymatic activity essential for the biological processes [165]. The history of blockbuster NP carriers in drug delivery is too short, while each carrier type may have different toxicity profiles over the other, and, in such scenarios, debating on the properties of NPs (size, shape, composition, charge, aggregation, etc.) related to toxicity seems promising.

NPs with smaller size and higher surface area easily interact with biomolecules and are distributed throughout the entire system and get metabolized. As discussed previously, NPs within 50–100 nm are internalized via clathrin-mediated endocytosis and seem to be an ideal size for receptor-mediated NP internalization [166]. NPs >150 nm are taken up by reticuloendothelial systems and filtered and further distribution is prevented, leading to elimination by the spleen and liver. A biodistribution study by Jong *et al.* intravenously administered different-sized gold NPs (10–250 nm) to rats and found that 10 nm NPs were present in all the major tissues, while 50–100 nm NPs were seen only in blood, liver, spleen, and lungs and not in the brain, heart, and kidneys, concluding that the increased surface area is the likely concern for the toxicity of smaller NPs [136]. Readers are encouraged to refer to other size-dependent toxicity reports of NPs for in-depth exposure [167–169].

Spherical NPs have the ability to endocytose easily and are less toxic over rod-shaped NPs due to ease of bending at cellular membranes. A study of zinc oxide NPs on human lung epithelial cell lines showed that smaller NPs with a rod shape were cytotoxic and caused severe mitochondrial damage and cell death within 72 h [170]. Another study in mice by Tak *et al.*

revealed that among rod-shaped, sphere and triangular silver NPs, holding similar surface charge (≈+30 mV) and size (≈50 nm), rod-shaped NPs had deeper, highest, and fastest penetration into the stratum corneum with higher concentration in blood, leading to systemic toxicity [171].

As reported in several literature, positively charged NPs favor binding to the biomolecules and are rapidly internalized, thus having higher toxicity over negative or neutral-charge NPs. This is because of higher ability of opsonization with positively charged NPs, leading to phagocytosis and precipitation of proteins [172]. Jeon *et al.* studied the impact of surface charge on cellular uptake of polyester NPs holding various functional groups in lung epithelial cell lines and demonstrated that positively charged NPs had twofold higher uptake against negatively charged NPs, while cellular toxicity was found to be dose dependent [173]. A comprehensive evaluation of various properties of NP on the human lung epithelial cell line reported that zinc NPs with positive charge entered cells faster relative to negatively charged titanium dioxide NPs and released zinc ion into cell fluid, causing cell damage. Moreover, these results were poorer with rod-shaped NPs than sphere, suggesting higher cellular toxicity with rod-shaped NPs [170]. Overall, it was obvious through several reports that size, shape, and charge are the key properties impacting toxicity of NPs, while other parameters will have minor influence in this direction; surface coating on NPs with biocompatible materials can help to control the NP properties related to toxicity.

5.6 Conclusion

Physicochemical and biological properties and efficiencies of the NPs including size, shape, durability, strengths, longer duration of action, and flexibility to accommodate many molecules attracted the interest of various researchers and scientists belonging to a variety of sectors including medicine, biology, and engineering. The NPs are widely used in the drug delivery applications like solubility enhancement, bioavailability enhancement, and controlled and sustained release profiling. NPs, in contrast to bulk materials, have unique properties. These properties influence applications of NPs to a great extent. Surface area and particle size are key factors in how materials interact with biological systems. Biological processes, such as endocytosis, cellular uptake, and the effectiveness of particle processing in the endocytic pathway, depend on the size of NPs. Apart from spherical NPs, nanorods, nanostars nanoplates, etc., have been synthesized. Shape of NPs influence their behavior in the biological system. Charge on the NPs can significantly influence opsonization, phagocytosis, their circulation in blood, and biodistribution. With features such as better drug loading and higher bioavailability, NPs provide an efficient way of drug delivery. With newer NP-based products being offered in the market, it is essential to focus more on their properties. No doubt, properties of NPs are alluring and open up a wide scope for exploring new applications, but as evident from literature, toxicity issues related with NPs also cannot be overlooked. It is crucial to characterize the physical–chemical properties of NPs in order to link biological and toxicological responses to these properties. This would definitely create avenues toward engineering NPs for their safer implementation in different fields.

References

1. Najahi-Missaoui, W., Arnold, R.D., Cummings, B.S., Safe nanoparticles: Are we there yet? *Int. J. Mol. Sci.*, 22, 385, 2020. doi: 10.3390/ijms22010385.
2. Al-Azayzih, A., Missaoui, W.N., Cummings, B.S., Somanath, P.R., Liposome-mediated delivery of the P21 activated kinase-1 (PAK-1) inhibitor IPA-3 limits prostate tumor growth *in vivo*. *Nanomed. Nanotechnol. Biol. Med.*, 12, 1231–1239, 2016, doi: 10.1016/j.nano.2016.01.003.
3. Puri, A., Loomis, K., Smith, B., Lee, J.-H., Yavlovich, A., Heldman, E., Blumenthal, R., Lipid-based nanoparticles as pharmaceutical drug carriers: From concepts to clinic. *Crit. Rev. Ther. Drug Carrier Syst.*, 26, 523–580, 2009, doi: 10.1615/CritRevTherDrugCarrierSyst.v26.i6.10.
4. Vauthier, C. and Bouchemal, K., Methods for the preparation and manufacture of polymeric nanoparticles. *Pharm. Res.*, 26, 1025–1058, 2009, doi: 10.1007/s11095-008-9800-3.
5. Bozzuto, G. and Molinari, A., Liposomes as nanomedical devices. *Int. J. Nanomedicine*, 10, 975, 2015.
6. Ganta, S., Talekar, M., Singh, A., Coleman, T.P., Amiji, M.M., Nanoemulsions in translational research—Opportunities and challenges in targeted cancer therapy. *AAPS PharmSciTech*, 15, 694–708, 2014.
7. Bhatia, S. and Bhatia, S., Nanotechnology and its drug delivery applications, in: *Natural Polymer Drug Delivery Systems Nanoparticles, Plants, and Algae*, pp. 1–32, 2016.
8. Yetisgin, A.A., Cetinel, S., Zuvin, M., Kosar, A., Kutlu, O., Therapeutic nanoparticles and their targeted delivery applications. *Molecules*, 25, 2193, 2020.
9. Wang, Q., Zhang, Y., Wu, L., Niu, S., Song, C., Zhang, Z., Lu, G., Qiao, C., Hu, Y., Yuen, K.-Y., Structural and functional basis of SARS-CoV-2 entry by using human ACE2. *Cell*, 181, 894–904, 2020.

10. Chavda, V.P., Vora, L.K., Vihol, D.R., COVAX-19® vaccine: Completely blocks virus transmission to non-immune individuals. *Clin. Complement. Med. Pharmacol.*, 1, 100004, 2021, doi: 10.1016/j.ccmp.2021.100004.
11. Chavda, V.P. and Apostolopoulos, V., COVID-19 vaccine design and vaccination strategy for emerging variants. *Expert Rev. Vaccines*, 21, 1359–1361, 2022, doi: 10.1080/14760584.2022.2112571.
12. Chavda, V.P., Teli, D., Balar, P.C., Vaghela, D., Solanki, H.K., Vaishnav, A., Vora, L., Potential anti-SARS-CoV-2 prodrugs activated by phosphorylation and their role in the aged population. *Molecules*, 28, 2332, 2023, doi: 10.3390/molecules28052332.
13. Mitchell, M.J., Billingsley, M.M., Haley, R.M., Wechsler, M.E., Peppas, N.A., Langer, R., Engineering precision nanoparticles for drug delivery. *Nat. Rev. Drug Discovery*, 20, 101–124, 2021.
14. Zhu, W., Wei, Z., Han, C., Weng, X., Nanomaterials as promising theranostic tools in nanomedicine and their applications in clinical disease diagnosis and treatment. *Nanomaterials*, 11, 3346, 2021.
15. Chavda, V.P., Balar, P.C., Patel, S.B., Nanotheranostics-based management of head and neck cancer. *Nanotheranostics*, 7, 202–209, 2023, doi: 10.7150/ntno.81724.
16. Chavda, V.P., Pandya, A., Kumar, L., Raval, N., Vora, L.K., Pulakkat, S., Patravale, V., Salwa, Duo, Y., Tang, B.Z., Exosome nanovesicles: A potential carrier for therapeutic delivery. *Nano Today*, 49, 101771, 2023. doi:10.1016/j.nantod.2023.101771.
17. Rizvi, S.A. and Saleh, A.M., Applications of nanoparticle systems in drug delivery technology. *Saudi Pharm. J.*, 26, 64–70, 2018.
18. Xiao, K., Luo, J., Fowler, W.L., Li, Y., Lee, J.S., Xing, L., Cheng, R.H., Wang, L., Lam, K.S., A self-assembling nanoparticle for paclitaxel delivery in ovarian cancer. *Biomaterials*, 30, 6006–6016, 2009.
19. Tu, L., Wang, G., Qi, N., Wu, W., Zhang, W., Feng, J., Multi-functional chitosan polymeric micelles as oral paclitaxel delivery systems for enhanced bioavailability and anti-tumor efficacy. *Int. J. Pharm.*, 578, 119105, 2020.
20. Qiu, N., Du, X., Ji, J., Zhai, G., A review of stimuli-responsive polymeric micelles for tumor-targeted delivery of curcumin. *Drug Dev. Ind. Pharm.*, 47, 839–856, 2021, doi: 10.1080/03639045.2021.1934869.
21. Jabri, T., Imran, M., Aziz, A., Rao, K., Kawish, M., Irfan, M., Malik, M.I., Simjee, S.U., Arfan, M., Shah, M.R., Design and synthesis of mixed micellar system for enhanced anticancer efficacy of paclitaxel through its co-delivery with naringin. *Drug Dev. Ind. Pharm.*, 45, 703–714, 2019.
22. Liu, Q., Lu, Y., Xiao, Y., Yuan, L., Hu, D., Hao, Y., Han, R., Peng, J., Qian, Z., Effects of docetaxel injection and docetaxel micelles on the intestinal barrier and intestinal microbiota. *Adv. Sci.*, 8, 2102952, 2021.
23. Sim, T., Kim, J.E., Hoang, N.H., Kang, J.K., Lim, C., Kim, D.S., Lee, E.S., Youn, Y.S., Choi, H.-G., Han, H.-K., Development of a docetaxel micellar

formulation using poly (ethylene glycol)–polylactide–poly (ethylene glycol) (PEG–PLA–PEG) with successful reconstitution for tumor targeted drug delivery. *Drug Deliv.*, 25, 1362–1371, 2018.

24. Zhang, X., Huang, Y., Li, S., Nanomicellar carriers for targeted delivery of anticancer agents. *Ther. Deliv.*, 5, 53–68, 2014.
25. Junnuthula, V., Kolimi, P., Nyavanandi, D., Sampathi, S., Vora, L.K., Dyawanapelly, S., Polymeric micelles for breast cancer therapy: Recent updates, clinical translation and regulatory considerations. *Pharmaceutics*, *14*, 1860, 2022.
26. Zheng, X., Xie, J., Zhang, X., Sun, W., Zhao, H., Li, Y., Wang, C., An overview of polymeric nanomicelles in clinical trials and on the market. *Chin. Chem. Lett.*, 32, 243–257, 2021.
27. Anumolu, S.S., Singh, Y., Gao, D., Stein, S., Sinko, P.J., Design and evaluation of novel fast forming pilocarpine-loaded ocular hydrogels for sustained pharmacological response. *J. Controlled Release*, *137*, 152–159, 2009.
28. Rubenicia, A.M.L., Cubillan, L.D., Sicam, V.A.D., Macabeo, A.P.G., Villaflores, O.B., Castillo, A.L., Intraocular pressure reduction effect of 0.005% latanoprost eye drops in a hyaluronic acid-chitosan nanoparticle drug delivery system in albino rabbits. *Transl. Vis. Sci. Technol.*, 10, 2–2, 2021.
29. Shah, B.M., Palakurthi, S.S., Khare, T., Khare, S., Palakurthi, S., Natural proteins and polysaccharides in the development of micro/nano delivery systems for the treatment of inflammatory bowel disease. *Int. J. Biol. Macromol.*, 165, 722–737, 2020.
30. Pan, B., Cui, D., Sheng, Y., Ozkan, C., Gao, F., He, R., Li, Q., Xu, P., Huang, T., Dendrimer-modified magnetic nanoparticles enhance efficiency of gene delivery system. *Cancer Res.*, 67, 8156–8163, 2007.
31. Zhao, L., Zhu, M., Li, Y., Xing, Y., Zhao, J., Radiolabeled dendrimers for nuclear medicine applications. *Molecules*, 22, 1350, 2017.
32. Chavda, V.P., Hossain, M.K., Beladiya, J., Apostolopoulos, V., Nucleic acid vaccines for COVID-19: A paradigm shift in the vaccine development arena. *Biologics*, 1, 337–356, 2021, doi: 10.3390/biologics1030020.
33. Khurana, A., Allawadhi, P., Khurana, I., Allwadhi, S., Weiskirchen, R., Banothu, A.K., Chhabra, D., Joshi, K., Bharani, K.K., Role of nanotechnology behind the success of MRNA vaccines for COVID-19. *Nano Today*, 38, 101142, 2021.
34. Chavda, V.P., Jogi, G., Dave, S., Patel, B.M., Nalla, L.V., Koradia, K., MRNA-based vaccine for COVID-19: They are new but not unknown! *Vaccines*, 11, 507, 2023, doi: 10.3390/vaccines11030507.
35. Chavda, V.P., Soni, S., Vora, L.K., Soni, S., Khadela, A., Ajabiya, J., MRNA-based vaccines and therapeutics for COVID-19 and future pandemics. *Vaccines*, 10, 2150, 2022, doi: 10.3390/vaccines10122150.
36. Rahim, M.A., Jan, N., Khan, S., Shah, H., Madni, A., Khan, A., Jabar, A., Khan, S., Elhissi, A., Hussain, Z., Recent advancements in stimuli responsive

drug delivery platforms for active and passive cancer targeting. *Cancers*, 13, 670, 2021.

37. Chen, J., Li, X., Zhao, X., Wu, Q., Zhu, H., Mao, Z., Gao, C., Doxorubicin-conjugated PH-responsive gold nanorods for combined photothermal therapy and chemotherapy of cancer. *Bioact. Mater.*, 3, 347–354, 2018.
38. Wu, W., Li, R., Bian, X., Zhu, Z., Ding, D., Li, X., Jia, Z., Jiang, X., Hu, Y., Covalently combining carbon nanotubes with anticancer agent: Preparation and antitumor activity. *ACS Nano*, 3, 2740–2750, 2009.
39. Villanueva-Flores, F., Castro-Lugo, A., Ramírez, O.T., Palomares, L.A., Understanding cellular interactions with nanomaterials: Towards a rational design of medical nanodevices. *Nanotechnology*, 31, 132002, 2020, doi: 10.1088/1361-6528/ab5bc8.
40. Qu, Z.G., He, X.C., Lin, M., Sha, B.Y., Shi, X.H., Lu, T.J., Xu, F., Advances in the understanding of nanomaterial–biomembrane interactions and their mathematical and numerical modeling. *Nanomedicine*, 8, 995–1011, 2013.
41. Wu, M., Guo, H., Liu, L., Liu, Y., Xie, L., Size-dependent cellular uptake and localization profiles of silver nanoparticles. *Int. J. Nanomedicine*, 14, 4247–4259, 2019.
42. Shi, W., Wang, J., Fan, X., Gao, H., Size and shape effects on diffusion and absorption of colloidal particles near a partially absorbing sphere: Implications for uptake of nanoparticles in animal cells. *Phys. Rev. E*, 78, 061914, 2008.
43. Shang, L., Nienhaus, K., Jiang, X., Yang, L., Landfester, K., Mailänder, V., Simmet, T., Nienhaus, G.U., Nanoparticle interactions with live cells: Quantitative fluorescence microscopy of nanoparticle size effects. *Beilstein J. Nanotechnol.*, 5, 2388–2397, 2014.
44. Rothen-Rutishauser, B., Bourquin, J., Petri-Fink, A., Nanoparticle-cell interactions: Overview of uptake, intracellular fate and induction of cell responses, in: *Biological Responses to Nanoscale Particles Molecular and Cellular Aspects and Methodological Approaches*, pp. 153–170, 2019.
45. Shi, X., Zhou, K., Huang, F., Wang, C., Interaction of hydroxyapatite nanoparticles with endothelial cells: Internalization and inhibition of angiogenesis in vitro through the PI3K/Akt pathway. *Int. J. Nanomedicine*, 12, 5781, 2017.
46. Penders, J., Stolzoff, M., Hickey, D.J., Andersson, M., Webster, T.J., Shape-dependent antibacterial effects of non-cytotoxic gold nanoparticles. *Int. J. Nanomedicine*, 12, 2457, 2017.
47. Woźniak, A., Malankowska, A., Nowaczyk, G., Grześkowiak, B.F., Tuśnio, K., Słomski, R., Zaleska-Medynska, A., Jurga, S., Size and shape-dependent cytotoxicity profile of gold nanoparticles for biomedical applications. *J. Mater. Sci. Mater. Med.*, 28, 1–11, 2017.
48. Gratton, S.E., Ropp, P.A., Pohlhaus, P.D., Luft, J.C., Madden, V.J., Napier, M.E., De Simone, J.M., The effect of particle design on cellular internalization pathways. *Proc. Natl. Acad. Sci.*, 105, 11613–11618, 2008.

49. Chu, Z., Zhang, S., Zhang, B., Zhang, C., Fang, C.-Y., Rehor, I., Cigler, P., Chang, H.-C., Lin, G., Liu, R., Unambiguous observation of shape effects on cellular fate of nanoparticles. *Sci. Rep.*, 4, 4495, 2014.
50. Nel, A.E., Mädler, L., Velegol, D., Xia, T., Hoek, E.M., Somasundaran, P., Klaessig, F., Castranova, V., Thompson, M., Understanding biophysicochemical interactions at the nano–bio interface. *Nat. Mater.*, 8, 543–557, 2009.
51. Christian, P., Von der Kammer, F., Baalousha, M., Hofmann, T., Nanoparticles: Structure, properties, preparation and behaviour in environmental media. *Ecotoxicology*, 17, 326–343, 2008.
52. Strambeanu, N., Demetrovici, L., Dragos, D., Lungu, M., Nanoparticles: Definition, classification and general physical properties, in: *Nanoparticles' Promises and Risks: Characterization, Manipulation, and Potential Hazards to Humanity and the Environment*, pp. 3–8, Springer, Switzerland, 2014.
53. Oake, A., Bhatt, P., Pathak, Y.V., Understanding surface Characteristics of Nanoparticles, in: *Surface Modification of Nanoparticles for Targeted Drug Delivery*, pp. 1–17, 2019.
54. Modena, M.M., Rühle, B., Burg, T.P., Wuttke, S., Nanoparticle characterization: What to measure? *Adv. Mater.*, 31, 1901556, 2019.
55. Li, N., Zhao, P., Astruc, D., Anisotropic gold nanoparticles: Synthesis, properties, applications, and toxicity. *Angew. Chem. Int. Ed.*, 53, 1756–1789, 2014.
56. Fröhlich, E., The role of surface charge in cellular uptake and cytotoxicity of medical nanoparticles. *Int. J. Nanomedicine*, 7, 5577, 2012. doi:10.2147/IJN.S36111.
57. Yu, T., Greish, K., McGill, L.D., Ray, A., Ghandehari, H., Influence of geometry, porosity, and surface characteristics of silica nanoparticles on acute toxicity: Their vasculature effect and tolerance threshold. *ACS Nano*, 6, 2289–2301, 2012.
58. Bhagyaraj, S.M. and Oluwafemi, O.S., Nanotechnology: The science of the invisible, in: *Synthesis of Inorganic Nanomaterials*, pp. 1–18, Elsevier, United Kingdom, 2018.
59. Quinten, M., *Optical Properties of Nanoparticle Systems: Mie and Beyond*, John Wiley & Sons, ISBN: ISBN 3-527-63315-4, Germany, 2010.
60. Kolhatkar, A., Jamison, A., Litvinov, D., Willson, R., Lee, T., Tuning the magnetic properties of nanoparticles. *Int. J. Mol. Sci.*, 14, 15977–16009, 2013, doi: 10.3390/ijms140815977.
61. Arruebo, M., Fernández-Pacheco, R., Ibarra, M.R., Santamaría, J., Magnetic nanoparticles for drug delivery. *Nano Today*, 2, 22–32, 2007, doi: 10.1016/S1748-0132(07)70084-1.
62. Garcia, M.A., Merino, J.M., Fernández-Pinel, E., Quesada, A., de la Venta, J., Ruíz González, M.L., Castro, G.R., Crespo, P., Llopis, J., González-Calbet, J.M. *et al.*, Magnetic properties of ZnO nanoparticles. *Nano Lett.*, 7, 1489–1494, 2007, doi: 10.1021/nl070198m.
63. El-Boubbou, K., Magnetic iron oxide nanoparticles as drug carriers: Clinical relevance. *Nanomedicine*, 13, 953–971, 2018, doi: 10.2217/nnm-2017-0336.

64. Walkey, C., Das, S., Seal, S., Erlichman, J., Heckman, K., Ghibelli, L., Traversa, E., McGinnis, J.F., Self, W.T., Catalytic properties and biomedical applications of cerium oxide nanoparticles. *Environ. Sci. Nano*, 2, 33–53, 2015, doi: 10.1039/C4EN00138A.
65. Nelson, B., Johnson, M., Walker, M., Riley, K., Sims, C., Antioxidant cerium oxide nanoparticles in biology and medicine. *Antioxidants*, 5, 15, 2016, doi: 10.3390/antiox5020015.
66. Wu, Q., Miao, W., Zhang, Y., Gao, H., Hui, D., Mechanical properties of nanomaterials: A review. *Nanotechnol. Rev.*, 9, 259–273, 2020, doi: 10.1515/ntrev-2020-0021.
67. Guo, D., Xie, G., Luo, J., Mechanical properties of nanoparticles: Basics and applications. *J. Phys. Appl. Phys.*, 47, 013001, 2013.
68. Albanese, A., Tang, P.S., Chan, W.C., The effect of nanoparticle size, shape, and surface chemistry on biological systems. *Annu. Rev. Biomed. Eng.*, 14, 1–16, 2012.
69. Williams, L.D. and Adams, W., *Nanotechnology Demystified*, McGraw Hill Professional, ISBN: ISBN 0-07-149053-1, New York, 2006.
70. Mitragotri, S., Anderson, D.G., Chen, X., Chow, E.K., Ho, D., Kabanov, A.V., Karp, J.M., Kataoka, K., Mirkin, C.A., Petrosko, S.H., Accelerating the Translation of Nanomaterials in Biomedicine. *ACS Nano*, 9, 6644–6654, 2015.
71. Mohanraj, V. and Chen, Y., Nanoparticles-a review. *Trop. J. Pharm. Res.*, 5, 561–573, 2006.
72. Chithrani, B.D., Ghazani, A.A., Chan, W.C., Determining the size and shape dependence of gold nanoparticle uptake into mammalian cells. *Nano Lett.*, 6, 662–668, 2006.
73. Hong, S., Choi, D.W., Kim, H.N., Park, C.G., Lee, W., Park, H.H., Protein-based nanoparticles as drug delivery systems. *Pharmaceutics*, 12, 604, 2020.
74. Liu, X., Huang, N., Li, H., Jin, Q., Ji, J., Surface and size effects on cell interaction of gold nanoparticles with both phagocytic and nonphagocytic cells. *Langmuir*, 29, 9138–9148, 2013.
75. Choi, H.S., Liu, W., Misra, P., Tanaka, E., Zimmer, J.P., Ipe, B.I., Bawendi, M.G., Frangioni, J.V., Renal clearance of quantum dots. *Nat. Biotechnol.*, 25, 1165–1170, 2007.
76. Dreaden, E.C., Austin, L.A., Mackey, M.A., El-Sayed, M.A., Size matters: Gold nanoparticles in targeted cancer drug delivery. *Ther. Deliv.*, 3, 457–478, 2012.
77. Redhead, H., Davis, S., Illum, L., Drug delivery in poly (lactide-co-glycolide) nanoparticles surface modified with poloxamer 407 and poloxamine 908: In vitro characterisation and *in vivo* evaluation. *J. Control. Release*, 70, 353–363, 2001.
78. Babaei, Z., Jahanshahi, M., Sanati, M., Fabrication and evaluation of gelatin nanoparticles for delivering of anti-cancer drug. *Int. J. Nanosci. Nanotechnol.*, 4, 23–30, 2008.

79. Guimarães, D., Cavaco-Paulo, A., Nogueira, E., Design of liposomes as drug delivery system for therapeutic applications. *Int. J. Pharm.*, 601, 120571, 2021.
80. Chavda, V.P., Balar, P.C., Patel, S.B., Interventional nanotheranostics in hepatocellular carcinoma. *Nanotheranostics*, 7, 128–141, 2023, doi: 10.7150/ntno.80120.
81. Xu, W., Ling, P., Zhang, T., Polymeric micelles, a promising drug delivery system to enhance bioavailability of poorly water-soluble drugs. *J. Drug Deliv.*, 2013, 340315, 2013.
82. Majumder, N., Das, N.G., Das, S.K., Polymeric micelles for anticancer drug delivery. *Ther. Deliv.*, 11, 613–635, 2020.
83. Cagel, M., Tesan, F.C., Bernabeu, E., Salgueiro, M.J., Zubillaga, M.B., Moretton, M.A., Chiappetta, D.A., Polymeric mixed micelles as nanomedicines: Achievements and perspectives. *Eur. J. Pharm. Biopharm.*, 113, 211–228, 2017.
84. Movassaghian, S., Merkel, O., Torchilin, V., Applications of polymer micelles for imaging and drug delivery. *Wiley Interdiscip. Rev. Nanomed. Nanobiotechnol.*, 7, 691–707, 2015.
85. Noriega-Luna, B., Godínez, L.A., Rodríguez, F.J., Rodríguez, A., de Larrea, G.Z.-L., Sosa-Ferreyra, C., Mercado-Curiel, R., Manríquez, J., Bustos, E., Applications of dendrimers in drug delivery agents, diagnosis, therapy, and detection. *J. Nanomater.*, 2014, 39–39, 2014.
86. Junghanns, J.-U.A. and Müller, R.H., Nanocrystal technology, drug delivery and clinical applications. *Int. J. Nanomedicine*, 3, 295–310, 2008.
87. Matea, C.T., Mocan, T., Tabaran, F., Pop, T., Mosteanu, O., Puia, C., Iancu, C., Mocan, L., Quantum dots in imaging, drug delivery and sensor applications. *Int. J. Nanomedicine*, 12, 5421, 2017.
88. Kinnear, C., Moore, T.L., Rodriguez-Lorenzo, L., Rothen-Rutishauser, B., Petri-Fink, A., Form follows function: Nanoparticle shape and its implications for nanomedicine. *Chem. Rev.*, 117, 11476–11521, 2017.
89. Millstone, J.E., Hurst, S.J., Métraux, G.S., Cutler, J.I., Mirkin, C.A., Colloidal gold and silver triangular nanoprisms. *Small*, 5, 646–664, 2009.
90. Truong, N.P., Whittaker, M.R., Mak, C.W., Davis, T.P., The importance of nanoparticle shape in cancer drug delivery. *Expert Opin. Drug Deliv.*, 12, 129–142, 2015.
91. Guo, J., Ma, L., Tan, T.Y.T., Ng, K.W., Loo, S.C.J., Zhao, X., Ng, S., Heng, B.C., Cytotoxicity of hydroxyapatite nanoparticles is shape and cell dependent. *Arch. Toxicol.*, 87, 6, 1037–1052, 2012.
92. Makvandi, P., Zarepour, A., Zheng, X., Agarwal, T., Ghomi, M., Sartorius, R., Zare, E.N., Zarrabi, A., Wu, A., Maiti, T.K., Non-spherical nanostructures in nanomedicine: From noble metal nanorods to transition metal dichalcogenide nanosheets. *Appl. Mater. Today*, 24, 101107, 2021.
93. Phan, H.T. and Haes, A.J., What does nanoparticle stability mean? *J. Phys. Chem. C*, 123, 16495–16507, 2019.

94. Young, K., Synthesis and galvanic replacement reaction of silver nanocubes in organic medium. *J. Young Investig.*, 2007.
95. Huang, L., Zhai, M.L., Long, D.W., Peng, J., Xu, L., Wu, G.Z., Li, J.Q., Wei, G.S., UV-Induced Synthesis, Characterization and formation mechanism of silver nanoparticles in alkalic carboxymethylated chitosan solution. *J. Nanopart. Res.*, 10, 1193–1202, 2008.
96. Chavda, V.P., Nalla, L.V., Balar, P., Bezbaruah, R., Apostolopoulos, V., Singla, R.K., Khadela, A., Vora, L., Uversky, V.N., Advanced phytochemical-based nanocarrier systems for the treatment of breast cancer. *Cancers*, 15, 1023, 2023, doi: 10.3390/cancers15041023.
97. Zhang, Y., Zhang, Y., Yin, L., Xia, X., Hu, F., Liu, Q., Qin, C., Lan, X., Synthesis and bioevaluation of iodine-131 directly labeled cyclic RGD-PEGylated gold nanorods for tumor-targeted imaging. *Contrast Media Mol. Imaging*, 2017, 1–10, 2017.
98. Sun, X., Huang, X., Yan, X., Wang, Y., Guo, J., Jacobson, O., Liu, D., Szajek, L.P., Zhu, W., Niu, G., Chelator-free 64Cu-integrated gold nanomaterials for positron emission tomography imaging guided photothermal cancer therapy. *ACS Nano*, 8, 8438–8446, 2014.
99. Kelly, K.L., Coronado, E., Zhao, L.L., Schatz, G.C., The optical properties of metal nanoparticles: The influence of size, shape, and dielectric environment. *J. Phys. Chem. B*, 107, 668–677, 2003.
100. Shuford, K.L., Ratner, M.A., Schatz, G.C., Multipolar excitation in triangular nanoprisms. *J. Chem. Phys.*, 123, 114713, 2005.
101. Pastoriza-Santos, I., Guerrero-Martinez, A., Silva, B., Liz-Marzan, L.M., Nanostars shine bright for you. *Curr. Opin. Colloid Interface Sci.*, 16, 118–127, 2011.
102. Pastoriza-Santos, I., Guerrero-Martinez, A., Silva, B., Liz-Marzan, L.M., Nanostars shine bright for you. *Curr. Opin. Colloid Interface Sci.*, 16, 118–127, 2011.
103. Mathaes, R., Winter, G., Besheer, A., Engert, J., Non-spherical micro-and nanoparticles: Fabrication, characterization and drug delivery applications. *Expert Opin. Drug Deliv.*, 12, 481–492, 2015.
104. Xu, S., Nie, Z., Seo, M., Lewis, P., Kumacheva, E., Stone, H.A., Garstecki, P., Weibel, D.B., Gitlin, I., Whitesides, G.M., Generation of monodisperse particles by using microfluidics: Control over size, shape, and composition. *Angew. Chem.*, 117, 734–738, 2005.
105. Palazzo, C., Ponchel, G., Vachon, J.J., Villebrun, S., Agnely, F., Vauthier, C., Obtaining nonspherical poly (alkylcyanoacrylate) nanoparticles by the stretching method applied with a marketed water-soluble film. *Int. J. Polym. Mater. Polym. Biomater.*, 66, 416–424, 2017.
106. Yin, Y. and Xia, Y., Self-assembly of spherical colloids into helical chains with well-controlled handedness. *J. Am. Chem. Soc.*, 125, 2048–2049, 2003.
107. Vanapalli, S.A., Iacovella, C.R., Sung, K.E., Mukhija, D., Millunchick, J.M., Burns, M.A., Glotzer, S.C., Solomon, M.J., Fluidic assembly and packing of

microspheres in confined channels. *Langmuir*, 24, 3661–3670, 2008, doi: 10.1021/la703840w.
108. Sau, T.K. and Rogach, A.L., Nonspherical noble metal nanoparticles: Colloid-chemical synthesis and morphology control. *Adv. Mater.*, 22, 1781–1804, 2010, doi: 10.1002/adma.200901271.
109. Thiagarajan, S., Sanmugam, A., Vikraman, D., Facile methodology of sol-gel synthesis for metal oxide nanostructures, in: *Recent Applications in Sol-Gel Synthesis*, U. Chandra, (Ed.), InTech, ISBN: ISBN 978-953-51-3245-5, Croatia, 2017.
110. Pillai, V., Kumar, P., Hou, M.J., Ayyub, P., Shah, D.O., Preparation of nanoparticles of silver halides, superconductors and magnetic materials using water-in-oil microemulsions as nano-reactors. *Adv. Colloid Interface Sci.*, 55, 241–269, 1995, doi: 10.1016/0001-8686(94)00227-4.
111. Lee, K.J., Jun, B.H., Kim, T.H., Joung, J., Direct synthesis and inkjetting of silver nanocrystals toward printed electronics. *Nanotechnology*, 17, 2424–2428, 2006, doi: 10.1088/0957-4484/17/9/060.
112. Sato-Berrú, R., Redón, R., Vázquez-Olmos, A., Saniger, J.M., Silver nanoparticles synthesized by direct photoreduction of metal salts. Application in surface-enhanced raman spectroscopy. *J. Raman Spectrosc.*, 40, 376–380, 2009, doi: 10.1002/jrs.2135.
113. Huang, C.-J., Wang, Y.-H., Chiu, P.-H., Shih, M.-C., Meen, T.-H., Electrochemical synthesis of gold nanocubes. *Mater. Lett.*, 60, 1896–1900, 2006, doi: 10.1016/j.matlet.2005.12.045.
114. Ahmed, M. and Narain, R., Rapid synthesis of gold nanorods using a one-step photochemical strategy. *Langmuir*, 26, 18392–18399, 2010, doi: 10.1021/la103339g.
115. Murphy, C.J. and Jana, N.R., Controlling the aspect ratio of inorganic nanorods and nanowires. *Adv. Mater.14*, 14, 1, 80–82, 2002.
116. Wang, Y., Camargo, P.H.C., Skrabalak, S.E., Gu, H., Xia, Y., A facile, water-based synthesis of highly branched nanostructures of silver. *Langmuir*, 24, 12042–12046, 2008, doi: 10.1021/la8020904.
117. Lu, L., Kobayashi, A., Tawa, K., Ozaki, Y., Silver nanoplates with special shapes: Controlled synthesis and their surface plasmon resonance and surface-enhanced raman scattering properties. *Chem. Mater.*, 18, 4894–4901, 2006, doi: 10.1021/cm0615875.
118. Kereselidze, Z., Romero, V.H., Peralta, X.G., Santamaria, F., Gold nanostar synthesis with a silver seed mediated growth method. *J. Vis. Exp.*, 59, 3570, 2012, doi: 10.3791/3570.
119. Mulenos, M.R., Lujan, H., Pitts, L.R., Sayes, C.M., Silver nanoparticles agglomerate intracellularly depending on the stabilizing agent: Implications for nanomedicine efficacy. *Nanomaterials*, 10, 1953, 2020, doi: 10.3390/nano10101953.
120. Patra, J.K., Das, G., Fraceto, L.F., Campos, E.V.R., del Pillar Rodriguez-Torres, M., Acosta-Torres, L.S., Diaz-Torres, L.A., Grillo, R., Swamy, M.K.,

Sharma, S. *et al.*, Nano based drug delivery systems: Recent developments and future prospects. *J. Nanobiotechnology*, 16, 71, 2018. doi:10.1186/s12951-018-0392-8.

121. Volpatti, L.R., Facklam, A.L., Cortinas, A.B., Lu, Y.-C., Matranga, M.A., MacIsaac, C., Hill, M.C., Langer, R., Anderson, D.G., Microgel encapsulated nanoparticles for glucose-responsive insulin delivery. *Biomaterials*, 267, 120458, 2021, doi: 10.1016/j.biomaterials.2020.120458.
122. Honary, S. and Zahir, F., Effect of zeta potential on the properties of nano-drug delivery systems-a review (part 2). *Trop. J. Pharm. Res.*, 12, 265–273, 2013, doi: 10.4314/tjpr.v12i2.20.
123. Foroozandeh, P. and Aziz, A.A., Insight into cellular uptake and Intracellular Trafficking of Nanoparticles. *Nanoscale Res. Lett.*, 13, 339, 2018, doi: 10.1186/s11671-018-2728-6.
124. Lamson, N.G., Berger, A., Fein, K.C., Whitehead, K.A., Anionic nanoparticles enable the oral delivery of proteins by enhancing intestinal permeability. *Nat. Biomed. Eng.*, 4, 84–96, 2019, doi: 10.1038/s41551-019-0465-5.
125. Yuan, Y.-Y., Mao, C.-Q., Du, X.-J., Du, J.-Z., Wang, F., Wang, J., Surface charge switchable nanoparticles based on zwitterionic polymer for enhanced drug delivery to tumor. *Adv. Mater.*, 24, 5476–5480, 2012, doi: 10.1002/adma.201202296.
126. Donahue, N.D., Acar, H., Wilhelm, S., Concepts of nanoparticle cellular uptake, intracellular trafficking, and kinetics in nanomedicine. *Adv. Drug Deliv. Rev.*, 143, 68–96, 2019, doi: 10.1016/j.addr.2019.04.008.
127. Duan, X. and Li, Y., Physicochemical characteristics of nanoparticles affect circulation, biodistribution, cellular internalization, and trafficking. *Small*, 9, 1521–1532, 2013, doi: 10.1002/smll.201201390.
128. Li, Y., Lian, Y., Zhang, L.T., Aldousari, S.M., Hedia, H.S., Asiri, S.A., Liu, W.K., Cell and nanoparticle transport in tumour microvasculature: The role of size, shape and surface functionality of nanoparticles. *Interface Focus*, 6, 20150086, 2016, doi: 10.1098/rsfs.2015.0086.
129. Asati, A., Santra, S., Kaittanis, C., Perez, J.M., Surface-charge-dependent cell localization and cytotoxicity of cerium oxide nanoparticles. *ACS Nano*, 4, 5321–5331, 2010, doi: 10.1021/nn100816s.
130. Huang, T., Zhang, T., Jiang, X., Li, A., Su, Y., Bian, Q., Wu, H., Lin, R., Li, N., Cao, H. *et al.*, Iron oxide nanoparticles augment the intercellular mitochondrial transfer–mediated therapy. *Sci. Adv.*, 7, eabj0534, 2021, doi: 10.1126/sciadv.abj0534.
131. Thorek, D.L.J. and Tsourkas, A., Size, charge and concentration dependent uptake of iron oxide particles by non-phagocytic cells. *Biomaterials*, 29, 3583–3590, 2008, doi: 10.1016/j.biomaterials.2008.05.015.
132. Calero, M., Gutiérrez, L., Salas, G., Luengo, Y., Lázaro, A., Acedo, P., Morales, M.P., Miranda, R., Villanueva, A., Efficient and safe internalization of magnetic iron oxide nanoparticles: Two fundamental requirements for biomedical applications. *Nanomed. Nanotechnol. Biol. Med.*, 10, 733–743, 2014. doi:10.1016/j.nano.2013.11.010.

133. Sharma, D., Maheshwari, D., Philip, G., Rana, R., Bhatia, S., Singh, M., Gabrani, R., Sharma, S.K., Ali, J., Sharma, R.K. *et al.*, Formulation and optimization of polymeric nanoparticles for intranasal delivery of lorazepam using box-Behnken design: *In vitro* and *in vivo* evaluation. *BioMed Res. Int.*, 2014, 1–14, 2014.
134. Patil, S., Sandberg, A., Heckert, E., Self, W., Seal, S., Protein adsorption and cellular uptake of cerium oxide nanoparticles as a function of zeta potential. *Biomaterials*, 28, 4600–4607, 2007. doi:10.1016/j.biomaterials.2007.07.029.
135. Jiang, J., Oberdörster, G., Biswas, P., Characterization of size, surface charge, and agglomeration state of nanoparticle dispersions for toxicological studies. *J. Nanopart. Res.*, 11, 77–89, 2009, doi: 10.1007/s11051-008-9446-4.
136. De Jong, W.H., Hagens, W.I., Krystek, P., Burger, M.C., Sips, A.J.A.M., Geertsma, R.E., Particle size-dependent organ distribution of gold nanoparticles after intravenous administration. *Biomaterials*, 29, 1912–1919, 2008, doi: 10.1016/j.biomaterials.2007.12.037.
137. Verrecchia, T., Spenlehauer, G., Bazile, D.V., Murry-Brelier, A., Archimbaud, Y., Veillard, M., Non-stealth (poly (lactic acid/albumin)) and stealth (poly (lactic acid-polyethylene glycol)) nanoparticles as injectable drug carriers. *J. Control. Release*, 36, 49–61, 1995, doi: 10.1016/0168-3659(95)00053-B.
138. Peracchia, M.T., Harnisch, S., Pinto-Alphandary, H., Gulik, A., Dedieu, J.C., Desmaële, D., d'Angelo, J., Müller, R.H., Couvreur, P., Visualization of in vitro protein-rejecting properties of PEGylated stealth® polycyanoacrylate nanoparticles. *Biomaterials*, 20, 1269–1275, 1999, doi: 10.1016/S0142-9612(99)00021-6.
139. Niidome, T., Yamagata, M., Okamoto, Y., Akiyama, Y., Takahashi, H., Kawano, T., Katayama, Y., Niidome, Y., PEG-modified gold nanorods with a stealth character for *in vivo* applications. *J. Control. Release*, 114, 343–347, 2006, doi: 10.1016/j.jconrel.2006.06.017.
140. Kang, S.I., Na, K., Bae, Y.H., Physicochemical characteristics and doxorubicin-release behaviors of PH/temperature-sensitive polymeric nanoparticles. *Colloids Surf. Physicochem. Eng. Asp.*, 231, 103–112, 2003, doi: 10.1016/j.colsurfa.2003.09.006.
141. Cong, Y., Pang, C., Dai, L., Banta, G.T., Selck, H., Forbes, V.E., Importance of characterizing nanoparticles before conducting toxicity tests. *Integr. Environ. Assess. Manage.*, 7, 502–503, 2011, doi: 10.1002/ieam.204.
142. Casals, E., Vázquez-Campos, S., Bastús, N.G., Puntes, V., Distribution and potential toxicity of engineered inorganic nanoparticles and carbon nanostructures in biological systems. *TRAC Trends Anal. Chem.*, 27, 672–683, 2008, doi: 10.1016/j.trac.2008.06.004.
143. Sharma, D., Maheshwari, D., Philip, G., Rana, R., Bhatia, S., Singh, M., Gabrani, R., Sharma, S.K., Ali, J., Sharma, R.K. *et al.*, Formulation and optimization of polymeric nanoparticles for intranasal delivery of lorazepam using box-behnken design: *In vitro* and *in vivo* evaluation. *BioMed. Res. Int.*, 2014, 1–14, 2014, doi: 10.1155/2014/156010.

144. McNamara, K. and Tofail, S.A.M., Nanoparticles in biomedical applications. *Adv. Phys. X*, 2, 54–88, 2017, doi: 10.1080/23746149.2016.1254570.
145. Dhamane, S.P. and Jagdale, S.C., Development of rifampicin loaded chitosan nanoparticles by 3 [2] full factorial design. *Res. J. Pharm. Technol.*, 13, 2545, 2020, doi: 10.5958/0974-360X.2020.00453.9.
146. Lu, B., Lv, X., Le, Y., Chitosan-modified PLGA nanoparticles for control-released drug delivery. *Polymers*, 11, 304, 2019, doi: 10.3390/polym11020304.
147. Bohr, A., Water, J., Beck-Broichsitter, M., Yang, M., Nanoembedded microparticles for stabilization and delivery of drug-loaded nanoparticles. *Curr. Pharm. Des.*, 21, 5829–5844, 2015, doi: 10.2174/138161282166615100812432 2.
148. Ali, M.I., Dayma, J., Moin, S., Nanoparticles for the enhancement of the therapeutic efficacy of herbal formulations, in: *Advances in Medical Diagnosis, Treatment, and Care*, R.K. Kesharwani, R.K. Keservani, A.K. Sharma, (Eds.), pp. 149–172, IGI Global, ISBN: ISBN 978-1-79984-453-2, United States, 2021.
149. Basha, S.K., Dhandayuthabani, R., Muzammil, M.S., Kumari, V.S., Solid lipid nanoparticles for oral drug delivery. *Mater. Today Proc.*, 36, 313–324, 2021, doi: 10.1016/j.matpr.2020.04.109.
150. Verma, S., Goand, U.K., Husain, A., Katekar, R.A., Garg, R., Gayen, J.R., Challenges of peptide and protein drug delivery by oral route: Current strategies to improve the bioavailability. *Drug Dev. Res.*, 82, 927–944, 2021, doi: 10.1002/ddr.21832.
151. Nasrollahzadeh, M., Sajadi, S.M., Sajjadi, M., Issaabadi, Z., An introduction to nanotechnology, in: *Interface Science and Technology*, vol. 28, pp. 1–27, Elsevier, ISBN: ISBN 978-0-12-813586-0, United Kingdom, 2019.
152. Venditti, I., Morphologies and functionalities of polymeric nanocarriers as chemical tools for drug delivery: A review. *J. King Saud. Univ. Sci.*, 31, 398–411, 2019, doi: 10.1016/j.jksus.2017.10.004.
153. Albanese, A., Tang, P.S., Chan, W.C.W., The effect of nanoparticle size, shape, and surface chemistry on biological systems. *Annu. Rev. Biomed. Eng.*, 14, 1–16, 2012, doi: 10.1146/annurev-bioeng-071811-150124.
154. Ridolfo, R., Tavakoli, S., Junnuthula, V., Williams, D.S., Urtti, A., van Hest, J.C.M., Exploring the impact of morphology on the properties of biodegradable nanoparticles and their diffusion in complex biological medium. *Biomacromolecules*, 22, 126–133, 2021, doi: 10.1021/acs.biomac.0c00726.
155. Kulkarni, M.B. and Goel, S., Microfluidic devices for synthesizing nanomaterials—a review. *Nano Express*, 1, 032004, 2020, doi: 10.1088/2632-959X/abcca6.
156. Sanyal, M.K., Datta, A., Hazra, S., Morphology of nanostructured materials. *Pure Appl. Chem.*, 74, 1553–1570, 2002, doi: 10.1351/pac200274091553.
157. Beddoes, C.M., Case, C.P., Briscoe, W.H., Understanding nanoparticle cellular entry: A physicochemical perspective. *Adv. Colloid Interface Sci.*, 218, 48–68, 2015, doi: 10.1016/j.cis.2015.01.007.

158. Cheng, D.-B., Qiao, Z.-Y., Xuan, L., Wang, H., Recent advances of morphology adaptive nanomaterials for anti-cancer drug delivery. *Prog. Nat. Sci. Mater. Int.*, 30, 555–566, 2020, doi: 10.1016/j.pnsc.2020.09.002.
159. Moghimi, S., Recent developments in polymeric nanoparticle engineering and their applications in experimental and clinical oncology. *Anticancer Agents Med. Chem.*, 6, 553–561, 2006, doi: 10.2174/187152006778699130.
160. Baer, D.R., Gaspar, D.J., Nachimuthu, P., Techane, S.D., Castner, D.G., Application of surface chemical analysis tools for characterization of nanoparticles. *Anal. Bioanal. Chem.*, 396, 983–1002, 2010, doi: 10.1007/s00216-009-3360-1.
161. Bankar, A., Joshi, B., Kumar, A.R., Zinjarde, S., Banana peel extract mediated novel route for the synthesis of palladium nanoparticles. *Mater. Lett.*, 64, 1951–1953, 2010, doi: 10.1016/j.matlet.2010.06.021.
162. Yang, L., Zhou, Z., Song, J., Chen, X., Anisotropic nanomaterials for shape-dependent physicochemical and biomedical applications. *Chem. Soc. Rev.*, 48, 5140–5176, 2019, doi: 10.1039/C9CS00011A.
163. Khan, I., Saeed, K., Khan, I., Nanoparticles: Properties, applications and toxicities. *Arab. J. Chem.*, 12, 908–931, 2019, doi: 10.1016/j.arabjc.2017.05.011.
164. Ingle, A.P., Gupta, I., Rai, M., Toxicological concerns of nanomaterials used in biomedical applications, in: *Nanobiotechnology in Diagnosis, Drug Delivery, and Treatment*, M. Rai, M. Razzaghi-Abyaneh, A.P. Ingle, (Eds.), pp. 375–398, Wiley, ISBN: ISBN 978-1-119-67173-2, United Kingdom, 2020.
165. Xia, T., Kovochich, M., Liong, M., Mädler, L., Gilbert, B., Shi, H., Yeh, J.I., Zink, J.I., Nel, A.E., Comparison of the mechanism of toxicity of zinc oxide and cerium oxide nanoparticles based on dissolution and oxidative stress properties. *ACS Nano*, 2, 2121–2134, 2008, doi: 10.1021/nn800511k.
166. Shin, S., Song, I., Um, S., Role of physicochemical properties in nanoparticle toxicity. *Nanomaterials*, 5, 1351–1365, 2015, doi: 10.3390/nano5031351.
167. Perde-Schrepler, M., Florea, A., Brie, I., Virag, P., Fischer-Fodor, E., Vâlcan, A., Gurzău, E., Lisencu, C., Maniu, A., Size-dependent cytotoxicity and genotoxicity of silver nanoparticles in cochlear cells *in vitro*. *J. Nanomater.*, 2019, 1–12, 2019, doi: 10.1155/2019/6090259.
168. Braakhuis, H.M., Cassee, F.R., Fokkens, P.H.B., de la Fonteyne, L.J.J., Oomen, A.G., Krystek, P., de Jong, W.H., van Loveren, H., Park, M.V.D.Z., Identification of the appropriate dose metric for pulmonary inflammation of silver nanoparticles in an inhalation toxicity study. *Nanotoxicology*, 10, 1–11, 2015, doi: 10.3109/17435390.2015.1012184.
169. Zhou, G., Li, Y., Ma, Y., Liu, Z., Cao, L., Wang, D., Liu, S., Xu, W., Wang, W., Size-dependent cytotoxicity of yttrium oxide nanoparticles on primary osteoblasts in vitro. *J. Nanopart. Res.*, 18, 135, 2016, doi: 10.1007/s11051-016-3447-5.
170. Hsiao, I.-L. and Huang, Y.-J., Effects of various physicochemical characteristics on the toxicities of ZnO and TiO2 nanoparticles toward human lung

epithelial cells. *Sci. Total Environ.*, 409, 1219–1228, 2011, doi: 10.1016/j.scitotenv.2010.12.033.

171. Tak, Y.K., Pal, S., Naoghare, P.K., Rangasamy, S., Song, J.M., Shape-dependent skin penetration of silver nanoparticles: Does it really matter? *Sci. Rep.*, 5, 16908, 2015, doi: 10.1038/srep16908.
172. Egbuna, C., Parmar, V.K., Jeevanandam, J., Ezzat, S.M., Patrick-Iwuanyanwu, K.C., Adetunji, C.O., Khan, J., Onyeike, E.N., Uche, C.Z., Akram, M. *et al.*, Toxicity of nanoparticles in biomedical application: Nanotoxicology. *J. Toxicol.*, 2021, 1–21, 2021, doi: 10.1155/2021/9954443.
173. Jeon, S., Clavadetscher, J., Lee, D.-K., Chankeshwara, S., Bradley, M., Cho, W.-S., Surface charge-dependent cellular uptake of polystyrene nanoparticles. *Nanomaterials*, 8, 1028, 2018, doi: 10.3390/nano8121028.

Part 2

NANOPARTICLES TO DELIVER ANTIGEN

6

Viral Vector–Based Nanoparticles

Suneetha Vuppu[1]*, Vivek P. Chavda[2], Toshika Mishra[1], Oishani Sengupta[1], Anand Sairam[1], Paridhi Soni[1], Mohit Joshi[1], Bhumi Bhalodiya[1] and Raj V.[3]

[1]Department of Biotechnology, School of Bio Sciences and Technology, Vellore Institute of Technology, Vellore, India
[2]Department of Pharmaceutics and Pharmaceutical Technology, L M College of Pharmacy, Navrangpura, Ahmedabad, Gujarat, India
[3]Department of Nano and Industrial Engineering Massachusetts Institute of Technology, Cambridge, MA, USA

Abstract

Viral vector–based nanoparticles have a pivotal role in revolutionizing the biopharmaceutics approach for detection and prophylactic measures of viral infectious diseases. In ancient times, interferons from vectors were used to treat diseases. Nanomedicine and viral vector–based nanoparticles have revolutionized the fields of pharmacology and medicine. Viral nanoparticles or viral vector–based nanoparticles include a wide range of nanoparticles and nanomaterials from bacteriophages and plant and mammalian viruses. These have been under research as vehicles of delivery in the past few years because of their valuable properties, safer utility, and easy manufacture. The coating of nanomaterials on the viral vector protects it from enzymatic digestion and host immune surveillance. The application of viral vector–based nanoparticles in nanomedicine is rapidly growing in various domains of therapeutics, including immunotherapies, drug delivery, vaccines, imaging, antimicrobial therapies, theranostics, photodynamic therapy, cardiovascular therapies, gene therapies, and cancer therapies. The viral vector–based nanoparticles are also used to design vaccines against numerous infections including viral infections. They are proven to increase the efficiency of current vaccines and treat infections that lack effective vaccines. This chapter focuses on nanoparticle-based viral vectors and their properties, mode of action, and applications in therapeutics in addition to recent advancements in the technology.

**Corresponding author*: vsuneetha@vit.ac.in; ORCID: https://orcid.org/0000-0002-4694-1944

Vivek P. Chavda and Vasso Apostolopoulos (eds.) Nanocarrier Vaccines: Biopharmaceutics-Based Fast Track Development, (195–240) © 2024 Scrivener Publishing LLC

Keywords: Viral vector–based nanoparticles, drug delivery, immunotherapy, imaging, theranostic, nanomedicine, applications, vaccines

6.1 Introduction

Nanoparticles are small materials of the size range of 1–100 nm. They are segregated into different categories based on their properties, shapes, or sizes. This structural uniqueness of the nanoparticles contributes to the exclusive physiochemical properties of the nanoparticles [1]. These properties also facilitate their applications in therapeutics and diagnostics, in addition to various industrial and environmental applications [1, 2].

Nanoparticles have important applications in modern medicine. In therapeutics, they are frequently utilized as a transporter for targeted gene or drug delivery to the afflicted cell, whereas, in the case of diagnostics, they are typically used in imaging techniques like fluorescence microscopy and magnetic resonance imaging (MRI) [2].

Studies have revealed that in cases of nanoparticles being used for therapeutic purposes, from the site of administration to the targeted tissue, nanoparticles will pass through a variety of physiological and cellular barriers that are naturally skilled in identifying and eliminating foreign particles. To overcome these limitations, numerous synthetic polymers are designed to modify the nanoparticles. However, biological vectors are efficient natural alternatives with minimal adverse effects and safer delivery. The viral vectors can widely be utilized for their favorable characteristics in drug delivery. This approach of employing biological vector systems for targeted drug delivery by hiding the nanoparticles is known as biomimetic nanotechnology. It involves three important steps: (1) isolation of biological vectors and removal of their genetic material, (2) generation of suitable nanoparticles, and (3) masking of nanoparticles with viral vectors producing viral nanoparticles (VNPs). VNPs are produced by sonication, electroporation, coextrusion, self-assembly, infusion, polymerization, genetic engineering, and bioconjugation sequentially at low-temperature conditions [3]. Alternatively, viral vector–based therapies exhibit limited efficiency in the sending of the desired gene to the target site. The immunogenicity of viral vectors also results in adverse reactions in patients. This can be controlled by introducing modifications on the virus surface using nanoparticle coating. The supplementation with nanoparticles enhances aggregation at the target site, evasion of the patient's immune response, and aversion of non-specific encounters with the serum complement or reticuloendothelial system. This approach is useful in targeted gene therapy [4].

In addition to the combination of viral vectors and nanoparticles for increased efficiency and efficacy in therapeutic delivery, the viruses can directly be used as nanomaterials as these are in the size range of 20–500 nm. These have several medical, biotechnological, and energy-based applications. The genetic material is removed from the viral capsid where the drug is incorporated. This approach is found to be efficient in the targeted delivery of cancer therapeutics [5].

The viruses have various features that make them favorable for targeted drug delivery. The sensitive drugs can be loaded inside the virus body, whereas the surface can be altered to help in immunogenicity and contact with other cellular molecules. Furthermore, VNPs also have utility in diagnostic techniques like imaging for the early detection of diseases [6]. In this chapter, we have discussed viral vector–based nanoparticles and their important applications extensively.

6.2 Characteristics of Viral Vector–Based Nanoparticles

Viral nanoparticles are compounds of tremendous interest in medical nano-biotechnology due to their very versatile nature. These are well-characterized, monodisperse constructions that can be mass-produced. There are several characteristic qualities of VNPs that make their structure be programmed and reprogrammed in a multitude of ways to fill their internal cavity with drug molecules and imaging reagents among other things. The exterior surface can be added with targeting ligands to let cell-specific transport of the above [6].

6.2.1 Structure of Viral Nanoparticles: Viral nanoparticles are highly defined three-dimensional structures. These have fixed arrays of viral coat proteins known as capsids that may be icosahedral (roughly spherical) or rod-shaped. The icosahedral particles have a size of 18–500 nm, while the rod-shaped or filamentous ones can have a length of around 2 μm [6, 7]. The diameters can range from 70 to 120 nm [4]. The VNP structure can be transformed by altering the template of nucleic acid coding for the viral proteins before synthesis and by chemically supplementing conjugates to precise amino acid side chains to make it more robust. For example, the mutated VNPs that are generated to have additional cysteine residues allow the covalent modifications using thiol-selective chemistry and labeling, as well as refinement, have been made easy by the integration of biotin-binding peptides, His6 tags, and the B domain of Staphylococcal protein A (SPAB) [6, 8]. The ones with singular protein hence have simpler structures. Apart from this, the capsid of the VNPs can be arranged as a monolayer, bilayer, or trilayer capsid [9].

6.2.2 Physicochemical Properties of Viral Nanoparticles: The viral capsid is naturally meant to protect the genetic material of the virus from being broken down by nucleases and from getting destroyed by physical hazards. These are therefore stable are resistant to temperature and pH extremes. This acts as an advantage when developing VNPs as it signifies the fact that they do not lose their integrity in a range of solvents and buffer mixtures and are capable of bearing the chemical modifications important for conjugation with targeting ligands or loading with cargos such as drugs or fluorophores [4, 6]. These are water-soluble particles as well [10].

6.2.3 Pharmacokinetics of Viral Nanoparticles: One of the most important factors that affect the transduction competence of the VNP in *in vivo* as well as *in vitro* conditions is the surface charge of the cell. The negatively charged glycosaminoglycans, which are expressed on the surface of the cell in large quantities, interact with cationic complexes of the nanoparticles with the help of electrostatic interaction. This subsequently increases the cellular uptake of nanocomplexes by the cell. Once the VNPs enter the cell, they are trafficked within the cell through different sections and enzymes situated in organelles where these are metabolized. The setting off of these endocytic processes is, in turn, induced by properties like stiffness and size of the VNP. Even though there may be an occurrence of different results, stiffer VNPs are usually more easily accepted, and both practical and theoretical investigations show that the endocytosis of stiff particles involves a lesser amount of energy. Additionally, VNPs that are less than 30 nm may not be proficient in adequately pushing the wrapping of the membrane enough to start the required processes. Several reports show that a decent cellular uptake and intracellular delivery of cargo happens when nanoparticles having diameters of approximately 50 nm are used [11]. Several pieces of evidence show that the main movement of the VNPs happens through the endolysosomal pathway, but the successful delivery of delicate molecules like that of nucleic acids suggests that alternate pathways are also involved. When it comes to the circulation and clearance profiles of the VNPs, these markedly differ in their composition. For example, non-PEGylated Potato virus X (PVX) showed a singular-phase plasma elimination pattern, whereas PEGylated PVX displayed double-phase breakdown from the bloodstream [10].

6.2.4 **Biocompatibility of Viral Nanoparticles**: Viral nanoparticles are non-toxic to mammalian cells. These are not known to cause any body weight changes. These have been observed not to induce any tumor necrosis factors, cause any type of inflammatory reactions, or cause tissue degradation of any kind. Thus, no chronic harmful reactions occurred in the experiments that may have caused animal misbehavior or fatality [10].

6.3 Applications

Some viruses, such as bacteriophages and plant viruses, are capable of entering mammal cells without inducing replication—due to this, nanoparticles based on viruses have been viewed as potentially useful in therapeutic functions, such as in cancer treatment or applications as vaccines in immunological applications. However, aside from uses in the immune system, they require further modifications to evade potential destruction by the immune system or to target specific tissues/cells.

One such method is the addition of ligands to the nanoparticles that match the receptors specifically expressed/overexpressed on particular types of cells. Another factor to be considered in modifying the selectivity toward target tissue is the size and shape of the ligands—two examples of compounds that have been utilized for this reason are folic acid and epidermal growth factor [6].

In drug delivery, various mechanisms and uses have been tested using plant viruses such as the Red Clover Necrotic Mosaic Virus and Cowpea Chlorotic Mottle Virus. One such method is the encapsulation of the chemical compounds within the capsid, which was used in the Cowpea Chlorotic Mottle Virus to administer oligodeoxynucleotides for immunotherapy against cancer. Binding the chemical/drug to the exterior of the particle's capsid has also been tested on the Red Clover Necrotic Mosaic Virus with the anticancer drug DOX. DOX was bound electrostatically with the use of Tris and phosphate buffer—the phosphate buffer, in particular, served to increase the amounts of DOX bound on the surface/exterior of the particle. The varied levels of DOX on the inside and outside of the molecule also affected the subsequent kinetics of the drug's release [12].

Another example of the combined use of therapeutic compounds in conjunction with viral-based nanoparticles is in the use of the compound cyclophosphamide along with cowpea mosaic virus–based particles that led to significant amounts of tumor cells being killed through augmenting the release and recognition of tumor-associated antigens and the presentation of antigens and allowing for immune cell invasion [13].

Virus-based nanoparticles have been utilized in the treatment and prevention of neurological disorders; for example, in the case of Alzheimer's disease, these were made to display the amyloid beta protein that has been indicated to have a role in the condition's development. Multiple test models were used—one Hepatitis B core antigen with the protein's N-terminal epitopes attached, another Papillomavirus model, and the bacteriophage Q-beta. The first has been tested on transgenic mice with Alzheimer's, showing its efficacy while the latter two models carried antigens and thus were able to induce the creation of antibodies, especially for the amyloid beta-protein in the host body [6].

Cardiovascular disease is the chief cause of death globally, and one of its main contributors/causative agents is blood clotting occurring in a vessel and restricting blood flow. In this regard, one potential solution that has been put forward is the application of plant virus–based nanoparticles for the delivery of thrombolysis-inducing agents. An example virus is the tobacco mosaic virus, which in structure is nanotube-like—this, along with its non-pathogenic nature to humans as well as its biocompatibility, has rendered it a suitable agent/vector for delivery of thrombolytic agents to the blocked area [14].

A lot of studies have focused on the expansion of novel and innovative therapeutics and diagnostics using nanotechnology-based approaches. The viral-vector nanomaterial formulations have proven to be an efficient and accurate method of delivery of drugs to the required site, with negligible side effects. The viral vector–based nanoparticles also promise a more accurate imaging technique when compared with the techniques currently in use [4]. The main characteristic of the virus exploited for its biomedical applications is its ability to protect and transport materials [12].

Another application of VNPs or VLPs has been in the vaccine delivery section. These vaccines contain essential genetic information regarding the antigens inside their own genome. The entry of the cell then subsequently starts the genomic processing, which then leads to the displaying the of

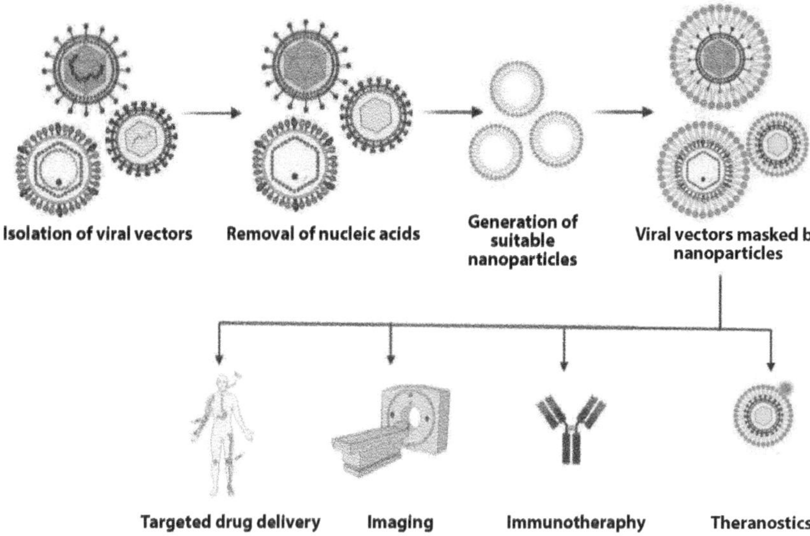

Figure 6.1 Steps for production of viral vector–based nanoparticles and their important applications in therapeutics and diagnosis (created by biorender.com).

encoded antigen on the surface of the cell. The intact replication machinery of the viral vector vaccine then starts making more and more of the viral vaccines. These replicated vaccines then are taken up by antigen-presenting cells (APCs) [12]. Since the construct of these vaccines is important to ensure maximum efficacy, nowadays, "smart" VLPs are being created that utilize immunoinformatic strategies and the correct identification of epitopes and involve artificial and genetic modifications [15].

The important biomedical applications (Figure 6.1) of viral-vector nanoparticles are discussed in this section.

6.3.1 Viral Nanoparticles for Drug Delivery

Advancements in nanotechnology, which involve the utilization of matter on a nanometer-size scale, unlocked the door to the growth of new-generation imaging instruments and targeted therapy with the heightened sensitivity of contrast agents and higher effectiveness of chemotherapy [16]. Viral nanoparticles include an assorted range of nanomaterials obtained from plant viruses, mammalian viruses, and bacteriophages. The application and evolution of VNPs and their no genome variations which are called virus-like particles (VLPs), for nanomedicine, is rising rapidly [12]. Viruses are deemed to be natural nucleic acid carriers because they guard

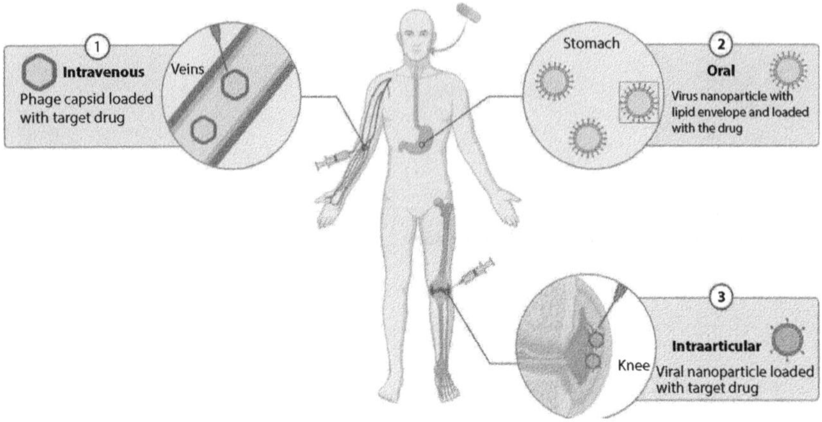

Figure 6.2 Different routes of drug delivery using VNPs (created by biorender.com).

and transport their cargo, and this is the key feature used for drug delivery (Figure 6.2). In the field of medicine, the cargo can be impregnated, encapsulated, absorbed, or conjugated to the interior and outer sides of the envelope protein interface using an amalgamation of chemistry and via attachment to numerous functional groups presented by the protein. This flexibility offers several options, which include reversible binding of active molecules, protection within protein matrices, and site-specific targeting [12].

The viral vector–based nanoparticles have applications as the mode of drug delivery in various therapeutic approaches for conditions like cancer, immunological disorders, microbial infections, and cardiovascular disorders. It is also used for the development of vaccines. This method is widely used in gene therapies for the correction of errors in nucleotide sequences [12].

6.3.1.1 *Antimicrobial Therapies*

The rapid rise in increased incidences of infectious diseases is caused by viruses, bacteria, fungi, and protozoans, contributing to increased morbidity and mortality. The greatest obstacle to the mitigation of infectious diseases is antimicrobial resistance, emphasizing the development of novel antimicrobial agents and management strategies like vaccination [17].

The viral vector–based nanoparticles provide an efficient alternative approach to therapeutics and vaccination as nanotherapies and nanovaccines. These particles facilitate targeted drug delivery to the infected organ minimizing any adverse effect on the neighboring healthy tissues. The viral vector–based nanoparticles carry the loaded drug molecules to the target diseased site and efficiently attach them to the host cells. This binding is facilitated by communication between the proteins, lipids, membranes, and carbohydrate constituents of the cell membranes. The VNP-based drug delivery carrier can evade the immune response. The approach also has advantages such as easy production, compatibility with the host system, targeted delivery, protection of the drug molecules from degradation, and biodegradability [17].

6.3.1.2 Cardiovascular Therapies

Cardiovascular diseases are a leading cause of death globally and are predicted to experience a significant increase in incidences by 2030. The viral vector–based nanoparticles contribute to an efficient, sensitive, and target-specific therapeutic approach for cardiovascular diseases. Studies report that over 50% of cases of cardiovascular diseases can be treated using nanotechnology. The nanoparticles are easily absorbed by the arteries and delivered to the target site where the drugs are continuously released over an extended period. The nanoparticles can also help in the rupture of plaques by causing apoptosis of macrophages and smooth muscle cells [18]. Similar to therapies for the treatment of cancer and genetic diseases, gene therapy can be employed as an efficient technology to target the affected cells and correct the errors in the nucleic acid sequence. The viral vector–based nanoparticles are suitable carriers for the gene to the target. It usually involves viruses like adenoviruses, adenovirus-associated viruses, retroviruses, and lentiviruses [19]. Different viral vector–based nanoparticles are presented in Table 6.1.

The utility of viral vector–based nanoparticles in immunotherapies is discussed in the later sections.

6.3.2 Viral Nanoparticles for Imaging

Viral nanoparticles are virus-based nanoparticle formulations that can act as the basis for new materials. Bacteriophages, plant or animal viruses, and infectious or non-infectious VNPs are all possible. VLPs are formed in heterologous systems but are not infectious. VNPs are self-assembling systems that produce symmetric, polyvalent, and monodisperse structures with a high degree of symmetry. Many different biochemical engineering techniques have been developed, enabling VNP/VLP formulations to transport

Table 6.1 Viral vector–based nanoparticles in drug delivery.

Sl. no.	Viral nanoparticle	Disease targeted	Mode of action	Comments	References
1.	Physalis mottle virus (PhMV)	Cancer	Synthesized a drug delivery system (DDS) by loading 6-maleimidocaproyl-hydrazone doxorubicin (DOX-EMCH) into the blank core of VLPs taken from Physalis mottle virus (PhMV) through a combination of chemical conjugation to cysteine residues and π–π stacking interactions with the attached doxorubicin molecule.	• Increases potency • Reduces immunogenicity • Increases cellular uptake • Allows specific cell targeting	[20–22]
2.	Johnson grass chlorotic stripe mosaic virus (JgCSMV)	Cancer	JgCSMV is easily transmitted by injecting infected crude plant extracts, purified virus preparations, andextracted RNA into embryos of already-soaked maize seeds	• JgCSMV is a plant RNA virus (genus Aureus virus, family Tombusviridae) • 30-nm icosahedral capsid made of 180 identical 41 kDa protein subunits packaging a molecule of ssRNA as the genome	[20, 23]

(*Continued*)

Table 6.1 Viral vector–based nanoparticles in drug delivery. (*Continued*)

Sl. no.	Viral nanoparticle	Disease targeted	Mode of action	Comments	References
3.	Hibiscus chlorotic ringspot virus (HCRSV)	Cancer	Folic acid was coupled onto the capsids to provide them with cancer-targeting capability.	• Bettered the uptake and cytotoxicity of doxorubicin in the ovarian cancer cells, OVCAR-3	[20, 24]
4.	Foot and mouth disease virus (FMDV)	Cancer	The antitumor efficacy *in vivo* was also determined in the a nude mouse xenograft model based on tumor volume/weight and histological changes. Natural FMDV specifically fastens to overexpressed integrin receptors in numerous cancer cells.	• The FMDV VLPs produced in E *Escherichia coli* was used for the first time as a delivery system of the antitumor drug doxorubicin (DOX).	[20, 25]

(*Continued*)

Table 6.1 Viral vector–based nanoparticles in drug delivery. (*Continued*)

Sl. no.	Viral nanoparticle	Disease targeted	Mode of action	Comments	References
5.	Hepatitis Bcore antigen	Cancer	HBV gets inside the host cell with the help of the sodium taurocholate cotransporting polypeptide pathway, eliminates the envelope protein, dispatches core particles, and makes a relaxed circular DNA (rcDNA) in the nucleus of the host cell.	• HBV genome is a partially double-stranded relaxed cyclic DNA molecule. • Encapsulated by the core antigen (HBcAg) and an envelope containing the surface antigen (HBsAg)	[20, 26–28]
6.	VP6-based VLPs	Cancer	Self-assembled RV VP6 particles combine with the small ubiquitin modifier (SUMO) domain to preserve their nanotube/ VLP structure in the gastrointestinal (GI) tract when provided orally and embody an apt biodegradable carrier.	• Biocompatible and biodegradable RV-VP6 structured protein assemblies in the form of nanotubes.	[20, 29]

(*Continued*)

Table 6.1 Viral vector–based nanoparticles in drug delivery. (*Continued*)

Sl. no.	Viral nanoparticle	Disease targeted	Mode of action	Comments	References
7.	Chimeric VLPs	Cancer	Chimeric hamsterpolyomavirus and SV40 VLPs incorporating T-cell epitopes from mucin 1 and influenza, respectively, can commence activation of epitope-specific CD8+ T cells.	• Chimeric VLPs have exhibited to be effective at inducing robust antibody responses • Their efficacy in generating cellular responses has not been comprehensively investigated.	[20, 30]
8.	Rous sarcoma virus	Cancer	RSV VLPs displaying hCC49 scFv were formed by the fusion of the transmembrane and cytoplasmic domains of hemagglutinin from influenza A (H1N1) virus and produced in silkworm larvae.	• RSV VLPs from silkworm larvae presented specific drug delivery to colon carcinoma cells *in vitro*.	[20, 31]

(*Continued*)

Table 6.1 Viral vector–based nanoparticles in drug delivery. *(Continued)*

Sl. no.	Viral nanoparticle	Disease targeted	Mode of action	Comments	References
9.	CCMV VLPs	Cancer	The CCMV VLPs are made from 90 homodimers and can be produced in *E. coli* or yeast	• They have beenpurified using size-exclusion chromatography or immobilized metal affinity chromatography, using coat proteins with hexahistidine extensions.	[20, 32]
10.	Tobacco Mosaic Virus (TMV)	Cancer	This plant virus is a safer alternative when compared with mammalian viruses. Usually, the virus has a hollow nanotube structure; however, it can be transformed into spherical forms by the application of heat. Both rod and sphere conformations can be used to deliver drugs to affected breast cancer cells.	• The formulation can be transported to the target tissue on intravenous administration. • Alterations and modifications of TMV viruses are easy to achieve by heat-induced processes.	[33, 34]

(Continued)

Table 6.1 Viral vector–based nanoparticles in drug delivery. (*Continued*)

Sl. no.	Viral nanoparticle	Disease targeted	Mode of action	Comments	References
11.	Viral nanoparticles as vehicles for photodynamic therapy (PDT)	Cancer, microbial infections, psoriasis, and atherosclerosis	Viral nanoparticles act as carriers or vehicles to deliver the photosensitizers to the target tissue where they get activated and induce cell death. The phage can genetically modify its coat proteins, which facilitates the delivery of photosensitizers.	• Filamentous, MS2, and Qβ phages are commonly used as nanocarriers in PDT.	[35]

(*Continued*)

Table 6.1 Viral vector–based nanoparticles in drug delivery. (*Continued*)

Sl. no.	Viral nanoparticle	Disease targeted	Mode of action	Comments	References
12.	Human Adenovirus serotypes (HAd)	Cardiovascular diseases	Viral nanoparticle vectors are widely used in gene therapy for cardiovascular diseases. These therapies can target pathways like lipid metabolism, immunoreaction and inflammation, and non-coding RNAs. The immunogenicity of the virus needs to be lowered as it elicits an immune response on introduction into the host body. The gene of interest is inserted into the viral genome, and when the virus reaches the target site in the host body, it releases its genetic material in the affected cells where the gene of interest gets integrated into the host genome.	• In addition to adenoviruses, other viruses that can be utilized in this approach include adenovirus-associated viruses and retroviruses. • The most favorable approach for delivery to the target atrial tissue in case of cardiovascular diseases is epicardial painting.	[36–39]

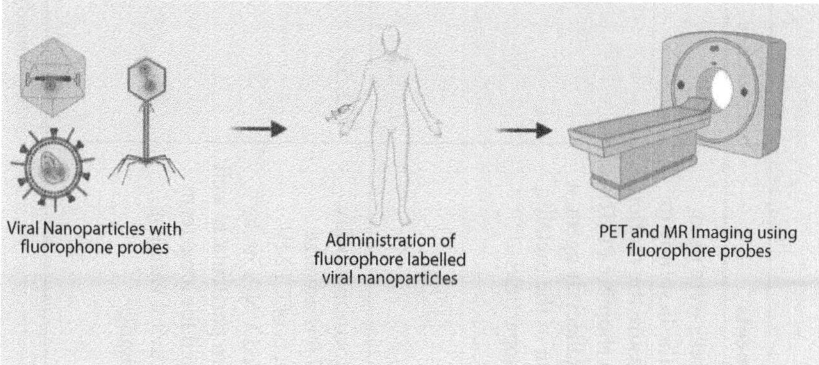

Figure 6.3 Viral nanoparticle-based imaging using fluorophore labeling (created by biorender.com).

sizable cargos of imaging tools or medications [4]. The discovery of non-toxic contrast compounds might help address current hurdles in quick and precise imaging of tissue microstructures and lesion characterization. This option is provided via nanoparticle technology [40]. The progress of imaging technologies and contrast agents is always driven by the pressing need for early illness identification and diagnosis [41].

A developing field of biomedical study called molecular imaging makes it possible to see, characterize, and quantify biological events in real time. VLPs can be more useful as imaging delivery agents than synthetic nanoparticles. Refer to Figure 6.3. They have the best signal-to-noise ratios with the least amount of toxicity at the target location. To visualize the biochemical pathways and disease mechanisms that are exhibited in the context of live organisms, a variety of imaging technologies, including optical imaging, computed tomography (CT), MRI, and positron emission tomography (PET), have been extensively used [12]. The use of viral-vector based nanoparticles in imaging is presented in Figure 6.3. Table 6.2 discusses different viral-vector based nanoparticles used in imaging.

6.3.2.1 *Nanoparticles are Used in PET/SPECT Scans*

18F-labeled DBCO-PEGylated mesoporous silica nanoparticles (MSNs) are used for imaging tumors [40, 42]. 125/124I-labeled anti-ICAM-1/PVPh-NP is used for detecting pulmonary inflammation [40, 43]. 64Cu-labeled CANF-comb nanoparticles are used for imaging natriuretic peptide clearance receptors in prostate cancer [40]. Refer to Table 6.2.

Table 6.2 Viral vector–based nanoparticles in imaging.

S. no.	Viral nanoparticle	Disease targeted	Mode of action	Comments	Reference
1.	Rhomboidal Pt (II) metallacycle-based hybrid VNPs	Cancer cells	The quantum dots encapsidation *in vitro* self-assembly system is used to trace the VNP when incubated with the living. Vero cells for cell imaging and visual transmission of Pt (II) metallacycle agent. Confirmation of the size is done by DLS (dynamic light scattering) measurement. Rhomboids have been shown to increase particle size and concentration.	Vero cells were used for the first time to measure the motion of a rhombic Pt (II) metallacycle agent, a form of the metallic nanoparticle.	[54]
2.	Cowpea mosaic virus (CPMV)–based VNP	Cancer therapeutics	A peptide ligand called E7p72 was added to CPMV to increase its affinity for epidermal growth factor-like domain 7, (EGFL7). Real-time intravital imaging revealed that the E7p72-CPMV particles exhibited great selectivity to EGFL7 and displayed binding to tumor neo-vasculature within 90 min.	The biocompatibility and few adverse effects of CPMV-based VNPs make them particularly well suited for long-term intravital vascular imaging.	[12, 55]

(*Continued*)

Table 6.2 Viral vector–based nanoparticles in imaging. (*Continued*)

S. no.	Viral nanoparticle	Disease targeted	Mode of action	Comments	Reference
3.	MS2 VLPs	Atherosclerotic plaques	The VLP is based on SV40 containing near infrared imaging quantum dots, a targeting peptide, and anticoagulant medication. Which CD (Carmit Diagnostic Imaging) stages the VLP could detect atherosclerosis in depended on the targeting peptide used.	VLPs can be added to long-lasting formulations employing metal–organic frameworks (MOFs), scaffolds, and slow-release polymers to reduce imaging times.	[12, 56]
4.	Tobacco Mosaic Virus VLPs	Breast cancer cells	PDA, a biopolymer inspired by mussels, was used to coat tobacco-based polydopamine-coated Gd-loaded TMV particles to enhance PTT, under the direction of multimodal magnetic resonance/photoacoustic imaging. At 60 MHz, the r1-relaxivity of the PDA particles was 80 mM1 s1, as opposed to 13.63 mg/L for particles that were not covered for Gd-TMV particles.	The PDA coating improves the MRI characteristics, grants PDA contrast, and makes PTT possible. Copper sulfide has only been encapsulated with HBc (hepatitis B core protein VLPs) for PTT.	[12, 52, 57, 58]

(*Continued*)

Table 6.2 Viral vector–based nanoparticles in imaging. (*Continued*)

S. no.	Viral nanoparticle	Disease targeted	Mode of action	Comments	Reference
5.	Canine parvovirus (CPV)	Tumor cells	Transferrin receptors (TfRs) are used by the canine parvovirus (CPV) to attach to and enter human and canine cells. The attachment of small molecules and transportation to tumor cells of CPV VLPs generated by the production of the CPV-VP2 capsid protein in a baculovirus expression system has been investigated. Six lysines per the VP2 subunit are potentially accessible for bioconjugation on the capsid exterior, according to structural modeling.	CPV-VLPs could be used as a new virus-based platform for the development of tumor-targeted medication delivery and image contrast agents. Under the right circumstances, it is possible to derivatize the lysine residues on them with small molecules, and doing so has no negative effects on their ability to attach to and internalize tumor cells.	[41, 59]

6.3.2.2 *Nanoparticles Used in Ultrasonic Tests*

Rattle-type MSN is used for ultrasound imaging agents [40, 44]. FA-PEG-CS and perfluorooctyl bromide nanocores are used for molecular tumor imaging agents [40, 45]. Exosome-like silica NP is used for exosome-like silica NPs [40, 46].

6.3.2.3 *Nanoparticles Utilized in CT Scans*

Liposomal iodine is used for imaging macrophage-rich atherosclerotic plaques and identifying the tumor vascular structure [40, 47, 48]. Tantalum oxide is used for producing greater imaging capability than iodine [40, 49]. AuNP is used for imaging brain malignant gliomas and enhancing radiotherapy [40, 50].

6.3.2.4 *Nanoparticles Employed in MRI Biomedical Applications*

Liposomal gadolinium is used for imaging the placenta as blood-pool contrast [40, 51]. FibPep-ION-Micelles are used for FibPep-ION-Micelles [40]. The Mn-SPIO micelle is used for high-power liver imaging contrast [52, 57].

6.3.2.5 *Illustrations of Nanoparticles Utilized in Fluorescence-Based Biological Applications*

Cationic oligofluorene-substrated POSS is used for imaging double-stranded DNA [40]. QD710-Cy7-PEGylated lipids are used for monitoring NP accumulation and dissociation kinetics in tumors [40, 53]. The perylenediimide-containing polysiloxane core and silica shell are used for detecting nanotoxicity in live cells [40].

6.3.3 Viral Nanoparticles for Immunotherapy

Viral nanoparticles are a broad category of naturally occurring nanomaterials that are generated from human, plant, and bacteriophage viruses. The use of VNPs and their genome-free counterparts, VLPs, in nanomedicine is expanding quickly. To achieve tissue selectivity, VLPs can be genetically or chemically coupled to targeted ligands as well as a variety of active components. VLPs are produced using biocompatible, biodegradable ingredients by scalable fermentation or molecular farming, VLPs can also be used for imaging and theranostics as well as cancer medicines,

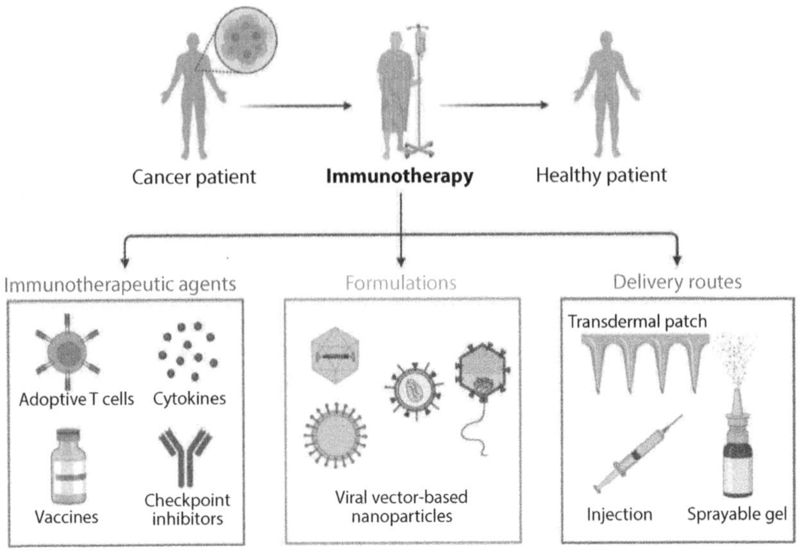

Figure 6.4 Details on the application of viral vector–based nanoparticle formulations in immunotherapy (created by biorender.com).

immunotherapies, vaccines, antimicrobial therapies, and gene therapies [12]. Refer to Figure 6.4.

VLPs are a very useful source of vaccines. VLPs mimic the conformation of native viruses harnessing their inherent immunogenicity but do not compromise on safety, as they cannot infectnor replicate because of the absence of a genome. A further degree of protection is provided by theinability of plant viruses and bacteriophages to multiply within mammalian cells, with or withouttheir viral genome. When displayed to the immune system of the host, VLPs are taken up by APCs inducing required and effective immune responses. Therefore, VLPs are durable contenders for use, as carriers to carry and present epitopes for vaccine design. Another emerging area in the use of VLPs is cancer immunotherapy—in which the inherent immune-stimulatory properties are utilized to reprogram the tumor microenvironment (TME) to launch antitumor immunity [12]. Refer to Table 6.3.

6.3.4 Viral Nanoparticles for Theranostic Applications

Several types of nanoparticle platforms are presently being established for applications in the medical field. Naturally occurring VNPs and VLPs are now the most studied because of their several benefits including production

Table 6.3 Viral vector nanoparticles in immunotherapy.

Sl. no.	Viral nanoparticles	Diseases targeted	Mode of action	Comments	References
1.	Poliovirus/ Rhinovirus chimera-PVSRIPO	Grade IV malignant glioma	Mammalian viruses for the longest time have been used as therapies in oncology. Oncolytic viruses directly lyse tumor cells leading to the release of tumor-associated antigens (TAAs). A potentially immunotherapeutic strainusing a recombinant Poliovirus/Rhinovirus chimera (PVSRIPO), which is currently in clinical trials against recurrent grade IV malignant glioma, is studied extensively. PVSRIPO infection of primary human dendritic cells (DCs) and macrophages produced persistent IFN-dominant activation and more resistance to cancer cell–mediated immunosuppression.	• Administration of viral nanoparticles into tumor microenvironments can trigger innate immune activation capable of instigating the transition to adaptive antitumor immunity. • A single injection of PVSRIPO elicited systemic antitumor cytotoxic T-cell (CTL) responses in B16F10.0-OVA-CD155 implanted C57BL/6 mice.	[12, 60]

(*Continued*)

Table 6.3 Viral vector nanoparticles in immunotherapy. (*Continued*)

Sl. no.	Viral nanoparticles	Diseases targeted	Mode of action	Comments	References
2.	Talimogene Laherparepvec (TVEC) derived from the type I herpes virus	Tumor cells	TVEC has two modes of action: 1) TVEC propagates tumor cells by dysregulating the protein kinase R (PKR) pathway 2) Activation of innate immune system response through the expression of proinflammatory molecules such as human granulocyte–macrophage colony-stimulating factor (GM-CSF)	• TVEC acts on the innate immune system and is recognized by pathogen-associated molecular pattern (PAMP) receptors. • Access of viral particles into the tumor cells leads to cell lysis, releasing TAAs, extracellularly for dendritic cells to process and activate T cytotoxic lymphocytes, which go onto attack additional tumor cells, in an expanding cyclic manner.	[61–65]

(*Continued*)

Table 6.3 Viral vector nanoparticles in immunotherapy. (*Continued*)

Sl. no.	Viral nanoparticles	Diseases targeted	Mode of action	Comments	References
				• TVEC is currently going through extended clinical trials in combination therapies and monotherapies for other cancers such as non-melanoma skin cancers, pancreatic cancer,and angiosarcomas	

(*Continued*)

Table 6.3 Viral vector nanoparticles in immunotherapy. (*Continued*)

Sl. no.	Viral nanoparticles	Diseases targeted	Mode of action	Comments	References
3.	Cow Pea Mosaic Virus (CPMV) VLPs	Cancer therapeutics	*In situ* vaccination of the CPMV demonstrated efficacy in the treatment of ovarian, colon, and breast tumor models. When vaccinated, the assimilation of neutrophils was observed. These activated neutrophils triggered a systemic antitumor response in the TME, which resulted in the uptake of administered CPMV VLPs.	• CPMV is a plant-based VLP, which is stable, non-toxic, modifiable with drugs and antigens, and can be mass-manufactured; they are the future of tumor treatment and are safe for humans.	[66–69]

(*Continued*)

Table 6.3 Viral vector nanoparticles in immunotherapy. (*Continued*)

Sl. no.	Viral nanoparticles	Diseases targeted	Mode of action	Comments	References
			CPMVs change the TME by engaging neutrophils and natural killer cells, macrophages are then reprogrammed from M2 to M1 antitumor macrophages. The innate immune cells then attack the tumor, and subsequent cell lysis allows APCs to process and present TAAs triggering tumor-specific T-cell responses. T-cell activation not only treats the injected tumor but also induces systemic antitumor immunity.	• Clinical trials of murine dermal melanoma, intracranial glioma, and ovarian cancer areunder process. • In addition to studies on mice, CPMV-based immunotherapeutic studies on canine patients are under trial.	

(*Continued*)

Table 6.3 Viral vector nanoparticles in immunotherapy. (*Continued*)

Sl. no.	Viral nanoparticles	Diseases targeted	Mode of action	Comments	References
4.	VLP-based vaccines for infections	Viral infections and diseases	Several VLP-based vaccines are being considered for treating various infections like human papillomavirus (HPV) and hepatitis B vaccines (HBV) as well as potential vaccines for influenza, Epstein–Barr virus, and malaria. The mechanism of action of different viral vaccine candidates can be of two types: The antigenic subunit is encapsulated. The antigenic subunit is displayed over the surface. The antigens are displayed on the cell surfaces priming both CD4+ and CD8+ T cells. The CD4+ T cells go on to activate B cells, which secrete antibodies. Activated CD8+ T cells induce a CTL response.	• Enterovirus 71 (EV71) SP70 epitope was used to tag HBc VLPs using a sortase A-mediated transpeptidation process, which produced powerful anti-EV71 antibodies and shielded Balb/c mice from a fatal challenge. Due to its simplicity, site specificity, high effectiveness, and adaptability, this chemoenzymatic site-specific technique has shown significant potential in VLP-based vaccine development.	[12]

(*Continued*)

Table 6.3 Viral vector nanoparticles in immunotherapy. (*Continued*)

Sl. no.	Viral nanoparticles	Diseases targeted	Mode of action	Comments	References
5.	CpG ODN-PROTEINS—nAVIDIN-based vaccines	Tumor cells	This nanovaccine primarily consists of nucleic acids (CpG ODN), proteins (such as a tumor-associated antigen, and neutravidin (nAvidin), which are the skeletal materials for constructing nanovaccine. It also contains carriers for loading CpG ODN and the tumor-associated antigen and dye molecules for tracing the vaccine *in vitro* and *in vivo*.	• According to animal studies, the nanovaccine exhibits exceptional anticancer activity, significantly slows the growth of tumors, and lengthens the survival ofmice with malignant tumors. • Nanovaccines produced by this method may mimic a virus and can treat cancer in a flexible, efficient, and economical way and have a lot of potential in clinical applications.	[70]

(*Continued*)

Table 6.3 Viral vector nanoparticles in immunotherapy. (*Continued*)

Sl. no.	Viral nanoparticles	Diseases targeted	Mode of action	Comments	References
			In vitro, the nanovaccine successfully increases dendritic cell maturation, enhances antigen cross-presentation, and increases cytokine production. It presents antigens to produce CD8+ T cells specific for tumor antigens and triggers a Th1- biased immune response; *in vivo* research shows the targeted lymph nodes. The most notable effect of a virus-like nanovaccine is its considerable inhibition of antigen-expressed tumor development and extension of the survival time of mice carrying the tumor.		

Figure 6.5 Viral vector nanoparticles for theranostic application (created by biorender.com).

in high titers, which is important for the diagnosis and treatment of the population. A wide spectrum of genetic and chemical engineering protocols has been authenticated that permit these particles to be at once loaded with imaging agents and drugs as cargo and thus be used for diagnosis and therapy together. This concept is known as theranostics, and this aspect holds supreme importance in the field of medical nanobiotechnology [8, 12]. Refer to Figure 6.5.

Table 6.4 below discusses the current viral and virus-like nanoparticles in the theranostic application of a variety of diseases.

6.4 Novel Advancements in Applications of Viral Nanoparticles

One novel use for VNPs has come in the field of bioimaging (i.e., fluorescence imaging)—specifically, the implantation of inorganic nanoparticles within these viruses without altering the structure or function of the particle's exterior appendages. This would allow for applications such as labeling.

Mineralization is one such method through which inorganic nano materials/probes can be inserted. It entails the attraction and trapping of these materials within the inner cavity via electrostatic methods such as ion affinity tags. In the case of the cowpea chlorotic mottle virus, the pores on the surface which allow for insertion of the inorganic material are controlled by a pH-dependent mechanism, swelling at pH above 6.5 and shrinking at pH below 6.5.

A second methodology that has been used is self-assembly, that is, the regulation of viral-based nanoparticle assembly through buffer conditions. In this method, the first and foremost consideration to be made is the surface coating of the inorganic nanoparticle to be inserted as it can affect the

Table 6.4 Viral vector–based nanoparticles in theranostics.

Sl. no.	Viral nanoparticles	Disease targeted	Mode of action	Comments	Reference
1.	Potato Virus X (PVX)	Breast and ductal Cancer	PVX was applied as a display stage for the TNF-related apoptosis-inducing ligand (TRAIL). The multivalent display it provided amplified its engagement and, subsequently, the activation of death receptors expressed on MDA-MB-231 breast cancer, HCC-38 primary ductal carcinoma, and BT-549 ductal carcinoma cell lines.	When the initial tumor volume was compared after therapy, the PVX-TRAIL showed no growth of tumor even by day 30 proving its retention of efficacy.	[12]

(Continued)

Table 6.4 Viral vector–based nanoparticles in theranostics. (*Continued*)

Sl. no.	Viral nanoparticles	Disease targeted	Mode of action	Comments	Reference
2.	Tobacco Mosaic Virus	Cardiovascular therapy	The conjugation of tissue plasminogen activator (tPA) to TMV formed the TMV-tPA complex. tPA is a known thrombolytic applied in acute myocardial infarctions to degrade thrombus obstruction. The brightness of the right carotid artery of C57BL/6 mice with a 540 nm green laser–produced thrombi. TMV-tPA was transported via tail vein injection inside the mice and sustained its effectiveness and performed better to free tPA by decreasing the average bleeding time from 858 s of free tPA to 429 s of the TMV-tPA complex.	• The TMV-tPA formulation exhibited an enhanced safety profile and the decreased bleeding showed that time is of significant importance as tPA does not explicitly collect at the unhealthy area.	[12]
3.	Influenza neuraminidase VLPs	Prostate cancer (PC-3)	The bioconjugation of TRAIL to influenza neuraminidase VLPs reduces the cell viability of PC-3 cell lines by 70–80%.	Studies have been conducted *in vitro* and are at a rudimentary stage.	[71]

(*Continued*)

Table 6.4 Viral vector–based nanoparticles in theranostics. (*Continued*)

Sl. no.	Viral nanoparticles	Disease targeted	Mode of action	Comments	Reference
4.	MS2 VLPs	Atherosclerotic plaques	MS2 VLPs are being engineered to transport near infrared imaging fluorescent dyes that enable imaging and anti-vascular cell adhesion protein (VCAM-1) antibodies to aid in the targeting of the inflammation sites. The modified VLPs were successful in targeting and detecting atherosclerotic plaques of a multitude of sizes in the descending aorta and aortic arch in ApoE knockout-type mice while the non-targeted controls revealed inadequate symptoms of fluorescence in the plaques.	• The primary organs where the signal accumulated were the liver, spleen, and kidneys. • The heart and pancreas along with fat had very limited quantities of agents persisting within the 24 h.	[56]

(*Continued*)

Table 6.4 Viral vector–based nanoparticles in theranostics. (*Continued*)

Sl. no.	Viral nanoparticles	Disease targeted	Mode of action	Comments	Reference
5.	Qβ-based VLPs	Lung Cancer	Qβ-based VLPs have been contrived for photothermal therapy (PTT) by the addition of sub7-nm sized gold nanoparticles on their exterior; the inner core was full of 500 doxorubicin DOX molecules. Once the 6-ns laser irradiation started, DOX was speedily discharged from the capsid without heating the bulk solution. This allowed a good amount of targeted cell killing of RAW264.7 macrophage and A549 lung cancer cells *in vitro*. Live-cell imaging showed that only cells contained by the irradiated area had morphological changes and detachment suggesting highly localized treatment abilities.	The PTT method displays unique benefits in oncotherapy such as high specificity, marginal invasiveness, and clear-cut spatial-temporal selectivity.	[12]

(*Continued*)

Table 6.4 Viral vector–based nanoparticles in theranostics. (*Continued*)

Sl. no.	Viral nanoparticles	Disease targeted	Mode of action	Comments	Reference
6.	P22 VLPs	Breast cancer	Photodynamic therapy (PDT) has been used with P22 VLPs to encapsulate cytochrome P450 enzymes (CYP), which changes the prodrug tamoxifen and wreathed with the photosensitizer protoporphyrin IX against estrogen receptor–positive breast cancer cells. The produced reactive oxygen species in synergy with enzymatic activity significantly improved the tamoxifen sensitivity *in vitro* and strongly stopped tumor cell growth.	The additional polyethylene glycol (PEG) coat of these VLPs makes them invisible (or difficult to detect) for macrophages, which makes the delivery of DOX much more efficient.	[72]

(*Continued*)

Table 6.4 Viral vector–based nanoparticles in theranostics. (*Continued*)

Sl. no.	Viral nanoparticles	Disease targeted	Mode of action	Comments	Reference
7.	Cowpea chlorotic mottle virus (CCMV)	Breast cancer	Programmable chimeric CCMV was built for directed delivery of drug with the association of CCMV capsid protein loaded with several molecules, i.e., CFSE/Alexa fluor as the imaging agent, folic acid as the targeting agent, and DOX as the anticancer drug. The complex showed exceedingly exact to folate receptors of the Michigan Cancer Foundation-7 (MCF7) cells (breast cancer cell lines).	The cytotoxicity of this complex was nearly 300% more than the conventional methods.	[73]

(*Continued*)

Table 6.4 Viral vector–based nanoparticles in theranostics. (*Continued*)

Sl. no.	Viral nanoparticles	Disease targeted	Mode of action	Comments	Reference
8.	Turnip Mosaic Virus (TuMV)	Inflammatory bowel disease (IBD)	The TuMV-derived VNPs conjugated and made functional with the Hsp60 peptide were shown to identify variances in Hsp60 autoantibody levels in peripheral blood. They also indicated that, in the IBD animal model, the greater quantities of the autoantibodies against Hsp60 protein were detected in the earlier non-pathological stage and lessened with the inflammation development assessed by the loss of weight and the rise of granulocyte levels in peripheral blood.	The differences in the autoantibody levels were untraceable by the traditional ELISA assay technique.	[74]

interactions between the virus' capsid proteins and the inorganic material. The primary goal is to mimic the viral genome's packaging—hence, coatings that are negatively charged are often employed as this mimic viral genetic material (due to its negative charge) and stimulate the viral-based nanoparticle assembly process. The effectiveness of this process and the assimilation of the inorganic molecule can vary, however, based on the choice of surface coating [75].

Novel advances have been made in the use of plant virus-based nanoparticles for the treatment of cancer as well, which show great promise. Certain plant viruses like the TMV have shorter half-lives (i.e., 3.5 min in mice that had no prior exposure to the virus) due to various immune system cells and antibodies surrounding the particle and preventing it from initiating any action. From this observation, one new modification came in the form of using a protein that would allow itself to enter the human body without being degraded by the immune system (i.e., serum albumin)—a study was carried out using this method in 2018, which showed that coating with serum albumin proved to be extremely effective in preventing antibody recognition, to the point that even antibodies specific to the TMV did not recognize it with the protein coating.

Another method that has arisen in recent years is the use of computer technology in designing nanoparticles for immunotherapy, which has reduced both the cost and time taken for the development of anticancer vaccines and allowed for development to be more efficient. Specifically, in the case of breast cancer, a vaccine was developed through the identification of a particular sequence from the protein and the selection of its epitopes that would be recognized by the immune system, from the database followed by combining the epitopes of interest for the vaccine and using a plant virus-based vector such as the TMV [76].

6.5 Limitations and Prospects of Viral Vector–Based Nanoparticle Approach

One of the main limitations in the development of viral vector–based nanoparticles is its cost. Most of the studies regarding these nanoparticles become difficult to complete due to the cost of the research. This also acts as a hindrance when it comes to the commercial availability of the same. There are also constraints when it comes to the addition of larger epitopes as they might hamper the assembly of the VLPs [77]. In different systems such as the enveloped VLPs, the expression of viral proteins can be different, which can affect the entire production process of the same. In cell-free systems, the scalability of the VNP is limited as well [9]. Another important challenge of this approach is that the procedure for preparation lacks standardization. Different companies and industries utilize different techniques for production and downstream processing raising questions on the regulatory and quality control aspect [78]. Even after all these shortcomings, VNPs have several advantages over generic drug delivery systems and hence are being explored more by the scientific world with so much interest. Many VNP-based approaches for different purposes like drug delivery, imaging, vaccination, gene therapy, and immunotherapy are pipelines.

6.6 Conclusion

Nanotechnology offers a more efficient solution to the problems related to the accuracy of therapeutic and diagnostic techniques, currently in practice. The viral vector–based nanoparticles are widely researched for their utility as drug or gene delivery vehicles in recent years due to their advantageous unique physical and chemical properties, safer use, and easy production. The alterations in the surface characteristics of the virus caused by nanoparticle coating protect it from enzymatic digestion and host immune monitoring. Immunotherapies, drug delivery, vaccines, imaging, antimicrobial therapies, theranostics, photodynamic therapy, cardiovascular therapies, gene therapies, and cancer therapies are just a few of the therapeutic areas where viral vector–based nanoparticles are being used.

Concerning what the future holds for virus-based nanoparticles, one of the technology's most prominent applications in the present and near future will likely be in the manufacturing of vaccines against human diseases such as hepatitis B or influenza through the use of plants and virus-based nanoparticles. An example of this exists in the case of the Hepatitis B virus where a TMV-based vector was used to produce large quantities of the surface antigen HBsAg in the plant species *Nicotiana benthamiana* in its correct form, allowing for the assembly of a large quantity of virus-based nanoparticles with the correct protein sequence [79]. Applications of VNPs will hopefully be successfully seen in precision medicine, which will include accurate drug dosage and treatment protocol for every individual for a variety of diseases. This will also include optimized ADME evaluations. However, more detailed research is required to facilitate easy access to these technologies by the global population including people from low-income countries.

The outbreak of the COVID-19 pandemic has resulted in rapid rise in the production of VNP-based vaccines. This has consequently led to an evolution in the domain of viral vector–based nanoparticle development [78]. The approach provides efficient therapeutic intervention as well as imaging techniques for diseases like cancers and cardiovascular diseases. The efficiency of this approach can be further enhanced by including it in combination therapy. The need of the hour is to modify the design of technology to overcome the limitations to enhance its application in therapeutics and diagnosis [3].

Acknowledgment

The authors want to express their gratitude to Dr. G. Viswanathan, Chancellor, VIT University, Vellore, for his constant support and encouragement.

References

1. Khan, I., Saeed, K., Khan, I., Nanoparticles: Properties, applications and toxicities. *Arab. J. Chem.*, 12, 7, 908–931, 2019.
2. Murthy, S.K., Nanoparticles in modern medicine: State of the art and future challenges. *Int. J. Nanomedicine*, 2, 2, 129, 2007.
3. Chen, L., Hong, W., Ren, W., Xu, T., Qian, Z., He, Z., Recent progress in targeted delivery vectors based on biomimetic nanoparticles. *Signal Transduction Targeting Ther.*, 6, 1, 1–25, 2021.
4. Kasala, D., Yoon, A.R., Hong, J., Kim, S.W., Yun, C.O., Evolving lessons on nanomaterial-coated viral vectors for local and systemic gene therapy. *Nanomedicine*, 11, 13, 1689, 2016.
5. Jeevanandam, J., Pal, K., Danquah, M.K., Virus-like nanoparticles as a novel delivery tool in gene therapy. *Biochimie*, 157, 38–47, 2019.
6. Koudelka, K.J., Pitek, A.S., Manchester, M., Steinmetz, N.F., Virus-based nanoparticles as versatile nanomachines. *Annu. Rev. Virol.*, 2, 1, 379, 2015.
7. Pokorski, J.K. and Steinmetz, N.F., The art of engineering viral nanoparticles. *Mol. Pharm.*, 8, 1, 29–43, 2011.
8. Yildiz, I., Shukla, S., Steinmetz, N.F., Applications of viral nanoparticles in medicine. *Curr. Opin. Biotechnol.*, 22, 6, 901–908, 2011.
9. Nooraei, S. *et al.*, Virus-like particles: Preparation, immunogenicity and their roles as nanovaccines and drug nanocarriers. *J. Nanobiotechnology*, 19, 1, 1–27, 2021.
10. Nkanga, C., II and Steinmetz, N.F., The pharmacology of plant virus nanoparticles. *Virology*, 556, 39–61, 2021.
11. Mitchell, M.J., Billingsley, M.M., Haley, R.M., Wechsler, M.E., Peppas, N.A., Langer, R., Engineering precision nanoparticles for drug delivery. *Nat. Rev. Drug Discovery*, 20, 2, 101–124, 2020.

12. Chung, Y.H., Cai, H., Steinmetz, N.F., Viral nanoparticles for drug delivery, imaging, immunotherapy, and theranostic applications. *Adv. Drug Deliv. Rev.*, 156, 214–235, Jan. 2020. doi: 10.1016/J.ADDR.2020.06.024.
13. Venkataraman, S., Apka, P., Shoeb, E., Badar, U., Hefferon, K., Plant virus nanoparticles for anti-cancer therapy. *Front. Bioeng. Biotechnol.*, 9, 817, 2021.
14. Pitek, A. S., *et al.*, Elongated plant virus-based nanoparticles for enhanced delivery of thrombolytic therapies. *Mol. Pharm.*, 14, 11, 3815–3823, 2017.
15. Seyfoori, A., *et al.*, Emerging advances of nanotechnology in drug and vaccine delivery against viral associated respiratory infectious diseases (VARID). *Int. Mol. Sci.*, 22, 13, 6937, 2021.
16. Steinmetz, N.F., Viral Nanoparticles in Drug Delivery and Imaging, Mol. *Pharm., 10*, 1, 1–2, 2013.
17. Yang, G., Chen, S., Zhang, J., Bioinspired and Biomimetic Nanotherapies for the Treatment of Infectious Diseases, Front. *Pharmacol.*, 10, 751, 2019.
18. Pala, R., Anju, V.T., Dyavaiah, M., Busi, S., Nauli, S.M., Nanoparticle-mediated drug delivery for the treatment of cardiovascular diseases. *Int. J. Nanomedicine*, 15, 3741, 2020.
19. Yan, C., Quan, X.-J., Feng, Y.-M., Nanomedicine for gene delivery for the treatment of cardiovascular diseases. *Curr. Gene Ther.*, 19, 1, 20, 2019.
20. Alvandi, N., Rajabnejad, M., Taghvaei, Z., Esfandiari, N., New generation of viral nanoparticles for targeted drug delivery in cancer therapy. *J. Drug Targeting*, 30, 2, 151–165, 2021.
21. Hu, H. and Steinmetz, N.F., Doxorubicin-loaded physalis mottle virus particles function as a pH-responsive prodrug enabling cancer therapy. *Biotechnol. J.*, 15, 12, e2000077, 2020.
22. Hu, H., Masarapu, H., Gu, Y., Zhang, Y., Yu, X., Steinmetz, N.F., Physalis mottle virus-like nanoparticles for targeted cancer imaging. *ACS Appl. Mater. Interfaces*, 11, 18213–18223, 20, 2019.
23. Alemzadeh, E. *et al.*, Enhanced anti-tumor efficacy and reduced cardiotoxicity of doxorubicin delivered in a novel plant virus nanoparticle. *Colloids Surf. B. Biointerfaces*, 174, 80–86, 2019.
24. Ren, Y., Sek, M.W., Lim, L.Y., Folic acid-conjugated protein cages of a plant virus: A novel delivery platform for doxorubicin. *Bioconjug. Chem.*, 18, 3, 836–843, 2007.
25. Yan, D. *et al.*, Foot-and-mouth disease virus-like particles as integrin-based drug delivery system achieve targeting anti-tumor efficacy. *Nanomedicine*, 13, 3, 1061–1070, 2017.
26. Will, H. *et al.*, Cloned HBV DNA causes hepatitis in chimpanzees. *Nature*, 299, 5885, 740–742, 1982. doi: 10.1038/299740a0.
27. Seto, W.K., Lo, Y.R., Pawlotsky, J.M., Yuen, M.F., Chronic hepatitis B virus infection. *Lancet*, 392, 10161, 2313–2324, Nov. 2018.
28. Miao, J. *et al.*, Advances in nanoparticle drug delivery systems for anti-hepatitis B virus therapy: A narrative review. *Int. J. Mol. Sci.*, 22, 20, 11227, 2021.

29. Shoja, Z., Jalilvand, S., Latifi, T., Roohvand, F., Rotavirus VP6: Involvement in immunogenicity, adjuvant activity, and use as a vector for heterologous peptides, drug delivery, and production of nano-biomaterials. *Arch. Virol.*, 167, 4, 1013–1023, 2022.
30. Pattinson, D.J. *et al.*, Chimeric virus-like particles and capsomeres induce similar CD8+ T cell responses but differ in capacity to induce CD4+ T cell responses and antibody responses. *Front. Immunol.*, 11, 564627, 1–13, 2020.
31. Kato, T., Yui, M., Deo, V.K., Park, E.Y., Development of rous sarcoma virus-like particles displaying hCC49 scFv for specific targeted drug delivery to human colon carcinoma cells. *Pharm. Res.*, 32, 11, 3699–3707, 2015.
32. Rohovie, M.J., Nagasawa, M., Swartz, J.R., Virus-like particles: Next-generation nanoparticles for targeted therapeutic delivery. *Bioeng. Transl. Med.*, 2, 1, 43–57, 2017.
33. Lin, R.D. and Steinmetz, N.F., Tobacco mosaic virus delivery of mitoxantrone for cancer therapy. *Nanoscale*, 10, 34, 16307, 2018.
34. Bruckman, M.A., Czapar, A.E., VanMeter, A., Randolph, L.N., Steinmetz, N.F., Tobacco mosaic virus-based protein nanoparticles and nanorods for chemotherapy delivery targeting breast cancer. *J. Control. Release*, 231, 103, 2016.
35. Lin, S., Liu, C., Han, X., Zhong, H., Cheng, C., Viral nanoparticle system: An effective platform for photodynamic therapy. *Int. J. Mol. Sci.*, 22, 4, 1–15, 2021.
36. Wolfram, J.A. and Donahue, J.K., Gene therapy to treat cardiovascular disease. *J. Am. Heart Assoc.*, 2, 4, e000119, 2013.
37. Pankajakshan, D., Agrawal, D.K., Pankajakshan, D., Agrawal, D.K., Clinical and translational challenges in gene therapy of cardiovascular diseases, in: *Gene Therapy-Tools and Potential Applications*, 2013.
38. Cao, G., Xuan, X., Zhang, R., Hu, J., Dong, H., Gene therapy for cardiovascular disease: Basic research and clinical prospects. *Front. Cardiovasc. Med.*, 8, 760140, 2021.
39. Schwartze, J.T., Havenga, M., Wam, B., Bradshaw, A.C., Nicklin, S.A., Adenoviral vectors for cardiovascular gene therapy applications: A clinical and industry perspective. *J. Mol. Med. (Berl)*, 100, 6, 875, 2022.
40. Han, X., Xu, K., Taratula, O., Farsad, K., Applications of nanoparticles in biomedical imaging. *Nanoscale*, 11, 3, 799–819, 2019.
41. Manchester, M. and Singh, P., Virus-based nanoparticles (VNPs): Platform technologies for diagnostic imaging. *Adv. Drug Deliv. Rev.*, 58, 14, 1505–1522, 2006.
42. Lee, S.B., Kim, H.L., Jeong, H.J., Lim, S.T., Sohn, M.H., Kim, D.W., Mesoporous silica nanoparticle pretargeting for PET imaging based on a rapid bioorthogonal reaction in a living body. *Angew. Chem. Int. Ed. Engl.*, 52, 40, 10549–10552, 2013.
43. Zern, B.J. *et al.*, Reduction of nanoparticle avidity enhances the selectivity of vascular targeting and PET detection of pulmonary inflammation. *ACS Nano*, 7, 3, 2461–2469, 2013.

44. Zhang, K. *et al.*, Double-scattering/reflection in a single nanoparticle for intensified ultrasound imaging. *Sci. Rep.*, 5, 8766, 1–11, 2015.
45. Hu, Y. *et al.*, Preparation and characterization of novel perfluorooctyl bromide nanoparticle as ultrasound contrast agent via layer-by-layer self-assembly for folate-receptor-mediated tumor imaging. *Biomed. Res. Int.*, 2016, 6381464, 2016.
46. Chen, F. *et al.*, Exosome-like silica nanoparticles: A novel ultrasound contrast agent for stem cell imaging. *Nanoscale*, 9, 1, 402–411, 2017.
47. Bhavane, R. *et al.*, Dual-energy computed tomography imaging of atherosclerotic plaques in a mouse model using a liposomal-iodine nanoparticle contrast agent. *Circ. Cardiovasc. Imaging*, 6, 2, 285–294, 2013.
48. Ghaghada, K.B. *et al.*, Evaluation of tumor microenvironment in an animal model using a nanoparticle contrast agent in computed tomography imaging. *Acad. Radiol.*, 18, 1, 20–30, 2011.
49. Steinmetz, N.F., Viral nanoparticles as platforms for next-generation therapeutics and imaging devices. *Nanomed. Nanotechnol. Biol. Med.*, 6, 5, 634–641, 2010.
50. Hainfeld, J.F., Smilowitz, H.M., O'connor, M.J., Dilmanian, F.A., Slatkin, D.N., Gold nanoparticle imaging and radiotherapy of brain tumors in mice. *Nanomedicine (Lond.)*, 8, 10, 1601, 2013.
51. Ghaghada, K.B. *et al.*, Pre-clinical evaluation of a nanoparticle-based blood-pool contrast agent for MR imaging of the placenta. *Placenta*, 57, 60–70, 2017.
52. Lu, J. *et al.*, Manganese ferrite nanoparticle micellar nanocomposites as MRI contrast agent for liver imaging. *Biomaterials*, 30, 15, 2919–2928, 2009.
53. Zhao, Y. *et al.*, Near-infrared fluorescence energy transfer imaging of nanoparticle accumulation and dissociation kinetics in tumor-bearing mice. *ACS Nano*, 7, 11, 10362–10370, 2013.
54. Li, R.H. *et al.*, Rhomboidal Pt(II) metallacycle-based hybrid viral nanoparticles for cell imaging. *Inorg. Chem.*, 60, 1, 431–437, 2021.
55. Leong, H.S. *et al.*, Intravital imaging of embryonic and tumor neovasculature using viral nanoparticles. *Nat. Protoc.*, 5, 8, 1406–1417, 2010.
56. Aanei, I.L., Huynh, T., Seo, Y., Francis, M.B., Vascular cell adhesion molecule-targeted MS2 viral capsids for the detection of early-stage atherosclerotic plaques. *Bioconjug. Chem.*, 29, 8, 2526–2530, 2018.
57. Hu, H., Yang, Q., Baroni, S., Yang, H., Aime, S., Steinmetz, N.F., Polydopamine-decorated tobacco mosaic virus for photoacoustic/magnetic resonance bimodal imaging and photothermal cancer therapy. *Nanoscale*, 11 19, 9760–9768, 2019.
58. Jia, Q. *et al.*, Biomineralization synthesis of HBc-CuS nanoparticles for near-infrared light-guided photothermal therapy. *J. Mater. Sci.*, 54, 20, 13255–13264, 2019.

59. Singh, P., Destito, G., Schneemann, A., Manchester, M., Canine parvovirus-like particles, a novel nanomaterial for tumor targeting. *J. Nanobiotechnology*, 4, 2, 2006.
60. Brown, M.C. *et al.*, Cancer immunotherapy with recombinant poliovirus induces IFN-dominant activation of dendritic cells and tumor antigen-specific CTLs. *Sci. Transl. Med.*, 9, 408, eaan4220, 1–16, 2017.
61. Conry, R.M., Westbrook, B., McKee, S., Norwood, T.G., Talimogene laherparepvec: First in class oncolytic virotherapy. *Hum. Vaccin. Immunother.*, 14, 4, 839, 2018.
62. Hughes, T., Coffin, R.S., Lilley, C.E., Ponce, R., Kaufman, H.L., Critical analysis of an oncolytic herpesvirus encoding granulocyte-macrophage colony stimulating factor for the treatment of malignant melanoma. *Oncolytic Virother.*, 3, 11, 2014.
63. Ott, P.A. and Hodi, F.S., Talimogene laherparepvec for the treatment of advanced melanoma. *Clin. Cancer Res.*, 22, 13, 3127–3131, 2016.
64. Chesney, J. *et al.*, Potential clinical and immunotherapeutic utility of talimogene laherparepvec for patients with melanoma after disease progression on immune checkpoint inhibitors and BRAF inhibitors. *Melanoma Res.*, 28, 3, 250, 2018.
65. Kelly, C.M. *et al.*, Objective response rate among patients with locally advanced or metastatic sarcoma treated with talimogene laherparepvec in combination with pembrolizumab: A phase 2 clinical trial. *JAMA Oncol.*, 6, 3, 402, 2020.
66. Lizotte, P.H. *et al.*, In situ vaccination with cowpea mosaic virus nanoparticles suppresses metastatic cancer. *Nat. Nanotechnol.*, 11, 3, 295, 2016.
67. Wang, C., Fiering, S.N., Steinmetz, N.F., Cowpea mosaic virus promotes anti-tumor activity and immune memory in a mouse ovarian tumor model. *Adv. Ther. (Weinh)*, 2, 5, 2019.
68. Kerstetter-Fogle, A. *et al.*, Plant virus-like particle in situ vaccine for intracranial glioma immunotherapy. *Cancers*, 11, 4, 515, 2019.
69. Wen, A.M. and Steinmetz, N.F., Design of virus-based nanomaterials for medicine, biotechnology, and energy. *Chem. Soc. Rev.*, 45, 15, 4074–4126, 2016.
70. Lu, Z. *et al.*, A biotin-avidin-system-based virus-mimicking nanovaccine for tumor immunotherapy. *J. Control. Release*, 332, 245–259, 2021.
71. Kaczmarczyk, S.J., Sitaraman, K., Young, H.A., Hughes, S.H., Chatterjee, D.K., Protein delivery using engineered virus-like particles. *Proc. Natl. Acad. Sci. (USA)*, 108, 41, 16998–17003, 2011.
72. Chauhan, K., Hernandez-Meza, J.M., Rodríguez-Hernández, A.G., Juarez-Moreno, K., Sengar, P., Vazquez-Duhalt, R., Multifunctionalized biocatalytic P22 nanoreactor for combinatory treatment of ER+ breast cancer. *J. Nanobiotechnology*, 16, 1, 1–14, 2018.
73. Wu, Y., Li, J., Shin, H.J., Self-assembled viral nanoparticles as targeted anticancer vehicles. *Biotechnol. Bioprocess Eng.*, 26, 1, 25–38, 2021.

74. Yuste-Calvo, C., López-Santalla, M., Garin, M., I, Ponz, F., Theranostic viral nanoparticles for autoimmune diseases. *J. Immunol. Biol.*, 5, 1, 149, 2020.
75. Zhang, W., Xu, C., Yin, G.Q., Zhang, X.E., Wang, Q., Li, F., Encapsulation of inorganic nanomaterials inside virus-based nanoparticles for bioimaging. *Nanotheranostics*, 1, 4, 358–368, 2017.
76. Shoeb, E. and Hefferon, K., Future of cancer immunotherapy using plant virus-based nanoparticles. *Future Sci. OA*, 5, 7, 401–2056, 2019.
77. Plummer, E.M. and Manchester, M., Viral nanoparticles and virus-like particles: Platforms for contemporary vaccine design. *Wiley Interdiscip. Rev. Nanomed. Nanobiotechnol.*, 3, 2, 174–196, 2011.
78. Capra, E., Gennari, A., Loche, A., Temps, C., Viral-vector therapies at scale: Today's challenges and future opportunities. *McKinsey, McKinsey & Company*, 1, 1–10, 2022. https://www.mckinsey.com/industries/life-sciences/our-insights/viral-vector-therapies-at-scale-todays-challenges-and-future-opportunities.
79. Venkataraman, S., Hefferon, K., Makhzoum, A., Abouhaidar, M., Combating human viral diseases: Will plant-based vaccines be the answer? *Vaccines*, 9, 7, 761, 2021.

7

Lipid-Based Nanoparticles

Sunny Shah[1*], Hardik Madhu[1], Moinuddin Soniwala[1], Dhaval Mori[1], Amit Vyas[1], Advaita Chauhan[2] and Bhupendra Prajapati[3]

[1]B. K. Mody Government Pharmacy College, Rajkot, India
[2]Long Island University, Brooklyn New York City, United States of America
[3]Shree S.K. Patel College of Pharmaceutical Education & Research, Faculty of Pharmacy, Ganpat University, Mehsana, Gujarat, India

Abstract
Lipid nanoparticles (LNPs) have gained significant attention in developing vaccines due to their unique properties, including biocompatibility, biodegradability, and the ability to encapsulate hydrophobic and hydrophilic molecules. They have been used to enhance the efficacy of vaccines by improving antigen delivery and presentation and facilitating targeted delivery to specific cells and tissues. Additionally, LNPs have been shown to have adjuvant effects, which can further enhance immune responses. These properties make LNPs promising tools for developing safe and effective vaccines against infectious diseases and cancers. The chapter overviews LNPs, including their types, synthesis methods, characteristics, and vaccine applications. The two main types of LNPs discussed are solid lipid nanoparticles and nanostructured lipid carriers. Various synthesis methods are also presented, including high-pressure homogenization, solvent emulsification–evaporation, and microfluidics. The characteristics of LNPs, including size and shape, surface charge, encapsulation efficiency, and stability, are also discussed. The chapter then delves into the applications of LNPs in vaccines, including enhancing immune responses, targeted delivery, and adjuvant effects. Finally, the challenges and future directions of LNPs are explored, including safety and toxicity concerns, stability issues, scale-up production challenges, and opportunities for future research.

Keywords: Toxicity, nanocarrier, ajuvant therapy, homogenization

**Corresponding author*: sunnyrshah@gmail.com; ORCID: https://orcid.org/0000-0002-0828-8059

Vivek P. Chavda and Vasso Apostolopoulos (eds.) Nanocarrier Vaccines: Biopharmaceutics-Based Fast Track Development, (241–274) © 2024 Scrivener Publishing LLC

7.1 Introduction

Lipid nanoparticles (LNPs) have a similar physical structure to liposomes. Still, they are particularly effective at encapsulating nucleic acids such as RNA and DNA, making them popular for non-viral gene delivery. LNPs, including liposomes and solid lipid nanoparticles (SLNs), can deliver drugs through different administration routes and successfully treat various diseases [1]. The first US Food and Drug Administration (FDA)-approved lipid-based drug formulation was marketed in 1995, and since then, more than a dozen therapeutics have been FDA-approved for multiple conditions [2]. Numerous vaccines are currently being tested in clinical trials, with some already authorized for emergency use, like Pfizer and Moderna COVID-19 vaccines. The most significant advances in medicine are taking place in nucleic acid delivery, which involves using LNPs to encapsulate different types of RNA, such as siRNA, miRNA, saRNA, or mRNA [3]. This technique can be used for gene silencing or activation and protein production. LNPs have been particularly effective in nucleic acid delivery, enabling gene silencing or activation and protein production. LNPs have received significant attention in drug discovery and delivery. These nanoparticles can transport hydrophobic and hydrophilic molecules to enhance the bioavailability of therapeutic compounds by controlling solubility, permeability, absorption, distribution, and metabolism. In addition, these nanoparticles show very low or no toxicity and increase the duration of drug action by extending the half-life and controlling drug release. Lipid nanosystems can include chemical modifications to avoid detection by the immune system (gangliosides or polyethylene glycol (PEG)) or to improve the therapeutic index. For example, PEG covalently linked to liposomes can reduce immunogenicity and antigenicity, protecting them from the training of the receptor's macrophage system to destroy foreign substances. Adding PEG can also alter physicochemical properties, reduce renal clearance and dosing frequency, and prolong circulation time [4].

Moreover, lipid nanoparticles can be prepared as pH-sensitive formulations to facilitate drug release in an acidic environment. They can also be associated with antibodies that recognize tumor cells or their receptors, such as folic acid (FoA). Nano drugs can also be combined with other therapeutic strategies to improve patients' responses. LNPs have been the subject of much research and interest in recent years. Two main types of LNPs are SLNs and nanostructured lipid carriers (NLCs). SLNs were developed to address the limitations of other nanoparticle carriers and offer benefits such as improved drug release and targeted delivery [5, 6]. NLCs are an advanced version of SLNs, with improved stability and capacity for drug loading. There are three structural models of NLCs. LNPs have potential applications in various fields, such as drug delivery, cosmetics, and clinical medicine [7, 8].

LNPs have emerged as promising candidates for vaccine delivery due to their unique properties, such as their ability to encapsulate a wide range of payloads, including both hydrophilic and hydrophobic molecules, and their ability to protect the encapsulated cargo from degradation. In particular, SLNs and NLCs have been extensively studied as potential vaccine delivery systems. These LNPs have been shown to enhance the immunogenicity of vaccines, including both prophylactic and therapeutic vaccines, by improving antigen stability, increasing uptake by antigen-presenting cells (APCs), and promoting a strong immune response [9]. Moreover, LNPs have the potential to overcome several limitations associated with traditional vaccine delivery systems, such as poor stability, rapid clearance from the body, and inadequate immune response. Overall, the use of LNPs as vaccine carriers represents a promising approach for the development of more effective and efficient vaccines and could significantly impact the future of immunization.

7.2 Types of Lipid-Based Nanoparticles

7.2.1 Solid Lipid Nanoparticles (SLNs)

SLNs are LNPs that consist of a solid lipid core stabilized by surfactants. They are typically 50–1,000 nm in size and have a spherical or elliptical shape [10]. SLNs have gained considerable attention in the field of drug delivery due to their biocompatibility, controlled release, and potential for improved bioavailability. SLNs have been used to deliver the hepatitis B vaccine, resulting in enhanced immune response and prolonged antigen release [11]. Influenza vaccine–loaded SLNs have been shown to improve vaccine stability, increase the shelf life of the vaccine, and enhance immune response [12]. There are fundamental differences between liposomes and SLNs. Liposomes and LNPs differ in their structures. Liposomes have a lipid bilayer membrane surrounding an aqueous cavity, while LNPs lack an aqueous cavity and have multilayer cores between lipid layers due to electrostatic interactions between cationic phospholipids and negatively charged nucleic acid substances [13]. LNPs and liposomes share similar lipid and cholesterol components, but ionizable lipids must be present in LNPs. Liposomes do not have strict requirements for the type of lipids used, and the ratio of each component, particularly the amount of cholesterol, is a significant distinguishing factor between liposomes and LNPs. The cholesterol content in the mRNA COVID-19 vaccines is significantly higher than in liposomes [14]. Liposomes and LNPs differ in their upstream manufacturing processes, but their downstream processes are similar. Liposomes are formed by combining lipid and aqueous phases and controlling particle size through homogenization or extrusion. LNPs are formed through a microfluidic mixing system with connectors that control particle size during the mixing and collision of the two phases.

7.2.2 Nanostructured Lipid Carriers (NLCs)

NLCs are a type of LNP that has gained considerable interest in vaccine delivery due to their unique advantages over other LNPs. NLCs are composed of a solid lipid matrix and a liquid lipid that form a nanostructure with an amorphous and imperfect crystalline structure [15, 16]. The amorphous and imperfect crystalline structure of NLCs allows for higher drug-loading capacity compared to other LNPs, thus enabling the delivery of a higher concentration of antigens in vaccines [17, 18]. NLCs exhibit better stability than other LNPs due to their improved structural integrity, which helps in preserving the vaccine components [19]. NLCs can release the vaccine components in a controlled manner, which ensures that the antigen remains at the site of injection for a longer duration, thereby enhancing the immune response [20]. NLCs are biocompatible and do not induce any significant toxicity or immune response. However, the preparation of NLCs requires a complex process that involves several steps, which makes it difficult to produce on a large scale. NLCs are susceptible to physicochemical instability, which may result in changes in particle size, drug release, and storage stability [21]. Although NLCs are generally considered safe, there is a possibility of toxicity due to the use of certain lipids or surfactants [22]. NLCs have been used as carriers for cancer vaccines due to their ability to deliver the vaccine components in a controlled manner and enhance the immune response against cancer cells [23, 24]. NLCs have been used for the delivery of various infectious diseases vaccines, such as influenza, hepatitis, and human papillomavirus (HPV) vaccines. NLCs have been explored as a carrier for allergy vaccines, with studies showing improved efficacy and safety.

7.3 Synthesis of Lipid-Based Nanoparticles

7.3.1 Introduction to Lipids

The choice of lipid used in SLNs is important as it can impact the stability, loading capacity, and release profile of the nanoparticles. Therefore, the selection of the appropriate lipid for a particular application depends on several factors, including the type of drug to be loaded, the desired release profile, and the intended route of administration.

Phospholipids: These are the most common type of lipids used in LNP formulations. Examples include 1,2-dimyristoyl-sn-glycero-3-phosphocholine (DMPC), 1,2-dioleoyl-sn-glycero-3-phosphocholine (DOPC), and 1,2-distearoyl-sn-glycero-3-phosphocholine (DSPC).

Cholesterol: This is often used in combination with phospholipids to improve the stability and rigidity of the LNP membrane.

Sphingolipids: These lipids are structurally similar to phospholipids but contain a sphingosine backbone instead of a glycerol backbone. Examples include ceramides and sphingomyelins.

Glycolipids: These are lipids that contain a carbohydrate moiety. Examples include monosialoganglioside (GM1) and galactosylceramide (GalCer).

Cationic lipids: These are lipids that contain a positively charged head group and are often used in gene delivery applications. Examples include 1,2-dioleoyl-3-trimethylammonium-propane (DOTAP) and N-[1-(2,3-dioleoyloxy)propyl]-N,N,N-trimethylammonium chloride (DOTMA) [25, 26].

PEGylated lipids: These are lipids that have a PEG moiety attached to the head group, which can improve the pharmacokinetics and biodistribution of the LNPs. Examples include distearoyl-phosphoethanolamine-PEG (DSPE-PEG) and cholesterol-PEG.

It is worth noting that LNP formulations can contain a combination of different types of lipids, and the choice of lipids can affect the stability, size, and drug-loading capacity of the resulting nanoparticles.

7.3.2 Methods for Formulating Lipid Nanoparticles

The synthesis of SLNs involves various methods that differ in their mechanism, efficiency, and product characteristics. Some of the different methods used for the synthesis of SLNs are as follows.

7.3.2.1 *High-Pressure Homogenization*

This method involves the application of high pressure and shear forces to the lipid mixture to reduce its particle size. The process starts by heating the lipid mixture, followed by homogenization using a high-pressure homogenizer. This method is widely used due to its efficiency, scalability, and reproducibility [27]. This technique involves high-pressure forces acting on the SLN dispersion to reduce the particle size, leading to the formation of stable nanoparticles. HPH can effectively reduce the size of SLNs, resulting in smaller and more uniform particles. This technique allows for precise control of particle size distribution. HPH can increase the loading of drugs in SLNs by reducing the particle size and increasing the surface area, leading to enhanced bioavailability and therapeutic efficacy. HPH is a scalable process that can be easily adapted to large-scale production. HPH does not require the use of organic solvents, which can reduce the toxicity and environmental impact of the SLN preparation process [28]. HPH requires specialized equipment, which can be expensive and may not be available in all laboratories. HPH can generate high shear forces that may cause damage to the lipid structure or the incorporated drug, leading to decreased stability and efficacy. The HPH process requires optimization of several parameters, such as pressure, temperature, and the number of cycles, which can be time-consuming and require significant expertise [29]. HPH-SLNs have been extensively investigated for vaccine delivery due to their ability to enhance antigen stability, uptake, and immune response. HPH-SLNs have also been investigated for the delivery of various drugs, including anticancer drugs, antifungal drugs, and antibiotics [30]. Thus, HPH is a promising method for the synthesis of SLNs for vaccine delivery, with several advantages such as high homogenization efficiency, improved drug loading, scalability, and reduced need for organic solvents. However, it also has some disadvantages such as equipment cost, shear forces, and process optimization.

7.3.2.2 *Solvent Emulsification–Evaporation*

This method involves dissolving the lipid and drug in an organic solvent, which is then emulsified in an aqueous phase. The solvent is then evaporated to obtain the SLNs [31]. This method is simple, cost-effective, and suitable for the encapsulation of hydrophobic drugs. Examples of SLNs synthesized using solvent emulsification–evaporation (SEE) include paclitaxel-loaded SLNs [32] and curcumin-loaded SLNs [33]. SEE is a widely used method for the synthesis of SLNs for vaccine delivery. The method involves the dissolution of a lipid in an organic solvent followed by the emulsification of this solution into an aqueous phase containing a surfactant. The solvent is then evaporated under reduced pressure, leading to the formation of SLNs. The method is simple and easy to scale up, making it suitable for large-scale production. The SLNs produced by this method have a narrow size distribution and a high drug-loading capacity. The method is compatible with a wide range of lipids and surfactants, allowing for flexibility in formulation design. The process can be modified to control the size and surface properties of the nanoparticles. The use of organic solvents can be toxic and pose a risk to health and the environment. The process requires careful optimization of various parameters, such as the choice of solvents, surfactants, and processing conditions, to ensure the stability and reproducibility of the nanoparticles [34]. The solvent removal step can lead to aggregation or fusion of the nanoparticles if not properly controlled. The solvent emulsification–evaporation method has been used to develop SLNs for the delivery of various vaccines, including influenza, hepatitis B, and malaria vaccines. SLNs prepared by this method have also been explored for the delivery of drugs, such as anticancer agents and antibiotics.

7.3.2.3 Microemulsion-Based Method

This method involves the formation of a microemulsion using a mixture of lipids, surfactants, cosurfactants, and water. The drug is then added to the microemulsion, and SLNs are obtained by removing the water phase. This method provides high drug-loading capacity and allows for the encapsulation of both hydrophilic and hydrophobic drugs. The microemulsion method is a widely used technique for the synthesis of SLNs for vaccine delivery. In this method, a microemulsion is formed using a mixture of oil, surfactant, cosurfactant, and water. The SLNs are then synthesized by adding a lipid phase to the microemulsion, followed by evaporation of the organic solvent. This method offers several advantages for the synthesis of SLNs [35]. The microemulsion method is a simple and cost-effective technique for the synthesis of SLNs. The equipment required for this method is readily available, and the process can be easily scaled up for commercial production. The microemulsion method offers high encapsulation efficiency (EE) of the vaccine antigen [36]. The small droplet size of the microemulsion leads to a high surface area-to-volume ratio, which facilitates the diffusion of the antigen into the lipid phase, resulting in high EE. The SLNs synthesized using the microemulsion method offer controlled release of the vaccine antigen. The small particle size and high surface area-to-volume ratio of the nanoparticles result in sustained release of the antigen, leading to a prolonged immune response. The microemulsion method offers excellent physical stability to the SLNs. The presence of a surfactant and a cosurfactant in the microemulsion stabilizes the lipid phase, preventing particle aggregation and ensuring long-term stability [37]. The microemulsion method requires the use of organic solvents, which can be toxic to living cells. The residual solvent can also affect the stability of the vaccine antigen. The microemulsion method has a limited drug-loading capacity due to the small size of the nanoparticles [38]. This can be a disadvantage for vaccines that require a high dose of antigen.

7.3.2.4 Hot-Melt Homogenization

This method involves the melting of the lipid and the drug at high temperatures, followed by homogenization to obtain the SLNs [39]. This method is suitable for the encapsulation of heat-stable drugs and provides SLNs with high drug-loading capacity. Hot-melt homogenization (HMH) is a widely used method for the preparation of SLNs. In this method, the lipid matrix is melted at a temperature above its melting point, and the drug is incorporated into the melted lipid. The mixture is then homogenized under high pressure and rapidly cooled to solidify the lipid matrix and form the nanoparticles [40]. The HMH method allows for the incorporation of a high concentration of hydrophobic drugs into the lipid matrix, resulting in a high drug-loading capacity. The HMH method can be easily scaled up for large-scale production of SLNs. SLNs prepared by the HMH method exhibit good physical stability due to the solidified lipid matrix. The particle size of SLNs can be controlled by varying the homogenization conditions, such as pressure and the number of homogenization cycles. The HMH method is not suitable for the encapsulation of hydrophilic drugs due to the hydrophobic nature of the lipid matrix. The high-temperature processing may cause thermal degradation of heat-labile drugs. The HMH method is widely used for the development of nanocarrier vaccines due to its ability to incorporate immunogens into the lipid matrix and produce stable and uniform nanoparticles. SLNs prepared by the HMH method have also shown potential in the delivery of anticancer drugs, as they can enhance drug bioavailability, reduce toxicity, and improve therapeutic efficacy. In conclusion, the HMH method is a promising approach for the preparation of nanocarrier vaccines. It offers several advantages, such as high drug-loading capacity, scalability, and stability, but it also has limitations, such as limited drug compatibility and high-temperature processing.

7.3.2.5 *Spray Drying*

This method involves the atomization of a lipid solution into droplets, which are then dried using a hot gas stream to obtain SLNs. This method is suitable for large-scale production and provides SLNs with good stability and uniform particle size. Spray drying is a method for synthesizing SLNs for vaccine delivery that involves the atomization of an emulsion or solution of lipid and active ingredients into a stream of hot air, causing rapid evaporation of the solvent and formation of nanoparticles [41]. The spray drying method can be used to produce large quantities of nanoparticles with a high degree of reproducibility and uniformity. The process is relatively simple, fast, and requires low amounts of organic solvents, making it an environmentally friendly approach. The resulting nanoparticles have good storage stability, physical integrity, and controlled drug release properties. This method can be limited by the tendency of some active compounds to degrade or lose their activity during the drying process. High temperatures and mechanical stresses during the process can cause changes in the physical and chemical properties of the nanoparticles, which can affect their efficacy and stability [42]. This method has also been applied in the delivery of other therapeutic agents, such as anticancer drugs and gene therapy. Overall, spray drying is a promising method for synthesizing SLNs for vaccine delivery, with potential applications in other fields of drug delivery.

7.3.2.6 *Solvent Injection Method*

Solvent injection is a method for the synthesis of SLNs that involves dissolving lipids and drug substances in a suitable organic solvent and injecting it into an aqueous phase containing a surfactant under controlled conditions of temperature, pressure, and mixing rate. This method relies on the rapid diffusion of the organic solvent into the aqueous phase, leading to the formation of nanosized lipid particles stabilized by a surfactant layer [43]. The main advantage of the solvent injection method is its simplicity and scalability, as it does not require high-energy input or specialized equipment. The method also allows for precise control of particle size, drug loading, and release kinetics by adjusting the composition of the organic and aqueous phases and the surfactant type and concentration. Moreover, the solvent injection method is compatible with a wide range of lipids and drug substances, including hydrophobic and hydrophilic molecules. However, the solvent injection method has some limitations and challenges, such as the potential toxicity of residual organic solvents and the need for rigorous purification and sterilization steps [44]. Additionally, the solvent injection method may lead to particle aggregation or instability under certain conditions, such as high surfactant concentrations or extreme pH values. The solvent injection method is a promising approach for the synthesis of SLNs and nanocarrier vaccines, with the potential for customization and optimization of particle properties and drug release profiles [45]. However, careful consideration must be given to the choice of lipids, surfactants, and organic solvents, as well as the purification and sterilization steps, to ensure the safety and efficacy of the final product.

7.3.2.7 Microfludics

Microfluidics is a method for producing LNPs that involves the use of microfabricated channels and chambers to mix and manipulate fluids at the microscale level. In this method, the lipid and aqueous phases are introduced into the microfluidic device, where they are mixed and processed to form the nanoparticles [46]. The principle of microfluidics is based on the laminar flow of fluids in microchannels, which allows for precise control of the flow rate, mixing, and reaction times. The size and morphology of the nanoparticles can be controlled by adjusting the parameters of the microfluidic device, such as the flow rate, channel geometry, and mixing conditions [47]. One of the main advantages of microfluidics is its ability to produce nanoparticles with a narrow size distribution and high reproducibility. The precise control of the flow rate and mixing conditions allows for the formation of homogenous nanoparticles with a high degree of control over their size, shape, and surface properties. Additionally, microfluidics is a scalable and high-throughput method that can produce large quantities of nanoparticles in a short period of time. However, there are also some disadvantages to microfluidics. The fabrication of microfluidic devices can be complex and expensive and requires specialized equipment and expertise. Additionally, the use of microfluidic devices can lead to the loss of materials and the formation of aggregates or impurities due to the high shear forces and stresses experienced by the fluids in the channels [48]. Microfluidics has a wide range of applications in the field of nanomedicine, including drug delivery and vaccine development. It can be concluded that microfluidics is a promising method for the production of LNPs for vaccine delivery. Its precise control over the size and morphology of the nanoparticles, as well as its scalability and high throughput, make it a valuable tool for the development of nanocarrier vaccines. However, the complexity and cost of the equipment and expertise required for microfluidics should be carefully considered when choosing a synthesis method.

7.4 Characterization of Lipid Nanoparticles

7.4.1 Size and Shape

LNPs are widely used as carriers for various types of drugs, including small molecules, nucleic acids, and proteins. One of the critical factors that determine the performance of LNPs is their size and shape. Size and shape are essential factors that influence the pharmacokinetics and pharmacodynamics of LNPs. The size of LNPs affects their biodistribution, cellular uptake, and clearance rate from the body. For example, LNPs with a size of 100 nm or smaller can evade the reticuloendothelial system and accumulate in tumor tissues through the enhanced permeability and retention (EPR) effect, leading to improved therapeutic efficacy [49]. Additionally, the shape of LNPs can affect their cellular uptake, as well as their interaction with biomolecules, such as proteins and lipids. Thus, controlling the size and shape of LNPs is crucial for optimizing their therapeutic efficacy and reducing potential side effects.

Several methods can be used to control the size and shape of LNPs. These methods include lipid composition, preparation method, and post-preparation modification. For example, the choice of lipids and their ratio can affect the size and shape of LNPs. Increasing the lipid concentration or adding a cosurfactant can reduce the particle size. Additionally, preparation methods such as sonication, microfluidization, and solvent evaporation can influence the size and shape of LNPs. Furthermore, post-preparation modification, such as coating LNPs with PEG, can affect their size and shape, as well as their stability and biodistribution [50].

Various analytical techniques can be used to characterize the size and shape of LNPs, such as dynamic light scattering (DLS), transmission electron microscopy (TEM), cryogenic transmission electron microscopy (Cryo-TEM), and atomic force microscopy (AFM).

Dynamic Light Scattering (DLS): DLS is a widely used technique to measure the hydrodynamic diameter of LNPs. In this method, light is scattered by the particles, and the intensity of the scattered light is analyzed to determine the size distribution of the particles. DLS is a non-invasive and rapid technique that can analyze a large number of samples at once. However, it has limitations in measuring the size of polydisperse and non-spherical particles accurately [51].

Transmission Electron Microscopy (TEM): TEM is a high-resolution imaging technique that provides detailed information on the size and shape of LNPs. In TEM, a beam of electrons is transmitted through the sample, and the resulting image is formed on a screen. TEM can provide information on the morphology, size, and shape of individual LNPs. However, it requires expensive equipment and skilled operators [52].

Cryogenic Transmission Electron Microscopy (Cryo-TEM): Cryo-TEM is an extension of TEM that allows the imaging of LNPs at cryogenic temperatures. In Cryo-TEM, LNPs are rapidly frozen to prevent the formation of ice crystals, which can damage the sample. Cryo-TEM can provide high-resolution images of LNPs, which can reveal their internal structure and morphology. However, it requires specialized equipment and expertise [53].

Atomic Force Microscopy (AFM): AFM is a high-resolution imaging technique that can provide information on the size and shape of LNPs. In AFM, a cantilever with a sharp tip is scanned over the sample surface, and the deflection of the cantilever is measured to obtain the topographical information of the sample. AFM can provide information on the surface morphology [54].

7.4.2 Surface Charge

Surface charge is an essential characteristic of LNPs that influences their interaction with biological systems and their therapeutic efficacy. The surface charge of LNPs affects their colloidal stability, cellular uptake, and interaction with biomolecules such as proteins and cell membranes. The surface charge of LNPs is determined by the nature of the lipid composition and the surrounding medium's pH and ionic strength. LNPs with a positive surface charge can interact with negatively charged cell membranes, leading to enhanced cellular uptake. On the other hand, LNPs with a negative surface charge can reduce non-specific interactions with cells and prolong their circulation time in the bloodstream.

Several methods can be used to control the surface charge of LNPs, including lipid composition, preparation method, and postpreparation modification. For example, the choice of lipids and their ratio can affect the surface charge of LNPs. Adding cationic lipids such as dioleoyltrimethylammonium propane (DOTAP) can increase the positive surface charge of LNPs. Similarly, adding anionic lipids such as phosphatidylserine (PS) can increase the negative surface charge of LNPs. Additionally, preparation methods such as sonication, microfluidization, and solvent evaporation can influence the surface charge of LNPs. Furthermore, postpreparation modification, such as coating LNPs with PEG, can affect their surface charge and improve their stability and biodistribution [55].

7.4.2.1 Analytical Techniques for Surface Charge Characterization

Various analytical techniques can be used to characterize the surface charge of LNPs, such as zeta potential measurement, electrophoresis, and isoelectric focusing.

7.4.2.2 Zeta Potential Measurement

Zeta potential is a measure of the electric potential at the shear plane of a particle in a suspension. Zeta potential measurement is a widely used technique to determine the surface charge of LNPs. In this method, the movement of LNPs in an electric field is measured, and the zeta potential is calculated using the electrophoretic mobility. Zeta potential measurement is a non-invasive and rapid technique that can analyze a large number of samples at once. However, it has limitations in measuring the surface charge of particles that are highly aggregated or non-spherical [56].

7.4.2.3 Electrophoresis

Electrophoresis is a technique that separates charged particles based on their charge and size under an applied electric field. In this method, LNPs are placed in a gel matrix, and an electric field is applied to separate them based on their surface charge. Electrophoresis can provide information on the surface charge and size distribution of LNPs. However, it requires specialized equipment and expertise [57].

7.4.2.4 Isoelectric Focusing

Isoelectric focusing is a technique that separates particles based on their isoelectric point (pI). In this method, LNPs are placed in a pH gradient gel, and an electric field is applied to separate them based on their pI. Isoelectric focusing can provide information on the surface charge and size distribution of LNPs. However, it requires specialized equipment and expertise [58].

Thus, controlling the surface charge of LNPs is crucial for optimizing their therapeutic efficacy and reducing potential side effects. Various methods can be used for surface charge control, such as lipid composition, preparation method, and postpreparation modification. Moreover, several analytical techniques can be used for surface charge characterization, such as zeta potential measurement, electrophoresis, and isoelectric focusing.

Choosing an appropriate method for surface charge characterization depends on the specific properties of LNPs and the research objectives.

7.4.3 Encapsulation Efficiency

EE is a critical parameter for LNPs that determines their therapeutic efficacy. EE is defined as the percentage of the drug or active compound that is encapsulated in LNPs. EE is a critical parameter for LNPs that influences their therapeutic efficacy, toxicity, and stability. High EE ensures that a maximum amount of drug or active compound is delivered to the target site, reducing the required dosage and potential side effects. Additionally, high EE ensures the stability of LNPs during storage and administration, reducing the likelihood of drug leakage and degradation [59].

7.4.3.1 Factors Affecting Encapsulation Efficiency

Several factors can affect the EE of LNPs, such as lipid composition, preparation method, drug properties, and storage conditions. For example, the choice of lipid composition can affect the EE of LNPs. Adding hydrophobic lipids such as cholesterol can increase the EE by providing a hydrophobic environment for the drug. Similarly, preparation methods such as sonication, microfluidization, and solvent evaporation can affect the EE by altering the size and surface properties of LNPs. Moreover, drug properties such as solubility, stability, and size can affect the EE by influencing the drug–lipid interaction. Finally, storage conditions such as temperature, light, and humidity can affect the EE by causing drug degradation and lipid oxidation [60].

7.4.3.2 Analytical Techniques for Encapsulation Efficiency Characterization

Various analytical techniques can be used to characterize the EE of LNPs, such as high-performance liquid chromatography (HPLC), ultraviolet-visible spectroscopy (UV-Vis), and fluorescence spectroscopy.

i. High-Performance Liquid Chromatography (HPLC): HPLC is a widely used technique for drug quantification and EE determination. In this method, the drug is extracted from LNPs using a suitable solvent, and the extracted drug is analyzed using HPLC. HPLC can provide accurate and precise measurements of drug concentration and EE. However, it requires specialized equipment and expertise.

ii. Ultraviolet-Visible Spectroscopy (UV-Vis): UV-Vis is a technique that measures the absorbance of light by molecules in the visible and ultraviolet range. In this method, the drug is extracted from LNPs, and the absorbance of the extracted drug is measured using UV-Vis. UV-Vis can provide a rapid and straightforward method for EE determination. However, it requires suitable chromophores or fluorophores in the drug molecule, limiting its applicability to some drugs.

iii. Fluorescence Spectroscopy: Fluorescence spectroscopy is a technique that measures the emission of light by molecules excited by a specific wavelength of light. In this method, the drug is extracted from LNPs, and the fluorescence of the extracted drug is measured using fluorescence spectroscopy. Fluorescence spectroscopy can provide a sensitive and specific method for EE determination. However, it requires suitable fluorophores in the drug molecule, limiting its applicability to some drugs.

Therefore, EE is a crucial parameter for LNPs that determines their therapeutic efficacy, toxicity, and stability. Several factors can affect the EE of LNPs, such as lipid composition, preparation method, drug properties, and storage conditions. Moreover, several analytical techniques can be used for EE characterization, such as HPLC, UV-Vis, and fluorescence spectroscopy. Choosing an appropriate method for EE characterization depends on the specific properties of LNPs and the research objectives.

7.4.4 Stability

LNPs have emerged as promising drug delivery systems due to their ability to encapsulate and deliver a wide range of drugs. However, the stability of LNPs is a critical parameter that affects their efficacy and shelf life. Stability is a critical parameter for LNPs that affects their therapeutic efficacy and shelf life. Stable LNPs can maintain their physicochemical properties, drug release, and targeting ability during storage and administration. However, unstable LNPs can undergo drug leakage, aggregation, and degradation, leading to reduced efficacy and potential toxicity.

7.4.4.1 *Factors Affecting Stability*

Several factors can affect the stability of LNPs, such as lipid composition, preparation method, storage conditions, and drug properties. For example, the choice of lipid composition can affect the stability of LNPs. Adding cholesterol can enhance the stability of LNPs by reducing their membrane fluidity and increasing their packing density. Similarly, preparation methods such as sonication and microfluidization can affect the stability of LNPs by altering their size, surface charge, and lipid packing. Moreover, storage conditions such as temperature, light, and humidity can affect the stability of LNPs by causing lipid oxidation, drug degradation, and physical instability. Finally, drug properties such as solubility, stability, and size can affect the stability of LNPs by influencing their drug–lipid interaction, drug release, and drug targeting.

7.4.4.2 *Analytical Techniques for Stability Characterization*

Various analytical techniques can be used to characterize the stability of LNPs, such as size distribution analysis, zeta potential analysis, and differential scanning calorimetry (DSC).

i. Size Distribution Analysis: Size distribution analysis is a technique that measures the size distribution of LNPs using DLS or nanoparticle tracking analysis (NTA). In this method, LNPs are dispersed in a suitable solvent, and the size distribution of LNPs is analyzed using DLS or NTA. Size distribution analysis can provide information about the stability and aggregation of LNPs. However, it requires specialized equipment and expertise.
ii. Zeta Potential Analysis: Zeta potential analysis is a technique that measures the surface charge of LNPs using electrophoretic mobility. In this method, LNPs are dispersed in a suitable electrolyte solution, and the zeta potential of LNPs is measured using a zeta potential analyzer. Zeta potential analysis can provide information about the stability and colloidal stability of LNPs. However, it requires careful selection of the electrolyte solution and expertise.
iii. Differential Scanning Calorimetry: DSC is a technique that measures the thermal behavior of LNPs by analyzing their heat flow under controlled temperature and pressure

conditions. In this method, LNPs are analyzed using DSC, and their thermal stability and lipid packing are evaluated. DSC can provide information about the thermal stability and physical stability of LNPs. However, it requires specialized equipment and expertise.

Stability is a critical parameter for LNPs that affects their therapeutic efficacy and shelf life. Several factors can affect the stability of LNPs, such as lipid composition, preparation method, storage conditions, and drug properties. Moreover, several analytical techniques can be used for stability characterization, such as size distribution analysis, zeta potential analysis, and DSC. Choosing an appropriate method for stability characterization depends on the specific properties of LNPs and the research objectives.

7.5 Applications of Lipid-Based Nanoparticles in Vaccines

7.5.1 Enhancement of Immune Response

LNPs have been extensively studied as adjuvants in vaccines to enhance the immune response against infectious diseases and cancers. LNPs can enhance the immune response by several mechanisms, such as antigen delivery, immune cell activation, and cytokine induction. LNPs can efficiently deliver antigens to APCs, such as dendritic cells, macrophages, and B cells, leading to enhanced antigen processing, presentation, and activation of T and B cells. Moreover, LNPs can activate immune cells, such as dendritic cells, macrophages, and natural killer cells, by releasing danger signals, such as lipopolysaccharides and CpG oligonucleotides, leading to increased cytokine production, phagocytosis, and antigen presentation [61]. Finally, LNPs can induce cytokine production, such as interferon-gamma and interleukin-12, leading to Th1 polarization, enhanced cytotoxic T lymphocyte activity, and long-term immune memory [62].

Examples of Vaccines Using LNPs for Immune Response Enhancement: The Moderna COVID-19 vaccine is a mRNA vaccine that uses LNPs as a delivery system and an adjuvant to enhance the immune response against SARS-CoV-2. The mRNA in the vaccine encodes the spike protein of SARS-CoV-2, which is encapsulated in LNPs and delivered to APCs. The LNPs in the vaccine also contain cholesterol and PEG, which enhance their stability and biodistribution [63]. The Pfizer-BioNTech COVID-19 vaccine is also an mRNA vaccine that uses LNPs as a delivery system and an adjuvant to enhance the immune response against SARS-CoV-2. The mRNA in the vaccine encodes the spike protein of SARS-CoV-2, which is encapsulated in LNPs and delivered to APCs. The LNPs in the vaccine also contain cholesterol and PEG, which enhance their stability and biodistribution [64]. The CureVac COVID-19 vaccine is an mRNA vaccine that uses LNPs as a delivery system and an adjuvant to enhance the immune response against SARS-CoV-2. The mRNA in the vaccine encodes the spike protein of SARS-CoV-2, which is encapsulated in LNPs and delivered to APCs. The LNPs in the vaccine also contain imidazole and cholesterol, which enhance their stability and immunogenicity [65]. The NanoFlu vaccine is a seasonal flu vaccine that uses LNPs as an adjuvant to enhance the immune response against influenza. The vaccine contains recombinant hemagglutinin antigen, which is encapsulated in LNPs and delivered to APCs. The LNPs in the vaccine also contain cholesterol and dimethyldioctadecylammonium bromide, which enhance their stability and immunogenicity [66].

Thus, LNPs have emerged as promising adjuvants in vaccines to enhance the immune response against infectious diseases and cancers. The use of LNPs in vaccines can enhance antigen delivery, immune cell activation, and cytokine induction, leading to enhanced immunogenicity and long-term immune memory. The examples of vaccines that use LNPs as adjuvants discussed in this essay demonstrate the potential of LNPs in vaccine development and highlight the importance of continued research in this area.

7.5.2 Targeted Delivery

LNPs have been widely investigated as a targeted drug delivery system due to their ability to encapsulate and protect therapeutic agents and selectively deliver them to specific cells or tissues. LNPs can achieve targeted delivery by several mechanisms, such as surface modification, passive targeting, and active targeting. Surface modification of LNPs with ligands, such as antibodies, peptides, and aptamers, can confer specificity and affinity for specific cells or tissues, such as cancer cells, inflamed tissues, and endothelial cells [67]. Passive targeting of LNPs can occur due to the EPR effect, where LNPs can accumulate in tumor tissues due to their leaky vasculature and impaired lymphatic drainage. Active targeting of LNPs can be achieved by conjugating ligands, such as FoA, transferrin, and hyaluronic acid, to the surface of LNPs, which can bind to receptors overexpressed on cancer cells or inflamed tissues, leading to enhanced uptake and accumulation [68].

7.5.2.1 *Cancer Immunotherapy*

LNPs have been investigated as a delivery system for cancer immunotherapy, such as checkpoint inhibitors and chimeric antigen receptor (CAR) T cells. LNPs can encapsulate and protect these agents from degradation and deliver them to cancer cells, leading to enhanced cytotoxicity and reduced systemic toxicity. Moreover, LNPs can be surface-modified with antibodies or peptides that target specific cancer cells, leading to enhanced efficacy and specificity [69].

7.5.2.2 *mRNA-Based Vaccines*

LNPs have been extensively studied as a delivery system for mRNA-based vaccines against infectious diseases and cancers. LNPs can encapsulate and deliver mRNA that encodes antigens or therapeutic proteins to immune cells, leading to enhanced immunogenicity and efficacy. Moreover, LNPs can be surface-modified with ligands that target specific immune cells, such as dendritic cells and B cells, leading to enhanced uptake and activation [70].

7.5.2.3 *Gene Therapy*

LNPs have been investigated as a delivery system for gene therapy, such as CRISPR/Cas9 and RNA interference. LNPs can encapsulate and deliver these agents to target cells, leading to gene editing or silencing. Moreover, LNPs can be surface-modified with ligands that target specific cells or tissues, such as neurons and liver cells, leading to enhanced specificity and efficacy [71].

It can be concluded that LNPs have emerged as a promising targeted drug delivery system due to their ability to encapsulate and protect therapeutic agents and selectively deliver them to specific cells or tissues. The mechanisms of targeted delivery by LNPs include surface modification, passive targeting, and active targeting. The examples of vaccines that use LNPs for targeted delivery discussed in this essay demonstrate the potential of LNPs in vaccine development and gene therapy and highlight the importance of continued research in this area.

7.5.3 Adjuvant Effects

Vaccines are one of the most effective interventions to protect against infectious diseases. However, some vaccines may require adjuvants to improve their efficacy, especially in vulnerable populations such as infants, the elderly, and immunocompromised individuals. Adjuvants are substances that are added to vaccines to enhance the immune response to the antigen, resulting in a more robust and long-lasting protective response. Adjuvants can be broadly classified into two categories: those that enhance the innate immune response, such as Toll-like receptor (TLR) agonists, and those that enhance the adaptive immune response, such as LNPs [72].

The mechanism of adjuvant effects can vary depending on the type of adjuvant used. TLR agonists, such as CpG oligonucleotides, flagellin, and lipopolysaccharides, activate the innate immune system by binding to specific receptors, resulting in the release of cytokines and chemokines. This, in turn, activates dendritic cells and macrophages, leading to the activation of antigen-specific T and B cells. LNPs, on the other hand, work by delivering the antigen to specific immune cells, such as dendritic cells and macrophages, resulting in enhanced antigen presentation and activation of antigen-specific T and B cells. LNPs can also act as depots, prolonging the release of antigen and enhancing the duration of the immune response [73].

7.5.3.1 mRNA COVID-19 Vaccines

The COVID-19 pandemic has highlighted the importance of rapid vaccine development and deployment. mRNA-based vaccines, such as the Pfizer-BioNTech and Moderna COVID-19 vaccines, have been shown to be highly effective at preventing severe disease and hospitalization. The mRNA vaccines utilize LNPs to deliver the mRNA encoding the SARS-CoV-2 spike protein to immune cells. The LNPs protect the mRNA from degradation and facilitate its uptake by immune cells, resulting in a robust and long-lasting immune response.

7.5.3.2 Human Papillomavirus (HPV) Vaccine

The HPV vaccine is a prophylactic vaccine that protects against HPV infection and related cancers. The vaccine contains virus-like particles (VLPs) that resemble the structure of the virus, along with an LNP adjuvant. The adjuvant enhances the immune response to the VLPs, resulting in long-lasting protection against HPV infection [74].

7.5.3.3 Influenza Vaccine

The influenza vaccine is an annual vaccine that protects against seasonal influenza. The vaccine contains inactivated or attenuated virus particles, along with an LNP adjuvant. The adjuvant enhances the immune response to the virus particles, resulting in improved efficacy of the vaccine [75].

Thus, LNPs are a promising approach for adjuvant development due to their ability to enhance antigen presentation and prolong the duration of the immune response. The COVID-19 pandemic has highlighted the importance of rapid vaccine development, and the use of LNPs in mRNA vaccines has shown promise in this regard. Future research in adjuvant development should continue to explore the use of LNPs to enhance the efficacy and duration of protective immune responses.

7.6 Challenges and Future Directions

7.6.1 Safety and Toxicity Concerns

Safety and toxicity concerns are major challenges in drug development and can lead to delays or even withdrawal of a drug from the market. Addressing safety and toxicity concerns is essential to ensure patient safety and maximize the benefit–risk ratio of a drug. One of the major challenges in addressing safety and toxicity concerns is the lack of understanding of the underlying mechanisms of toxicity. Toxicity can occur due to a variety of reasons, such as off-target effects, dose-dependent effects, or idiosyncratic reactions. In some cases, toxicity may only be identified after a drug is on the market and has been administered to a large number of patients. Furthermore, toxicity can be complex and may involve multiple organ systems, making it difficult to identify the cause of toxicity.

7.6.1.1 Preclinical Safety Evaluation

Preclinical safety evaluation is an essential step in drug development to identify potential safety concerns before a drug is tested in humans. Preclinical safety evaluation includes *in vitro* and *in vivo* testing to assess the safety profile of a drug candidate. In vitro assays can be used to evaluate the potential for a drug candidate to cause cytotoxicity, genotoxicity, or other adverse effects. *In vivo* testing in animals can provide valuable information on the pharmacokinetics and pharmacodynamics of a drug candidate, as well as identify potential safety concerns [76].

7.6.1.2 Human Pharmacology Studies

Human pharmacology studies are conducted to evaluate the safety and tolerability of a drug candidate in healthy volunteers. These studies can provide information on the pharmacokinetics and pharmacodynamics of a drug candidate in humans, as well as identify any potential safety concerns [77].

7.6.1.3 *Postmarketing Surveillance*

Post-marketing surveillance is a critical component of drug safety monitoring and involves ongoing monitoring of the safety profile of a drug after it is on the market. Postmarketing surveillance can identify previously unrecognized safety concerns and can provide valuable information on the long-term safety of a drug [78].

7.6.1.4 *Adverse Event Reporting*

Adverse event reporting is a key mechanism for identifying safety concerns and involves the reporting of adverse events by healthcare professionals and patients. Adverse event reporting can provide early warning signals of potential safety concerns and can facilitate the rapid identification and investigation of safety concerns [79].

Safety and toxicity concerns are major challenges in drug development, and addressing these concerns is essential to ensure patient safety and maximize the benefit–risk ratio of a drug. Preclinical safety evaluation, human pharmacology studies, postmarketing surveillance, and adverse event reporting are essential strategies for addressing safety and toxicity concerns. The continued development and refinement of these strategies are essential to ensure the safety and efficacy of drugs and to improve patient outcomes.

7.6.2 Stability Issues

Stability is a critical attribute of pharmaceutical products and can affect the efficacy, safety, and shelf life of a drug. Stability issues can arise due to a variety of factors, such as degradation, oxidation, hydrolysis, or physical instability. Addressing stability issues is essential to ensure the quality, safety, and efficacy of a drug. One of the major challenges in addressing stability issues is the complex nature of drug degradation. Degradation can occur due to a variety of factors, such as exposure to light, heat, moisture, or pH changes. Furthermore, degradation can result in the formation of impurities, which can affect the safety and efficacy of a drug. Identifying the underlying mechanisms of drug degradation can be challenging, and developing strategies to mitigate stability issues can be time-consuming and costly [80].

7.6.2.1 *Formulation Optimization*

Formulation optimization is a key strategy for addressing stability issues and involves the selection of appropriate excipients and the optimization of the drug product's physicochemical properties. The use of stabilizing excipients, such as antioxidants, chelating agents, or pH modifiers, can mitigate stability issues and improve the shelf life of a drug product. Furthermore, the optimization of the drug product's physicochemical properties, such as particle size, crystal form, or polymorphism, can improve stability and reduce the risk of degradation.

7.6.2.2 *Analytical Method Development*

Analytical method development is critical for identifying stability issues and monitoring the stability of a drug product over time. Analytical methods, such as HPLC, can be used to identify degradation products and quantify impurities. Stability-indicating methods can also be developed to detect changes in drug product stability over time.

7.6.2.3 *Accelerated Stability Studies*

Accelerated stability studies can be used to evaluate the stability of a drug product over a short period of time by subjecting the drug product to stress conditions, such as high temperature and humidity. Accelerated stability studies can provide valuable information on the degradation pathways of a drug product and can be used to develop strategies to mitigate stability issues.

7.6.2.4 *Quality by Design (QbD)*

QbD is a systematic approach to drug development that involves the identification of critical quality attributes (CQAs) and the development of a control strategy to ensure the quality of the drug product. QbD can be used to identify stability as a CQA and develop a control strategy to mitigate stability issues.

It can be concluded that stability issues are a significant challenge in drug development, and addressing these concerns is essential to ensure the quality, safety, and efficacy of a drug. Formulation optimization, analytical method development, accelerated stability studies, and QbD are essential strategies for addressing stability issues. Continued development and refinement of these strategies are essential to ensure the stability of drug products and improve patient outcomes.

7.6.3 Scale-Up Production Challenges

Scaling up production is a critical step in the development of pharmaceutical products, as it involves increasing production volumes from the laboratory scale to the commercial scale. Scaling up production can be challenging, as it requires careful consideration of several factors, such as equipment design, process optimization, and regulatory compliance [81].

7.6.3.1 Equipment Design

One of the major challenges in scaling up production is equipment design. Equipment used in laboratory-scale production may not be suitable for commercial-scale production. Commercial-scale equipment must be designed to handle larger volumes of raw materials, maintain consistent process conditions, and meet regulatory compliance requirements. Designing and installing commercial-scale equipment can be expensive and time-consuming.

7.6.3.2 Process Optimization

Process optimization is critical in scaling up production. Process parameters must be optimized to ensure product quality, consistency, and yield. Factors such as raw material quality, processing time, temperature, and pressure must be carefully monitored and controlled to achieve optimal results. Process optimization requires extensive testing, data analysis, and collaboration between process development and manufacturing teams.

7.6.3.3 Regulatory Compliance

Scaling up production requires compliance with regulatory requirements, such as current Good Manufacturing Practices (cGMPs). cGMP regulations ensure that the drug product is safe, effective, and of high quality. Compliance with cGMP regulations involves extensive documentation, testing, and validation, which can be time-consuming and expensive.

Strategies for addressing scale-up production challenges

i. Early involvement of manufacturing teams
   Involving manufacturing teams early in the development process can help address scale-up production challenges. Manufacturing teams can provide valuable insights into equipment design, process optimization, and regulatory

compliance. Collaborating with manufacturing teams can help identify potential issues and develop solutions to address them.

ii. Process development and optimization
Process development and optimization are critical in scaling up production. Process development teams must identify critical process parameters, optimize process conditions, and develop robust control strategies. Process optimization involves extensive testing, data analysis, and collaboration between process development and manufacturing teams.

iii. Technology transfer
Technology transfer is the process of transferring knowledge and technology from the laboratory scale to the commercial scale. Technology transfer involves documentation, training, and collaboration between process development and manufacturing teams. Ensuring that technology transfer is well documented and well executed is essential to address scale-up production challenges.

iv. Continuous process improvement
Continuous process improvement is essential in addressing scale-up production challenges. Continuous improvement involves evaluating and optimizing the production process over time to improve efficiency, reduce waste, and improve product quality. Continuous process improvement requires a culture of collaboration, data analysis, and process monitoring.

Scaling up production is a critical step in the development of pharmaceutical products. Equipment design, process optimization, and regulatory compliance are major challenges in scaling up production. Addressing these challenges requires early involvement of manufacturing teams, process development and optimization, technology transfer, and continuous process improvement. Effective strategies for addressing these challenges are essential to ensure the quality, safety, and efficacy of pharmaceutical products.

7.6.4 Opportunities for Future Research

LNPs have emerged as a promising tool for the delivery of vaccines. They are biodegradable and biocompatible and can encapsulate a variety of antigens and adjuvants. Moreover, LNPs can be modified to target specific cells, tissues, or organs.

7.6.4.1 Novel Antigen and Adjuvant Formulations

LNPs have the potential to encapsulate a wide range of antigens and adjuvants, which can be delivered to APCs to stimulate an immune response. There is a need to develop novel antigen and adjuvant formulations that can elicit robust and long-lasting immune responses. For example, the use of neoantigens and personalized vaccines can be explored using LNPs [82].

7.6.4.2 Targeted Delivery

LNPs can be modified to target specific cells, tissues, or organs. Targeted delivery can enhance the efficacy and safety of vaccines. There is a need to develop targeted LNPs that can deliver vaccines to specific cells, such as dendritic cells or T cells, to improve vaccine efficacy.

7.6.4.3 Manufacturing Process Optimization

The manufacturing process of LNPs is critical for their efficacy and safety. There is a need to optimize the manufacturing process to ensure consistent quality, reproducibility, and scalability. Moreover, there is a need to develop large-scale manufacturing processes that can produce LNPs in sufficient quantities to meet the demands of global vaccination programs.

7.6.4.4 Immunological Mechanisms

LNPs can stimulate both innate and adaptive immune responses. However, the immunological mechanisms underlying LNP-mediated immune responses are not well understood. There is a need to investigate the immunological mechanisms underlying LNP-mediated immune responses to improve the design and development of LNPs.

7.6.4.5 Opportunities for Future Research

COVID-19 Vaccine Development

The COVID-19 pandemic has highlighted the urgent need for effective and safe vaccines. LNPs have been used in the development of COVID-19 vaccines, such as the Pfizer-BioNTech and Moderna vaccines. There is a need to further investigate the efficacy and safety of LNPs in COVID-19 vaccine development and to develop novel LNPs that can elicit broader and more robust immune responses.

Cancer Vaccines

LNPs have the potential to deliver cancer vaccines to stimulate an immune response against tumor cells. There is a need to investigate the efficacy and safety of LNPs in cancer vaccine development and to develop targeted LNPs that can deliver cancer vaccines to tumor cells.

Personalized Vaccines

LNPs can be used to deliver personalized vaccines, such as neoantigen vaccines, which are tailored to an individual's tumor mutations. There is a need to investigate the efficacy and safety of LNPs in personalized vaccine development and to develop manufacturing processes that can produce personalized vaccines in a timely and cost-effective manner.

LNPs have emerged as a promising tool for the delivery of vaccines. There are several areas for further research and development of LNPs, such as the development of novel antigen and adjuvant formulations, targeted delivery, manufacturing process optimization, and investigation of immunological mechanisms. Opportunities for future research include COVID-19 vaccine development, cancer vaccines, and personalized vaccines. The continued investigation and development of LNPs in vaccine development have the potential to revolutionize vaccination strategies and improve public health.

7.7 Conclusion

LNPs have emerged as a game-changer for future immunization strategies. The unique properties of LNPs, such as their biocompatibility, biodegradability, and ability to encapsulate a wide range of antigens and adjuvants, make them an ideal tool for the delivery of vaccines. Moreover, LNPs can be modified to target specific cells, tissues, or organs, which can enhance the efficacy and safety of vaccines.

LNPs have already been successfully used in the development of COVID-19 vaccines, and there is immense potential for their use in other vaccine development programs. The future development of LNPs in vaccines requires further research and optimization of manufacturing processes to ensure consistent quality, reproducibility, and scalability.

LNPs hold promise for the development of personalized vaccines, cancer vaccines, and targeted vaccines for infectious diseases. The continued development of LNPs in vaccine delivery has the potential to revolutionize immunization strategies and improve public health. Therefore, it is crucial to invest in further research and development of LNPs in vaccine delivery to harness their full potential for future immunization strategies.

References

1. De, A. and Ko, Y.T., A tale of nucleic acid–ionizable lipid nanoparticles: Design and manufacturing technology and advancement. *Expert Opin. Drug Deliv.*, 20, 1, 75–91, Jan. 2023. doi: 10.1080/17425247.2023.2153832.
2. Jia, Y. *et al.*, Approved nanomedicine against diseases. *Pharmaceutics*, 15, 3, 774, Feb. 2023. doi: 10.3390/pharmaceutics15030774.
3. Dammes, N. and Peer, D., Paving the road for RNA therapeutics. *Trends Pharmacol. Sci.*, 41, 10, 755–775, Oct. 2020. doi: 10.1016/j.tips.2020.08.004.
4. Das, A. and Ali, N., Nanovaccine: an emerging strategy. *Expert Rev. Vaccines*, 20, 10, 1273–1290, Oct. 2021. doi: 10.1080/14760584.2021.1984890.
5. Zao, C.-X., Xu, L., Wang, X., Liu, Y., Yang, G., Falconer, R.J., Lipid nanoparticles for drug delivery. *Adv. NanoBiomed Res.*, 2, 2, 2100109, Feb. 2022. doi: 10.1002/anbr.202100109.
6. Dhiman, N., Awasthi, R., Sharma, B., Kharkwal, H., Kulkarni, G.T., Lipid nanoparticles as carriers for bioactive delivery. *Front. Chem.*, 9, Apr. 2021. doi: 10.3389/fchem.2021.580118.
7. Abdel-Mageed, H.M., Abd El Aziz, A.E., Mohamed, S.A., AbuelEzz, N.Z., The tiny big world of solid lipid nanoparticles and nanostructured lipid carriers: An updated review. *J. Microencapsul.*, 39, 1, 72–94, Jan. 2022. doi: 10.1080/02652048.2021.2021307.
8. Tenchov, R., Bird, R., Curtze, A.E., Zhou, Q., Lipid nanoparticles from liposomes to mRNA vaccine delivery, a landscape of research diversity and advancement. *ACS Nano*, 15, 11, 16982–17015, Nov. 2021. doi: 10.1021/acsnano.1c04996.
9. Liu, G., Zhu, M., Zhao, X., Nie, G., Nanotechnology-empowered vaccine delivery for enhancing CD8+ T cells-mediated cellular immunity. *Adv. Drug Deliv. Rev.*, 176, 113889, Sep. 2021. doi: 10.1016/j.addr.2021.113889.
10. Eskandari, Z., Bahadori, F., Celik, B., Onyuksel, H., Targeted nanomedicines for cancer therapy, from basics to clinical trials. *J. Pharm. Pharm. Sci.*, 23, 1, 132–157, May 2020. doi: 10.18433/jpps30583.
11. Luzuriaga, M.A., Shahrivarkevishahi, A., Herbert, F.C., Wijesundara, Y.H., Gassensmith, J.J., Biomaterials and nanomaterials for sustained release vaccine delivery. *Wiley Interdiscip. Rev. Nanomed. Nanobiotechnol.*, 13, 6, e1735, Nov. 2021. doi: 10.1002/wnan.1735.
12. Lim, M. *et al.*, Engineered nanodelivery systems to improve dna vaccine technologies. *Pharmaceutics*, 12, 1, 30, Jan. 2020. doi: 10.3390/pharmaceutics12010030.
13. Ferhan, A.R., Park, S., Park, H., Tae, H., Jackman, J.A., Cho, N.J., Lipid nanoparticle technologies for nucleic acid delivery: A nanoarchitectonics perspective. *Adv. Funct. Mater.*, 32, 37, 2203669, Sep. 2022. doi: 10.1002/adfm.202203669.
14. Uddin, M.N. and Roni, M.A., Challenges of storage and stability of mrna-based covid-19 vaccines. *Vaccines*, 9, 9, 1033, Sep. 2021. doi: 10.3390/vaccines9091033.

15. Zhong, Q. and Zhang, L., Nanoparticles fabricated from bulk solid lipids: Preparation, properties, and potential food applications. *Adv. Colloid Interface Sci.*, 273, 102033, Nov. 2019. doi: 10.1016/j.cis.2019.102033.
16. Salvi, V.R. and Pawar, P., Nanostructured lipid carriers (NLC) system: A novel drug targeting carrier. *J. Drug Deliv. Sci. Technol.*, 51, 255–267, Jun. 2019. doi: 10.1016/j.jddst.2019.02.017.
17. Musielak, E., Feliczak-Guzik, A., Nowak, I., Synthesis and potential applications of lipid nanoparticles in medicine. *Mater. (Basel).*, 15, 2, 682, Jan. 2022. doi: 10.3390/ma15020682.
18. Viegas, C., Seck, F., Fonte, P., An insight on lipid nanoparticles for therapeutic proteins delivery. *J. Drug Deliv. Sci. Technol.*, 77, 103839, Nov. 2022. doi: 10.1016/j.jddst.2022.103839.
19. Khan, M.S. *et al.*, Lipid-based colloidal nanoparticles for applications in targeted vaccine delivery. *Nanoscale Adv.*, 5, 2023. doi: 10.1039/d2na00795a.
20. Jazayeri, S.D., Lim, H.X., Shameli, K., Yeap, S.K., Poh, C.L., Nano and microparticles as potential oral vaccine carriers and adjuvants against infectious diseases. *Front. Pharmacol.*, 12, Jun. 2021. doi: 10.3389/fphar.2021.682286.
21. Trenkenschuh, E. and Friess, W., Freeze-drying of nanoparticles: How to overcome colloidal instability by formulation and process optimization. *Eur. J. Pharm. Biopharm.*, 165, 345–360, Aug. 2021. doi: 10.1016/j.ejpb.2021.05.024.
22. Ahmad, J., Rizwanullah, M., Amin, S., Warsi, M.H., Ahmad, M.Z., Barkat, M.A., Nanostructured lipid carriers (NLCs): Nose-to-brain delivery and theranostic application. *Curr. Drug Metab.*, 21, 14, 1136–1143, Dec. 2020. doi: 10.2174/1389200221666200719003304.
23. Aldosari, B.N., Alfagih, I.M., Almurshedi, A.S., Lipid nanoparticles as delivery systems for RNA-based vaccines. *Pharmaceutics*, 13, 2, 1–29, Feb. 2021. doi: 10.3390/pharmaceutics13020206.
24. He, R., Zang, J., Zhao, Y., Dong, H., Li, Y., Nanotechnology-based approaches to promote lymph node targeted delivery of cancer vaccines. *ACS Biomater. Sci. Eng.*, 8, 2, 406–423, Feb. 2022. doi: 10.1021/acsbiomaterials.1c01274.
25. Garcês, A., Amaral, M.H., Lobo, J.M.S., Silva, A.C., Formulations based on solid lipid nanoparticles (SLN) and nanostructured lipid carriers (NLC) for cutaneous use: A review. *Eur. J. Pharm. Sci.*, 112, 159–167, 2018. doi: 10.1016/j.ejps.2017.11.023.
26. Weber, S., Zimmer, A., Pardeike, J., Solid lipid nanoparticles (SLN) and nanostructured lipid carriers (NLC) for pulmonary application: A review of the state of the art. *Eur. J. Pharm. Biopharm.*, 86, 1, 7–22, 2014. doi: 10.1016/j.ejpb.2013.08.013.
27. Yadav K, K.S., Kale, High pressure homogenizer in pharmaceuticals: Understanding its critical processing parameters and applications. *J. Pharm. Innov.*, 15, 4, 690–701, Dec. 2020. doi: 10.1007/s12247-019-09413-4.
28. Khairnar, S.V. *et al.*, Review on the scale-up methods for the preparation of solid lipid nanoparticles. *Pharmaceutics*, 14, 9, 1886, Sep. 2022. doi: 10.3390/pharmaceutics14091886.

29. Soni, G., Kale, K., Shetty, S., Gupta, M.K., Yadav, K.S., Quality by design (QbD) approach in processing polymeric nanoparticles loading anticancer drugs by high pressure homogenizer. *Heliyon*, 6, 4, e03846, Apr. 2020. doi: 10.1016/j.heliyon.2020.e03846.
30. Duong, V.-A., Nguyen, T.-T.-L., Maeng, H.-J., Recent advances in intranasal liposomes for drug, gene, and vaccine delivery. *Pharmaceutics*, 15, 1, 207, Jan. 2023. doi: 10.3390/pharmaceutics15010207.
31. Duong, V.A., Nguyen, T.T.L., Maeng, H.J., Preparation of solid lipid nanoparticles and nanostructured lipid carriers for drug delivery and the effects of preparation parameters of solvent injection method. *Molecules*, 25, 20, 4781, 2020. doi: 10.3390/molecules25204781.
32. Chaurawal, N., Misra, C., Raza, K., Lipid-based nanocarriers loaded with taxanes for the management of breast cancer: Promises and challenges. *Curr. Drug Targets*, 23, 6, 544–558, Apr. 2021. doi: 10.2174/1389450122666210824144304.
33. Araya-Sibaja, A.M. *et al.*, Curcumin loaded and co-loaded nanosystems: A review from a biological activity enhancement perspective. *Pharm. Nanotechnol.*, 9, 2, 85–100, Feb. 2020. doi: 10.2174/2211738508666201228150659.
34. Cun, D., Zhang, C., Bera, H., Yang, M., Particle engineering principles and technologies for pharmaceutical biologics. *Adv. Drug Deliv. Rev.*, 174, 140–167, Jul. 2021. doi: 10.1016/j.addr.2021.04.006.
35. Shirodkar, R.K., Kumar, L., Mutalik, S., Lewis, S., Solid lipid nanoparticles and nanostructured lipid carriers: Emerging lipid based drug delivery systems. *Pharm. Chem. J.*, 53, 5, 440–453, Aug. 2019. doi: 10.1007/s11094-019-02017-9.
36. Jain, A.K. and Thareja, S., Solid lipid nanoparticles, in: *Nanotechnology in the Life Sciences*, pp. 221–249, 2020.
37. Chirio, D. *et al.*, Development of solid lipid nanoparticles by cold dilution of microemulsions: Curcumin loading, preliminary in vitro studies, and biodistribution. *Nanomaterials*, 9, 2, 230, Feb. 2019. doi: 10.3390/nano9020230.
38. Samimi, S., Maghsoudnia, N., Eftekhari, R.B., Dorkoosh, F., Lipid-based nanoparticles for drug delivery systems, in: *Characterization and Biology of Nanomaterials for Drug Delivery: Nanoscience and Nanotechnology in Drug Delivery*, pp. 47–76, Elsevier, 2018. doi: 10.1016/B978-0-12-814031-4.00003-9.
39. Souto, E.B. *et al.*, SLN and NLC for topical, dermal, and transdermal drug delivery. *Expert Opin. Drug Deliv.*, 17, 3, 357–377, Mar. 2020. doi: 10.1080/17425247.2020.1727883.
40. Musielak, E., Feliczak-Guzik, A., Nowak, I., Optimization of the conditions of solid lipid nanoparticles (SLN) synthesis. *Molecules*, 27, 7, 2202, Mar. 2022. doi: 10.3390/molecules27072202.
41. Amoabediny, G. *et al.*, Overview of preparation methods of polymeric and lipid-based (niosome, solid lipid, liposome) nanoparticles: A comprehensive review. *Int. J. Polym. Mater. Polym. Biomater.*, 67, 6, 383–400, 2018. doi: 10.1080/00914037.2017.1332623.

42. Sarabandi, K., Jafari, S.M., Mohammadi, M., Akbarbaglu, Z., Pezeshki, A., Heshmati, M.K., Production of reconstitutable nanoliposomes loaded with flaxseed protein hydrolysates: Stability and characterization. *Food Hydrocoll.*, 96, 442–450, Nov. 2019. doi: 10.1016/j.foodhyd.2019.05.047.
43. Patel, D., Patel, B., Thakkar, H., Lipid based nanocarriers: Promising drug delivery system for topical application. *Eur. J. Lipid Sci. Technol.*, 123, 5, 2000264, May 2021. doi: 10.1002/ejlt.202000264.
44. Shah, S., Dhawan, V., Holm, R., Nagarsenker, M.S., Perrie, Y., Liposomes: Advancements and innovation in the manufacturing process. *Adv. Drug Deliv. Rev.*, 154–155, 102–122, 2020. doi: 10.1016/j.addr.2020.07.002.
45. Sivadasan, D., Sultan, M.H., Madkhali, O., Almoshari, Y., Thangavel, N., Polymeric lipid hybrid nanoparticles (Plns) as emerging drug delivery platform—a comprehensive review of their properties, preparation methods, and therapeutic applications. *Pharmaceutics*, 13, 8, 1291, Aug. 2021. doi: 10.3390/pharmaceutics13081291.
46. Jia, F., Gao, Y., Wang, H., Recent advances in drug delivery system fabricated by microfluidics for disease therapy. *Bioengineering*, 9, 11, 625, Oct. 2022. doi: 10.3390/bioengineering9110625.
47. Niculescu, A.G., Chircov, C., Bîrcă, A.C., Grumezescu, A.M., Nanomaterials synthesis through microfluidic methods: An updated overview. *Nanomaterials*, 11, 4, 864, Mar. 2021. doi: 10.3390/nano11040864.
48. Niculescu, A.G., Chircov, C., Bîrcă, A.C., Grumezescu, A.M., Fabrication and applications of microfluidic devices: A review. *Int. J. Mol. Sci.*, 22, 4, 1–26, Feb. 2021. doi: 10.3390/ijms22042011.
49. Di, J. *et al.*, Biodistribution and non-linear gene expression of mRNA LNPs affected by delivery route and particle size. *Pharm. Res.*, 39, 1, 105–114, Jan. 2022. doi: 10.1007/s11095-022-03166-5.
50. Veiga, N., Diesendruck, Y., Peer, D., Targeted lipid nanoparticles for RNA therapeutics and immunomodulation in leukocytes. *Adv. Drug Deliv. Rev.*, 159, 364–376, 2020. doi: 10.1016/j.addr.2020.04.002.
51. Jia, Z., Li, J., Gao, L., Yang, D., Kanaev, A., Dynamic light scattering: A powerful tool for in situ nanoparticle sizing. *Colloids Interfaces*, 7, 1, 15, Feb. 2023. doi: 10.3390/colloids7010015.
52. Franken, L.E., Grünewald, K., Boekema, E.J., Stuart, M.C.A., A technical introduction to transmission electron microscopy for soft-matter: Imaging, possibilities, choices, and technical developments. *Small*, 16, 14, 1906198, Apr. 2020. doi: 10.1002/smll.201906198.
53. Watt, J., Huber, D.L., Stewart, P.L., Soft matter and nanomaterials characterization by cryogenic transmission electron microscopy. *MRS Bull.*, 44, 12, 942–948, 2019. doi: 10.1557/mrs.2019.285.
54. Takechi-Haraya, Y., Usui, A., Izutsu, K., II, Abe, Y., Atomic force microscopic imaging of mRNA-lipid nanoparticles in aqueous medium. *J. Pharm. Sci.*, 112, 3, 648–652, Mar. 2022. doi: 10.1016/j.xphs.2022.11.026.

55. Guevara, M.L., Persano, S., Persano, F., Lipid-based vectors for therapeutic mRNA-based anti-cancer vaccines. *Curr. Pharm. Des.*, 25, 13, 1443–1454, 2019. doi: 10.2174/1381612825666190619150221.
56. Duan, Y. *et al.*, A brief review on solid lipid nanoparticles: Part and parcel of contemporary drug delivery systems. *RSC Adv.*, 10, 45, 26777–26791, 2020. doi: 10.1039/d0ra03491f.
57. Liu, M., Wen, J., Sharma, M., Solid lipid nanoparticles for topical drug delivery: Mechanisms, dosage form perspectives, and translational status. *Curr. Pharm. Des.*, 26, 27, 3203–3217, Aug. 2020. doi: 10.2174/1381612826666200526145706.
58. Kumar, R., Guttman, A., Rathore, A.S., Applications of capillary electrophoresis for biopharmaceutical product characterization. *Electrophoresis*, 43, 1–2, 143–166, Jan. 2022. doi: 10.1002/elps.202100182.
59. de Oliveira, I.F. *et al.*, Cutting-edge advances in therapy for the posterior segment of the eye: Solid lipid nanoparticles and nanostructured lipid carriers. *Int. J. Pharm.*, 589, 119831, Nov. 2020. doi: 10.1016/j.ijpharm.2020.119831.
60. Lopes, C., Cristóvão, J., Silvério, V., Lino, P.R., Fonte, P., Microfluidic production of mRNA-loaded lipid nanoparticles for vaccine applications. *Expert Opin. Drug Deliv.*, 19, 10, 1381–1395, Oct. 2022. doi: 10.1080/17425247.2022.2135502.
61. Tursi, N.J., Xu, Z., Kulp, D.W., Weiner, D.B., Gene-encoded nanoparticle vaccine platforms for in vivo assembly of multimeric antigen to promote adaptive immunity. *Wiley Interdiscip. Rev. Nanomed. Nanobiotechnol.*, 15, 4, Feb. 2023. doi: 10.1002/wnan.1880.
62. Nguyen, K.G. *et al.*, Localized interleukin-12 for cancer immunotherapy. *Front. Immunol.*, 11, Oct. 2020. doi: 10.3389/fimmu.2020.575597.
63. Park, K.S., Sun, X., Aikins, M.E., Moon, J.J., Non-viral COVID-19 vaccine delivery systems. *Adv. Drug Deliv. Rev.*, 169, 137–151, Feb. 2021. doi: 10.1016/j.addr.2020.12.008.
64. Noor, R., mRNA vaccines as an efficient approach for the rapid and robust induction of host immunity against SARS-CoV-2. *SN Compr. Clin. Med.*, 4, 1, 88, Apr. 2022. doi: 10.1007/s42399-022-01168-3.
65. Verbeke, R., Lentacker, I., De Smedt, S.C., Dewitte, H., The dawn of mRNA vaccines: The COVID-19 case. *J. Control. Release*, 333, 511–520, May 2021. doi: 10.1016/j.jconrel.2021.03.043.
66. Corder, B.N., Bullard, B.L., Poland, G.A., Weaver, E.A., A decade in review: A systematic review of universal influenza vaccines in clinical trials during the 2010 decade. *Viruses*, 12, 10, 1186, Oct. 2020. doi: 10.3390/v12101186.
67. Xu, Y., Fourniols, T., Labrak, Y., Préat, V., Beloqui, A., Des Rieux, A., Surface modification of lipid-based nanoparticles. *ACS Nano*, 16, 5, 7168–7196, May 2022. doi: 10.1021/acsnano.2c02347.
68. Khalil, I.A., Sato, Y., Harashima, H., Recent advances in the targeting of systemically administered non-viral gene delivery systems. *Expert Opin. Drug Deliv.*, 16, 10, 1037–1050, Oct. 2019. doi: 10.1080/17425247.2019.1656196.

69. Hou, Y. *et al.*, Recent advance in nanomaterials for cancer immunotherapy. *Chem. Eng. J.*, 435, 134145, May 2022. doi: 10.1016/j.cej.2021.134145.
70. Huang, X., Kong, N., Zhang, X., Cao, Y., Langer, R., Tao, W., The landscape of mRNA nanomedicine. *Nat. Med.*, 28, 11, 2273–2287, Nov. 2022. doi: 10.1038/s41591-022-02061-1.
71. Kazemian, P., Yu, S.Y., Thomson, S.B., Birkenshaw, A., Leavitt, B.R., Ross, C.J.D., Lipid-nanoparticle-based delivery of CRISPR/Cas9 genome-editing components. *Mol. Pharm.*, 19, 6, 1669–1686, Jun. 2022. doi: 10.1021/acs.molpharmaceut.1c00916.
72. Kozak, M. and Hu, J., The integrated consideration of vaccine platforms, adjuvants, and delivery routes for successful vaccine development. *Vaccines*, 11, 3, 695, Mar. 2023, doi: 10.3390/vaccines11030695.
73. Pulendran, B., Arunachalam, P.S., O'Hagan, D.T., Emerging concepts in the science of vaccine adjuvants. *Nat. Rev. Drug Discovery*, 20, 6, 454–475, Jun. 2021. doi: 10.1038/s41573-021-00163-y.
74. de Moura, I.A., Silva, A.J.D., de Macêdo, L.S., da Conceição Viana Invenção, M., de Sousa, M.M.G., de Freitas, A.C., Enhancing the effect of nucleic acid vaccines in the treatment of HPV-related cancers: An overview of delivery systems. *Pathogens*, 11, 12, 1444, Nov. 2022. doi: 10.3390/pathogens11121444.
75. Pilkington, E.H. *et al.*, From influenza to COVID-19: Lipid nanoparticle mRNA vaccines at the frontiers of infectious diseases. *Acta Biomater.*, 131, 16–40, Sep. 2021. doi: 10.1016/j.actbio.2021.06.023.
76. Vlatkovic, I., Non-immunotherapy application of lnp-mrna: Maximizing efficacy and safety. *Biomedicines*, 9, 5, 530, May 2021. doi: 10.3390/biomedicines9050530.
77. Luxi, N. *et al.*, Allergic reactions to COVID-19 vaccines: Risk factors, frequency, mechanisms and management. *BioDrugs*, 36, 4, 443–458, Jul. 2022. doi: 10.1007/s40259-022-00536-8.
78. Cole, A., Webster, P., Van Liew, D., Salas, M., Aimer, O., Malikova, M.A., Safety surveillance and challenges in accelerated COVID-19 vaccine development. *Ther. Adv. Drug Saf.*, 13, 204209862211164, Jan. 2022. doi: 10.1177/20420986221116452.
79. Moghimi, S.M., Allergic reactions and anaphylaxis to LNP-based COVID-19 vaccines. *Mol. Ther.*, 29, 3, 898–900, Mar. 2021. doi: 10.1016/j.ymthe.2021.01.030.
80. Cheng, F. *et al.*, Research advances on the stability of mRNA vaccines. *Viruses*, 15, 3, 668, Mar. 2023. doi: 10.3390/v15030668.
81. Liu, X. and Meng, H., Consideration for the scale-up manufacture of nanotherapeutics—a critical step for technology transfer. *View*, 2, 5, 20200190, Oct. 2021. doi: 10.1002/viw.20200190.
82. Bolhassani, A., Lipid-based delivery systems in development of genetic and subunit vaccines. *Mol. Biotechnol.*, 65, Dec. 2022. doi: 10.1007/s12033-022-00624-8.

8

Nanoparticle-Based mRNA Vaccines: Are We One Step Closer to Targeted Cancer Therapy?

Lakshmi Vineela Nalla[1]*, Siva Nageswararao Gajula[2] and Vivek P. Chavda[3]

[1]*Department of Pharmacology and Toxicology, GITAM School of Pharmacy, GITAM (Deemed to be University), Rushikonda, Visakhapatnam, Andhra Pradesh, India*
[2]*Department of Pharmaceutical Analysis, GITAM School of Pharmacy, GITAM (Deemed to be) University, Rushikonda, Visakhapatnam, Andhra Pradesh, India*
[3]*Department of Pharmaceutics and Pharmaceutical Technology, L.M. College of Pharmacy, Ahmedabad, Gujarat, India*

Abstract

Messenger RNA (mRNA) has recently become a compelling alternative for treating and preventing cancers with substantial therapeutic outcomes. *In vitro* transcription of mRNA has been developed to mimic natural mRNA's structure for protein replacement therapy, cancer immunotherapy, and vaccination. The systemic distribution of mRNA for safer human use is hampered by various unresolved structural, biological, and technical issues. To address the outstanding challenges and realize the full potential of mRNA therapeutics, advancements in mRNA design and manufacturing and the choice of creative delivery vehicles are required. Despite tremendous attempts to improve mRNA therapeutic intracellular delivery, challenges linked to numerous applications via various pathways persist. Vaccine nanotechnology has considerable promise in treating cancer. Nanotechnologies have the properties required to achieve the ultimate aim of cancer vaccines: ultimately mobilize the immune system's efficacy as a live therapy to detect tumor antigens and eliminate tumor cells. This chapter provides the current state of mRNA treatments, improvements in delivery methods, current translational clinical tractability concerns, and the likelihood of resolving any mRNA-related issues. This also offers new approaches for targeted cancer therapy

**Corresponding author*: lnalla@gitam.edu; vineelavinni154@gmail.com

Vivek P. Chavda and Vasso Apostolopoulos (eds.) Nanocarrier Vaccines: Biopharmaceutics-Based Fast Track Development, (275–304) © 2024 Scrivener Publishing LLC

by summarizing the most recent uses and developments of nanoparticle-based mRNA vaccines in malignancies.

Keywords: mRNA vaccines, cancer therapy, nanoparticles, nanotechnology, *in vitro*-transcribed (IVT) mRNA, delivery system

8.1 Introduction

A ground-breaking category of potent medications to target "undruggable" genes and their related products has been developed due to the increased interest in messenger RNA (mRNA), creating a novel model for treating diseases ranging from COVID-19 to cancers [1]. In cancer treatment, mRNA offers promising therapeutic possibilities for replacing and restoring defective genes in tumor cells that have lost their functionality. Restoring the expression of unmutable cardinal genes via mRNA significantly inhibits the development of cancer cells [2]. Also, CRISPR/Cas9 mRNA brings a fresh perspective to cancer gene therapy by enabling site-specific targeted genome editing to inactivate overexpressed genes linked to tumors specifically [3]. mRNA is a natural biomolecule that serves as a bridge for converting genetic data from genes into useful proteins. It is neither a genetic molecule that lasts forever nor a stable, functional product. Compared to the majority of currently available drugs, the distinct properties of mRNA, such as its temporary nature and adaptability, provide patients with a great deal of optimism for complete therapeutic applications in cancer treatment, protein replacement therapy, and immunization [4]. With the aid of mRNA, the recipient's immune systems can be quickly educated to identify and eradicate cancers. Since its discovery in 1961, mRNA has been extensively investigated [5]. This research has revealed mRNA's structure and biological function, its molecular translational and metabolic routes, and its function in protein synthesis. Several teams attempted in the early 1970s to give naked mRNA directly and use various techniques to convert it into a target protein across multiple cell types [6–8]. The first fruitful translation of rabbit globin mRNA into mouse spleen cells was then reported in 1978 utilizing unilamellar liposomes [9]. Despite the isolated mRNA's effective translation into a protein, many unanswered questions remain. Production cost, poor cellular internalization, biological instability, and immunogenicity hindered the clinical implementation of mRNA treatments. As the first step in developing mRNA therapeutics, the first preclinical research of *in vitro*-transcribed (IVT) mRNA in a mouse model was published in 1990 [10]. After 5 years, the mRNA polynucleotide vaccine was created, and it significantly inhibited the growth of tumors and elicited an immune response against the Carcinoembryonic Antigen (CEA), a tumor-associated antigen [11]. One area of research that has recently received the most attention is finding an ideal way to improve the mRNA vaccines' stability. The best strategy to transfer mRNA is using a substance that can prevent deterioration and encourage efficient cell absorption following injection.

The production of novel materials with 1–100-nm size is called "nanotechnology" per the materials science criteria. Still, further research has broadened the definition of these novel materials [12]. Nanomedicine uses nanoparticles in medical settings for disease diagnosis, detection, control, and therapy [13]. Nanomaterials, which differ from bulk materials in that they have unique physical and chemical characteristics, including a large surface-area-to-volume ratio, strong reactivity, and ultra-small size, can be employed to eliminate the drawbacks of conventional therapeutic reagents [14]. Nanomaterials can be used to design medications with various activities and are similar in size to biomolecules [15]. There is a lot of potential for medical applications of nanotechnology. Due to delicate designs and alterations, nano-drugs have different drug release patterns, a longer half-life, a higher drug loading, and lower cytotoxicity to healthy tissue. They are more bioavailable and selective than conventional drug formats [16]. Studies show that mRNA vaccines based on nanocarriers are frequently used to treat diseases.

A severe disease like cancer places a financial strain on individuals, families, communities, and health systems on a global scale [17]. Microwave therapy, immunotherapy, chemotherapy, radiotherapy, and surgery are some options that can be employed to treat these severe diseases, considerably enhancing the effect of therapeutics and patient quality of life [18]. Unfortunately, due to severe side effects and ineffective treatment, the abovementioned treatments are limited for many individuals. Due to recent advancements in the study of adjuvants and vaccine delivery methods, nanotechnology has become increasingly significant in the field of vaccination [19]. Also, certain mRNA vaccines based on nanocarriers have been effectively used to treat coronavirus disease (COVID-19) [20], substantially increasing public confidence in developing mRNA cancer vaccines. Moreover, research has shown that mRNA vaccines based on nanocarriers may be used as cancer vaccines [21–23]. More research and development of nanocarrier-based mRNA vaccines for cancer might help address mRNA vaccine instability and improve drug treatment targeting [24].

8.2 Use of mRNA in Vaccines: Advantages and Challenges

The extraordinary development in mRNA technology opens up clinical applications for mRNA treatments in regenerative medicine, immunization, genomic editing, cellular reprogramming, and protein replacement therapy. Researchers are drawn to mRNA because of its distinctive properties. The first issue is that it cannot assimilate into the host cell genome, which is a typical issue with DNA and viral therapies. Thus, insertional mutagenesis is not a possibility. Second, unlike other DNA therapeutics that necessitate nuclear envelope collapse during cell division to enter the nucleus, mRNA is translated into the cytosol and does not need to enter the nucleus to perform at its best [25]. Third, mRNA's impact on cells is momentary. Once it has accomplished its purpose, normal physiological mechanisms eliminate it, improving control over pharmacokinetics and dose [26]. Finally, it lessens the chance of tolerance development [27].

Except for mRNA production, there are no cell culture contaminants or protein aggregations because the proteins are produced in the host cells [25]. The cost of producing mRNA is minimal, and the reaction yield is high. Furthermore, it has a greater safety profile because it is manufactured using just animal or cellular components and no hazardous chemicals. Rapid production and scaling of mRNA make it the perfect tool for combating possible pandemics [28]. Protein replacement therapy can bypass the cold chain logistics typically associated with protein products by taking advantage of the ability to store mRNA in the lyophilized form [29]. Traditional recombinant technology has inherent disadvantages, such as long-term storage, cold chain infrastructure for transit, quality control, and the need for manufacturing facilities. Messenger RNA treatments can readily address these issues with protein products used in protein replacement therapy. Other unique properties of mRNA in regenerative medicine include the potential to create stem cells by stimulating the production of homing proteins. Another advantage of mRNA expression over surface modification with proteins or DNA treatment, which may have negative effects owing to the technique's durability and rupture of the cell surface membrane, is that it is transitory [30]. Pluripotency is induced by mRNA, which also prevents residual expression of transgenes caused by insertional mutagenesis when used with viral vectors. Because of the possibility of non-specific editing, mRNA's transient activity makes it suitable for gene editing without the unfavorable and ongoing generation of nucleases [25].

Notwithstanding the benefits of using mRNA to create vaccines and medicines, several issues must be resolved to boost their efficacy. First, estimating the amount of RNA required to provide a therapeutic effect might be challenging, especially in protein replacement therapy [25]. Second, IVT mRNA can act as an intrinsic adjuvant and stimulate the immune system. This property is beneficial for mRNA vaccines but unsuitable for protein replacement applications. Regarding vaccines, IVT mRNA only stimulates $CD8^+$ T cells via cell-mediated immunity, resulting in a weaker immunological response [31]. A recent study, however, shows that mRNA immunization can activate $CD4^+$ T cells. Finding an adjuvant that will not prevent mRNA uptake is another difficult task [27]. Due to blood-borne nucleases and pattern-recognition receptors, mRNA has a brief half-life *in vivo*. Moreover, it is more vulnerable to oxidation than DNA. Hence, a delivery method is needed for many mRNA treatments. Clinical trials have produced moderate outcomes compared to what was anticipated from preclinical research, making it hard to translate mRNA therapics to clinical use [32].

8.3 How Do mRNA Vaccines Work?

Clinical trials have examined a few mRNA vaccines for non-cancerous disorders, particularly in treating COVID-19. Numerous preclinical and clinical research has investigated the use of mRNA vaccines for anticancer purposes, either by direct injection or adopted transfer on dendritic cells (DCs) *in vitro* [33, 34]. The fundamental idea behind using mRNA as a cancer vaccine is that it can carry a target transcript that encodes one or more tumor-specific antigens (TSAs) or tumor-associated antigens (TAAs) into the host cell's cytoplasm, particularly in antigen-presenting cells (APCs), where it will subsequently be expressed as the target antigen [35]. The expressed antigens can be displayed on APC's cellular surface via a major histocompatibility complex, thereby triggering antitumor immunity [36]. The corresponding antigens will be obtained from mRNA through translation when introduced into host cells, mimicking cellular and humoral immunity comparable to viral infection [37]. By raising T cells' antigenic activity, mRNA vaccines help the host's ability to fight off viruses and tumors [38, 39]. In most studies investigating T-cell immunity triggered by antigen-encoded mRNA, researchers utilize autologous DCs loaded with mRNA *in vitro* and reintroduce them into patients. Although DC function *in vitro* needs intricate customized vaccination techniques, which significantly impede their application in people, clinical trials have proved the safety and feasibility of this strategy. So, there is a demand for fresh vaccination strategies that can specifically target DCs *in vivo* [27].

Nanomaterials are frequently used as drug carriers. mRNA therapy provides several benefits over treatments using proteins (including antibodies), viral systems, oligonucleotides, DNA, and small chemicals [40]. mRNA vaccines can concurrently mediate inhibition patterns and stimulation and replace or express faulty proteins, broadening the spectrum of their potential indications in contrast to oligonucleotides and most small molecular pharmacological targets. There is no risk of genome integration with mRNA vaccines compared to DNA vaccines and lentiviral vectors because they need to penetrate the cytoplasmic ribosomal translation machinery rather than the nucleus. Moreover, antigen expression following mRNA vaccination is temporary, preventing T-cell weakening brought on by chronic antigen contact [41]. mRNA vaccines have the benefits of low cost, quick production and purification over protein vaccinations, and simple synthesis [42].

Nanomaterial-based mRNA vaccines are a promising area of research and development in immunotherapy. These vaccines utilize nanomaterials as delivery vehicles for mRNA molecules, enabling targeted and efficient delivery to cells and tissues. The fundamental concept behind nanomaterial-based mRNA vaccines is to use messenger RNA (mRNA) to instruct cells to produce specific proteins that can trigger an immune response. The mRNA encodes the instructions for synthesizing a particular antigen, which, when expressed by the cells, can stimulate the immune system to recognize and target specific disease-associated targets, such as cancer cells. Nanomaterials, typically nanoparticles, serve as carriers for the mRNA molecules. These nanoparticles can comprise various materials, including lipids, polymers, or inorganic materials, such as gold or silica. The choice of nanomaterial is crucial, as it affects the mRNA's stability, biocompatibility, and delivery efficiency. The nanomaterials used in these vaccines offer several advantages. They can protect the fragile mRNA from degradation and enzymatic attack, ensuring its stability during storage and transportation.

Furthermore, nanomaterials can facilitate the targeted delivery of mRNA to specific cells or tissues, enhancing the vaccine's efficacy while minimizing off-target effects. Developing nanomaterial-based mRNA vaccines for various diseases, including cancer, is an active area of research. These vaccines hold great potential for personalized medicine, as the mRNA can be tailored to target specific antigens or mutations associated with individual patient's tumors. Additionally, the modular nature of mRNA allows for rapid adaptation to emerging variants or new targets, providing flexibility in vaccine design. While nanomaterial-based mRNA vaccines offer promising prospects, there are still challenges to overcome. These include optimizing the delivery efficiency, ensuring the long-term stability of mRNA, minimizing potential toxicities, and addressing potential immunogenicity concerns associated with repeated administration. Rigorous preclinical and clinical studies must evaluate these vaccines' safety, efficacy, and immunogenicity profiles.

8.4 Nanocarriers for mRNA Delivery

In the early years, the limited utilization of naked mRNA stemmed from the absence of an effective delivery mechanism despite its ability to generate encoded proteins upon initial injection [10]. Naked mRNA encounters challenges in traversing cell membranes and is susceptible to degradation by extracellular RNases. To successfully function *in vivo*, mRNA delivery systems must overcome various barriers outside and inside cells. These obstacles include enhancing mRNA's ability to enter target tissues, evade degradation by nucleases in physiological fluids, circumvent clearance by the mononuclear phagocytosis system, prevent elimination through glomerular filtration after systemic injection, and evade clearance by the mononuclear phagocytosis system. Thus, the development of efficient mRNA delivery mechanisms becomes crucial. Encapsulating mRNA in a delivery vehicle enables it to reach target cells through various methods depending on the properties of the delivery platform and the specific cell type involved. Endocytosis and micropinocytosis, for example, are two mechanisms by which mRNA given by lipid nanoparticles (LNPs) may be absorbed. Polyplexes enter cells by caveolae-mediated endocytosis, whereas lipoplexes enter via clathrin-mediated endocytosis or cell membrane fusion [43]. mRNA delivery system optimization is critical for the development of mRNA medicines.

The choice of a nanocarrier depends on several factors, including the specific mRNA payload, the target cell or tissue, desired release kinetics, and safety considerations. The goal is to develop nanocarriers that efficiently deliver mRNA to the target site, protect it from degradation, facilitate cellular uptake, and effectively translate the mRNA into proteins. Ongoing research focuses on optimizing the design of nanocarriers, improving their stability, enhancing target specificity, minimizing toxicity, and enabling large-scale production. Clinical trials are being conducted to evaluate the safety and efficacy of nanocarrier-mediated mRNA delivery in various applications, including vaccines, gene therapies, and cancer treatments. Nanocarriers are critical to protecting and delivering mRNA molecules to target cells or tissues. Lipid, polymeric, and inorganic nanoparticles are among the most commonly studied nanocarriers for mRNA delivery. Continued research in this field aims to overcome challenges and further advance the development of safe and effective nanocarriers for mRNA-based therapeutics. Various types of nanocarriers have been investigated for mRNA delivery, including lipid nanoparticles, polymeric nanoparticles, and inorganic nanoparticles.

8.4.1 Liposomes and RNA Lipoplexes

Lipids have always been considered promising substrates for mRNA transfer because they are biodegradable and selectively electrical. They have the opposite electrical charge that RNA does. As a result, cationic lipids are the favored method of transporting nucleic acids. Liposomes, spherical lipid bilayers that enclose a chamber capable of containing aqueous solutions, often incorporate cationic lipids like DOTMA and DOTAP [44, 45] to bind RNA that have a negative charge, along with specific helper lipids, thus forming the lipid bilayer structure.

Cationic liposomes combined with RNA are called RNA lipoplexes (RNA LPX). Liposomes are lipid-based nucleic acid carriers that are used as transfection agents *in vitro*. Their positive surface charges enable them to create complexes (continuous bilayer structures or in the form of spherical structures) with negatively charged nucleic acids. Because of its efficacy in delivering DNA, LPX was initially employed for RNA delivery in 1989 and is currently often used for mRNA vaccine distribution. In creating RNA-LPX, RNA is diluted with liposomes in a sodium chloride solution and ethanol using a predetermined charge ratio [46].

Target specificity might change if the ratio of cationic lipids to RNA changes. The spleen may be delivered systematically if the concentration of cationic lipids is reduced. When administered intravenously, this enhanced mRNA-LPX could be specifically targeted to the spleen DCs (IL-6, IL-12, and cytokines, such as TNF) and act as a vaccine for cancer [46]. This formulation treats melanoma by delivering tumor-associated antigens (TAAs) [47] and autoimmune encephalomyelitis by providing autoantigens [48]. There are numerous ways to give LPX mRNA, with intravenous injection being the most popular. The most efficient delivery and good immunological organ specificity were demonstrated by intravenous injection. The systematic distribution of cancer vaccines is the principal application for LPX mRNA, and several projects are undergoing clinical testing [49].

8.4.2 Lipid Nanoparticles

Lipid nanoparticles (LNPs) are nanoscale structures composed of amphipathic phospholipid molecules that self-assemble to form lipid vesicles. With controlled release characteristics, outstanding biocompatibility, and minimal toxicity, LNPs are attractive nanocarriers for delivering nucleic acids [50]. LNPs are essential ingredients in both mRNA medicines and vaccinations. LNPs contain important characteristics that may impact the effectiveness of immune activation, including adjustable size, shape, and charge. LNP modification can produce the best immune reactions [51]. LNPs can deliver numerous antigens and adjuvants simultaneously when used as nanovaccines. Moreover, LNPs' membrane surfaces can show antigens with improved natural conformation representation.

LNPs have demonstrated excellent potential for the development of nanovaccines in a variety of clinical and preclinical applications. As was already established, LNPs are essential for effectively delivering mRNA vaccines by shielding them from nucleases. According to a recent assessment, numerous other LNP-mRNA formulations are currently undergoing clinical trials to prevent and treat severe dangers to human health, like cancer, viruses, and genetic disorders [43]. Cationic lipids, ionizable lipids, and other lipid types can also be used as LNP components. Furthermore, modification allows for the functionalization of lipids, including PEGlyation, which increases the flexibility and potency of LNPs for vaccine production [4].

8.4.3 Polymer-Based Nanoparticles

Colloidal systems of polymeric nanoparticles have a broad size range (10-1000 nm) [52]. Polymeric nanoparticles can be injected into the core and coupled to the surface and have good immunogenicity and stability for effective antigen encapsulation and presentation. Polymeric nanoparticles can be self-adjuvant, have an adjustable size, and are typically solid [53]. Polymeric nanoparticles can increase the phagocytosis or endocytosis efficiency of antigen uptake by APCs [54]. Natural polymeric nanoparticles like dextran and chitosan and manufactured polymeric nanomaterials like PLGA and PLA are both helpful for creating nanovaccines. Natural polymeric nanoparticles are cost-effective, soluble in water, and highly biocompatible. For instance, the linear cationic polysaccharide chitosan can be utilized to administer vaccines. Chitosan, with its cationic charge and bioadhesive properties, is a suitable option for gene delivery and coating other polymeric nanoparticles to enhance adhesion and immunogenicity [55]. Chitosan can also be modified based on purpose by adding functional groups [56]. Synthetic polymeric nanoparticles typically have higher repeatability and better control over molecular weight compositions and degradation rates than natural polymers [57]. For instance, PLGA nanoparticles are highly biodegradable and have adjustable characteristics. PLGA and PEG can be linked to form a polymeric micelle for improved T-cell responses and hydrophobic peptide antigen delivery [58].

8.4.4 Hybrid Nanoparticles

Polymeric nanoparticles (NPs) and liposomes are the most often used nanocarriers for mRNA delivery, and both have shown remarkable effectiveness in preclinical and clinical studies. Although liposomes and polymeric nanoparticles have demonstrated efficacy in mRNA delivery, they also possess limitations, including challenges in escaping the endosomal compartment within cells, instability in complex biological environments, and toxicity. Various research groups have proposed hybrid composites comprising peptide molecules, inorganic nanoparticles, polymers, or lipids to address these issues. By harnessing the advantages of each carrier independently, these hybrid systems aim to provide potential solutions to overcome the challenges above. The created hybrid nanoparticles improve the transfection effectiveness of the desired protein while lowering the toxicity of the currently used vectors. For instance, after systemic injection in a mouse, a combination system of L-histidine-(N,N-di-n-hexadecylamine) cationic liposomes (lipopolyplexes) and PEGylated cationic polymer histidylated polylysine considerably slowed the growth of B16 melanoma. The rapid liberation of encoded mRNA is facilitated by leveraging the pH-dependent release properties of histidylated cationic polymer and lipids within the acidic compartment of early endosomes. This pathway triggers the generation of cytotoxic T lymphocytes (CTLs) that play a pivotal role in eliciting antitumor immune responses [59]. To enhance this process, the authors introduced 11 mol% of -D-mannopyranosyl-N-dodecylhexadecanamide (Man-lipid) injections into cationic histidine-(N,N-di-n-hexadecylamine) ethylamide cationic liposomes (lipopolyplexes). Mannosylated liposomes (Man11-Lip100) were created due to this technique to increase the therapeutic potency of MART-1 mRNA-loaded vaccinations and the transfection of DCs [60]. After intravenous treatment, splenic DCs (CD11c + cells) that were prompted to express EGFP were four times more numerous and larger than non-mannosylated liposomes.

8.5 Nanoparticle-Based mRNA Vaccines in Cancer Therapy

Human mortality is still primarily caused by cancer [61–63]. Developing an anticancer vaccine is a crucial first step to treating this widespread disease with customized medicine. Full clarification of the cancer pathophysiology is complex despite great efforts. Cancerous cells typically develop due to various environmental and genetic causes mutating healthy cells. Thus, unlike infectious diseases, cancer is often heterologous and challenging to avoid. The development of vaccines is vital for eradicating leftover malignant cells and inhibiting tumor recurrence and metastasis since even tiny numbers of malignant cells can cause relapse. Nanomedicine can battle cancer by utilizing the nanomaterials listed in this study to enhance tumor vaccine delivery, *in situ* tumor antigen exposure, or the presentation of the immune activation program. Several nanomaterials have been investigated as effective delivery systems for tumor vaccinations.

Nanoparticle-based mRNA vaccines offer a higher level of selectivity compared to alternative treatments. By minimizing the non-specific absorption of therapeutic substances in healthy tissues and utilizing nanomaterials for drug delivery, the potential for side effects can be reduced. As a result, these vaccines are particularly suitable for individuals with compromised physical conditions who may be unable to tolerate conventional treatment. Moreover, targeted therapy is a successful approach to treating tumors. The inherent carriers hold substantial promise for delivering therapeutic nanoparticles due to their favorable minimal toxicity and biocompatibility. Anticancer medicines can be wrapped in nanocarriers as an effective drug delivery method to the lesion.

8.5.1 Breast Cancer

Breast cancer (BC), a highly heterogeneous disease, is the second most common neoplasia and the leading cause of mortality in women [64–67]. Research has shown that 60–70% of these individuals die due to the condition. Findings also indicate that from 2014 to 2018, breast cancer in women gradually increased (0.5% annually). The World Health Organization estimates that 2.3 million females were diagnosed with BC in 2020 and that

685,000 fatalities were recorded worldwide. Early breast cancer detection can lower treatment costs and mortality rates. Thus, it is critical to develop accurate breast cancer diagnosis techniques. Recent advancements in analytical techniques, especially LC-MS [68–70], enable the identification and quantification of analytes even at low concentrations [71–73]. These advancements help diagnose diseases through biomarkers [74–76].

On the other hand, using the immune system to destroy cancerous cells is a revolutionary therapeutic approach. Active immunotherapies can reset the immune system to an antitumor surveillance state, protecting against neoplastic tissue. A promising area of medical study is RNA therapy to prevent and treat breast cancer. RNA-based medications can potentially be effective drug regulators against cancer cells by altering the expression of particular proteins. These characteristics make high specificity and a low chance of off-target effects possible. Recently, Liu *et al.* discovered that MUC1-based mRNA vaccination effectively induces a robust cytotoxic T lymphocyte (CTL) response against triple-negative breast cancer (TNBC). The combination of mRNA vaccines with ipilimumab, an anti-cytotoxic T lymphocyte-associated protein 4 (CTLA-4) monoclonal antibody (mAb), yielded enhanced T-cell immunity compared to either the single mRNA vaccine or the mAb alone.

Furthermore, combining lipid nanoparticle-based mRNA with chemotherapeutic medicines can enhance the therapeutic effect against aggressive targets, particularly in triple-negative breast cancer (TNBC) cases.

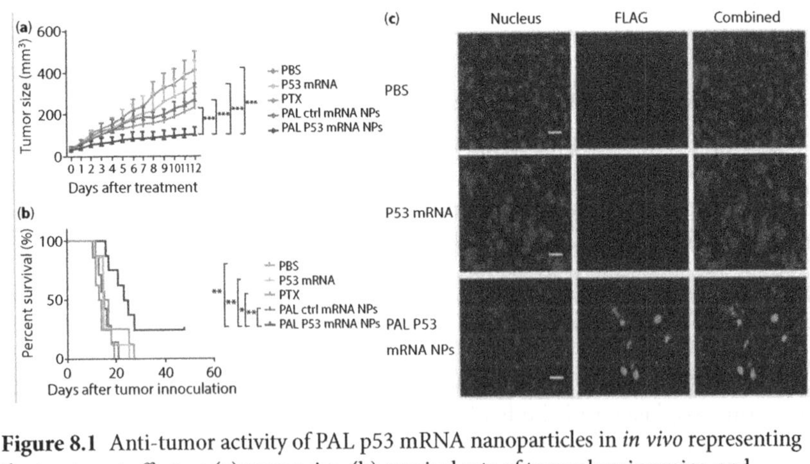

Figure 8.1 Anti-tumor activity of PAL p53 mRNA nanoparticles in *in vivo* representing the treatment effect on (a) tumor size, (b) survival rate of tumor bearing mice, and (c) immunohistochemistry of p53 expression in isolated tumors.

Zhang *et al.* developed PAL P53 mRNA NPs by conjugating paclitaxel with amino lipid nanoparticles (PAL) via an ester bond, including p53 mRNA [77–79] and incorporating modified amino lipid analogs, and significantly improved paclitaxel's stability, encapsulation efficiency, and release profile. The interaction between paclitaxel and p53 mRNA demonstrated anticancer activity in an orthotopic TNBC model, leading to an extended median survival time (Figure 8.1).

8.5.2 Colorectal Cancer

The number of new colorectal cancer cases and fatalities in 2018 was predicted to be above 1.8 million and 881,000, respectively [80]. Most stage I or stage II colorectal cancer patients today receive surgical treatment, and total mesenterectomy is the go-to surgical surgery for colorectal cancer. For patients with metastatic colorectal cancer, a standard recommendation often involves the utilization of systemic chemotherapy or a combination of targeted biologics over surgery and chemotherapy for stage III colorectal cancer patients [81].

Precision therapy has been employed to treat colorectal cancer in recent years due to the tight connection between the occurrence of colorectal cancer

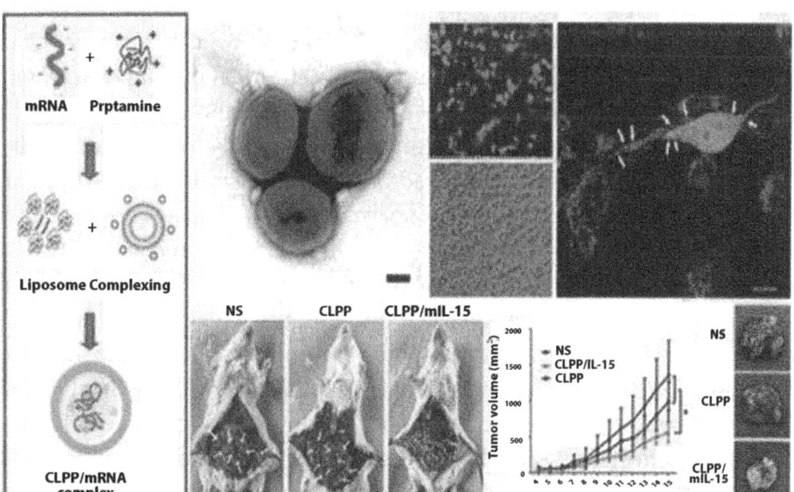

Figure 8.2 Represents the characterization and colocalization studies of CLPP/mIL-15 complex using TEM and immunocytochemistry respectively. The anti-tumor activity of CLPP/mIL-15 complex inhibited the metastasis of abdominal cavity and C26 colon carcinoma.

and heterogeneity [82]. LNP-based mRNA vaccines offer special benefits for treating colorectal cancer [83]. A liposome/protamine system (CLPP) nano-delivery system was developed by Lei *et al.* to deliver IL-15-encoded mRNA for colon cancer gene therapy [84]. The results demonstrated that the ready CLPP system could deliver IVTIL-15mRNA into CT26 colorectal cancer cells safely and efficiently. Local and systemic treatment of the CLPP/mIL-15 combination revealed substantial anticancer effects through multiple pathways in a C26 colorectal cancer cell model of subcutaneous, abdominal, and lung metastasis. In the CT26 abdominal metastasis model, subcutaneous metastatic model, and lung metastasis model, 70%, 55%, and 69% inhibition rates, respectively, were found following local or systemic treatment of the CLPP/mIL-15 combination. They are all quite reliable and secure. Following immunization, lymphocyte activity was markedly boosted, dramatically increasing the number of $CD8^+$ T cells. These results demonstrated the great therapeutic potential of the CLPP/mIL-15 complex in the immune gene therapy of colorectal cancer, demonstrating the suitability of CLPP for mRNA transport and the potential of the CLPP/mIL-15 complex for cancer immune gene therapy (Figure 8.2).

8.5.3 Lung Cancer

The most often diagnosed cancer and the main reason for cancer-related fatalities is lung cancer [80]. The most prevalent form of lung cancer, non-small cell lung cancer (NSCLC), is detected in 40% of cases with stage IV metastatic disease. Treatment for lung cancer typically involves a combination of chemotherapy, surgery, or radiotherapy, depending on the kind and stage of malignant tumors. Unfortunately, due to inadequate early detection, most patients with lung cancer have their disease discovered at an advanced stage, with local tumor invasion or distant metastases. Therefore, they are not candidates for surgical intervention. Patients with stage IV lung cancer have an average survival duration of 4 months as the disease advances. Hence, it is essential to step in as soon as possible. Advanced NSCLC is now mostly treated with systemic cytotoxic therapy, although chemotherapy's therapeutic effects have peaked, necessitating the development of novel treatment modalities [85]. The standard treatment for advanced lung cancer is systemic chemotherapy, which increases patients' chances of survival and quality of life [86]. Platinum-based medications are a part of the typical first-line chemotherapy regimen for lung cancer [87]. However, the dose-limiting adverse effects of platinum-based chemotherapy include peripheral neuropathy, intestinal damage, anemia, nephrotoxicity, and cardiotoxicity.

Other symptoms include nausea, restlessness, and exhaustion. Patients with molecular anomalies addressed by new drugs have a median survival rate of only approximately 2–3 years, but the survival rate of standard chemotherapy combinations has stabilized [88].

Limited preclinical trials are investigating LNP-based mRNA vaccines for lung cancer. In a Phase Ib clinical study conducted by Papachristofilou *et al.*, a combination of local radiation therapy with or without pemetrexed, along with BI1361849 (an immunotherapeutic comprising five antigens encoded by CV9202), demonstrated good tolerance, minimal side effects, and the ability to trigger a targeted immune response in patients with stage IV non-small cell lung cancer (NSCLC). Immunization increased functional CD4+ and CD8+ T cells in 40% of patients, exceeding baseline levels by at least two-fold [89]. This therapy proved that BI1361849 (CV9202) and immune checkpoint inhibitors benefited NSCLC patients. Similarly, in a clinical phase I/IIa study conducted by Sebastian *et al.*, it was observed that antigen-specific immunotherapy (CV9201) was well tolerated and led to an enhanced immune response in patients diagnosed with stage IIIB/IV NSCLC [90]. In a promising phase I/II research (NCT03164772) [91], the effectiveness and safety of mRNA vaccines (BI 1361849) in combination with anti-CTLA-4 (tremelimumab) and anti-PD-L1 (durvalumab) checkpoint inhibitors were evaluated for the treatment of NSCLC. These studies recommended the value of immune checkpoint inhibitors in conjunction with mRNA-based immunotherapy for treating NSCLC.

8.5.4 Glioma Tumor

Neurotumors are tumors that affect either the central nervous system or the peripheral nervous system. After a certain level of nerve injury, poor nerve healing plays a significant role in developing malignant tumors [92]. Central nervous system tumors comprise a small minority of neurotumor cases, but their management is complex. Due to the central nervous system's deep placement, surgical access to the tumor site is difficult. As a result, there is an urgent need to investigate the regulatory mechanism and look for glioma tumor biomarkers because the existing treatment still has many limitations.

The major hurdle in managing brain tumors (BT) is the molecule's capability to cross the blood–brain barrier (BBB). Often, because of the distortion of BBB, the blood–tumor barrier is also formed, which adds to the hurdles of transportation of therapeutic moiety into the brain. Several mechanisms, including passive diffusion, endocytosis, carrier-mediated transportation, and active transport, are used to move the chemical into the BBB or BTB. Seldom is the process known as passive diffuse reliant on concentration gradient [93]. Inorganic nanoparticles such as silica, carbon, and metal oxides are involved in the imaging techniques of BT. Inorganic nanoparticles have been shown in multiple studies to be effective enough to cross the BBB. Still, their undesired drug delivery and severe degradability have prevented them from being used in clinical trials [94]. Exosomes are the most rapidly growing field, upholding various advantages over other nanoparticles [95]. Innovative exosome research has been used in several investigations, including those involving RNA therapies.

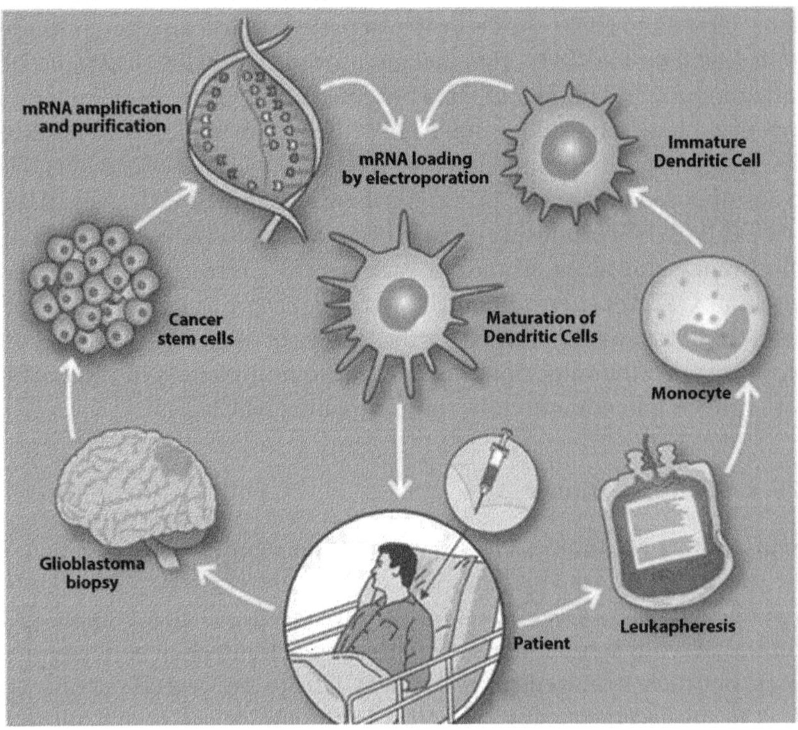

Figure 8.3 An overview of DCs production to target GSCs. *Left circle*: After performing the biopsy of the glioblastoma patient, the isolated CSCs tumorsphere cultures should be processed for RNA purification once the sphere reached the tertiary sphere formation. The RNA was amplified to mRNA using the strand switch method. Right circle: Prior to starting radio-chemotherapy, the patient underwent leukapheresis to extract monocytes. Autologous monocytes were then developed into immature dendritic cells in vitro. Centre: Electroporation was used to transfect dendritic cells with autologous glioblastoma stem cell mRNA. Below: After EORTC-recommended chemo-radiotherapy, matured dendritic cells expressing glioblastoma stem cell antigens were injected intradermally five times for induction throughout the first three weeks and thereafter monthly for up to 18 vaccinations.

Exosomes carrying mRNA suppressed glioma tumor growth by preventing and reducing vasculogenic mimicry, migration, and angiogenesis [96].

mRNA can be employed to address genetic diseases or facilitate tissue repair due to its ability to express specific functional proteins. Additionally, it can be utilized in immunotherapy to generate antigens, antibodies, or receptors. Despite not being developed enough, these treatments provide many research possibilities. A single-cell suspension was created by Vik-Mo *et al.* after they successfully extracted brain tumor biopsy tissues [97]. In this study, autologous glioma stem cells (GSCs) were cultivated to form tumor spheres *in vitro*. Subsequently, GSC-mRNA was amplified and transfected into autologous dendritic cells derived from monocytes. These dendritic cells, carrying autologous GSC-mRNA, were then utilized to elicit an immune response explicitly targeting the patient's GSCs. Compared to somatic brain progenitor cells, GSCs exhibited higher levels of telomerase activity and survivin, an apoptosis-inhibiting protein. After receiving therapy with a DC vaccine that targets GSCs in solid tumors, patients with glioma had a 2.9-fold improvement in progression-free survival. Specific-induced lymphocyte proliferation was detected in all seven patients after stimulation with tumorsphere lysate, hTERT, or survivin peptides (Figure 8.3).

8.5.5 Other Tumors

Provenge (Sipuleucel-T), an immune cell-based vaccination, became the first therapeutic cancer vaccine approved by the US Food and Drug Administration in 2010 [98]. It was specifically developed for patients with hormone-resistant prostate cancer. Huang *et al.* identified CD247, FCGR1A, and TRRAP as potential antigens for mRNA vaccines against cholangiocarcinoma in patients with immune subtype 2 (IS2) tumors. This discovery laid the foundation for developing anti-cholangiocarcinoma vaccines and identifying suitable targets for vaccination [99]. Also, according to Huang *et al.*, patients with the IS4 and IS5 tumors could benefit from the anti-pancreatic cancer mRNA vaccine, which contained the antigens TMOD3, ADAM9, WNT7A, EFNB2, TPX2, and MET [100]. Prostate cancer can also be treated with anticancer vaccinations CV9103 and CV9104 [101].

8.6 Clinical Trials

With the validation of the data from preclinical, *in vitro*, or *ex vivo* techniques, it is crucial to carry out clinical research on humans as a final stage

Table 8.1 Clinical trials of mRNA-based vaccines for various cancers.

Other Identifiers (IDs)	Disease	Number of enrolled	Route of administration	NCT number
mRNA-2416-P101	• Ovarian cancer • Refractory/Relapsed (R/R) solid tumor	79	Intertumoral (IT)	NCT03323398
mRNA-2752-P101 2022-001597-55	• R/R solid tumor • Breast cancer • Urothelial cancer	264	IT	NCT03739931
180074 18-C-0074	• Hepatocellular carcinoma • Colon cancer • Melanoma • Gastrointestinal cancer • Genitourinary cancer	5	Intramuscular (IM)	NCT03480152
CV-9103-001	• Prostate cancer	48	Intradermal (ID)	NCT00831467
Pro00003877	• Malignant neoplasms of the brain	42	ID	NCT00639639

before commercial use of any therapeutic product. Refer to Table 8.1 for the details of national clinical trials (NCTs).

8.6.1 Considerations for Clinical Translation

For the intended application, the characteristics of nanoparticle-mRNA formulations must be precisely specified. To get the best therapeutic results, nanoparticle-mRNA formulations may require different features than they do as medicines or vaccines, such as differing biodistribution profiles. Therapeutics are directed at specific organs, whereas vaccines must interact with immune cells. Nanoparticle-mRNA formulations should therefore be created in accordance with biomedical requirements. Good manufacturing practice (GMP) is essential to transform nanoparticle-mRNA systems from bench to bedside to ensure medication purity, therapeutic benefits, and factors like safety profiles and storage conditions.

The first stage in the GMP synthesis of mRNA is creating a linearized DNA template. The mRNA is subsequently translated *in vitro* using the DNA template as a guide in the presence of an RNA polymerase and ribonucleoside triphosphates. DNase digestion removes the remaining DNA template, then chemical or enzymatic processes cap the mRNA units. Eventually, double-stranded RNA, free nucleotides, shortened nucleic acid fragments, enzymes, and precipitation procedures based on microbeads are used to purify the mRNA. The pure mRNA can be filtered for sterility, dissolved in a storage solution, and frozen for long-term storage [25, 32, 102, 103]. Various strategies have been investigated to enhance the structural components of mRNA to promote stability and translational effectiveness.

To choose the most suitable mRNA constructs for cancer therapy, it is crucial to comprehend the molecular mechanisms underlying cancer development and identify critical cancer-associated genes involved in tumor growth and metastasis. Once these genes are identified, mRNA can be modified to enhance gene expression or induce immune responses against the tumor. It is essential to explore sequential optimization and modifications within the Cap structures, Untranslated Region (UTRs), poly-A tails, and the inclusion of modified bases to develop stable and well-tolerated next-generation mRNA therapeutics with optimal translational efficiency. Investigating the structure-activity relationship (SAR) of *in vitro* transcribed (IVT) mRNA plays a vital role in this process. To receive speedy clinical approval, it may be necessary to guarantee the industrial flexible purification processes, standard quality control, and impurity content during the IVT manufacture. Also, by simply altering the coding sequence, researchers may concentrate on using the same essential mRNA-building component to treat various disorders. It is crucial to allocate increased efforts towards developing cost-effective synthetic approaches and manufacturing technologies to enable high-yield production of mRNA pharmaceuticals at an affordable price.

Several transporters have been investigated in preclinical and clinical investigations to deliver mRNA for cancer treatment based on the current situation. In contrast to animal research, there are fewer clinical approvals and a lower rate. Thus, some areas need to be the focus of scientists, including: (i) nanocarriers made from FDA-approved materials to accelerate rapid succession; (ii) significant advancements in formulation techniques, the biodegradability of nanocarriers, and commercially viable industrial manufacture; (iii) sufficient stability of nanoparticle-mRNA complexes in systemic circulation and effective targeting ability to prevent medication loss and off-target distribution; (iv) explanation of the molecular routes of cellular internalization and release mechanism from delivery payloads at endosomal stages since they are essential to the pharmacological action of pharmaceuticals; and (v) analysis of data from animal and human clinical studies to optimize the dose and toxicity of nanocarriers.

Lipid nanoparticles play a more valuable role among mRNA vaccine carrier systems. Previously, reverse-phase evaporation and thin-film hydration created lipid nanoparticle–nucleic acid formulations [104]. By using extrusion methods, the diameters of lipid nanoparticles were further homogenized [104]. The technique of rapid mixing is commonly employed to formulate lipid nanoparticle-mRNA complexes. It combines an ethanol phase containing the lipid components with an aqueous phase containing the mRNA molecules under a controlled flow rate and pH conditions. This method can produce lipid nanoparticle–mRNA formulations with exceptional encapsulation efficiency and uniform size distribution, ensuring the repeatability and scalability of the manufacturing process [105, 106]. As mRNA is easily degraded, formulation buffers must be free of ribonuclease contamination [25, 32]. Enrichment can enhance the final mRNA concentration after the lipid nanoparticle–mRNA formulations undergo additional processes to eliminate organic solvents and residual components. Subsequently, the filtered and frozen lipid nanoparticle-mRNA formulations undergo tests compliant with good manufacturing practice (GMP). These tests encompass evaluations of physical attributes like charges, particle sizes, mRNA encapsulation, and compendial tests, including osmolality, particulate matter, bacterial endotoxins, sterility, and other quality assessments.

The long-term stability of lipid nanoparticle–mRNA formulations is influenced by storage conditions, such as lyophilized storage, freezing, or aqueous storage, and the choice of cryoprotectants, such as mannitol, trehalose, or sucrose. These factors need careful consideration before their clinical application [107]. For example, when lipid nanoparticle–mRNA formulations are stored in liquid nitrogen, adding 5% w/v trehalose or sucrose can maintain mRNA delivery efficacy for at least 3 months [107]. Notably, the approved COVID-19 mRNA vaccines are stored at freezing temperatures with the addition of sucrose. However, there are differences in their handling and storage requirements. While BNT162b2 needs to be thawed and diluted with saline before administration, mRNA-1273 can be stored at 15°C to 20°C and administered immediately after thawing [106].

Lipid nanoparticle–mRNA compositions' safety profile correlates with their mRNA molecules and lipid constituents. After systemic or local injection, lipid components may stimulate the human immune system; for instance, PEG-lipids may cause hypersensitivity reactions by activating the complement system. Moreover, by accelerating blood clearance, anti-PEG antibodies may cause quick systemic clearance of laterally given PEGylated nanoparticles. The drug encapsulated in PEGylated nanoparticles' bioavailability and biodistribution may vary due to rapid blood clearance, leading to adverse effects. Several natural and synthetic polymers, which are now being tested in clinical studies, have been researched as PEG substitutes to alleviate safety concerns [108, 109]. It has also been noted that ionizable and cationic lipids enhance the release of pro-inflammatory cytokines and reactive oxygen species. The complement system and Toll like receptors may contribute to the activation of the innate immune system, although the immunogenicity of these lipids is not yet fully known. Depending on the dose, the cytotoxicity of lipid compounds, cell types, and lipid characteristics is also a safety concern. Lipid nanoparticles have been used *in vivo* to cause liver and lung damage in mice. This damage may be caused by the materials' cytotoxicity and the production of pro-inflammatory factors [110, 111]. Biodegradable lipids may be used to increase the biocompatibility of lipid nanoparticles.

By accumulating enough human data from current clinical trials of nanoparticle-based mRNA vaccines and translational research, the scientific community may launch next-generation nanoparticle–mRNA treatments to eliminate cancer.

8.7 Conclusion

Instances of insertional mutagenesis are less likely because of mRNA's distinct and fleeting expression. Synthetic mRNA has drawn much attention from biology, medicine, and material science researchers since the discovery of *in vitro* transcribed mRNA. Immunotherapies, cell reprogramming, protein replacement therapies, and gene editing have all undergone revolutionary changes due to various mRNA therapeutic applications. Significant advancements further overcome the instability and immunogenicity of mRNA therapeutics in mRNA design, structure, and chemical changes. The extraordinary COVID-19 pandemic has ushered in a new era of mRNA vaccines. However, some significant issues must be resolved before the technology can treat other genetic illnesses like cancer. The main challenge is finding realistic and scalable mRNA synthesis methods that work with current pharmaceutical technologies. Clinically applicable mRNA transporters, appropriate systemic circulation stability, and immunogenicity resolution are requirements for effective mRNA therapies.

In recent years, there has been growing interest in developing nanomaterial-based mRNA vaccines for cancer treatment. These innovative vaccines utilize nanoparticles as delivery systems for mRNA, aiming to stimulate an immune response against cancer cells. Nanoparticle-based mRNA vaccines offer several advantages in cancer treatment. Using nanoparticles as carriers enhances the stability and protection of mRNA molecules, ensuring their efficient delivery to target cells. Additionally, nanoparticles can be engineered to have specific properties, such as improved cellular uptake and controlled release, enhancing the vaccine's effectiveness. Clinical trials investigating nanoparticle-based mRNA vaccines for cancer treatment have been conducted and are underway. These trials aim to assess these novel vaccines' safety, dosage, and effectiveness. By leveraging the potential of mRNA technology and nanomaterials, researchers and clinicians strive to develop targeted and personalized cancer therapies that harness the patient's immune system to combat the disease. However, it is crucial to note that nanoparticle-based mRNA vaccines for cancer treatment are rapidly evolving.

In conclusion, nanoparticle-based mRNA vaccines hold promise as a potential breakthrough in cancer treatment. By combining the advantages of mRNA technology and nanomaterials, these vaccines aim to enhance the immune response against cancer cells, offering new possibilities for targeted and personalized therapies. Clinically advanced nanomaterials improve mRNA-based nanotherapeutics for the treatment of cancers. As a result of their apparent advantages, lipid nanoparticle-based mRNA vaccines have demonstrated promise in the treatment and prevention of tumors. They are anticipated to be a viable strategy in several cancer therapies. Furthermore, it is conceivable that LNP-based mRNA vaccines will soon play a significant role in illness diagnosis and prognosis. Ongoing research and clinical trials will continue to shed light on the safety, efficacy, and practicality of these innovative approaches, potentially paving the way for more effective cancer treatments in the future.

References

1. Chavda, V.P. *et al.*, mRNA-based vaccine for COVID-19: They are new but not unknown! *Vaccines*, 11, 3, 507, 2023.
2. Xiao, Y. *et al.*, Combining p53 mRNA nanotherapy with immune checkpoint blockade reprograms the immune microenvironment for effective cancer therapy. *Nat. Commun.*, 13, 1, 758, 2022.
3. Zhang, D. *et al.*, Enhancing CRISPR/Cas gene editing through modulating cellular mechanical properties for cancer therapy. *Nat. Nanotechnol.*, 17, 7, 777–787, 2022.
4. Hajj, K.A. and Whitehead, K.A., Tools for translation: Non-viral materials for therapeutic mRNA delivery. *Nat. Rev. Mater.*, 2, 10, 1–17, 2017.
5. Brenner, S., Jacob, F., Meselson, M., An unstable intermediate carrying information from genes to ribosomes for protein synthesis. *Nature*, 190, 576–581, 1961.
6. Lane, C.D., Marbaix, G., Gurdon, J.B., Rabbit haemoglobin synthesis in frog cells: The translation of reticulocyte 9 s RNA in frog oocytes. *J. Mol. Biol.*, 61, 1, 73–91, 1971.
7. Maeyer-Guignard, J.D., Maeyer, E.D., Montagnier, L., Interferon messenger RNA: Translation in heterologous cells. *Proc. Natl. Acad. Sci.*, 69, 5, 1203–1207, 1972.
8. Furusawa, M. *et al.*, Injection of foreign substances into single cells by cell fusion. *Nature*, 249, 449, 450, 1974.
9. Dimitriadis, G.J., Translation of rabbit globin mRNA introduced by liposomes into mouse lymphocytes. *Nature*, 274, 5674, 923, 924, 1978.
10. Wolff, J.A. *et al.*, Direct gene transfer into mouse muscle *in vivo*. *Science*, 247, 4949, 1465–1468, 1990.
11. Conry, R.M. *et al.*, Characterization of a messenger RNA polynucleotide vaccine vector. *Cancer Res.*, 55, 7, 1397–1400, 1995.
12. Liu, Y., Miyoshi, H., Nakamura, M., Nanomedicine for drug delivery and imaging: A promising avenue for cancer therapy and diagnosis using targeted functional nanoparticles. *Int. J. Cancer*, 120, 12, 2527–2537, 2007.
13. Kim, B., Rutka, J., Chan, W., Effect of Coumarate 3-zhydroxylase Down regulation on lignin structure. *Nanomed. N. Engl. J. Med.*, 363, 2434–2443, 2010.
14. Zhang, L. *et al.*, Nanoparticles in medicine: Therapeutic applications and developments. *Clin. Pharmacol. Ther.*, 83, 5, 761–769, 2008.
15. Cullis, P.R. and Hope, M.J., Lipid nanoparticle systems for enabling gene therapies. *Mol. Ther.*, 25, 7, 1467–1475, 2017.

16. Cheng, Z. *et al.*, Nanomaterials for cancer therapy: Current progress and perspectives. *J. Hematol. Oncol.*, 14, 1, 1–27, 2021.
17. Siegel, R., Miller, K., Fuchs, H., Jemal, A., Cancer statistics, 2020. *CA Cancer J. Clin.*, 70, 1, 7–30, 2020.
18. Wang, Y. *et al.*, Application of artificial intelligence to the diagnosis and therapy of colorectal cancer. *Am. J. Cancer Res.*, 10, 11, 3575, 2020.
19. Chauhan, G. *et al.*, Nanotechnology for COVID-19: Therapeutics and vaccine research. *ACS Nano*, 14, 7, 7760–7782, 2020.
20. Garcia-Beltran, W.F. *et al.*, COVID-19-neutralizing antibodies predict disease severity and survival. *Cell*, 184, 2, 476–488.e11, 2021.
21. Billingsley, M.M. *et al.*, Ionizable lipid nanoparticle-mediated mRNA delivery for human CAR T cell engineering. *Nano Lett.*, 20, 3, 1578–1589, 2020.
22. Zhu, G. *et al.*, Efficient nanovaccine delivery in cancer immunotherapy. *ACS Nano*, 11, 3, 2387–2392, 2017.
23. Verbeke, R. *et al.*, Broadening the message: A nanovaccine co-loaded with messenger RNA and α-GalCer induces antitumor immunity through conventional and natural killer T cells. *ACS Nano*, 13, 2, 1655–1669, 2019.
24. Zhong, X. *et al.*, Facile synthesis of porous germanium-iron bimetal oxide nanowires as anode materials for lithium-ion batteries. *Nano Res.*, 11, 3702–3709, 2018.
25. Sahin, U., Karikó, K., Türeci, Ö., mRNA-based therapeutics—Developing a new class of drugs. *Nat. Rev. Drug Discovery*, 13, 10, 759–780, 2014.
26. Petsch, B. *et al.*, Protective efficacy of *in vitro* synthesized, specific mRNA vaccines against influenza A virus infection. *Nat. Biotechnol.*, 30, 12, 1210–1216, 2012.
27. Pollard, C. *et al.*, Challenges and advances towards the rational design of mRNA vaccines. *Trends Mol. Med.*, 19, 12, 705–713, 2013.
28. Bahl, K. *et al.*, Preclinical and clinical demonstration of immunogenicity by mRNA vaccines against H10N8 and H7N9 influenza viruses. *Mol. Ther.*, 25, 6, 1316–1327, 2017.
29. Kwon, H. *et al.*, Emergence of synthetic mRNA: *In vitro* synthesis of mRNA and its applications in regenerative medicine. *Biomaterials*, 156, 172–193, 2018.
30. Kauffman, K.J., Webber, M.J., Anderson, D.G., Materials for non-viral intracellular delivery of messenger RNA therapeutics. *J. Control. Release*, 240, 227–234, 2016.
31. Painter, M.M. *et al.*, Rapid induction of antigen-specific CD4+ T cells is associated with coordinated humoral and cellular immunity to SARS-CoV-2 mRNA vaccination. *Immunity*, 54, 9, 2133–2142.e3, 2021.
32. Pardi, N. *et al.*, mRNA vaccines—A new era in vaccinology. *Nat. Rev. Drug Discovery*, 17, 4, 261–279, 2018.
33. Islam, M.A. *et al.*, Adjuvant-pulsed mRNA vaccine nanoparticle for immunoprophylactic and therapeutic tumor suppression in mice. *Biomaterials*, 266, 120431, 2021.

34. Rittig, S.M. *et al.*, Long-term survival correlates with immunological responses in renal cell carcinoma patients treated with mRNA-based immunotherapy. *Oncoimmunology*, 5, 5, e1108511, 2016.
35. Cobb, M., Who discovered messenger RNA? *Curr. Biol.*, 25, 13, R526–R532, 2015.
36. Van der Jeught, K. *et al.*, Dendritic cell targeting mRNA lipopolyplexes combine strong antitumor T-cell immunity with improved inflammatory safety. *ACS Nano*, 12, 10, 9815–9829, 2018.
37. Yokokawa, H. *et al.*, Induction of humoural and cellular immunity by immunisation with HCV particle vaccine in a non-human primate model. *Gut*, 67, 2, 372–379, 2018.
38. Palucka, K. and Banchereau, J., Cancer immunotherapy via dendritic cells. *Nat. Rev. Cancer*, 12, 4, 265–277, 2012.
39. Topalian, S.L., Weiner, G.J., Pardoll, D.M., Cancer immunotherapy comes of age. *J. Clin. Oncol.*, 29, 36, 4828, 2011.
40. Buschmann, M.D. *et al.*, Nanomaterial delivery systems for mRNA vaccines. *Vaccines*, 9, 1, 65, 2021.
41. Hovav, A.-H. *et al.*, Duration of antigen expression *in vivo* following DNA immunization modifies the magnitude, contraction, and secondary responses of CD8+ T lymphocytes. *J. Immunol.*, 179, 10, 6725–6733, 2007.
42. Storjohann, A., The shifted number system for fast linear algebra on integer matrices. *J. Complex.*, 21, 4, 609–650, 2005.
43. Hou, X. *et al.*, Lipid nanoparticles for mRNA delivery. *Nat. Rev. Mater.*, 6, 12, 1078–1094, 2021.
44. Felgner, P.L. *et al.*, Lipofection: A highly efficient, lipid-mediated DNA-transfection procedure. *Proc. Natl. Acad. Sci.*, 84, 21, 7413–7417, 1987.
45. Stamatatos, L. *et al.*, Interactions of cationic lipid vesicles with negatively charged phospholipid vesicles and biological membranes. *Biochemistry*, 27, 11, 3917–3925, 1988.
46. Kranz, L.M. *et al.*, Systemic RNA delivery to dendritic cells exploits antiviral defence for cancer immunotherapy. *Nature*, 534, 7607, 396–401, 2016.
47. Krienke, C. *et al.*, A noninflammatory mRNA vaccine for treatment of experimental autoimmune encephalomyelitis. *Science*, 371, 6525, 145–153, 2021.
48. Sahin, U. *et al.*, An RNA vaccine drives immunity in checkpoint-inhibitor-treated melanoma. *Nature*, 585, 7823, 107–112, 2020.
49. Qureischi, M. *et al.*, mRNA-based therapies: Preclinical and clinical applications. *Int. Rev. Cell Mol. Biol.*, 372, 1–54, 2022.
50. Samaridou, E., Heyes, J., Lutwyche, P., Lipid nanoparticles for nucleic acid delivery: Current perspectives. *Adv. Drug Deliv. Rev.*, 154, 37–63, 2020.
51. Bachmann, M.F. and Jennings, G.T., Vaccine delivery: A matter of size, geometry, kinetics and molecular patterns. *Nat. Rev. Immunol.*, 10, 11, 787–796, 2010.
52. Kang, H. *et al.*, Size-dependent EPR effect of polymeric nanoparticles on tumor targeting. *Adv. Healthc. Mater.*, 9, 1, 1901223, 2020.

53. Simón-Vázquez, R., Peleteiro, M., González-Fernández, Á., Polymeric nanostructure vaccines: Applications and challenges. *Expert Opin. Drug Deliv.*, 17, 7, 1007–1023, 2020.
54. Das, A. and Ali, N., Nanovaccine: An emerging strategy. *Expert Rev. Vaccines*, 20, 10, 1273–1290, 2021.
55. Wu, D. *et al.*, Chitosan-based colloidal polyelectrolyte complexes for drug delivery: A review. *Carbohydr. Polym.*, 238, 116126, 2020.
56. Butkovich, N. *et al.*, Advancements in protein nanoparticle vaccine platforms to combat infectious disease. *Wiley Interdiscip. Rev. Nanomed. Nanobiotechnol.*, 13, 3, e1681, 2021.
57. Karlsson, J., Vaughan, H.J., Green, J.J., Biodegradable polymeric nanoparticles for therapeutic cancer treatments. *Ann. Rev. Chem. Biomol. Eng.*, 9, 105–127, 2018.
58. Rietscher, R. *et al.*, Antigen delivery via hydrophilic PEG-b-PAGE-b-PLGA nanoparticles boosts vaccination induced T cell immunity. *Eur. J. Pharm. Biopharm.*, 102, 20–31, 2016.
59. Mockey, M. *et al.*, mRNA-based cancer vaccine: Prevention of B16 melanoma progression and metastasis by systemic injection of MART1 mRNA histidylated lipopolyplexes. *Cancer Gene Ther.*, 14, 9, 802–814, 2007.
60. Perche, F. *et al.*, Enhancement of dendritic cells transfection *in vivo* and of vaccination against B16F10 melanoma with mannosylated histidylated lipopolyplexes loaded with tumor antigen messenger RNA. *Nanomed. Nanotechnol. Biol. Med.*, 7, 4, 445–453, 2011.
61. Rihan, M. *et al.*, Pyruvate kinase M2: A metabolic bug in re-wiring the tumor microenvironment. *Cancer Microenviron.*, 12, 149–167, 2019.
62. Rihan, M. *et al.*, Boronic acid derivative activates pyruvate kinase M2 indispensable for redox metabolism in oral cancer cells. *Bioorg. Med. Chem. Lett.*, 59, 128539, 2022.
63. Gajula, S.N.R., Naga, D., Bale, J., Nageswara, S., Nanjappan, S.K., Analytical and omics approaches in the identification of oxidative stress-induced cancer biomarkers, in: *Handbook of Oxidative Stress in Cancer: Mechanistic Aspects*, pp. 1–24, 2020.
64. Chavda, V.P. *et al.*, Advanced phytochemical-based nanocarrier systems for the treatment of breast cancer. *Cancers*, 15, 4, 1023, 2023.
65. Nalla, L.V. and Khairnar, A., Empagliflozin mediated miR-128-3p upregulation promotes differentiation of hypoxic cancer stem-like cells in breast cancer. *Eur. J. Pharmacol.*, 943, 175565, 2023.
66. Nalla, L.V. *et al.*, Targeting specificity protein 1 with miR-128-3p overcomes TGF-β1 mediated epithelial-mesenchymal transition in breast cancer: An *in vitro* study. *Mol. Biol. Rep.*, 49, 7, 6987–6996, 2022.
67. Nalla, L.V., Kalia, K., Khairnar, A., Self-renewal signaling pathways in breast cancer stem cells. *Int. J. Biochem. Cell Biol.*, 107, 140–153, 2019.

68. Gajula, S.N.R. and Nanjappan, S., Metabolomics: A recent advanced omics technology in herbal medicine research, in: *Medicinal and Aromatic Plants*, pp. 97–117, Elsevier, Cambridge, Massachusetts, United States, 2021.
69. Rao Gajula, S.N. *et al.*, Cytochrome P450 enzymes: A review on drug metabolizing enzyme inhibition studies in drug discovery and development. *Bioanalysis*, 13, 17, 1355–1378, 2021.
70. Gajula, S.N.R., Nadimpalli, N., Sonti, R., Drug metabolic stability in early drug discovery to develop potential lead compounds. *Drug Metab. Rev.*, 53, 3, 459–477, 2021.
71. Anchi, P. *et al.*, Nimbolide loaded sustained release microparticles as single-dose formulations for effective management of arthritis. *J. Drug Deliv. Sci. Technol.*, 75, 103638, 2022.
72. Sahoo, S.K. *et al.*, Bioevaluation of quinoline-4-carbonyl derivatives of piperazinyl-benzothiazinones as promising antimycobacterial agents. *Arch. Pharm.*, 355, 11, 2200168, 2022.
73. Sahoo, S.K. *et al.*, Identification of nitrofuranylchalcone tethered benzoxazole-2-amines as potent inhibitors of drug resistant Mycobacterium tuberculosis demonstrating bactericidal efficacy. *Bioorg. Med. Chem.*, 64, 116777, 2022.
74. Satti, S. *et al.*, Anxiolytic-and antidepressant-like effects of Bacillus coagulans Unique IS-2 mediate via reshaping of microbiome gut-brain axis in rats. *Neurochem. Int.*, p, 105483, 2023.
75. Gajula, S.N.R. *et al.*, LC-MS/MS: A sensitive and selective analytical technique to detect COVID-19 protein biomarkers in the early disease stage. *Expert Rev. Proteomics*, 20, 5–18, 2023 (just-accepted).
76. Dandekar, M.P. *et al.*, Multi-strain probiotic formulation reverses maternal separation and chronic unpredictable mild stress-generated anxiety- and depression-like phenotypes by modulating gut microbiome–brain activity in rats. *ACS Chem. Neurosci.*, 13, 13, 1948–1965, 2022.
77. Zhang, X. *et al.*, Local and systemic delivery of mRNA encoding survivin-T34A by lipoplex for efficient colon cancer gene therapy. *Int. J. Nanomedicine*, p, 2733–2751, 2019.
78. Zhang, F. *et al.*, Genetic programming of macrophages to perform antitumor functions using targeted mRNA nanocarriers. *Nat. Commun.*, 10, 1, 3974, 2019.
79. Zhang, C. *et al.*, Chemotherapy drugs derived nanoparticles encapsulating mRNA encoding tumor suppressor proteins to treat triple-negative breast cancer. *Nano Res.*, 12, 855–861, 2019.
80. Bray, F. *et al.*, Global cancer statistics 2018: GLOBOCAN estimates of incidence and mortality worldwide for 36 cancers in 185 countries. *CA Cancer J. Clin.*, 68, 6, 394–424, 2018.
81. Papamichael, D. *et al.*, Treatment of colorectal cancer in older patients: International Society of Geriatric Oncology (SIOG) consensus recommendations 2013. *Ann. Oncol.*, 26, 3, 463–476, 2015.

82. Zamarin, D. *et al.*, Localized oncolytic virotherapy overcomes systemic tumor resistance to immune checkpoint blockade immunotherapy. *Sci. Transl. Med.*, 6, 226, 226ra32–226ra32, 2014.
83. Lapworth, K. *et al.*, Impulsivity and positive psychotic symptoms influence hostility in methamphetamine users. *Addict. Behav.*, 34, 4, 380–385, 2009.
84. Lei, S. *et al.*, Efficient colorectal cancer gene therapy with IL-15 mRNA nanoformulation. *Mol. Pharm.*, 17, 9, 3378–3391, 2020.
85. Steven, A., Fisher, S.A., Robinson, B.W., Immunotherapy for lung cancer. *Respirology*, 21, 5, 821–833, 2016.
86. Yano, T. *et al.*, Therapeutic strategy for postoperative recurrence in patients with non-small cell lung cancer. *World J. Clin. Oncol.*, 5, 5, 1048, 2014.
87. Momcilovic, M. *et al.*, Targeted inhibition of EGFR and glutaminase induces metabolic crisis in EGFR mutant lung cancer. *Cell Rep.*, 18, 3, 601–610, 2017.
88. Maemondo, M., Park, K., Kobayashi, K., First-SIGNAL: First-line single-agent iressa versus gemcitabine and cisplatin trial in never-smokers with adenocarcinoma of the lung. *N. Engl. J. Med.*, 362, 2380–2388, 2010.
89. Papachristofilou, A. *et al.*, Phase Ib evaluation of a self-adjuvanted protamine formulated mRNA-based active cancer immunotherapy, BI1361849 (CV9202), combined with local radiation treatment in patients with stage IV non-small cell lung cancer. *J. Immunother. Cancer*, 7, 1–14, 2019.
90. Sebastian, M. *et al.*, A phase I/IIa study of the mRNA-based cancer immunotherapy CV9201 in patients with stage IIIB/IV non-small cell lung cancer. *Cancer Immunol. Immunother.*, 68, 799–812, 2019.
91. Khan, P. *et al.*, RNA-based therapies: A cog in the wheel of lung cancer defense. *Mol. Cancer*, 20, 1–24, 2021.
92. Neumeister, M.W. and Winters, J.N., Neuroma. *Clin. Plast. Surg.*, 47, 2, 279–283, 2020.
93. Sanchez-Covarrubias, L. *et al.*, Transporters at CNS barrier sites: Obstacles or opportunities for drug delivery? *Curr. Pharm. Des.*, 20, 10, 1422–1449, 2014.
94. Yang, Z. *et al.*, Pharmacological and toxicological target organelles and safe use of single-walled carbon nanotubes as drug carriers in treating Alzheimer disease. *Nanomed. Nanotechnol. Biol. Med.*, 6, 3, 427–441, 2010.
95. Gajula, S.N.R. and Nalla, L.V., Small extracellular vesicle biomarkers in breast cancer: A real-time snapshot for early diagnosis and prognosis? *Future Sci.*, 15, 367–37, 2023.
96. Zhang, Z. *et al.*, MicroRNA-29a-3p delivery via exosomes derived from engineered human mesenchymal stem cells exerts tumour suppressive effects by inhibiting migration and vasculogenic mimicry in glioma. *Aging*, 13, 4, 5055, 2021.
97. Vik-Mo, E.O. *et al.*, Therapeutic vaccination against autologous cancer stem cells with mRNA-transfected dendritic cells in patients with glioblastoma. *Cancer Immunol. Immunother.*, 62, 1499–1509, 2013.

98. Cheever, M.A. and Higano, C.S., PROVENGE (sipuleucel-T) in prostate cancer: The first FDA-approved therapeutic cancer vaccine therapeutic prostate cancer vaccine. *Clin. Cancer Res.*, 17, 11, 3520–3526, 2011.
99. Huang, X. *et al.*, Identification of tumor antigens and immune subtypes of cholangiocarcinoma for mRNA vaccine development. *Mol. Cancer*, 20, 1–17, 2021.
100. Huang, X. *et al.*, Identification of tumor antigens and immune subtypes of pancreatic adenocarcinoma for mRNA vaccine development. *Mol. Cancer*, 20, 1–18, 2021.
101. Rausch, S. *et al.*, mRNA vaccine CV9103 and CV9104 for the treatment of prostate cancer. *Hum. Vaccines Immunother.*, 10, 11, 3146–3152, 2014.
102. F. DeRosa *et al.*, Methods for purification of messenger RNA. Google Patents, 2021.
103. F. Von Der Mülbe *et al.*, Method for producing RNA. Google Patents, 2018.
104. MacLachlan, I., Liposomal formulations for nucleic acid delivery, in: *Antisense Drug Technology*, pp. 255–288, CRC Press, Boca Raton, Florida, United States, 2007.
105. Leung, A.K. *et al.*, Microfluidic mixing: A general method for encapsulating macromolecules in lipid nanoparticle systems. *J. Phys. Chem. B*, 119, 28, 8698–8706, 2015.
106. Kim, J. *et al.*, Self-assembled mRNA vaccines. *Adv. Drug Deliv. Rev.*, 170, 83–112, 2021.
107. Zhao, P. *et al.*, Long-term storage of lipid-like nanoparticles for mRNA delivery. *Bioact. Mater.*, 5, 2, 358–363, 2020.
108. Lila, A.S.A., Kiwada, H., Ishida, T., The accelerated blood clearance (ABC) phenomenon: Clinical challenge and approaches to manage. *J. Control. Release*, 172, 1, 38–47, 2013.
109. Knop, K. *et al.*, Poly (ethylene glycol) in drug delivery: Pros and cons as well as potential alternatives. *Angew. Chem. Int. Ed.*, 49, 36, 6288–6308, 2010.
110. Li, S. *et al.*, Effect of immune response on gene transfer to the lung via systemic administration of cationic lipidic vectors. *Am. J. Physiol.-Lung Cell. Mol. Physiol.*, 276, 5, L796–L804, 1999.
111. Dokka, S. *et al.*, Oxygen radical-mediated pulmonary toxicity induced by some cationic liposomes. *Pharm. Res.*, 17, 521–525, 2000.

9

Protein Delivery by Nanoparticles

Harshita Gauraha[1], Ankita Bhadoriya[1], Rupesh K. Gautam[1*] and Dinesh Kumar Mishra[2†]

[1]Indore Institute of Pharmacy, IIST Campus, Rau, Indore (M.P.), India
[2]Department of Pharmacy, Guru Ghasidas Vishwavidyalaya (A Central University) Koni, Bilaspur (C.G.), India

Abstract
Protein therapy has a significant influence on the treatment of several serious human illnesses. Many clinical proteins, including enzymes, growth regulators, hormones, and mediators, are immunogenic and have limited biological stability due to their fundamentally fragile structure and susceptibility to enzymatic degradation, which may restrict their positive advantages and, in some cases, their utility. The most effective and direct approach to overcome this limitation is to deliver protein by nanotechnology, which delivers the proteins into the target cell to exchange the dysfunctional protein and keep the balance of the organism. These delivery systems include soluble polymers, microparticles, microcapsules, cells, lipoproteins, liposomes, micelles, and dendrimers. These carriers can be targeted by conjugating them with certain proteins or antibodies targeting a particular area of interest. They are sensitive to pH and temperature changes, and have a slow rate of degradation. This chapter outlines the benefits and drawbacks of commercially available methods to deliver proteins and those still in development. Also, it offers a concentrated summary of major advancements in nanoformulations development technology for the administration of clinical proteins as well as potential prospects.

Keywords: Nanotechnology, proteins and peptides delivery, nanocarriers, nanoparticles

9.1 Introduction

The term "protein" originates from the Greek word "proteios," meaning "prime". Proteins are the major functional and organizational unit needed

**Corresponding author*: drrupeshgautam@gmail.com
†*Corresponding author*: dineshdops@gmail.com

Vivek P. Chavda and Vasso Apostolopoulos (eds.) Nanocarrier Vaccines: Biopharmaceutics-Based Fast Track Development, (305–344) © 2024 Scrivener Publishing LLC

to perform immunological, circulatory, and equilibrium processes. They are the essential biological large molecules found in living organisms, constituting about 20% of the body's overall mass and more than 50% of a cell's dry weight [1]. Large molecules having secondary and tertiary structures that are naturally occurring include proteins and peptides. These large molecules have a strong ability to regulate a variety of physiological processes. They are used for bodybuilding as well as essential biological processes like transport, regulation, and the production of enzymes and hormones [2].

The function of the protein is greatly influenced by its structure. Because its structure is influenced by its environment, factors like temperature, pH, and salt content will have an impact on both the protein's structure and its ability to perform its intended function. Protein structural abnormalities can cause molecular diseases that significantly alter metabolic processes [3]. All proteins are polymers of long chain amino acids. Proteins can interact with other proteins or biomolecules to create complex biomolecules that have capabilities that the individual components are unable to achieve [4].

With advances in biotechnology, formulation scientists are becoming more interested in protein- and peptide-based therapies. Due to their target-specific binding characteristics, these are these are highly powerful and low in toxicity. Proteins and peptides are part of a new class of prospective therapies that are currently on the market [5]. In biological membranes, proteins are present in a variety of receptors and channel that aid in the movement of substances throughout the body. Interaction of polypetides with several metabolic pathways trigger chemical focusing [6].

Proteins and peptides are most likely to be delivered by parenteral route. Parenteral delivery has a number of disadvantages compared to non-invasive delivery techniques such as intranasal, bronchial, and topical, which necessitates repeated injections due to a short *in vivo* half-life that results in poor patient compliance. Plasma concentration levels must be maintained with frequent dosage administration [7]. Numerous chemical alteration strategies have been reported as solutions to these restrictions. To overcome these limitations, distribution of proteins and peptides through nanostructured delivery vehicles can enhance absorption [8]. Protein treatments provide the benefits of higher selectivity, more activity, and less toxicity compared to conventional small-molecular medications, which continue to lead the pharmaceutical industry. The effective transport of several clinical proteins (such as enzymes and cytokines) to specific diseased locations is hindered by their sensitivity to enzymatic decomposition, short systemic half-lives, and low skin permeation [9].

9.2 Major Challenges in Protein Delivery

Recently, substantial techniques have been researched to deliver proteins and peptides orally. Considerable attempts have been made for the traditional distribution of tiny molecules [10]. However, designing an oral formulation that is effective for protein and peptide distribution through the oral route presents countless difficulties for pharmaceutical researchers. Shorter serum half-life, proteolytic protein and peptide breakdown, ion permeability, bigger molecule, deposition of polypeptides, capacity to trigger a cell-mediated immunological response, and adsorption via cell surface are a few of the challenging characteristics [11].

Peptides and proteins found it challenging for exogenous administration due to their physicochemical characteristics, particularly when administered orally [12]. Less than 1% of proteins and peptides are bioavailable orally. Hepatic and intestinal metabolism, intestinal membrane permeability, solubility, and molecular size are the obstacles to orally given peptides [13].

The interaction with mucin fiber, or adhesion, between the mucoadhesive system and mucous layer, allows for a 4–5 hour stay in the colon. This is due to the mucin's update time in intestinal mucus, which is 50–270 minutes. There are absorption issues with the transdermal method for the administration of proteins and peptides [14]. The nasal route has drawbacks such as nasal discomfort, a small drug delivery volume, rapid nasal epithelium renewal, an acidic pH, and endo- and exopeptidases. Protein and peptide administration through the mouth results in mucosal irrita bility, decreased permeability, and a bitter taste [15].

The following are the issues and difficulties pharmaceutical companies are currently experiencing with drug development and delivery:

- **2.1 Low Solubility:** It is one of the main issues when creating a specific pharmaceutical formulation. Poor solubility affects the drug's bioavailability. As a result, it poses a substantial obstacle for each unique chemical entity that scientists and companies discover [16].
- **2.2 Low Bioavailability:** The percentage of a drug dose that is available in blood or for systemic circulation is known as bioavailability. It is a significant factor of drug pharmacokinetics. Drug bioavailability is 100% when injected intravenously; however, due to partial absorption, when supplied orally, a drug's bioavailability lowers. Therefore,

bioavailability must be taken into account for administration other than intravenous [17].

2.3 Low Efficacy: The maximum effect a medicine dose may have is known as efficacy. To be extremely successful, drugs must have high binding affinities to their targets. The affinity of the drug's molecule is used to describe this [18]. The maximum reaction will be reduced if there is insufficient affinity between the medication and the target molecule. Low efficacy of drug molecules is one of the key problems that prolongs the duration of therapy for a serious disease [19].

2.4 Fast Excretion: The process of removing drugs from the body through urinary or renal organs, like the kidneys, is termed as excretion or elimination. Fast drug elimination reduces the effectiveness of the medicine since the optimum required dose of the drug molecule is not received by the targeted organs [20].

2.5 Drug Level at Specific Site: In some areas of the organ, such as the tumor cells, a specific level of drug accumulation is required for proper and effective therapy. The fraction of the drug required zone does not persist in these areas. Chemotherapeutic drugs for the treatment of cancer are connected with a lack of optimal drug accumulation [21].

Many medication delivery methods have been created in recent years by scientists and researchers, and some of them are still being worked on today [22]. Modern drug delivery techniques include soluble polymers, microparticles, microcapsules, cells, lipoproteins, liposomes, micelles, dendrimers, hydrogels, and carbon nanotubes. These carriers can be targeted by conjugating them with certain antibodies targeting a particular area of interest. They are sensitive to pH and temperature changes, and have a slow rate of degradation. Recently, highly developed and engineered nanomaterials have been used to combat these issues [23].

9.3 Nanotechnology

The science and technology that deal with the development of specially engineered materials at nanoscale is termed as nanotechnology. The Greek term "nano" means dwarf. A nanometer is one billionth (10^{-9}) of a meter [24]. In addition, one nanometer is equivalent to the sum of the diameters of three atoms.

Researchers have used nanotechnology to create novel nanocarriers and nanodrug delivery systems for the medication molecules are successfully delivered to the location of action. This system is founded on a technique that provides a certain dosage of a medication for an extended length of time to a certain bodily part with disease [25].

The highly integrated nature of this nanodrug delivery system demands collaboration among experts from a variety of fields, including chemists, biologists, and engineers [26]. The development of selective drug delivery, in which the drug is only effective on the targeted specific body part (for instance, in malignant tissues), and prolonged release formulations, in which the drug is released from preparations over time in a controlled way, are two ongoing attempts to deliver polypeptides [27].

Nanotechnology has shown a lot of potential for use in the medical field. Numerous nanomedicines have previously received approval for use in clinical settings, and many more are still being tested. Particularly, nanoparticles with sizes typically between 10 and 30 nm, including liposomes, micelles, polymer nanoparticles, and inorganic nanomaterials, have significant benefits as drug carriers [28]. Structure and different modifications that can help in the delivery of protein nanoparticle are shown in Figure 9.1.

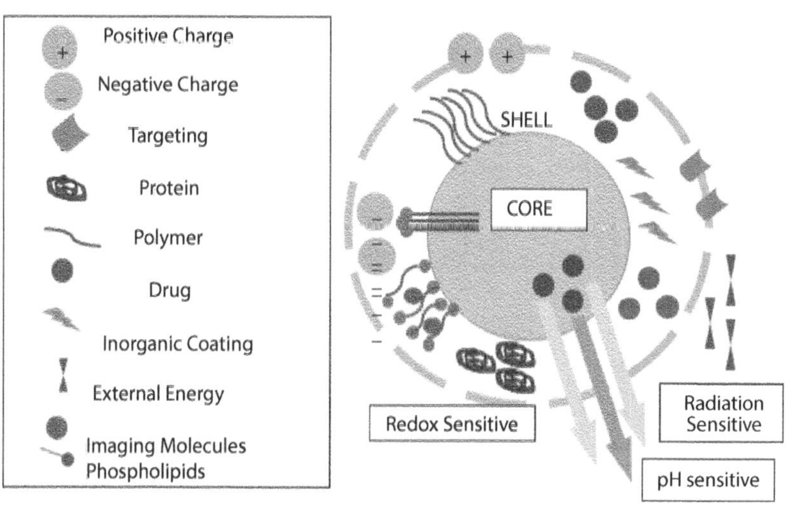

Figure 9.1 Schematic representation of core–shell nanostructures with several possible modifications.

9.4 Nanoparticles

Nanoparticles (NP) are a fast-growing area, which includes the development of structures in the 5 to 1000 nm size range. The method of preparation decides the encapsulation or entrapment of protein into the nanoparticles [29]. Reticuloendothelial and phagocytic clearance can be avoided easily by submicron-sized nanoparticles. This eases their penetration deep into tissues through fine capillaries and is taken up easily by the targeted cells. Additionally, they can overcome the barriers and have excellent biocompatibility in contrast to larger microparticles [30].

The nanoparticles ought to be made with good therapeutic impact in mind in addition to protein trapping. Therefore, before developing, it is crucial to understand more about intracellular delivery systems. Nanoparticles start moving inside cells after being taken in. They are first sent to endosomes, where they later combine with lysosomes (Figure 9.2). Following their release from lysosomes, nanoparticles spread to the therapeutic target, such as the nucleus and other organelles, in the cytoplasm [31].

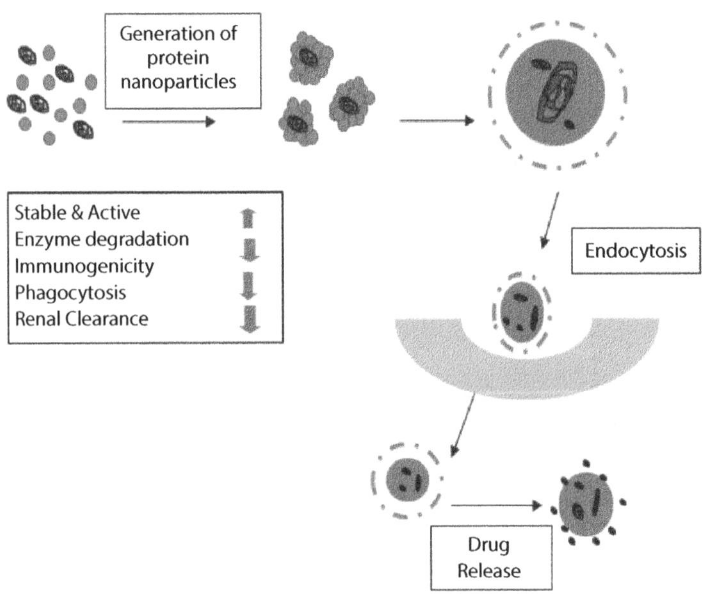

Figure 9.2 Delivery of insoluble proteins by endocytosis.

Pharmaceutically active chemicals and biomolecules, such as neoplastic agents and clinical polypeptides, have prominently used nanostructures as the vehicle. Proteins are a promising option for the polymeric materials typically used in the production of nanoparticles because of their safety [32].

In comparison to other nanocarriers, polymeric nanoparticles have grown in popularity for the delivery of proteins and peptides. Amphiphilic peptides, linear and cyclic peptides, as well as -helical and -sheet peptides, can all self-assemble into nanostructures. Thus, understanding these molecules and their participating pathways is necessary for drug development. Therefore, the small size of nanoparticles makes them available to penetrate the biological system and affect the normal cell functioning [33].

9.4.1 Nanocarriers

Active substances in medicinal, veterinary, or agricultural applications can be made better and more effective by using nanocarrier-based delivery systems (as shown in Figure 9.3), especially when those ingredients are unstable, sparingly soluble, or have side effects. Numerous medications

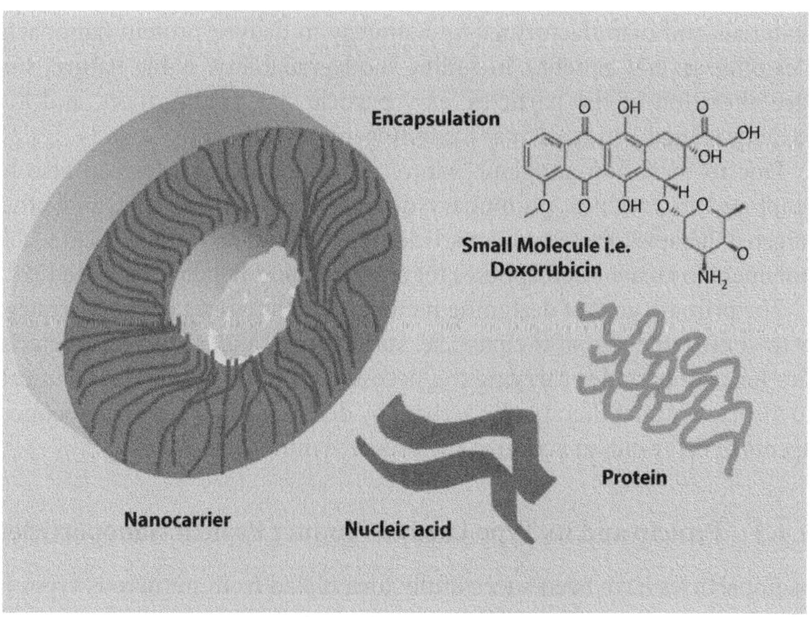

Figure 9.3 Structure of nanocarriers.

have low solubility, instability, or off-target effects after systemic administration, which limits their effectiveness. For instance, the toxic effects of cancer treatments on healthy cells and the formation of resistant cells that have oncogenes for efflux transporters and proteins involved in multidrug resistance are two reasons why cancer therapy frequently fails [34].

The potential of off-target harm increases because larger dosages must be administered to maintain the drug concentration within the therapeutic range due to the tumor's low active component bioavailability [35]. By using the enhanced permeability and retention (EPR) effect, a well-known event based on the combination of leaky vasculature and inadequate lymphatic drainage at the tumor site, nanocarriers can solve this difficulty by delivering active components [36].

The accumulation of active substances in nanocarriers also decreases the rate of clearance through phagocytosis and renal excretion, extending the active ingredient's circulation time, and hence, its therapeutic duration [37].

9.4.2 Protein Nanocarrier

Proteins can be used as the basic components for the development of nanoparticles since they have special uses and characteristics in biological materials and manufacturing. As a strategy to deliver, protein nanoparticles offer several benefits, including biodegradability, stable nature, surface alteration of the particles, easy particle size modification, and low toxicity-related issues, such as immunogenicity [38].

Due to their non-antigenic nature, protein nanoparticles can also be employed in vaccines, pulmonary delivery, cancer treatment, and other targeted therapies. Protein nanoparticles can be combined with biodegradable polymers to create microspheres for sustained and regulated release [39].

The primary goal of designing nanoparticles as a drug delivery strategy is to control the size of the particle, surface area, and surface characteristics for nanoparticles carrying the necessary amounts of pharmaceuticals to display optimal therapeutic activity by delivering active pharmaceutical agents to carry out an action at the pre-determined site [40].

9.4.3 Protein and Its Type Used to Produce Protein Nanoparticles

Nanoparticles have been successfully formulated from numerous types of proteins including water-soluble proteins (e.g., bovine and human serum albumin) and insoluble proteins (e.g., zein and gliadin). Some of them are shown in Figure 9.4.

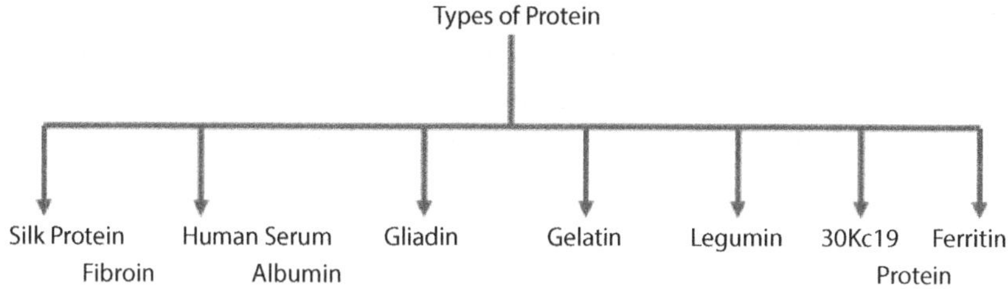

Figure 9.4 Types of protein.

9.4.3.1 *Silk Protein Fibroin*

Fibroin, which makes up 65% to 85% of the total protein in silk fibers, is the predominant protein found there. In general, the outer sericin is eliminated using a thermochemical degumming procedure using Na_2CO_3 to extract the fibroin that is frequently obtained from silk fiber made by the *Bombyx mori* silkworm. Because the separated fibroin is insoluble, LiBr or $CaCl_2$ is used to create a soluble version known as regenerating fibroin [41].

Due to its versatility, dynamic power, high stability, low immunogenicity, biodegradability, and biocompatibility, as well as its abundance and lower cost, fibroin is commonly used to create nanoparticles. Fibroin nanoparticles have a negative charge on their zeta potential. By covering it with a positively charged polymer, such as PEI, chitosan, or EDC, the surface can be used as a cross-linker to convert it into a positively charged moiety. Several factors, such as fibroin MW, crystalline nature, encapsulation ability, and manufacturing conditions, may have an impact on FNP attributes, such as medium size, size distribution, surface zeta potential, drug encapsulation, release profile, and stability of particle formation [42].

Since they can address a number of shortcomings of low-molecular-weight medications, fibroin nanoparticles have been the subject of numerous investigations to be distributed and used in the delivery of pharmaceutical formulations. All fibroin nanostructures having small sized nanoparticles loaded into them have enhanced drug-treatment effectiveness, increased capture efficiency, tunable protracted release profile, higher drug solubilization, stabilization, inhibition of drug decomposition, and lower adverse effects [43].

Advantages

Silk Protein Fibroin is advantageous as they are highly stable with good mechanical strength and flexibility. Moreover, it has a low immunogenicity, and is biodegradable and biocompatible.

Disadvantages
Their major disadvantage is that sericin may cause immunogenic reactions and antiparallel ß-sheet crystal conformation, referred to as silk-II form, rendering it as insoluble filament leading to its degradation.

9.4.3.2 Human Serum Albumin

Human serum albumin (HSA) is a globular protein that is mostly found in the circulatory system. It contains up to about 585 amino acids and has a molecular weight (MW) of 66 kDa. HSA is made up of three main units: I, II, and III, each of which also includes subunits A and B. HSA contains subunits IIA and IIIA, respectively, include Sudlow's sites I and II, the two principal binding sites. HSA is well-known for its ability to act as a carrier for other chemicals and helps improve solubility of hydrophobic molecules into blood. HSA is employed to deliver various chemicals to particular biological tissues. Additionally, HSA is extremely pH stable (stabilized in the pH range of 4 to 9) [44].

Several medications can be loaded onto HSA for transport through the circulatory system because of its high ligand-binding capabilities. Additionally, protein nanoparticle preparation—in particular, HSA—seems to be a useful approach for avoiding the unfavorable interactions with serum that frequently occur following intravenous administration of complexes. A pathologic calcification in the media of large and intermediate arteries that are linked to the heart is one application where HSA nanoparticles have been frequently used for their treatment [45].

Advantages
They are highly stable, soluble in physiological fluids, biodegradable and shows low toxicity, immunogenicity, and have both accessibility and readiness.

Disadvantage
The only drawback of this protein is its cost.

9.4.3.3 Gliadin

Gliadin, which is obtained by extracting wheat gluten with 70% ethanol, is one of the principal proteins present in it. Each single-chain polypeptide chain in gliadin has an average weight (MW) of 25 to 100 kDa, and the chains are joined together by intramolecular disulfide bonds. Except in cases of severe pH, it shows less solubility in aqueous solutions. The hydrophobic interactions and disulfide linkages that result in the protein's distinctive folded shape are what cause its low water solubility. Because of this, gliadin nanoparticles can be utilized to deliver medications that are amphiphilic and hydrophobic in a regulated manner [46]. Utilizing proteins that are insoluble in water has the benefit of not requiring an additional curing step to preserve the integrity of goods that are based on water. Consequently, gliadin is acceptable for use because of its good biological compatibility, biodegradable nature, non-toxicity, and stability. It can also be a polymer that can be affixed to the mucosal membrane and is appropriate for oral and local medication delivery systems because it adheres to the mucosa. Due to their affinity for the upper GIT tract but poor anchoring to other gastrointestinal tracts, they are polymers that are good for targeted medication administration [47].

Advantages
This protein shows high compatibility with its biological environment; highly stable, biodegradable, and non-toxic and non-immunogenic.

Disadvantages
This protein has many advantages, but have the major limitations of larger particle size and rapid degradation.

9.4.3.4 Gelatin

A natural water-soluble polymer known as denatured protein gelatin can be made by hydrolyzing collagen in basic or acidic environments, or by thermally or enzymatically degrading collagen. It is also the first proteinaceous substance which helps in creating nanoparticles. Gelatin is a soluble protein in water at temperatures above 35°C to 40°C, and has an average molecular weight of 20 to 220 kDa [48]. On the basis of collagen hydrolysis process, there are two different forms of gelatin (type A and type B) offered commercially. Several biologically active materials, including DNA and RNA, bovine serum albumin (BSA), bone morphogenetic protein 2 (BMP-2), and basic fibroblast growth factor (bFGF), are commonly encapsulated in gelatin nanoparticles. However, gelatin has a weak mechanical structure and a quick rate of nanoparticle formation. Therefore, when gelatin is used to make nanoparticles, it needs to be mechanically, biologically, or molecularly cross-linked by a range of copolymers, such as glutaraldehyde (GA), to strengthen it mechanically and to decrease disintegration rate and solubility in aqueous solutions [49].

Advantages
This is a useful protein because of its high compatibility with biological fluids, biodegradable in nature, and with safe and easy bridging property.

Disadvantages
The limitations of this protein are low mechanical strength and speedy degradation.

9.4.3.5 Legumin

Legumin, which is a member of the 11S globulin protein family, is one of the essential storage proteins found in soybean seeds (*Pisum sativum* L.). Legumin is made up of six subunits and has a molecular mass between 300 and 400 kDa. It is also abundant in amino acids that contain sulphur [50]. Legumin-derived nanoparticles have properties such as bioadhesion and have a significant surface area, indicating a greater potential for interaction with biological surfaces. They are also smaller in size, highly stable, and with minimal antigenicity; however, further detailed study is required to raise the poor yield that is associated with them and verify their use in biomedical applications [51].

Advantages
This protein is bioadhesive in nature with a large surface area and smaller particle size, which make it an ideal choice in formulations.

Disadvantages
It gives very low yield, which limits its usage.

9.4.3.6 30Kc19 Protein Obtained from Silkworm Hemolymph

The 30K protein family, which includes the 30Kc6, 30Kc12, 30Kc19, 30Kc21, and 30Kc23, is a group of similar-looking proteins found in the hemolymph of silkworms (*Bombyx mori*). The 30K proteins play important functions in cell development and viability in many cells, affect enzyme stability, and have about 30 kDa of molecular weight. Making small-sized nanoparticles only from the 30Kc19 protein has proven to be difficult [52]. In contrast to HSA nanoparticles, 30Kc19-HSA nanoparticles, which contain 50% of 30Kc19, demonstrated greater intracellular transport efficiency and activity of the loaded protein. These results imply that HSA combined with 30Kc19 is more suited than pure HSA as a drug delivery nanoparticle material. Furthermore, it was demonstrated that the 30Kc19 - the efficacy of cargo proteins inside a cell transportation was compared peptides -penetrating Pep-c19 cell, showing that the helix domain can be employed as a component of protein nanoparticles. Helix domain can be used to construct nanoparticles even when used singly [53].

Advantages

- ❖ Highly stable
- ❖ Increased cell growth and feasibility
- ❖ Biodegradable
- ❖ Non-immunogenicity
- ❖ Low toxicity
- ❖ Enzyme stabilization
- ❖ Penetrability

Disadvantages

- ❖ Small size and yield when using only 30Kc19α

9.4.3.7 Ferritin

Microorganisms, plants, and mammals all contain ferritin, an iron-storing protein, first discovered in 1937 by Laufberger. It is a hollow protein glob ule with a molecular weight (MW) of 474 kDa and that consists of 24 subunits. Its 6-nm inorganic core is made of hydrated iron oxide ferrihydrite, and its spherical polypeptide shell surrounds it (Apoferritin) [54].

Advantages
It is a highly biodegradable protein with better pH and thermal stability.

Disadvantages
This is a highly expensive protein.

9.5 Methods of Preparation

Protein nanoparticles can be created using one of three techniques: chemical, physical, or self-assembly (Figure 9.5 and Table 9.1). Emulsion and complicated coacervation techniques are widely utilized as chemical procedures. Electrospray technology and a nano spray drying process are two examples of physical approaches [55].

9.5.1 Chemical Methods

9.5.1.1 Emulsion/Solvent Extraction

Polymeric nanoparticles are routinely produced using these methods as shown in Figure 9.6. It has also been applied to the production of

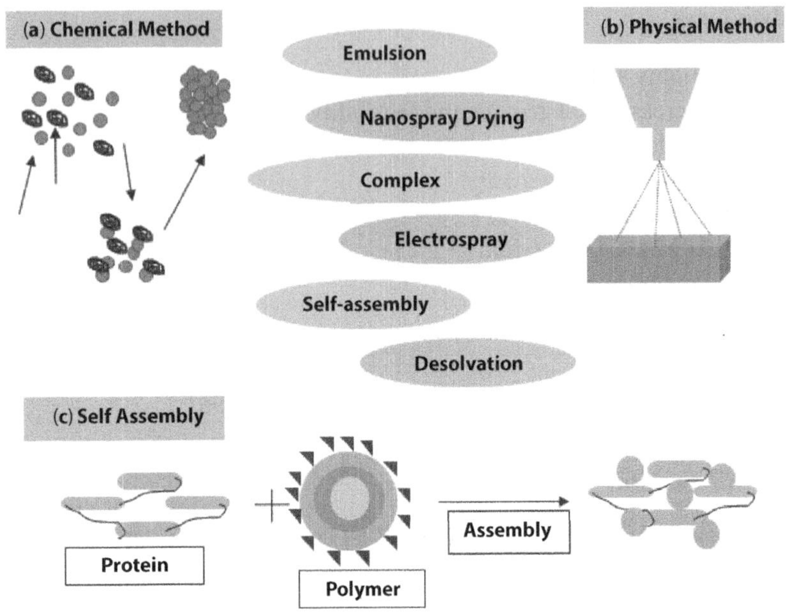

Figure 9.5 Diagrammatic representation of different methods of protein nanoparticles preparation.

Table 9.1 Descriptions of different methods of protein nanoparticles preparation.

Method	Definition
1. Chemical method	Uses a chemical procedure to create protein nanoparticles, and incorporates sophisticated coacervation and emulsion
2. Physical method	After breaking down proteins into nanosized particles, it incorporates physical aggregation together with an electrospray approach and a nano spray drying method.
3. Self-assembly method	It is a technique for creating nanoparticles that uses the desolvation method as well as protein agglutination on its own.

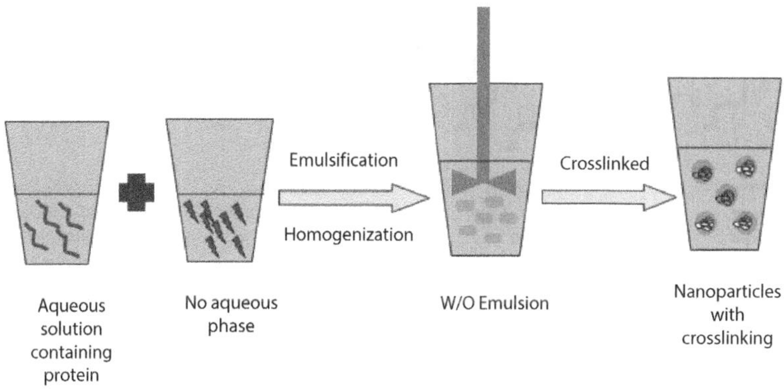

Figure 9.6 Emulsion/solvent extraction.

proteinaceous nanoparticles. A mixture of two or more immiscible liquids that occurs when the liquids dispersed into each other forms an emulsion. A polymeric solution (in an organic solvent) or proteinaceous solution (in an aqueous buffer) is mechanically agitated or sonicated to generate nanoparticles, forming an emulsion system in its most alkaline form. The amount of protein and the ratio of the water to oil phase are key components for protein nanoparticles. Hence, obtained protein nanoparticles are purified either by addition of a cross-linking agent or thermally by adding a W/O emulsion to an oil, warmed at a temperature of 100°C or higher [56].

Emulsion Process: After an emulsion system has been created by dispersion by mechanical agitation or ultrasonic waves, this process is a way to create nanoparticles by using a solvent or by eliminating it. Additionally, a double emulsion technique is employed [57].

Complex Coacervation Process: Using a sophisticated coacervation process, nanoparticles are created by interacting with other polymers and other proteins to make cationic or anionic proteins [57].

Advantages

This process is highly stable with an efficient encapsulation capacity, and the reaction conditions can control the shape and size of nanoparticles.

Disadvantages

The major limitations of this process are larger particles are obtained, as compared to desolvation, and they also need surfactants due to thermodynamic instability.

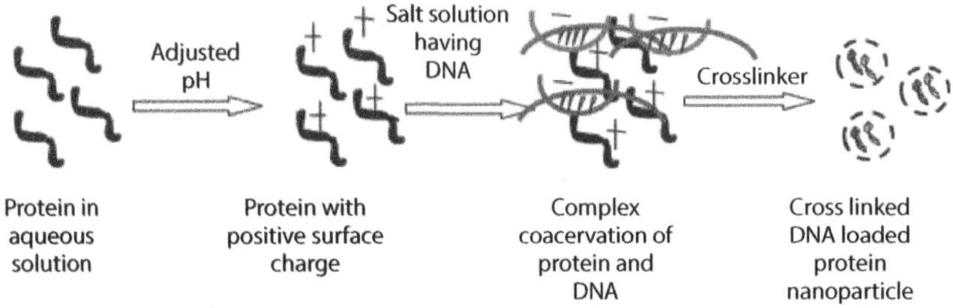

Figure 9.7 Protein nanoparticles by complex coacervation method.

Polyelectrolyte Complexation/Complex Coacervation Method

Despite being amphoteric with several charged functional groups, proteins can be either cationic or anionic by changing a number of factors, including the protein's pH. Different polymeric electrolytes may interact electrostatically with the charged protein as illustrated in Figure 9.7. To help with the regulated transmission of DNA and bioactive medications, stable biocompatible nanoparticles and coacervates may be made using the pH-dependent electron transfer between proteins and other polymers. Typically, oligonucleotides with the shape of lengthy strings of therapeutic DNA/RNA molecules are converted to nanoparticles by employing cationic protein polymers to generate a compound containing anion oxygen atoms in a phospholipid backbone [58].

For instance, Truong-Le *et al.* trapped DNA in gelatin nanoparticles by using salt-induced complex coacervation process. At pH 5, the gelatin had a positive charge and was combined with DNA to produce complex coacervates. The loading efficiency was between 25% and 30%, and the particle sizes ranged from 200 to 700 nm. DNA was trapped inside the protein matrix during the coacervation process. Hydrophobic interactions or hydrogen bonds were equally effective at enhancing protein–polymer synthesis as electrostatic interactions. The quantitative determination of proteinaceous and polymeric complex is thought to be determined by the equilibrium between the protein–polymer interaction and polymer–polymer interaction. When complex coacervates are formed, nanoparticle size and loading effectiveness can be modified through ultrasonic processing [58].

9.5.2 Physical Method

9.5.2.1 Nano Spray Drying

In the utilization of nano spray drying technique, nanoparticles in liquid samples are processed. With heated nitrogen and carbon dioxide gas flowing in the nozzle's spray direction, liquid samples are sprayed into chambers. Nanoparticles are gathered in a chamber by an electrode at the bottom. As the sprayed droplets descend to the chamber's bottom, electrodes cause them to accumulate an electrostatic charge. This process for producing tiny protein particles is systematic, rapid, and affordable [59].

As hydrophilic medications can be enclosed in these spray-dried nanoparticles, drug delivery systems are one use for spray drying. Because solvent evaporation helps retain nanoparticle droplets at a consistent temperature, utilizing nanoparticles enables the technique to be used on specimens that are sensitive to heat. This technique for producing nanoparticles has the advantage that the size of the particles can be changed by adjusting factors like the nozzle's size and spraying rate. To stabilize polymer particles when using protein nanoparticles, surfactant additives are typically required [60].

9.5.2.2 *Electrospraying Technique*

It is a technique for atomizing liquids. It can be used for manipulating materials at the submicron scale. The protein solution needs to be exposed to a high voltage to spray the liquid jet stream through the nozzle and produce an aerosolized droplet using this method. Colloidal-sized protein nanoparticles have been gathered and aerosolized as droplets. Using this methodology, it is straightforward to incorporate drugs and nucleic acids into highly efficient nanoparticles. Solvent evaporation can be used to cre ate solid particles. Based on the kind of medication to be delivered, there are differences in the diameter of needle gauge, voltage, rate of flow, and operating distance [60].

The fundamental principle underlying electrospraying is to mix polymer solutions at high voltage so that the polymers are delivered by syringes as nanoparticles. In actuality, this is a method that is used to create molecules with nanostructures that combat similarity with electrical radiation method. Gong *et al.* studied that micellar nanoparticles embedded with protein based on the exact interaction between the major amino group of albumin and acetaldehyde to construct a core-shell nanomicelle. It was mixed, dissolved in water, and then put through a 0.45-m pore size fine porous filter [59].

9.5.2.3 *Self-Assembly*

Individual protein chains can spontaneously form nanosized aggregates known as protein micelles that are dissolved in a solution higher than the critical micelle concentration (CMC) at the critical solution temperature (CMT).

The micellar chains can be stabilized by the solidification process by creating a bridge between them. Through the hydrophobic modification process, the hydrophilic protein albumin can acquire amphiphilic properties. When added to aqueous solutions, hydrophobically altered proteins can self-assemble into micelle nanoparticles. The dialysis approach was used to successfully incorporate protein micelle containing paclitaxel, with a drug loading efficiency of roughly 33.1% and a high encapsulation efficiency of 90.5% [61].

9.5.2.4 *Desolvation*

Desolvation is the method most frequently used to create protein-based nanoparticles. With the addition of desolvating chemicals, such as ethanol and acetone, to protein solutions that contain medications, the desolvation method makes it simple to produce nanoparticles. Proteins are altered by desolvating chemicals, making them less soluble, which results in precipitation in the form of protein nanoparticles. When a particle reaches a certain size, the process of creating new particles gradually increases the number of particles of that same size [62].

Bridging agents like GA are used to join nanoparticles after they have been formed. Protein amount, desolvating agent addition rate, pH, and temperature have the biggest effects on particle size. It is possible to create smaller nanoparticles, particularly at high pH and low protein concentration. By applying a desolvating chemical to make the protein less soluble, the protein is concentrated, turning the desolvation process into a self-assembly approach rather than directly producing nanoparticles [61].

9.6 Nanoformulations Available for Protein and Peptide Delivery

In protein delivery, nanotechnology can:

- Provide protection to protein against premature decomposition inside the body;
- Extend the half-life of proteins in the bloodstream that have poor oral bioavailability; and
- Improve the safety and effectiveness of biological treatments by controlling sustained release that can keep drug concentrations in the therapeutic range and target damaged tissues, cells, and internal compartments.

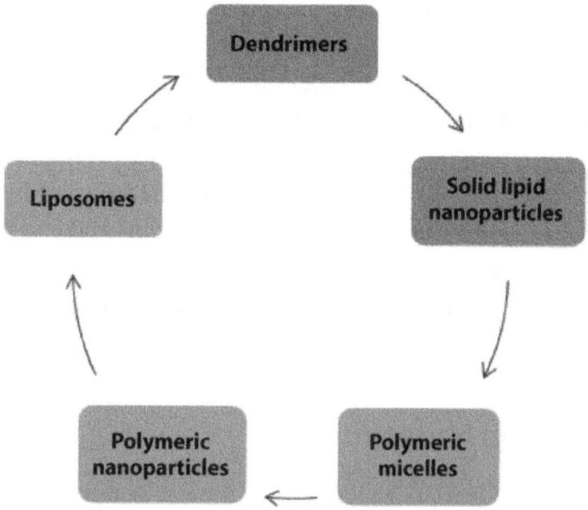

Figure 9.8 Types of nanocarrier systems.

Nanoparticles (< 100 nm) are effective drug delivery systems because they can deliver medications to the site of action in a regulated manner at a lower dose and dosing frequency. This reduces the toxicity and/or negative consequences of conventional medicines, including their quick clearance, nonspecific distribution, limited bioavailability, and irregular drug release [63].

Clinical nanostructures can be liposomes, nanoparticles made of gold, iron, or silica, formulations based on polymers, or formulations based on micelles. Increased intestinal absorption into the bloodstream, protection from severe environments, site-specific targeting, and controlled release into a target site are all functions of the nanocarrier. Drug delivery technologies are therefore becoming more and more crucial. Among them, numerous nanomaterials, including liposomes, polymers, dendrimers, and magnetic nanoparticles, as shown in (Figure 9.8), are utilized as drug delivery vehicles [64].

9.6.1 Dendrimers

To get past multiple biological and physical barriers, they serve as carriers for the transfer of bioactive chemicals to the desired location within the cell. It has been discovered that dendrimers' physical and chemical characteristics vary in a number of ways that enhance their interactions with biological molecules such proteins, peptides, genes, and lipid biomembranes. Dendrimers' structural development gives it a unique vehicle for the transport of peptide medicines to the brain [64].

9.6.2 Liposomes

It is an essential element of the drug delivery system that lowers drug toxicity and dose dumping, and is crucial for the spreading of genetic material. To transmit proteins and peptides, liposomes have been developed as a unique therapeutic strategy. Liposomes' surfaces can be altered with ligands or antibodies to improve tissue targeting by making it more recognizable to types of cells. A unique method of delivering proteins by using biodegradable polymer and lipid has been described as liposomes in alginate (LIA) [65].

9.6.3 Solid Lipid Nanoparticles

Solid lipid nanoparticles (SLNs), which can be as small as 30 nm or as large as 1000 nm, are made from lipids that break down. Particle sizes of SLNs that can bypass spleen and liver filtration range from 120 nm to 200 nm. They have many benefits over other nanoformulations, including improved bioavailability, simple large-scale manufacture, minimal acute or chronic toxicity, the capacity to incorporate both hydrophilic and lipophilic medicines, reduced solvent use, and protection for legally responsible phytoconstituents (Figure 9.9) [63].

9.6.4 Polymeric Nanoparticles

Drug delivery using polymeric nanoparticles (with sizes < 100 nm) is being addressed here. They maintain the system's thermodynamic equilibrium and can quickly pass through biological barriers. Polymeric nanoparticles are used to transport proteins and peptides as they can be encapsulated

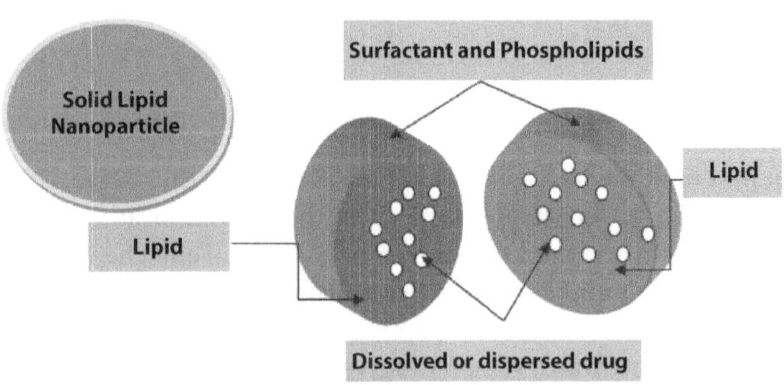

Figure 9.9 Solid lipid nanoparticles.

effectively and have the capacity to prevent medications from deteriorating and premature release before reaching the target region. Biodegradable polymeric nanoparticles are thought to be effective at delivering proteins and peptides because they are either attached to the surface as nanospheres or to the inner core as nanocapsules [66].

9.6.5 Polymeric Micelles

Polymeric micelles have extended circulation duration when intravenously administered due to their hydrophilic moiety and smaller particle size. A polymeric micelle that has been loaded with medication has been shown to be a useful tool in the pharmaceutical treatment of tumors. Decreased lymphatic drainage and increased vascular permeability are two advantages of polymeric micelles, and as a result, more deposition happens in the tissues around inflammatory or cancerous tumors [64].

Polymeric micelles have been created to treat neurodegenerative illnesses like Parkinson's and Alzheimer's, HIV-associated dementia, and brain stroke. Polymeric micelles offer the advantage of carrying the physiologically active substance to the brain without having to cross the blood-brain barrier. [65]. To circumvent the BBB, polymeric micelles have been employed in two different ways:

- In the first method, polymeric micelles are packed with molecules of ligand or antibodies that can cause transcytosis across the blood–brain barrier (BBB), which is made up of endothelial cells in microvessels.
- The second method reduces medication efflux and increases BBB permeability by adding pluronic block copolymer to a polymeric micelle.

9.7 Clinical Trials and Market-Approved Nanoparticles

Lipid-based nanoparticles, polymeric nanoparticles, micellar nanoparticles, nanostructures, inorganic nanoparticles (metals/metal oxides and other inorganic nanomaterials), and proteinaceous nanoparticles are some of the FDA-approved nanodrugs that are now used in clinical care (listed in Table 9.2 and the report of clinical trials and marketed nanoparticles are shown in Figure 9.10 and Figure 9.11).

Table 9.2 FDA-approved protein nanoparticles.

Drug/company	Active ingredients	Nanoparticle size	Specific treatment	Approval year
Liposomal nanoparticles				
Onivyed/Merrimark	Irinotecan	90–130 nm	Pancreatic cancer	2015–2016
Vyxeos/Jazz Pharmaceuticals	Daunorubicin/cytarabine	100–200 nm	Acute myeloid leukemia	2017–2018
Onpattro/Alnylam	Transthyretin-directed small interfering RNA	80 nm	Hereditary Transthyretin-mediated amyloidosis	2018
Proteinaceous nanoparticles				
Abraxane/Celgene	Albumin-bound Paclitaxel	130 nm	Pancreatic cancer	FDA 2013, Europe 2005
			Breast cancer	2005–2008
Viral nanoparticles				
Zolgensma/Novartis	AAV9 containing SMN1 gene	22 nm	Spinal muscular atrophy	2019

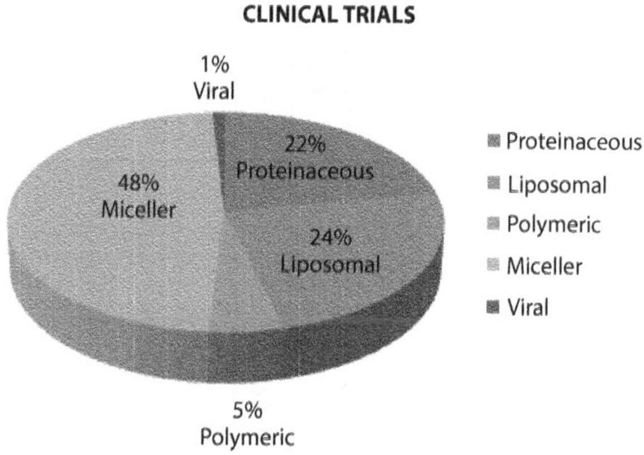

Figure 9.10 Clinical trials using different nanoparticles.

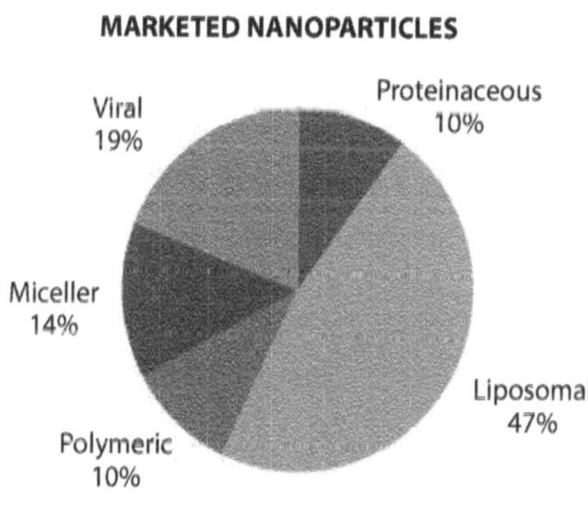

Figure 9.11 Marketed nanoparticles formulation.

9.7.1 Nanomedicine and Proteins in the Field of Treatment and Diagnosis

In medicine and therapy, various polypeptides, including insulin, vaccines, antibodies, and numerous conjugated proteins, have been used. This is also one of the key study fields. The creation of brand-new medication delivery systems through nanomedicine aims to enhance their performance and characteristics. On the other hand, one of the biggest problems with nano pharmaceutical drug delivery systems is how different blood proteins interact with them and build a protein crown (called a protein corona) around the nanostructures, which has a significant impact on how well they work in the end [67].

Table 9.3 Marketed nanoparticles for protein delivery.

Trade name	Generic name	Composition	Description/indications	Company
Sandostatin Lar	Octreotide Acetate	PLGA	Acromegaly	Novartis
PegIntron	PEG-α-interferon-2b	PEGylated proteins	Treats hepatitis C and melanoma	Schering Plough
Oncaspar	PEG-Lasparaginase	PEGylated proteins	Treats acute lymphoblastic leukemia	Enzon
Pegasys	PEG-α-interferon-2a	PEGylated proteins	Treats hepatitis C and hepatitis B	Roche
Somavert	Pegvisomant	PEGylated proteins	The growth hormone receptor antagonist is used in the treatment of acromegaly	Pfizer

These consist of complement systems, immune system proteins, and antibodies. To enhance the effectiveness of nanomedicines, significant efforts are being made to regulate how drug formulations interact with various proteins, particularly immune system proteins. Several peptides and proteins, including as cell-penetrating peptides (CPPs), antibodies, and phage peptides, have also been considered to target drug-containing nanoparticles to target tissue or tissues, such as tumor tissues. Finally, one of the newest medication delivery techniques uses protein nanocarriers for effective drug delivery. Albumin nanoparticles containing anti-cancer medication are found in the medicine Abraxane [68] (other marketed products are listed in Table 9.3).

Proteins have a key role in both the diagnosis of certain diseases and the creation of sensors that can be used to detect other diseases. Several nano biosensors are being explored to detect various proteins, immune responses, allergens, and biomarkers. Antibodies and enzymes are used to detect the illness. Examples include the use of antibodies to identify various viral illnesses and the production of glucose nano biosensors using the enzyme glucose oxidase [68].

9.8 Characterization of Protein Nanoparticles

There are different methods to characterize protein nanoparticles such as particle size, morphology, particles structure, drug loading as shown in Figure 9.12.

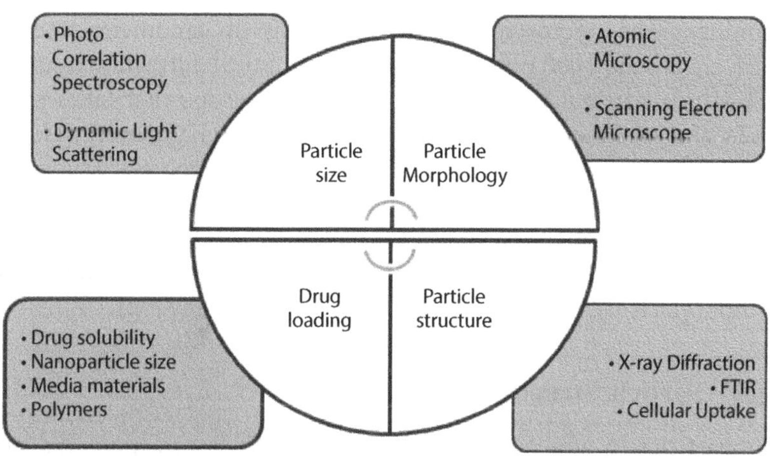

Figure 9.12 Pictorial diagram for the characterization of protein nanoparticles.

9.8.1 Particle Size and Polydispersity

The key features of nanoparticle systems are particle size and size distribution. Nanoparticles are appropriate for a variety of biological objectives because they display substantially greater absorption within the cell with their smaller sizes, compared to microparticles and their mobility. Measurement of particle size using Photon Correlation Spectroscopy (PCS) or Dynamic Light Scanning (DLS) is the quickest and most popular technique [69].

In the industrial context, photon correlation spectroscopy (PCS) is the technique of choice for submicron mouth analysis. Particles that are evenly scattered in liquid mediums make up the samples examined by PCS equipment. The PCS laser travels through the Brownian motion, where particles continue to have under these circumstances, and measures the speed of this motion. The PCS calculates sample PI and average particle size. A tried-and-true method for determining particle sizes, from nanometer to micrometer, is the DLS hypothesis. The assumption that tiny particles in suspension move in random patterns are used by this theory. At the same temperature, larger particles travel more slowly than smaller ones [69, 70].

9.8.2 Particle Morphology

Electronic, diagnostic, and therapeutic applications can be revolutionized by changing the physicochemical characteristics of materials at the nanoscale. Atomic force microscopy (AFM) and scanning electron microscopy (SEM) are two methods for structural determination of nanoparticles. AFMs or scan force microscopes (SFMs) are extremely high-resolution scanning probe microscopes with resolutions of nanometer fractions, which are around 1000 times greater than the optical diffraction limit [71].

SEM is a particular kind of electronic microscope that takes photographs of a specimen's surface by scanning it with a high-energy electron beam in a raster injection pattern. SEM possesses the necessary submicron size range nanometer resolution, which is essential for assessing particle form. As a result of the electron's interactions with the sample's atoms, a signal is created that comprises details about the sample, including its composition, electrical conductivity, and surface topography [72].

9.8.3 Particle Structure/Surface Charge

Nanoparticles in aqueous solutions can have their zeta potential measured, which is a typical method of determining the surface charge.

The distribution of the nanoparticles can also be determined using the polydispersity index. Particle interaction is crucial for the stability of colloidal systems. This interaction is quantified by predicting stability using the zeta potential data. Zeta potential is a metric for particle rebound. Additionally, as electrostatic repulsion stabilizes the majority of water-soluble colloid systems, the more repulsion between the particles, the less probable it is that they will approach one another and create cohesiveness. It has been found that nanoparticles with zeta potentials greater than 30 mV (+/) are stable for deposition because clotting is inhibited by the surface charges [73].

9.8.4 Drug Loading and Release

The soluble nanoparticle spotting for drug delivery should be able to incorporate an abundance of drugs into a limited amount of medium. Two methods are used to attach the medication to nanoparticles. The first technique involves simultaneously loading nanoparticles with drugs and combining (or adding) nanoparticles with drugs when developing nanoparticles. After making the nanoparticles, the second method involves a high amount of drug incorporation into them. Drug loading capacity varies depending on the drug solubility, size of the nanoparticles, media type, and the polymers. The effects of drug loading differ based on the solubility of the drug, the size of the nanoparticles, the media type, and the polymers. Major roles are played by polymer solubility in substances, which is correlated with content, molecular weight, and drug-polymer interaction [72, 73].

9.9 Applications of Protein Nanoparticles

The use of protein nanostructures as vehicles for the administration of proteins, medications, and peptides via various routes has been studied in more detail. Applications for protein nanoparticles in biomedicine are shown in Figure 9.13.

9.9.1 Routes

9.9.1.1 *Oral Route*

For all pharmacological applications, oral administration is the first-line approach since it offers a variety of benefits, including patient convenience and compliance and the prevention of contamination issues and

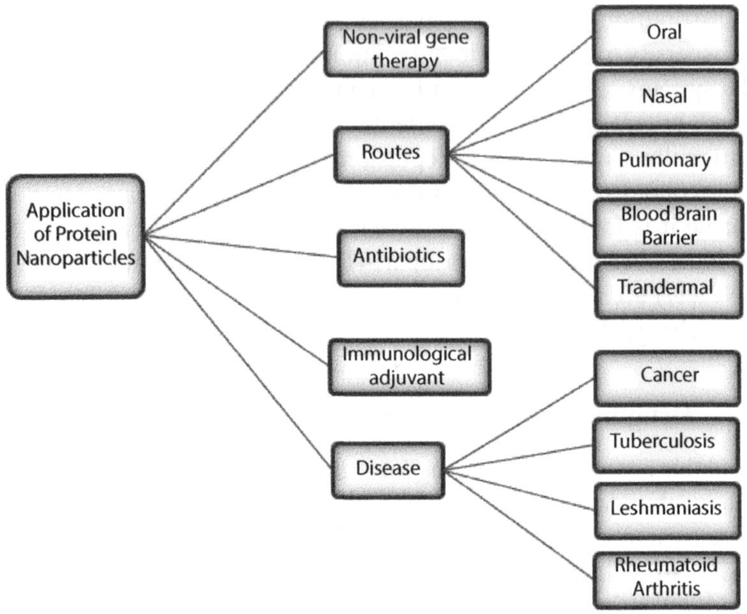

Figure 9.13 Applications of protein nanoparticles.

infections. When consumed orally, proteins must, however, withstand the GI tract's challenging enzymatic and acidic conditions and should trespass the intestinal epithelial barrier before entering the circulation and reaching the target locations [74]. One of the most preferred carriers for protein drug delivery through oral administration are nanoparticles, and it is due to their capacity to preserve proteins from enzymatic activity and hydrolytic deterioration as well as their ability to successfully cross epidermal barriers via selective protein transport, adsorption-mediated transport, paracellular transport, and recognition by microfold cells (M cells) in Peyer's patches. Recently, a lot of time and energy have been put into creating nanotechnology-based oral protein delivery systems, especially insulin [75].

For instance, Sonaje *et al.* used the ionic-gelatin approach to create a pH-sensitive nanoparticle system made of chitosan, poly (-glutamic acid), insulin, sodium tripolyphosphate, and magnesium sulfate ($MgSO_4$), and then tested its oral administration capability for pH-dependent insulin delivery. Enteric-coated capsules (Eudragit®S100 and L100-55) were used to cover the nanoparticles to prevent them from being degraded by the acidic environment of the stomach [76].

9.9.1.2 *Nasal and Pulmonary Route*

For protein delivery, the use of the respiratory system, which includes the nasal and pulmonary routes, has numerous benefits. Compared to the gastrointestinal tract, the local proteolytic action in each delivery route is relatively low. It is easy to trigger large immune responses in each route due to the serologically active mucosal-associated lymphoid tissue, which comprises multiple lymphoid follicles (B-cell regions), macrophages, and dendritic cells. In addition, each of the techniques require lesser medicine doses than the oral route [77]. In light of this, the respiratory tract has quite a potential, specifically in terms of antigenic protein delivery techniques for immunization. The upper respiratory tract is protected in a mucus layer and the airway epithelium is tightly closed via tight junctions, which limit the entry of protein molecules into the respiratory tract [78].

Due to their desired size for local deposition, potential to control drug dissolution rates, and capacity to move protein molecules throughout the epidermal region, including the mucosal barrier, while shielding them from agglomeration, nanoformulations are one of the most suitable approaches to regulate the administration of inhaled proteins [77]. An intranasal vaccine delivery approach has been created by Nochi *et al.*, specifically the use of a cationic pullulan hydrogel (nanogel) composed of cholesteryl that is nanometer-sized (cCHP). The physiologically bridged nanogels (40 nm) enclose proteins in the molecular chains predominantly through hydrophobic contacts, and they increase the efficiency of antigen transport via association with the anion exchange into the epithelium. It was recently stated that pulmonary immunization with nanoparticles turned into getting used to encourage powerful effector memory CD8+T cells (TEM)-biased responses [79].

9.9.1.3 *Blood–Brain Barrier Route*

Protein nanoparticles have the capacity to penetrate the blood–brain barrier, which is inaccessible to conventional drugs administered intravenously. Loperamide, tubocurarine, and doxorubicin are some of the medications attached to protein nanoparticles [80, 81].

9.9.1.4 *Transdermal Delivery*

As proteins and other big or hydrophilic molecules absorb very slowly through the stratum corneum, commonly known as the horny layer, transdermal patches are a noninvasive substitute for injectables; however, they are mostly used for administering minute hydrophobic medications. Transdermal administration of drugs necessitates the passage of drug molecules either via epithelium and surrounding lipids (intracellular transport), across epithelium (intercellular transport), or through skin appendages like hair follicles, scabs, sweat and sebaceous glands. In the stratum corneum, lipidic membrane comprises of both lipophilic and hydrophilic regions covering the intercellular gaps [82].

To improve skin penetration qualities, Choi *et al.* encapsulated proteins into chitosan-conjugated, Pluronic-based nanocarrier (nanogel) and assessed the effectiveness of its transdermal protein delivery. Diacrylate Pluronic F127 and acrylate chitosan were used to create a nanogel (70 nm), and many proteins like insulin, BSA, and galactosidase were then enclosed within the nanogel by leveraging the large volume transition properties of the Pluronic [82, 83].

9.9.2 Non-Viral Gene Therapy

Gelatin nanoparticles that have been cationized have the potential to be a new, efficient method of delivering non-viral genes. The main advantages of gelatin-loaded nanoparticles are their being extremely less toxic to cells, their straightforward preparation, and affordable price [84].

9.9.3 Immunological Adjuvant

Immunological adjuvants, like gelatin nanoparticles, are used to boost cellular and humoral responses to antigens. Gelatin-DNA nanosphere coacervate was employed by Sundar *et al.* (2010) as a means of delivering the CFTR gene to the human tracheal epithelial cells [85].

9.9.4 Antibiotics

Protein nanoparticles exhibit effective effects when used in antibiotics. After interacting with protein nanoparticles, antibiotics either become more effective or become less harmful. Nanoparticles loaded with amoxicillin and gliadin (AGNP) demonstrated anti-*helicobacter pylori* activity, while AGNP had a lower dosage requirement for overall eradication [85, 86].

9.9.5 Diseases

9.9.5.1 Tuberculosis (TB)

Mycobacterium tuberculosis is the bacterium that causes it, and, if it is not treated promptly, can be fatal. Although there are TB medications on the market, nanotechnology is an emerging technique in this area since it allows for a controlled release to the infected cells for a prolonged length of time using an efficient delivery mechanism. According to cytotoxicity experiments, rifampicin-loaded (RIF-loaded) gelatin nanoparticles injure healthy cells less than the free drug, and make the medicine more targetable by using a lower dose. In comparison to normal RIF, RIF-gelatin-loaded nanoparticles were discovered to have high deposition capacity in several organs [87].

The RIF nanoparticles will also enhance the pharmacokinetics of the drug and provide prolonged plasma levels by boosting mean residence time and area under the curve. As an effective and secure delivery system for the treatment of tuberculosis, when isoniazid-loaded mannosylated gelatin nanoparticles were injected intravenously, a considerable reduction of colony forming units of bacteria were seen in the lungs and spleen of TB-infected mice [88].

9.9.5.2 Cancer Therapy

Typically, chemotherapy and radiotherapy are used to treat a group of disorders known as cancer. Even though targeted medication delivery with nanoparticles is being investigated, standard therapy still has several negative side effects, including drug toxicity and drug resistance [89]. *In vitro* and *in vivo* drug solubility of paclitaxel-loaded gelatin nanoparticles in aqueous media is rate-limiting. The targeting of bladder tumors by intravenous paclitaxel gelatin nanoparticles was beneficial, and the amount absorbed by the body was reduced. Paclitaxel-loaded gelatin nanoparticles demonstrate rate-limiting drug solubility *in vitro* and *in vivo* in aqueous media. Gelatin nanoparticles containing paclitaxel delivered intravenously demonstrated effective bladder tumor targeting and a reduction in systemic absorption [90].

The drug dilution issue is resolved by the paclitaxel-loaded gelatin nanoparticles' consistent drug release, and because of the sustained drug level, the frequency of treatments may be reduced. Breast cancer treatment with doxorubicin-polyethyleneimine (DOX-PEI) loaded in human serum albumin (HSA) nanoparticles has been proven to be effective. HSA that had been loaded with DOX-PEI showed reduced cytotoxicity and improved biocompatibility. Additionally, under transfection circumstances, only 80% of the cells were transfected with HSA nanoparticles that included bovine serum albumin that was tetramethyl rhodamine-conjugated [91].

9.9.5.3 *Leishmaniasis*

The protozoan parasite *Leishmania donovani* is the cause of the microbiological disease leishmaniasis. A polyene antibiotic called amphotericin B (AmB) is used to treat visceral leishmaniasis. Functionalized gelatin nanoparticles loaded with AmB minimize cytotoxicity and improve absorption by macrophages. Gelatin nanoparticles coated in 1,2-diacyl-sn-glycerol-3-phospho-1-serine and loaded with AmB exhibit significant levels of AmB accumulation in the liver and spleen, which improve anti-leishmanial activity both *in vitro* and *in vivo* [92].

9.9.5.4 *Rheumatoid Arthritis*

An autoimmune condition called rheumatoid arthritis (RA) affects the body's joints, impairs mobility, damages joints, and lowers quality of life. To overcome the specificity associated with inflamed tissues and lengthen the half-life, methotrexate (MTX) is frequently utilized for the treatment of RA. The deterioration of cartilage and synovial fibroblasts was reduced by MTX-human serum albumin (HSA) [93].

9.10 Conclusion

One of the carriers with the most potential in today's drug delivery methods are nanoparticles. Above all, protein nanoformulations stand out by virtue of its several factors, such as renewable resources availability, low cost, biocompatibility, and biodegradability, as well as many active moieties that allow them to load sizable quantities of drugs and the potential to be linked to targeting groups. They are considered when directing protein-loaded nanocarriers to a certain desired site or area.

The complex coacervation and desolvation processes are used in the production of nanoparticles. Until now, nano spray drying is in its infancy stage; alternative physical and chemical procedures also have their own drawbacks, such as limited flowing tendency or the requirement of surfactant removal. To overcome these limitations, active research must be done in the future. By changing aspects including size, shape, and surface charge, protein-loaded nanoformulations lead to efficient protein delivery. Furthermore, the features of the nanoformulations, concentration, and drug all affect how well proteins load and release drugs. There is a lack of evidence comparing the usage of protein nanoparticles with other approaches, even though it has already led to some exciting results, and has shown significant promise for the future.

9.11 Future Developments

More and more are understood about different proteins and their functions in regulatory networks of human health and illness because of the rapid progress in the field of molecular biology. Pharmaceutical experts are focusing on developing nanotechnology-based formulations to meet the growing market demand for numerous therapeutic medicines and moving beyond the drawbacks of traditional protein and peptide delivery. Due to their small particle size, excellent encapsulation, and targeting ability, nanoparticles give these bio-macromolecules the potential for improved therapeutic outcomes. Bioactives, when delivered at low doses and with precision, reduce toxicity. A progressive rise in the number of clinical trials, patents, and commercialized nanoformulations for protein and peptide delivery have been observed during the last several years.

References

1. Kianfar, E., Protein nanoparticles in drug delivery: Animal protein, plant proteins and protein cages, albumin nanoparticles. *J. Nanobiotechnology*, 19, 159, 2021.
2. Martínez-López, A., Cristina, P. *et al.*, Protein-based nanoparticles for drug delivery purposes. *Int. J. Pharm.*, 581, 119289, 2020.
3. Baimanov, D., Cai, R., Chen, C., Understanding the chemical nature of nanoparticle–protein interactions. *Bioconjug. Chem.*, 30, 7, 1923–37, 2019.
4. Cao, S.J., Xu, S., Wang, H.M., Ling, Y. *et al.*, Nanoparticles: Oral delivery for protein and peptide drugs. *AAPS PharmSciTech*, 20, 5, 190, 2019.
5. Joseph, M., Trinh, H.M., Mitra, A.K., Peptide and protein-based therapeutic agents, in: *Emerging Nanotechnologies for Diagnostics, Drug Delivery and Medical Devices*, pp. 145–167, 2017.
6. Shaji, J. and Patole, V., Protein and peptide drug delivery: Oral approaches. *Ind. J. Pharm. Sci.*, 70, 3, 269–277, 2008.
7. Patel, A., Patel, M., Yang, X., Mitra, A.K., Recent advances in protein and peptide drug delivery: A special emphasis on polymeric nanoparticles. *Protein Pept. Lett.*, 21, 11, 1102–1120, 2014.
8. Homayun, B., Lin, X., Choi, H.J., Challenges and recent progress in oral drug delivery systems for biopharmaceuticals. *Pharmaceutics*, 11, 3, 129, 2019.

9. Aguirre, T.A., Teijeiro-Osorio, D., Rosa, M., Coulter, I.S. *et al.*, Current status of selected oral peptide technologies in advanced preclinical development and in clinical trials. *Adv. Drug Deliv. Rev.*, 106, 223–241, 2016.
10. Kompella, U.B. and Lee, V.H., Delivery systems for penetration enhancement of peptide and protein drugs: Design considerations. *Adv. Drug Deliv. Rev.*, 46, 1-3, 211–245, 2001.
11. Desai, N., Challenges in development of nanoparticle-based therapeutics. *AAPS J.*, 14, 2, 282–95, 2012.
12. Burnham, N.L., Polymers for delivering peptides and proteins. *Am. J. Hosp. Pharm.*, 51, 210–218, 1994.
13. Nischan, N. and Hackenberger, C.P., Site-specific PEGylation of proteins: Recent developments. *J. Org. Chem.*, 79, 10727–10733, 2014.
14. Chaurasia, G., A review on pharmaceutical preformulation studies in formulation and development of new drug molecules. *Int. J. Pharm. Sci. Res.*, 7, 6, 2313, 2016.
15. Tanaka, H., Oasa, S., Kinjo, M., Tange, K. *et al.*, Temperature and pH sensitivity of a stabilized self-nanoemulsion formed using an ionizable lipid-like material via an oil-to-surfactant transition. *Colloids Surf. B Biointerfaces*, 151, 95–101, 2017.
16. Blanco, E., Shen, H., Ferrari, M., Principles of nanoparticle design for overcoming biological barriers to drug delivery. *Nat. Biotechnol.*, 33, 9, 941–951, 2015.
17. Safari, J. and Zarnegar, Z., Advanced drug delivery systems: Nanotechnology of health design a review. *J. Saudi Chem. Soc.*, 18, 2, 85–99, 2014.
18. Artigas, M.A., Fani, A.A., Belloso, O.M., Improving the shelf life of low-fat cut cheese using nanoemulsion-based edible coatings containing oregano essential oil and mandarin fiber. *Food Control*, 76, 1–12, 2017.
19. Yan, M., Du, J., Gu, Z., Liang, M. *et al.*, A novel intracellular protein delivery platform based on single-protein nano capsules. *Nat. nanotechnol.*, 5, 48–53, 2010.
20. Ragelle, H., Danhier, F., Préat, V., Langer, R., Anderson, D.G., nanoparticle-based drug delivery systems: A commercial and regulatory outlook as the field matures. *Expert Opin. Drug Deliv.*, 14, 1–14, 2016.
21. Rother, M., Nussbaumer, M.G., Renggli, K., Bruns, N., Protein cages and synthetic polymers: A fruitful symbiosis for drug delivery applications, bionanotechnology and materials science. *Chem. Soc. Rev.*, 45, 22, 6213–6249, 2016.
22. Zhao, M., Biswas, A., Hu, B., Joo, K.I. *et al.*, Redox-responsive nanocapsules for intracellular protein delivery. *Biomaterials*, 32, 22, 5223–5230, 2011.
23. Naik, A. and Palanivelu, D., Nanoparticles in protein formulation: Challenges, current trends and future perspectives. *Novel Approaches Drug Designing & Dev.*, 2, 1, 15–20, 2021.
24. Pathak, Y., Recent developments in nanoparticulate drug delivery systems, in: *Drug Delivery Nanoparticles Formulation and Characterization*, pp. 1–7, Informa Healthcare, USA, 2009.

25. Verma, R.K. and Garg, S., Current status of drug delivery technologies and future directions. *Pharm. Technol.*, 25, 1–14, 2001.
26. Han, Y., Lin, Z., Zhou, J., Yun, G., Polyphenol-mediated assembly of proteins for engineering functional materials. *Angew. Chem. Int. Ed.*, 59, 15618–15625, 2020.
27. Verma, D., Gulati, N., Kaul, S., Mukherjee, S., Nagaich, U., Protein based nanostructures for drug delivery. *J. Pharm.*, 2018, 9285854, 1–18, 2018.
28. Chenthamara, D., Subramaniam, S., Ramakrishnan, S.G., Krishnaswamy, S. *et al.*, Therapeutic efficacy of nanoparticles and routes of administration. *Biomater. Res.*, 23, 20, 1–29, 2019.
29. Sassi, A.B., Nagarkar, R., Hamblin, P., Vaccines and cancer therapies, in: *Novel Approaches and Strategies for Biologics*, M. Singh, and M. Salnikova, (Eds.), pp. 199–217, Elsevier Inc, Elsevier, London, UK, Waltham, MA, USA, San Diego, CA, 92101-4495. 2015.
30. Nischan, N. and Hackenberger, C.P., Site-specific PEGylation of proteins: Recent developments. *J. Org. Chem.*, 79, 10727–10733, 2014.
31. Kim, J., Cao, L., Shvartsman, D., Silva, E.A. *et al.*, Targeted delivery of nanoparticles to ischemic muscle for imaging and therapeutic angiogenesis. *Nano Lett.*, 11, 694–700, 2011.
32. Dombu, C.Y. and Betbeder, D., Airway delivery of peptides and proteins using nanoparticles. *Biomaterials*, 34, 516–525, 2013.
33. Wang, M., Alberti, K., Sun, S., Arellano, C.L., Combinatorially designed lipid-like nanoparticles for intracellular delivery of cytotoxic protein for cancer therapy. *Angew. Chem. Int. Ed. Engl.*, 53, 2893–2898, 2014.
34. Gu, Z., Biswas, A., Zhao, M., Tang, Y., Tailoring nanocarriers for intracellular protein delivery. *Chem. Soc. Rev.*, 40, 3638–3655, 2011.
35. Scaletti, F., Hardie, J., Lee, Y.W., Luther, D.C., Protein delivery into cells using inorganic nanoparticle–protein supramolecular assemblies. *Chem. Soc. Rev.*, 47, 3421–3432, 2018.
36. Qin, X., Yu, C., Wei, J., Li, L., Rational design of nanocarriers for intracellular protein delivery. *Adv. Mater.*, 31, 1–32, 2019.
37. Lv, J., Tan, E., Wang, Y., Fan, Q., Tailoring guanidyl-rich polymers for efficient cytosolic protein delivery. *J. Control. Release*, 320, 412–420, 2020.
38. Spicer, C.D., Jumeaux, C., Gupta, B., Stevens, M.M., Peptide and protein nanoparticle conjugates: Versatile platforms for biomedical applications. *Chem. Soc. Rev.*, 47, 3574–3620, 2018.
39. Lv, J., Fan, Q., Wang, H., Cheng, Y., Polymers for cytosolic protein delivery. *Biomaterials*, 218, 1–12, 2019.
40. Stewart, M.P., Langer, R., Jensen, K.F., Intracellular delivery by membrane disruption: Mechanisms, strategies, and concepts. *Chem. Rev.*, 118, 7409–7531, 2018.
41. Ho, Y.T., Poinard, B., Yeo, E.L.L., Kah, J.C.Y., An instantaneous colorimetric protein assay based on spontaneous formation of a protein corona on gold nanoparticles. *Analyst*, 140, 4, 1026–36, 2015.

42. Numata, K. and Kaplan, D.L., Silk-based delivery systems of bioactive molecules. *Adv. Drug Del. Rev.*, 62, 1497–1508, 2010.
43. Ki, C.S., Park, Y.H., Jin, H.-J., Silk protein as a fascinating biomedical polymer: Structural fundamentals and applications. *Macromol. Res.*, 17, 935–942, 2009.
44. Jahanban-Esfahlan, A., Dastmalchi, S., Davaran, S., A simple improved desolvation method for the rapid preparation of albumin nanoparticles. *Int. J. Biol. Macromol.*, 91, 703–709, 2016.
45. Fasano, M., Curry, S., Terreno, E., Galliano, M. *et al.*, The extraordinary ligand binding properties of human serum albumin. *IUBMB Life*, 57, 787–796, 2005.
46. Joye, I.J., Nelis, V.A., McClements, D.J., Gliadin-based nanoparticles: Fabrication and stability of food-grade colloidal delivery systems. *Food Hydrocoll.*, 44, 86–93, 2015.
47. Wu, W., Kong, X., Zhang, C., Hua, Y., Improving the stability of wheat gliadin nanoparticles–Effect of gum arabic addition. *Food Hydrocoll.*, 80, 78–87, 2018.
48. Abrams, D., Huang, Y., McQuarrie, S., Roa, W. *et al.*, Optimization of a two-step desolvation method for preparing gelatin nanoparticles and cell uptake studies in 143B osteosarcoma cancer cells. *J. Pharm. Pharm. Sci.*, 9, 124–132, 2006.
49. Madkhali, O., Mekhail, G., Wettig, S.D., Modified gelatin nanoparticles for gene delivery. *Int. J. Pharm.*, 554, 224–234, 2019.
50. Mirshahi, T., Irache, J., Nicolas, C., Mirshahi, M. *et al.*, Adaptive immune responses of legumin nanoparticles. *J. Drug Targeting*, 10, 625–631, 2002.
51. Cho, Y.-H. and Jones, O.G., Assembled protein nanoparticles in food or nutrition applications. *Adv. Food Nutr. Res.*, 88, 47–84, 2019.
52. Park, J.H., Park, H.H., Choi, S.S., Park, T.H., Stabilization of enzymes by the recombinant 30Kc19 protein. *Process. Biochem.*, 47, 164–169, 2012.
53. Park, H.H., Woo, Y.H., Ryu, J., Lee, H.J., Park, J.H., Park, T.H., Enzyme delivery using protein-stabilizing and cell-penetrating 30Kc19α protein nanoparticles. *Process. Biochem.*, 63, 76–83, 2017.
54. Wang, G. and Uludag, H., Recent developments in nanoparticle-based drug delivery and targeting systems with emphasis on protein-based nanoparticles. *Expert Opin. Drug Deliv.*, 5, 499–515, 2008.
55. Dora, C.P., Singh, S.K., Sanjeev, K., Ashok, D. *et al.*, Development and characterization of nanoparticles of glibenclamide by solvent displacement method. *Acta Poloniae Pharm. Drug Res.*, 67, 3, 283–290, 2010.
56. Elzoghby, A.O., Samy, W.M., Elgindy, N.A., Protein-based nanocarriers as promising drug and gene delivery systems. *J. Control. Release*, 161, 38–49, 2012.
57. Zhang, W. and Zhong, Q., Microemulsions as nanoreactors to produce whey protein nanoparticles with enhanced heat stability by sequential enzymatic cross-linking and thermal pretreatments. *J. Agric. Food Chem.*, 57, 19, 9181–9189, 2009.

58. Ren, X., Hou, T., Liang, Q., Zhang, X., Hu, D., Xu, B., Chen, X., Chalamaiah, M., Ma, H., Effects of frequency ultrasound on the properties of zein-chitosan complex coacervation for resveratrol encapsulation. *Food Chem.*, 279, 223–230, 2019.
59. Lee, S.H., Heng, D., Ng, W.K., Chan, H.-K., Tan, R.B., Nano spray drying: A novel method for preparing protein nanoparticles for protein therapy. *Int. J. Pharm.*, 403, 192–200, 2011.
60. Bock, N., Woodruff, M.A., Hutmacher, D.W., Dargaville, T.R., Electrospraying, a reproducible method for production of polymeric microspheres for biomedical applications. *Polymers*, 3, 131–149, 2011.
61. Batrakova, E.V., Bronich, T.K., Vetro, J.A., Kabanov, A.V., Polymer micelles as drug carriers, in: *Nanoparticulates as Drug Carriers*, pp. 57–93, 2006.
62. Sabra, S.A., Elzoghby, A.O., Sheweita, S.A., Haroun, M., et.al., Self-assembled amphiphilic zein-lactoferrin micelles for tumor targeted co-delivery of rapamycin and wogonin to breast cancer. *Eur. J. Pharm. Biopharm.*, 128, 156–169, 2018.
63. Ganesan, P., Ramalingam, P., Karthivashan, G., Ko, Y.T. *et al.*, Recent developments in solid lipid nanoparticle and surface-modified solid lipid nanoparticle delivery systems for oral delivery of phyto-bioactive compounds in various chronic diseases. *Int. J. Nanomedicine*, 13, 1569–1583, 2018.
64. Araújo, R.V., Santos, S.D., Ferreira, E., II, Giarolla, J., New advances in general biomedical applications of PAMAM dendrimers. *Molecules*, 23, 11, 2849, 2018.
65. Tunki, L., Kulhari, H., Sistla, R., Pooja, D., Dendrimer-based targeted drug delivery, in: *Pharmaceutical Applications of Dendrimers*, pp. 107–129, 2020.
66. Yang, J., Bahreman, A., Daudey, G., Bussmann, J. *et al.*, Drug delivery via cell membrane fusion using lipopeptide modified liposomes. *ACS Cent. Sci.*, 2, 621–630, 2016.
67. Ozboyaci, M., Kokh, D., Wade, R., Three steps to gold: Mechanism of protein adsorption revealed by Brownian and molecular dynamics simulations. *Phys. Chem. Chem. Phys.*, 18, 15, 10191–200, 2016.
68. Corbo, C., Molinaro, R., Tabatabaei, M., Farokhzad, O.C. *et al.*, Personalized protein corona on nanoparticles and its clinical implications. *Biomater. Sci.*, 5, 3, 378–87, 2017.
69. Boyd, R.D., Pichaimuthu, S.K., Cuenat, A., New approach to inter-technique comparisons for nanoparticle size measurements; Using atomic force microscopy, nanoparticle tracking analysis and dynamic light scattering. *Colloids Surf. Physicochem. Eng. Asp.*, 387, 35–42, 2011.
70. Hoo, C.M., Starostin, N., West, P., Mecartney, M.L., A comparison of atomic force microscopy (AFM) and dynamic light scattering (DLS) methods to characterize nanoparticle size distributions. *J. Nanopart. Res.*, 10, 89e96, 2008.

71. Bummer, P.M., Chemical considerations in protein and peptide stability, in: *Protein Formulation and Delivery*, p. 7e42, Informa Healthcare, Boca Raton, 2007.
72. Reddy, L.H., Arias, J.L., Nicolas, J., Couvreur, P., Magnetic nanoparticles: Design and characterization, toxicity and biocompatibility, pharmaceutical and biomedical applications. *Chem. Rev.*, 112, 5818e78, 2012.
73. Kirby, B.J. and Hasselbrink, E.F., Zeta potential of microfluidic substrates: 1. Theory, experimental techniques, and effects on separations. *Electrophoresis*, 25, 187e202, 2004.
74. Woitiski, C.B., Carvalho, R.A., Ribeiro, A.J., Neufeld, R.J. *et al.*, Strategies toward the improved oral delivery of insulin nanoparticles via gastrointestinal uptake and translocation. *BioDrugs*, 22, 223–237, 2008.
75. Shan, W., Zhu, X., Liu, M., Li, L. *et al.*, Overcoming the diffusion barrier of mucus and absorption barrier of epithelium by self-assembled nanoparticles for oral delivery of insulin. *ACS Nano*, 9, 2345–2356, 2013.
76. Pridgen, E.M., Alexis, F., Kuo, T.T., Levy-Nissenbaum, E. *et al.*, Transepithelial transport of Fc-targeted nanoparticles by the neonatal fc receptor for oral delivery. *Sci. Transl. Med.*, 5, 213, 2013.
77. Nochi, T., Yuki, Y., Takahashi, H., Sawada, S. *et al.*, Nanogel antigenic proteindelivery system for adjuvant-free intranasal vaccines. *Nat. Mater.*, 9, 572–578, 2010.
78. Li, A.V., Moon, J.J., Abraham, W., Suh, H. *et al.*, Generation of effector memory T cell-based mucosal and systemic immunity with pulmonary nanoparticle vaccination. *Sci. Transl. Med.*, 2013.
79. Thornton, E.E., Looney, M.R., Bose, O., Sen, D. *et al.*, Spatiotemporally separated antigen uptake by alveolar dendritic cells and airway presentation to T cells in the lung. *J. Exp. Med.*, 209, 1183–1199, 2009.
80. Guixer, B., Arroyo, X., Belda, I., Sabidó, E. *et al.*, Chemically synthesized peptide libraries as a new source of BBB shuttles. Use of mass spectrometry for peptide identification. *J. Pept. Sci.*, 22, 9, 577–591, 2009.
81. Ulbrich, K., Hekmatara, T., Herbert, E., Kreuter, J., Transferrin-and transferrin-receptor-antibody-modified nanoparticles enable drug delivery across the blood–brain barrier (BBB). *Eur. J. Pharm. Biopharm.*, 71, 251–256, 2009.
82. Veiseh, O., Tang, B.C., Whitehead, K.A., Anderson, D.G., Langer, R. *et al.*, Managing diabetes with nanomedicine: Challenges and opportunities. *Nat. Rev. Drug Discovery*, 14, 45–57, 2015.
83. Nose, K., Pissuwan, D., Goto, M., Katayama, Y. *et al.*, Gold nanorods in an oil-base formulation for transdermal treatment of type 1 diabetes in mice. *Nanoscale*, 4, 3776–3780, 2012.
84. Huang, J., Shu, Q., Wang, L., Wu, H., Wang, A.Y., Mao, H., Layer-by-layer assembled milk protein coated magnetic nanoparticle enabled oral drug delivery with high stability in stomach and enzyme-responsive release in small intestine. *Biomaterials*, 39, 105–113, 2015.

85. Sundar, S., Kundu, J., Kundu, S.C., Biopolymeric nanoparticles. *Sci. Technol. Adv. Mater.*, 11, 1–13, 2010.
86. Kundu, J., Chung, Y.-I., Kim, Y.H., Tae, G., Kundu, S.C., Silk fibroin nanoparticles for cellular uptake and control release. *Int. J. Pharm.*, 388, 1–2, 242–250, 2010.
87. Saraogi, G.K., Gupta, P., Gupta, U.D., Jain, N.K. *et al.*, Gelatin nanocarriers as potential vectors for effective management of tuberculosis. *Int. J. Pharm.*, 385, 1–2, 143–149, 2010.
88. Saraogi, G., Sharma, K., Joshi, B., Mannosylated gelatin nanoparticles bearing isoniazid for effective management of tuberculosis. *J. Drug Targeting*, 19, 3, 219–227, 2011.
89. Yu, M., Wu, J., Shi, J., Nanotechnology for protein delivery: Overview and perspective. *J. Control. Release*, 240, 24–37, 2015.
90. Lu, Z., Yeh, K., Wang, J. *et al.*, Paclitaxel gelatin nanoparticles for intravesical bladder cancer therapy. *J. Urol.*, 185, 1478–1483, 2011.
91. Paul, A., Shao, W., Prakash, S., Abbasi, S. *et al.*, Cationic albumin nanoparticles for enhanced drug delivery to treat breast cancer: Preparation and *in vitro* assessment. *J. Drug Deliv.*, 8, 1–8, 2012.
92. Yasmin, R., Shah, M., Khan, S.A., Ali, R., Gelatin nanoparticles: A potential candidate for medical applications. *Nanotechnol. Rev.*, 6, 2, 191–207, 2017.
93. Ren, K., Dusad, A., Dong, R., Quan, L. *et al.*, Albumin as a delivery carrier for rheumatoid arthritis. *J. Nanomed. Nanotechnol.*, 4, 1–4, 2013.

Part 3

ROUTE OF ADMINISTRATION

10

Oral Vaccine Delivery: Current Status

Pankti C. Balar[1], Vidhi A. Modh[2], Sanjay P. Chauhan[2], Hetvi K. Solanki[1], Rajeshkumar K. Patel[2*], Dasharath M. Patel[2] and Maharshi B. Padya[2]

[1]Pharmacy Section, L.M. College of Pharmacy, Ahmedabad, Gujarat, India
[2]Graduate School of Pharmacy, Gujarat Technological University, Gandhinagar, India

Abstract
Vaccines are preparations of dead or weakened pathogens, such as bacteria or viruses, that, when administered, induce the formation of antibodies or cellular protection against the pathogen. Thanks to the innovative biotechnology techniques used in both the initial and final analyses of these products, the successful and currently rapid advancement of various vaccine types and monoprophylaxis through mass vaccination have proven to be a successful approach to control such viral infections. In contrast to conventional medications, which may have an overall negative effect on the body, nanovaccines are vaccinations made of nanoparticles (NPs) and are rapidly developing as an entirely novel category of vaccines that specifically target the spot in the body where the sickness or infection began. Due to the same nanoscale sizes of pathogens and nanoscale materials, the body's immune system responds to stimuli appropriately, resulting in favorable cellular and humoral responses of the immune system. To address the many challenges in vaccine manufacturing, targeted dispersion of NPs produces long-term release together with enhanced stability and antibody response. To get beyond the physiological and immunological barriers that protect the body's internal environment from outside substances, an oral vaccination is necessary. This chapter's emphasis is on nanovaccines that are employed in oral vaccine delivery.

***Keywords*:** Vaccines, oral vaccines, GIT barriers, antigens, antibody

**Corresponding author*: rajeshpatel@gtu.edu.in

Vivek P. Chavda and Vasso Apostolopoulos (eds.) Nanocarrier Vaccines: Biopharmaceutics-Based Fast Track Development, (347–366) © 2024 Scrivener Publishing LLC

10.1 Introduction

The delivery of a vaccine encourages the creation of antibodies or immune systems against the pathogen by using a dead or weakened pathogen, such as a weakened bacterium, virus, or component of the pathogen's structure. A particle of substance with a diameter between 1 and 100 nanometers (nm) is referred to as a nanoparticle. To offer effective immunization through improved targeting and by inducing a reaction of antibodies at the cellular level, the invention of nanocarrier-based vaccines has begun to get a lot of interest. Effective vaccine delivery methods that can shield the encapsulated antigen from the hostile *in vivo* environment and maintain a sustained release have been developed with the use of nanotechnology. This helps to activate the immunotherapy features of the vaccines. Since parenteral vaccine delivery triggers potent cellular and systemic immune responses, it is now the most used approach [1]. The parenteral route does, however, have some intrinsic drawbacks, such as low patient compliance since it hurts when it is administered, the inability to self-medicate, and the absence of mucosal immune responses. Therefore, there is an urgent need to create a non-invasive method of administering vaccines. In comparison to parenteral, nasal, rectal, and vaginal routes, the oral route is regarded as the best way to administer vaccines since it is non-invasive, simple to administer, capable of mass vaccination, has a wide range of safety characteristics, and is simple to produce and store. Refer to Figure 10.1. Oral vaccinations are preferred because parenteral vaccines are more difficult to produce and scale up since they require strict environmental control and purity. Researchers are looking into vaccine administration for therapeutic, as well as preventative, purposes due to the wide immune response (i.e., mucosal and systemic) associated with the oral route [2].

In contrast to conventional medications, nanovaccines are vaccinations made of nanoparticles and are rapidly developing as an entirely novel type of vaccine that concentrates on the spot in the body where the sickness or infection began. Intranasal, intravenous, transdermal, oral, and other administration methods are all simple ways to distribute nanovaccines, which can also be modified to traverse the blood–brain barrier [3]. Unlike conventional subunit vaccines, these vaccine formulations have longer thermal stability at ambient temperature and do not need a cold chain [4]. The storage durability of nanovaccines greatly lowers the cost of medical storage and enables worldwide distribution to far-off places. Nanoparticles are capable of retaining antigen protein stability and functioning for prolonged periods of time in addition to thermal stability, which is sought to trigger a suitable immune response. Additionally, they can be "pathogen-like" for

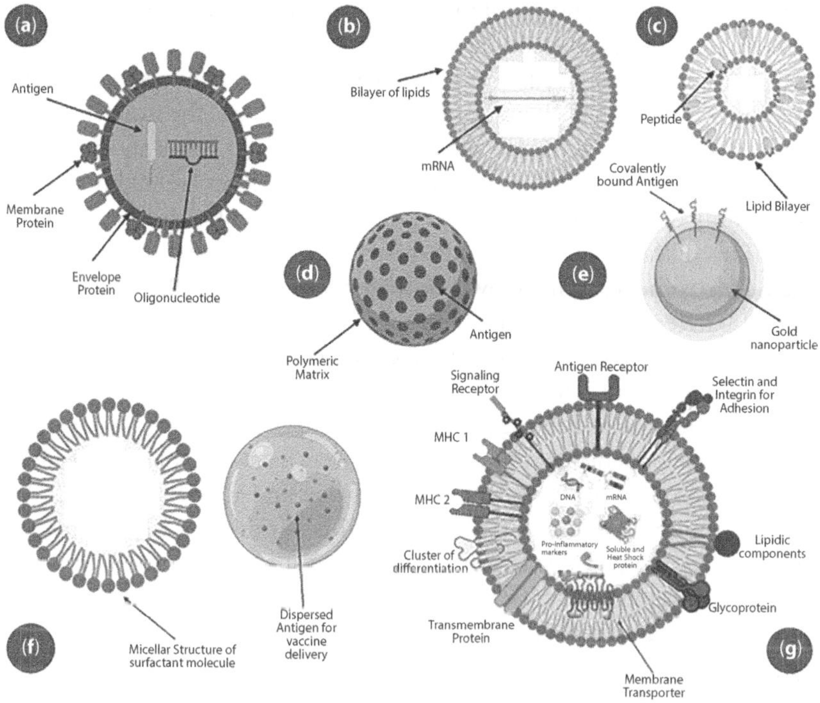

Figure 10.1 A schematic presentation of the use of different nanoparticles. Adopted under CC BY 4 from [5].

effective absorption by being linkers functionalized with substances. The particles break down to release the encapsulated antigen over time after being ingested by immune cells. Typically, biocompatible polymeric materials that are both natural and artificial are used to create particles, creating a "plug and play" environment. Due to the variety of materials employed in their synthesis, nanoparticles can have their form, size, surface charge, hydrophobicity, and eventually their actions, which are self-adjuvating, altered [4].

10.2 Need for Oral Vaccines

The inherent instability and returning to their virulent state, thereby rendering the antigen challenging to deliver, are major obstacles in the development of live vaccines. Flu, hepatitis A, and other inactivated or dead vaccinations and toxoids, including lockjaw and diphtheria, typically require many doses and result in a poorer immune response. Additionally, protein subunit vaccines do have certain drawbacks, such as low stability and the capacity to modify protein structure, which may prevent the production of some neutralizing antibodies. Furthermore, the isolated protein has not sufficiently activated the immune system to result in immunization. Similar to this, DNA and protein antigen vaccines are highly unstable and readily disintegrate in the biological environment. Additionally, many infectious disease vaccines are either ineffective or poorly formulated, and their effectiveness is limited by their inability to successfully overcome biological barriers and reach the target areas. The development of oral vaccines has lately looked into the nanotechnology-based technique to tackle such difficulties [6]. The skin and mucosal membranes act as physical barriers and first line of defense against pathogenic infiltrations. As they often contact with outside surroundings, infections have the ability to infiltrate the mucosal membranes of the aerodigestive and urogenital tracts. As a result, immunoglobulin A (IgA) synthesis is stimulated, which can more specifically stop pathogens from penetrating the mucosa at the infection site. The development of systemic immunoglobulin G (IgG), which can prevent pathogen reproduction at later stages of infections and provide total defense, must be stimulated in addition to local mucosal immunity [7]. The bulk of today's vaccinations, however, are primarily given by parenteral ways via subcutaneous or intramuscular injection, and they normally only work by strongly inducing local mucosal immune responses. The first goal is to create an efficient vaccination technique that can create a robust immune barrier at both mucosal and systemic locations. These barriers, which are typically employed for systemic usage, can be bypassed by the development of efficient mucosal immunity. The majority of mucosal vaccines primarily induce immunity by using live or attenuated versions of pathogens [8]. In recent decades, advanced nanotechnology has considerably accelerated the development of oral vaccine delivery. Although a great deal of work has been made with nanocarriers for oral vaccination administration, the majority are still in the lab and relatively few have started clinical trials. Safety, effectiveness, cost, and scalability need to be carefully examined to transfer NP-mediated oral immunization from the lab to the bedside.

10.3 Nanoparticles as an Oral Vaccine Delivery System

To get beyond the physiological and immunological barriers that protect the body's internal environment from outside substances, oral vaccination is necessary. While peptidases found in the GI tract serve as the primary

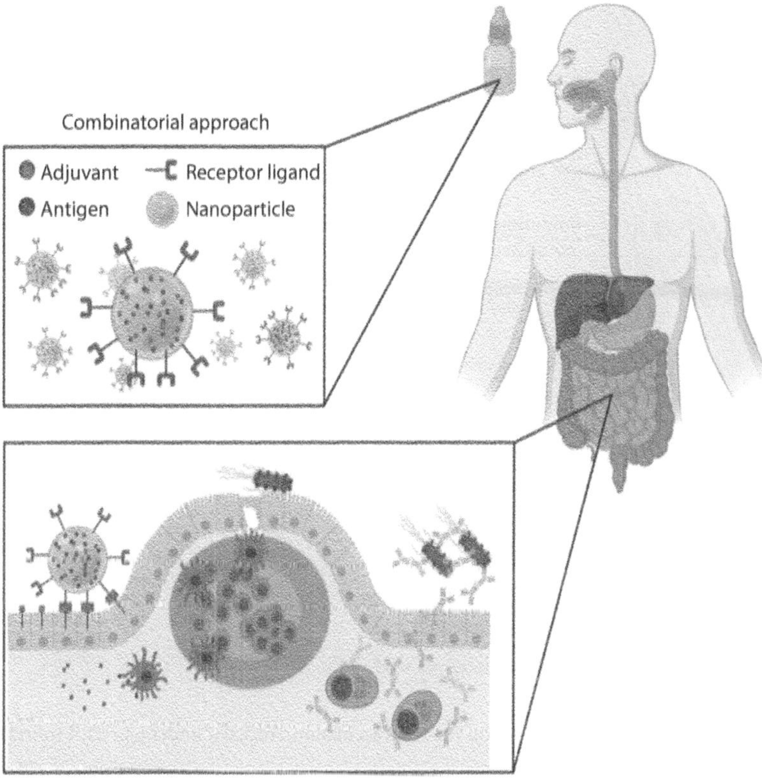

Figure 10.2 Various novel means by which an oral vaccination can be loaded in the body. Adopted under CC BY 4 from [9].

enzyme that enzymatically destroys foreign material and stops their invasion, the GI epithelium produces a tight connection membrane barrier that is semipermeable and regulates absorption. The effectiveness of the vaccinations is influenced by the type of epitope and delivery methods. The mucosa is where most infections first enter the body; it is crucial to immunize the mucosa to trigger a defensive mucosal immune response. Refer Figure 10.2 for understanding various novel approaches by with the oral vaccination can be delivered.

The sublingual (SL) mucosa is one of the finest possibilities for mucosal immunization despite various hurdles to vaccine penetration caused by its distinct architecture and physiology. The creation of potent SL vaccines against infectious illnesses will be facilitated by the efficient and targeted delivery of adjuvants and antigens to the proper mediators of immunity in the SL tissue [10]. Refer to Figure 10.3. An effective immunological response to influenza antigenic substances co-delivered sublingually with

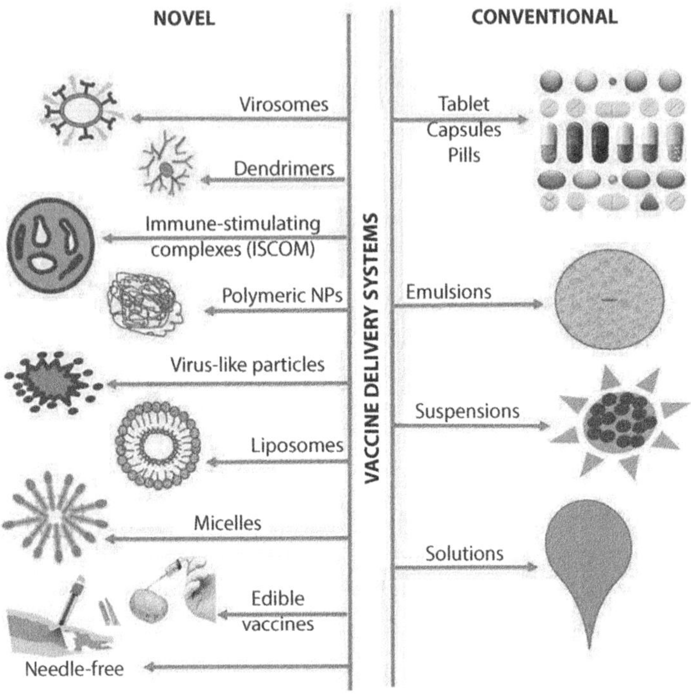

Figure 10.3 Various delivery systems for oral vaccination. Adopted under CC BY 4 from [11].

the synthetic toll-like receptor-4 agonist, CRX-601, was seen. In an SL murine influenza vaccination model, the potential of liposome modification using PEG copolymers (Pluronics), phospholipid-PEG conjugates, and chitosan to elicit an immunological response was assessed. Pluronic copolymers were less successful than phospholipid-PEG conjugates at producing stable, surface-neutral liposomes. In comparison to unmodified liposomes, SL immunization with surface-transformed liposomes bearing CRX-601 adjuvant produced much better flu-specific responses [10].

To carry antigens to the target location and properly shield vaccines/antigens from the GIT's adverse surroundings, the carrier system must be perfect. Currently, majority of oral vaccinations are administered as live, weakened, or disabled organisms mixed with adjuvants, and they may have the potential to revert to an active form, especially in individuals with inadequate defenses, ultimately hurting the host. For sufficient exposure to antigen-presenting cells (APCs), vaccines can be encased in nanoparticles or connected to their surfaces. The particle size, surface charge, surface area, high-loading capacity, protection in stomach fluid, and improved penetration ability through the mucosal barrier of the gut are only some of the unusual controllable characteristics of

NPs [12]. The various means to ensure the efficacy of nanocarrier-mediated oral vaccinations are discussed here.

10.3.1 Lipid Nanoparticles

Due to its capacity for encasing both hydrophobic and hydrophilic components, liposomes have become extremely important in the medication delivery industry. These are made up of phospholipids that self-assemble into an exterior lipid bilayer and an interior aqueous core. While hydrophobic components can be included in the lipid bilayer or at the intersection of an inner and outer layer, hydrophilic antigens can be effectively retained inside the core. Liposomes are a useful technique for delivering antigens and immunostimulants because they may entrap both hydrophilic and lipophilic components. Unmodified liposomes are readily disintegrated by lipase produced by the pancreas and break down in an acidic pH. Additionally, intestinal bile salts may weaken the phospholipid bilayer, causing liposomes to lyse. The entrapped substance is released early as a result of this lysis [13]. The instability of liposomes in the GIT can be controlled by changing their formulation. Biosomes are created when biodegradable and compatible bile salts are added to the liposome composition. Bile salts aid in protecting liposomes from oxidation in the GIT by stabilizing them [14].

The outer oily or aqueous phase is used to disperse w/o or o/w emulsions to create several emulsions for the oral administration of vaccines. These different emulsions are further strengthened by the introduction of exogenous surfactants. This formulation's oral administration revealed increased macrophage and APC uptake in mucosal lymph nodes, which strengthened the immune response by increasing systemic antibody production. One study showed how to include OVA into a multiple emulsion with squalene oil as the basis (w/o/w). Emulsions are inexpensive in addition to their other advantages [15, 16].

10.3.2 Polymeric Nanoparticles

These are extremely small antigen carriers with the capacity to enclose, adsorb, or combine any antigen on or within them. Over the course of 20 years, many polymers have been investigated for their possible application in the delivery of vaccines. Encapsulating vaccines in polymers protects antigens and regulates the release at the site since biological molecules are delicate and susceptible to degradation while traveling through the GIT. To improve active targeting, these polymeric nanoparticles' surfaces can be modified. Polyester, specifically poly lactic acid (PLA) and poly lactic-co-glycolic acid, is the most common kind of synthetic polymer used to manufacture biodegradable nanoparticle-based vaccines that allow persistent antigen release with a single dosage (PLGA) [17].

Carreno *et al.* created an oral vaccination by encapsulating *Salmonella typhi* porin proteins in PLGA and considering oral administration. Because it stopped antigens from deteriorating in the stomach's pH after oral administration, this encapsulation demonstrated better effectiveness. Antigens are shielded from the hostile stomach environment by pH-sensitive polymers, which only allow for antigen releases in the intestine. By adding M-cell-specific compounds aimed at the periphery of chitosan nanoparticles, the oral transfer of antigens can be enhanced [18, 19].

10.3.3 Vesicular Delivery

Exosomes are round and vesicular structures that are related to endosomes, and are composed of a variety of proteins and lipid bilayers. They have diameters between 40 and 160 nm. They are released by a wide range of cell types, and they are extensively dispersed throughout all physiological tissues and fluids. Exosomes are used during intercellular communication as a means of transferring materials from donor cells to recipient cells. Exosomes are employed in the therapy of a number of diseases because of their innate capacity to regulate intricate metabolic processes. They can be altered to deliver different pharmacological and bioactive substances at certain sites. Exosomes can deliver various proteins to destination cells, changing the biological processes of the cells in the process. Exosome targeting activity in exosome donor cells can be produced via direct surface alteration and manipulation [20, 21].

When colloidal particles self-assemble, microcapsular structures called colloidosomes are produced. Among other mechanical properties, these particles regulate colloidosome size, permeability, and strength. Depending on the kind of emulsion utilized in their creation, various kinds of colloidosomes may be identified. They can be emulsion-based colloidosomes containing or without emulsions. A number of approaches, like thermal annealing, gel trapping, covalent crosslinking, and many more, have been used to study the creation of colloidosomes [22].

Phytosomes are aqueous-based phytomedicine-phospholipid complexes that are colloidal, tiny vesicles with micellar or hexagonal nanostructures, comparable to liposomes. For the enteral route of drug administration, phytosomes are desirable carriers due to their unique characteristics, including their nanometric size, amphiphilic nature, high entrapment efficiency, and high stability [23]. Attaching drugs to phospholipid molecules, a crucial part of phytosomes, has a number of benefits, including the prevention of drug leakage from the carrier and delayed, sustained drug release that boosts the efficacy of certain therapies [24]. Sphingosome development is based on challenging stability issues with liposomes. Liposome instability is brought on by the phospholipid ester link's hydrolysis. This hydrolysis can be stopped by using lipids with ether or amide links in place of ester bonds [25].

10.4 Advantages of Oral Nanovaccines

- Compared to alternative delivery methods, oral vaccination is the most appealing since it is non-invasive, safe, and straightforward to administer, demonstrating high patient compliance and clinical viability [8].
- Oral vaccinations are needle-free and can significantly lower the hazards associated with needle-stick injuries that could result in the cross-transmission of blood-borne diseases owing to the reuse of infected or improperly discarded syringes and needles [26].
- It is a convenient and simple method of remote immunization.
- The cost of producing oral vaccinations can be reduced by avoiding the rigorous sterile conditions required for parenteral formulation.
- Oral vaccinations might minimize the need for a cold chain, resulting in less material needing to be stored and transported [27].

10.5 Drawbacks and Disadvantages of Oral Nanovaccines

- The GI system has several physiochemical and biological barriers that oral antigens must overcome.
- The entire procedure of immunization varies with various factors such as particle charge, density, size, molecular weight, bond rotation, elasticity, and polymorphism.
- The intestinal epithelium's mucus-secreting layers act as a biological barrier and aid in the digestion of food for nutritional absorption as well as defense against the ingestion of harmful substances.

- The stomach's extremely acidic environment, the GI tract's shift in pH, and the presence of certain proteolytic enzymes are all factors in the destruction of proteins.
- All of these have an impact on the transport of biomolecules that are prone to denaturation and degradation, such as peptides or antigenic proteins.
- Another issue with the creation of oral vaccines is that in comparison to parenteral formulation, a larger dosage of antigen is required to trigger an immune response.
- A higher dosage has a greater effect on tolerance [28].

10.6 Barriers in Oral Vaccines Delivery

10.6.1 Physiological Barrier

Enzyme-catalyzed hydrolysis and GI tract pH variations have an effect on the durability of oral vaccinations. They lead to the degradation and irreversible conformational alterations of the oral vaccine antigens. Denatured vaccines further result in precipitate in the GI-lumen, a decrease in intestinal binding, and the presentation of dormant epitopes to PP, all of which have the potential to elicit an unwanted immunological response [26]. The GI-transit time has an impact on the oral vaccination's GI stability as well. Given the varying gastric-emptying durations, which can be as short as 2 hours while fasting or as long as 16 hours when fed, the prolonged residency of the vaccine in the stomach may result in proteolytic digestion or denaturation under the extremely low stomach pH (pH 2–4). According to Meynell *et al.*, exposure to *Streptococcus pneumoniae* R36a led to an increase in M cells [27]. Due to the potential side effects such food allergies and inflammatory illnesses, the attention has shifted to developing better strategies for targeting already-existing M cells. M cell-specific moieties can help vaccines specifically target M cells [29].

The principal location for the transport or absorption of antigens is the GI tract (gastrointestinal tract). The GI tract's structure is crucial to the effectiveness of oral vaccination formulation. Because of the GI membrane's usual shape, orally administered drugs must get past a number of physiological and immunological barriers. The primary physiological barrier is the intestinal epithelium. Mucus-secreting layers cover it, aiding in nutritional digestion and absorption, as well as providing defense against pathogen invasion.

To maintain these conditions, the GI tract must release proteolytic enzymes and bile salts for fat and protein digestion, as well as maintain a pH that is varied throughout the tract and includes a very acidic environment (stomach). Oral vaccinations may deteriorate and lose their immunogenicity under such conditions. Second, their absorption is restricted by the condensed epithelial cell surface and thick mucus layer. Furthermore, delivering antigens to a particular area to cause long-lasting and mucosal secretions is challenging. Due to the aforementioned considerations, oral vaccinations need a higher dosage of antigens than typical injectable methods do to produce powerful, long-lasting mucosal immunity [30].

10.6.2 Immunological Barrier

Oral vaccinations must not only overcome physiological barriers to be successful; they must also retain their safety and efficacy while avoiding oral tolerance. In terms of establishing mucosal and humoral immunity, the oral poliovirus vaccine (OPV) is an excellent form of vaccination. However, the transition from parenteral inactivated poliovirus vaccine (IPV) to oral poliovirus vaccine (OPV) has highlighted the importance of a safety concern in the development of an oral vaccine. According to multiple studies using animal models, oral tolerance may be evident in activation-induced cell death (deletion), anergy, and most recently, the generation of regulatory T cells [31].

Microfold cells (M cells) transport antigens through the gastrointestinal epithelium, where they are then taken up by dendritic cells (DC) via dendrites. Antigens are processed and presented by dendritic cells to stimulate immature T lymphocytes. The naive T cells also activate B cells, cause them to change into IgA+ plasma B cells, and cause them to generate antibodies that are specific to the antigen. The mucosal membrane's surface is then reached by the antibodies.

10.7 Currently Licensed Oral Vaccines

In challenging gastrointestinal settings, formulation solutions using innovative lipid and polymer-based nano/microparticle platforms maintain the stability and antigenicity of vaccines. Affixed ligands may precisely target immune cell absorption, and particles' physicochemical characteristics can be altered to delay the release of antigens. These are workable methods for administering oral vaccinations successfully. The most crucial stage is to develop a vaccination strategy that can establish a strong immune barrier both mucosally and systemically. Majority of adjuvants with medicinal approval are mostly systemically given and ineffective at promoting mucosal immunity. Most mucosal vaccines work to instill immunity largely by utilizing live or attenuated microorganisms. In addition, it is envisaged that the production of oral vaccinations will not need to adhere to the same strict and sterile standards as parenteral versions, which will reduce expenses. The oral vaccine delivery approach can also result in mucosal specific antigen IgA antibodies in the blood in addition to systemically activating antigen-specific antibodies like IgG [32]. The simplest way to take therapeutics is orally. Compared to alternative mucosal locations for vaccination, the oral route provides a number of benefits: By avoiding "needle phobia" and being painless, oral vaccines can improve patient compliance to vaccinations. Finally, oral vaccinations are easy to do, are self-administered, and can increase the capacity for mass immunization. To reduce needle-stick injuries during traditional delivery techniques, qualified medical professionals are required [4]. By avoiding cold-chain constraints, oral vaccines might cut down on storage and travel. Cold-chain travel could be eliminated with novel carrier mediated vaccine delivery. Deliveries without needles could also cost less to transmit iatrogenic blood-borne pathogens. The manufacture of oral vaccinations might not require sterile conditions like parenteral counterparts, resulting in cheaper production costs. IgG and mucosal IgA antibodies may be systemically activated by oral vaccinations in the blood. Despite their many benefits, oral immunizations must overcome physical and immunological restrictions, including ineffective mucosal permeability, limited immunogenicity, and mucosal antigen instability. In a study by Chen *et al.*, their goal was to create dry-coated vaccinations that will result in a large number of antigen-presenting cells [33]. NP technology has the potential to address many of these important problems. First, a test example demonstrates that dry-coated NPs for the influenza vaccine maintain their stability for at least 6 months at 23°C, possibly reducing the significant burden of the cold chain in vaccination campaigns in temperate and subtropical regions. Second, NPs offer great dose sparing (30 times) when flu vaccines are administered selectively to thousands of the skin's APCs. If successfully implemented in humans, this dose-sparing method will significantly reduce the cost of immunization, especially for more expensive vaccines (like Gardasil, a vaccine from Merck against HPV) [33]. Refer to Table 10.1.

Table 10.1 Marketed oral vaccines.

Name	Targeted diseases	Incorporated antigen	Type of formulations	Frequency and dosage of administration	Manufacturer	Reference
Polio Sabin™	Polio	Live attenuated poliomyelitis virus Sabin strains type 1	Solution	One dose of 0.1 ml, three doses, taken at least 4 weeks apart	Multiple	[34]
Dukoral	Cholera	Inactivated whole – cell *Vibrio cholera* Inaba, O1 Ogawa, classic and El Tor strains	Suspension	Two doses are taken orally	Valneva Sweden AB (Solna, Sweden)	[35]
Rotateq™		RV5, which consists of five live human –bovine reassortant rotavirus strains (G1, G2, G3, G4, and PIA)	Solution	Three doses, 4–10 weeks apart	Merck (West Point, PA, USA)	[36]
Rotarix™	Rotavirus	Attenuated human rotavirus (G1P strain)	Suspension	Two or three doses	GlaxoSmithKline (Wavre, Belgium)	[37]

10.8 Descriptions of Licensed Oral Vaccines

10.8.1 Oral Polio Vaccine

It was the first mucosal vaccination to be commercialized successfully. Each of the three infection serotypes is represented by a live attenuated poliovirus strain in the oral polio vaccination. It takes three doses to produce immunity on the mucosal and humoral levels, providing protection. The vaccination contains an antibody that aids in preventing the poliovirus from spreading to the neurological system, protecting children from paralysis at a young age. The intestinal mucosa, which is the main location for poliovirus entrance and multiplication, has a local immunological reaction as a result. This local gut reaction response aids in preventing the spread of infection from person to person [38].

10.8.2 Live Oral Typhoid Vaccine

Salmonella typhi, which is spread by contaminated food or water, causes typhoid fever. Typhoid fever is therefore more prevalent in industrial areas and less developed nations. Oral live attenuated Ty21a vaccine is one of the two approved typhoid fever vaccines [39]. The current formulation comes in two forms: a liquid solution and an enteric-coated capsule. Three or four doses are taken on different days [40].

10.8.3 Cholera Vaccine

A severe diarrheal illness called cholera is spread through feces contaminating food and drink. *Vibrio cholera*, which is acquired by consumption of tainted food and water, colonizes the epithelial lining of the gut and causes copious diarrhea that, if left untreated, kills within hours. Recombinant cholera toxin B offers defense against several serotypes, is secure, and offers 2 years of cholera protection [41].

The second and most recently approved vaccination on a global scale is the oral live attenuated cholera vaccine CVD-10-HgR (VAXCHORA, PaxVax, USA), which contains a genetically modified strain of *V. cholera* called the Inaba strain. An earlier CVD 103-HgR vaccine (Orochol, Mutachol) was reformulated into this one [42].

10.8.4 Rotavirus Vaccine

Infants and kids under the age of 5 who have diarrheal mortality are typically affected by rotavirus. The virus is a three-layered particle with a variety of antigens. Human rotavirus illness is mostly brought on by five serotypes. One is a monovalent attenuated human rotavirus vaccine (RotaRix) and the other is a pentavalent bovine-human rotavirus vaccine (Rotateq). Given that Rotarix requires fewer dosages and hence requires less storage space, it has a better cost-effective ratio [43].

10.8.5 Oral Adenovirus Vaccine

Oral-dosing permits targeted aseptic infection in the lower intestine tract, in contrast to upper respiratory tract immunity [44]. Human adenoviruses (AdV) infrequently lead to illness. More severe respiratory tract infections and acute respiratory diseases are brought on by AdV-4 and AdV-7 [45]. Only the live oral AdV-4 and AdV-7 vaccine, approved in the US, is given to prospective military members. Although long-term immunity has not been evaluated, placebo-controlled clinical studies have demonstrated the vaccine's great safety and antibody response. Serum samples from 2011 co-administered live oral AdV-4 and AdV-7 vaccination recipients were collected over a 6-year period to examine the durability of the antibody response [46]. At 6 years after immunization, geometric mean titers (GMT) for the AdV-4 and AdV-7 vaccine component groups were comparable to those from earlier years. Neither AdV-4 nor AdV-7 patients had decreasing neutralizing antibody (NAb) titers of greater than 5%. Several patients showed a four-fold increase in NAb titers against AdV-4 or AdV-7 at various points after vaccination, indicating homotypic or heterotypic re-exposure. This study demonstrated that the live oral AdV-4 and AdV-7 immunization provided durable immunity to these viruses [47].

10.9 Conclusion and Future Prospect

With the exception of live-attenuated or inactivated microorganism-based immunizations that employ nanoparticles, no vaccine has ever been delivered orally previously. Antigens are stabilized and protected from an adverse gastrointestinal environment by NPs, which also increase immunogenicity. The potent immune response of NPs may be explained by the fact that their size, which is comparable to viruses', makes them nanoscale. By ligand-specific targeting to M cells, epithelial dendritic cells, and enterocytes, the oral administration's poor antigen penetration across the intestine's mucosal barrier can be overcome. It has been proposed that using nanotechnology, which not only makes it possible to target intestinal immune cells specifically, but also reduces harmful side effects, would help solve the problems caused by the GIT. Nanoparticle-based oral immunizations can be made by selecting a suitable carrier into which vaccines may be injected that can better activate an immune response. The current analysis concludes that delivering vaccinations orally as opposed to parenterally offers much higher benefits. However, the oral route for vaccine delivery also has a number of drawbacks, such as the degradation of antigens under unfavorable conditions. Thanks to nanotechnology, we now have access to carriers that can get beyond the drawbacks of vaccinations given orally. The only question that remains is whether they can be produced using economically sensible ways to enable mass immunization at a low cost, as oral immunization has proven, which is especially helpful in underdeveloped nations.

References

1. Chen, Y.C., Cheng, H.F., Yang, Y.C., Yeh, M.K., Nanotechnologies applied in biomedicines' vaccine, in: *Micro and Nanotechnologies for Biotechnology*, S.G. Stanciu, (Ed.), InTech, Taiwan 2016, [Internet] [cited 2023 Apr 15]. Available from: http://www.intechopen.com/books/micro-and-nanotechnologies-for-biotechnology/nanotechnologies-applied-in-biomedicines-vaccine.
2. D'Souza, M.J., Tawde, S.A., Akalkotkar, A., Chablani, L., D'Souza, M., Chiriva-Internati, M., Nanotechnology in vaccine delivery, in: *Molecular Vaccines*, M. Giese, (Ed.), Springer International Publishing, Cham, 2014, [Internet] [cited 2023 Apr 15]. p. 727–41. Available from: https://link.springer.com/10.1007/978-3-319-00978-0_19.
3. Chavda, V.P., Baviskar, K.P., Vaghela, D.A., Raut, S.S., Bedse, A.P., Nasal sprays for treating COVID-19: A scientific note. *Pharmacol. Rep.*, 75, 2, Feb. 27, 2023. [Internet] [cited Mar. 5, 2023], Available from: https://link.springer.com/10.1007/s43440-023-00463-7.
4. Maina, T.W., Grego, E.A., Boggiatto, P.M., Sacco, R.E., Narasimhan, B., McGill, J.L., Applications of nanovaccines for disease prevention in cattle. *Front. Bioeng. Biotechnol.*, 8, 608050, Dec. 11, 2020.

5. Bezbaruah, R., Chavda, V.P., Nongrang, L., Alom, S., Deka, K., Kalita, T. *et al.*, Nanoparticle-based delivery systems for vaccines. *Vaccines*, 10, 11, 1946, Nov. 17, 2022.
6. Kour, P., Rath, G., Sharma, G., Goyal, A.K., Recent advancement in nanocarriers for oral vaccination. *Artif. Cells Nanomed. Biotechnol.*, 46, sup3, 1102–14, Nov. 12, 2018.
7. Banerjee, A., Qi, J., Gogoi, R., Wong, J., Mitragotri, S., Role of nanoparticle size, shape and surface chemistry in oral drug delivery. *J. Control. Release*, 238, 176–85, Sep. 2016.
8. Zhu, Q., Talton, J., Zhang, G., Cunningham, T., Wang, Z., Waters, R.C. *et al.*, Large intestine–targeted, nanoparticle-releasing oral vaccine to control genitorectal viral infection. *Nat. Med.*, 18, 8, 1291–6, Aug. 2012.
9. Van der Weken, H., Cox, E., Devriendt, B., Advances in oral subunit vaccine design. *Vaccines*, 9, 1, 1, Dec. 22, 2020.
10. Oberoi, H.S., Yorgensen, Y.M., Morasse, A., Evans, J.T., Burkhart, D.J., PEG modified liposomes containing CRX-601 adjuvant in combination with methylglycol chitosan enhance the murine sublingual immune response to influenza vaccination. *J. Control. Release*, 223, 64–74, Feb. 2016.
11. Mangla, B., Javed, S., Sultan, M.H., Ahsan, W., Aggarwal, G., Kohli, K., Nanocarriers-assisted needle-free vaccine delivery through oral and intranasal transmucosal routes: A novel therapeutic conduit. *Front. Pharmacol.*, 12, 757761, Jan. 11, 2022.
12. Chavda, V.P., Balar, P.C., Patel, S.B., Nanotheranostics-based management of head and neck cancer. *Nanotheranostics*, 7, 2, 202–9, 2023.
13. Chavda, V.P., Balar, P.C., Patel, S.B., Interventional nanotheranostics in hepatocellular carcinoma. *Nanotheranostics*, 7, 2, 128–41, 2023.
14. Mühlberg, E., Burtscher, M., Umstätter, F., Fricker, G., Mier, W., Uhl, P., Trends in liposomal nanocarrier strategies for the oral delivery of biologics. *Nanomedicine*, 16, 20, 1813–32, Aug. 2021.
15. Shaker, D.S., Ishak, R.A.H., Ghoneim, A., Elhuoni, M.A., Nanoemulsion: A review on mechanisms for the transdermal delivery of hydrophobic and hydrophilic drugs. *Sci. Pharm.*, 87, 3, 17, 2019 Jul 12.
16. Chavda, V.P., Balar, P.C., Teli, D., Davidson, M., Bojarska, J., Apostolopoulos, V., Antibody–biopolymer conjugates in oncology: A review. *Molecules*, 28, 6, 2605, Mar. 13, 2023.
17. Perry, S.L. and McClements, D.J., Recent advances in encapsulation, protection, and oral delivery of bioactive proteins and peptides using colloidal systems. *Molecules*, 25, 5, 1161, Mar. 5, 2020.
18. Carreño, J.M., Perez-Shibayama, C., Gil-Cruz, C., Printz, A., Pastelin, R., Isibasi, A. *et al.*, PLGA-microencapsulation protects Salmonella typhi outer membrane proteins from acidic degradation and increases their mucosal immunogenicity. *Vaccine*, 34, 35, 4263–9, Jul. 2016.

19. Yoo, M.K., Kang, S.K., Choi, J.H., Park, I.K., Na, H.S., Lee, H.C. *et al.*, Targeted delivery of chitosan nanoparticles to Peyer's patch using M cell-homing peptide selected by phage display technique. *Biomaterials*, 31, 30, 7738–47, Oct. 2010.
20. Zani-Ruttenstock, E., Antounians, L., Khalaj, K., Figueira, R.L., Zani, A., The role of exosomes in the treatment, prevention, diagnosis, and pathogenesis of COVID-19. *Eur. J. Pediatr. Surg.*, 31, 04, 326–34, Aug. 2021.
21. Chavda, V.P., Valu, D.D., Parikh, P.K., Tiwari, N., Chhipa, A.S., Shukla, S. *et al.*, Conventional and novel diagnostic tools for the diagnosis of emerging SARS-CoV-2 variants. *Vaccines*, 11, 2, 374, Feb. 6, 2023.
22. Attia, M.S., Abdel-Mottaleb, M.S.A., Mohamed, E.H., Smart nanovesicles for drug delivery, in: *Systems of Nanovesicular Drug Delivery*, Elsevier, 2022, [Internet] [cited 2023 Apr 15]. p. 367–85. Available from: https://linkinghub.elsevier.com/retrieve/pii/B9780323918640000176.
23. Gholap, A.D., Sayyad, S.F., Hatvate, N.T., Dhumal, V.V., Pardeshi, S.R., Chavda, V.P. *et al.*, Drug delivery strategies for avobenzone: A case study of photostabilization. *Pharmaceutics*, 15, 3, 1008, Mar. 21, 2023.
24. Barani, M., Sangiovanni, E., Angarano, M., Rajizadeh, M.A., Mehrabani, M., Piazza, S. *et al.*, Phytosomes as innovative delivery systems for phytochemicals: A comprehensive review of literature. *Int. J. Nanomedicine*, 16, 6983–7022, Oct. 2021.
25. Dhiman, N., Awasthi, R., Sharma, B., Kharkwal, H., Kulkarni, G.T., Lipid nanoparticles as carriers for bioactive delivery. *Front. Chem.*, 9, 580118, Apr. 23, 2021.
26. Jazayeri, S.D., Lim, H.X., Shameli, K., Yeap, S.K., Poh, C.L., Nano and microparticles as potential oral vaccine carriers and adjuvants against infectious diseases. *Front. Pharmacol.*, 12, 682286, Jun. 2, 2021.
27. Cordeiro, A.S., Patil-Sen, Y., Shivkumar, M., Patel, R., Khedr, A., Elsawy, M.A., Nanovaccine delivery approaches and advanced delivery systems for the prevention of viral infections: From development to clinical application. *Pharmaceutics*, 13, 12, 2091, Dec. 5, 2021.
28. Pumchan, A., Sae-Ueng, U., Prasittichai, C., Sirisuay, S., Areechon, N., Unajak, S., A novel efficient piscine oral nano-vaccine delivery system: Modified halloysite nanotubes (HNTs) preventing streptococcosis disease in tilapia (oreochromis sp.). *Vaccines*, 10, 8, 1180, Jul. 25, 2022.
29. Meynell, H.M., Thomas, N.W., James, P.S., Holland, J., Taussig, M.J., Nicoletti, C., Up-regulation of microsphere transport across the follicle-associated epithelium of Peyer's patch by exposure to *Streptococcus pneumoniae* R36a. *FASEB J.*, 13, 6, 611–9, Apr. 1999.
30. Chavda, V.P., Patel, A.B., Vora, L.K., Singla, R.K., Shah, P., Uversky, V.N. *et al.*, Nitric oxide and its derivatives containing nasal spray and inhalation therapy for the treatment of COVID-19. *Curr. Pharm. Des.*, 28, 46, 3658–70, Dec. 2022.

31. Mowat, A.M.I., Anatomical basis of tolerance and immunity to intestinal antigens. *Nat. Rev. Immunol.*, 3, 4, 331–41, Apr. 2003.
32. Li, M., Kaminskas, L.M., Marasini, N., Recent advances in nano/microparticle-based oral vaccines. *J. Pharm. Investig.*, 51, 4, 425–38, Jul. 2021.
33. Chen, X., Fernando, G.J.P., Crichton, M.L., Flaim, C., Yukiko, S.R., Fairmaid, E.J. *et al.*, Improving the reach of vaccines to low-resource regions, with a needle-free vaccine delivery device and long-term thermostabilization. *J. Control. Release*, 152, 3, 349–55, Jun. 2011.
34. Bandyopadhyay, A.S., Garon, J., Seib, K., Orenstein, W.A., Polio vaccination: Past, present and future. *Future Microbiol.*, 10, 5, 791–808, May 2015.
35. Lucas, M.E.S., Deen, J.L., von Seidlein, L., Wang, X.Y., Ampuero, J., Puri, M. *et al.*, Effectiveness of mass oral cholera vaccination in Beira, Mozambique. *N. Engl. J. Med.*, 352, 8, 757–67, Feb. 24, 2005.
36. Mo, Z., Mo, Y., Li, M., Tao, J., Yang, X., Kong, J. *et al.*, Efficacy and safety of a pentavalent live human-bovine reassortant rotavirus vaccine (RV5) in healthy chinese infants: A randomized, double-blind, placebo-controlled trial. *Vaccine*, 35, 43, 5897–904, Oct. 2017.
37. Steele, A.D., Neuzil, K.M., Cunliffe, N.A., Madhi, S.A., Bos, P., Ngwira, B. *et al.*, Human rotavirus vaccine rotarix™ provides protection against diverse circulating rotavirus strains in African infants: A randomized controlled trial. *BMC Infect. Dis.*, 12, 1, 213, Dec. 2012.
38. Oldenkamp, H.F., Vela Ramirez, J.E., Peppas, N.A., Re-evaluating the importance of carbohydrates as regenerative biomaterials. *Regen. Biomater.*, 6, 1, 1–12, Feb. 1, 2019.
39. Levine, M.M., Ferreccio, C., Abrego, P., Martin, O.S., Ortiz, E., Cryz, S., Duration of efficacy of Ty21a, attenuated salmonella typhi live oral vaccine. *Vaccine*, 17, S22–7, Oct. 1999.
40. Black, R.E., Levine, M.M., Ferreccio, C., Clements, M.L., Lanata, C., Rooney, J. *et al.*, Efficacy of one or two doses of Ty21a salmonella typhi vaccine in enteric-coated capsules in a controlled field trial. *Vaccine*, 8, 1, 81–4, Feb. 1990.
41. Sánchez, J. and Holmgren, J., Virulence factors, pathogenesis and vaccine protection in cholera and ETEC diarrhea. *Curr. Opin. Immunol.*, 17, 4, 388–98, Aug. 2005.
42. Holmgren, J., Svennerholm, A.M., Jertborn, M., Clemens, J., Sack, D.A., Salenstedt, R. *et al.*, An oral B subunit: Whole cell vaccine against cholera. *Vaccine*, 10, 13, 911–4, Jan. 1992.
43. Vesikari, T., Rotavirus vaccination: A concise review. *Clin. Microbiol. Infect.*, 18, 57–63, Oct. 2012.
44. Top, F.H., Control of adenovirus acute respiratory disease in U.S. army trainees. *Yale J. Biol. Med.*, 48, 3, 185–95, Jul. 1975.
45. Chavda, V.P., Bezbaruah, R., Valu, D., Patel, B., Kumar, A., Prasad, S. *et al.*, Adenoviral vector-based vaccine platform for COVID-19: Current status. *Vaccines*, 11, 2, 432, Feb. 13, 2023.

46. Krishnan, A., Gangadaran, P., Chavda, V.P., Jogalekar, M.P., Muthusamy, R., Valu, D. *et al.*, Convalescent serum-derived exosomes: Attractive niche as COVID-19 diagnostic tool and vehicle for mRNA delivery. *Exp. Biol. Med. (Maywood)*, 247, 14, 1244–52, Jul. 2022.
47. Collins, N.D., Adhikari, A., Yang, Y., Kuschner, R.A., Karasavvas, N., Binn, L.N. *et al.*, Live oral adenovirus type 4 and type 7 vaccine induces durable antibody response. *Vaccines*, 8, 3, 411, Jul. 23, 2020.

11

Nanovaccines for Mucosal Immunity

Shashi Kiran Misra[1], Anupria Kapoor[1] and Kamla Pathak[2]*

[1]School of Pharmaceutical Sciences, Chhatrapati Shahu Ji Maharaj University, Kanpur, India
[2]Faculty of Pharmacy, Uttar Pradesh University of Medical Sciences Saifai, Etawah, India

Abstract
Elevated morbidity and mortality rates associated with mucosal infections have become alarming globally. Recently, the SARS-CoV-2 pandemic has portrayed some harsh images of respiratory mucosal infections all over the world. The existence of nonspecific suboptimal mucosal vaccines against COVID-19 critically presents challenges to managing and saving the lives of those who suffer. The mammalian mucosal immune system contains a typical system of cellular tissues and soluble molecular components that performs a defense mechanism against invaders (pathogens) and malignant cells. Mucosal nanovaccines have emerged as a prospective first-line of defense approach for devastating pathogens by eliciting both site-specific and systemic immunity or responses. Mucosal nanovaccines not only check pathogenic infections at the mucosal site, and boost immunogenicity but also combat associated systemic disorders precisely due to their superior surface area, augmented bioavailability, enriched antigen-invading cell environment at the mucosal site, and thus shield the entry portal (oral and nasal) for antigens. Mucosal immunization elicits promising immunomodulation by turning up and down proteins involved in certain allergies and autoimmune disorders. This chapter outlines different types of mucosal immune responses and advanced applications of nanovaccines developed by polymeric nanoparticles, carbon nanotubes, inorganic particles, and virus-like biomimetic particles against pathogenic infections, autoimmune mucosal disorders, and so they can aid in chemotherapy and prophylactics as well.

Keywords: Respiratory mucosal infection, cell environment, mucosal immunity, antigen, first-line defense, nanovaccines, polymeric nanoparticles, biomimetic

**Corresponding author*: kamlapathak5@gmail.com

Vivek P. Chavda and Vasso Apostolopoulos (eds.) Nanocarrier Vaccines: Biopharmaceutics-Based Fast Track Development, (367–404) © 2024 Scrivener Publishing LLC

11.1 Introduction

Vaccination has an important role in the induction of immunity for centuries now. It has contributed immensely to maintaining the immunity of the community. Killed, weakened (attenuated), or a fraction of pathogens are chosen to provoke or stimulate immunity in the body. It is safe, ingenious, and is the most valuable measurement that reduces risks of infection by activating the body's natural defense mechanism. The body's defense or immunity memorizes the pathogen and response after each attack. Hence, it is so far the finest approach to the elimination of infection. However, significant surges of drug resistance and drug-related toxicity require robust therapy to protect mankind. Traditional vaccine-carrying subunits sometimes fail to produce long-lasting protection and maintenance of body immunity against noxious pathogens. Moreover, attenuated vaccines need stringent storage facilities in maintaining vaccine effectiveness [1]. One of the major issues concerned with the traditional vaccine is that it cannot be applied to immunocompromised individuals and against some fatal diseases such as HIV, tuberculosis, and hepatitis C. Further, subunit vaccines containing soluble proteins show shorter half-life and are less immunogenic. That's why they require multiple booster dosages to ensure long-term immunity. Table 11.1 compiles some salient features of the different kinds of vaccines available for the development of humoral and cellular immunities [2].

Currently, nanovaccines are emphasized as a novel approach to vaccination. In this sequence, microemulsions, nanogels, liposomes, micelles, virosomes, and dendrimers have offered successful strategies for targeting antigens and limited issues related to traditional vaccines. Myriad functions such as target delivery, stimulation of immunity, antigen presentation, and induction of T cells, all are responded to by nanovaccines while combating a disease or cancer [3]. Nanomaterials and nanocarriers, i.e., gold/silica nanoparticles, carbon-based materials, virus particles, and mucoadhesive polymers, are targeted for the formulation of nanovaccines. Preferred needle-free routes for painless and effective nanovaccinations are oral, nasal, and topical microprojection patches [4].

Similar to conventional vaccines, these nanovaccines also induce humoral and cell-mediated immunities. These are more efficient and convenient to use as they can be administered through nasal drops. These consist of nanoscaled particles and portray an emerging class of vaccines for the prevention and/or spread of infections. In contrast to traditional vaccines, nanovaccines aim the specific part of the body with persisting infection or disease

Table 11.1 Salient features of developed vaccines of different routes.

Properties	Oral solutions	Traditional vaccines	Microneedles	Nanovaccines
Self-administration	Yes	No	Yes	No
Administration mode	Oral solution	Needle	Microneedle	Injection or inhalation
Storage	Cold chain	Cold chain	Room temperature	Formulation dependent
Onset time	Slow	Rapid	Rapid	Rapid
Safety	High	Poor	High	High
Use	High	Moderate	High	High
Patient compliance	High	Poor	Better	Better
Pain	Painless	Painful	Painless	Both painful and painless

without disturbing the body's immune system. Convenient usage and better therapeutic efficacy make nanovaccines a promising and effective approach for curing devastating diseases and infections. Further, these vaccines do not need a cold chain as they have prolonged thermal stability at room temperature. This storage stability encourages global delivery to distant areas and lessens medical storage costs [5]. Furthermore, nanovaccines are efficient in maintaining the functionality and stability of admixed antigen proteins that are required to elicit the desired immunity [6]. Figure 11.1 summarizes the stimulation of cellular, humoral, and mucosal immunities in the body.

Nanovaccines containing customized shapes and sizes of pathogens can be recognized better and easily subjected to uptake by immune cells. Further, functionalization or surface modification with alpha-1,2-linked dimannopyranoside makes an antigen-like structure that can be efficiently subjected to uptake by immune cells [7]. After internalization, the encapsulated antigen releases out from natural or synthetic biocompatible polymeric adjuvants and remains there for a long period. The ability of the innate immune system to distinguish alarm signals (danger) initiates the activation of immune cells. These alarm signals may be either exogenous (microorganism) or endogenous (dead cells) patterns that are specifically identified by pattern recognition receptors. The selection of polymers plays a crucial role in provoking

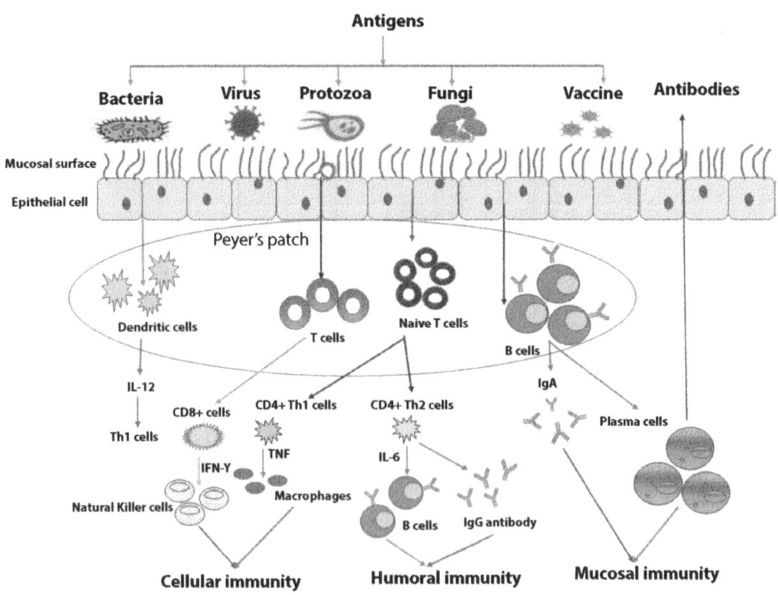

Figure 11.1 Schematic representation of the development of various immunities in the body.

immunogenicity. Biocompatible and biodegradable natural polysaccharides (chitosan), synthetic polyanhydrides (1,6-bis(p-carboxyphenoxy) hexane), poly(diaminosulphides), and polyesters (PLGA) are widely employed for nanovaccine formulations. These polymeric materials enable high antigenic payloads, sustained release, and stability and initiate cell-mediated immunity at local and systemic sites. Additionally, these polymers facilitate the modulation of particles' physical characteristics (size, shape, surface charges, hydrophobicity, and hydrophilicity) that significantly influence their cellular internalization process [8].

Literature envisages that submicron-sized particles were easily internalized through antigen-presenting cells and provoked profound immunity whereas particles greater than 10 μm faced an issue for internalization by the same cells [9]. Different research concluded that smaller size particles were effectively internalized through antigen-presenting cells (APCs) owing to ease of transportation across biological barriers, i.e., epithelia [10].

Gutierro *et al.* developed nanovaccines of bovine serum albumin containing poly (d,l-lactic-co-glycolic acid) possessing diverse particle sizes, i.e., 200 nm, 500 nm, and 1 μm. The formulations were subjected to subcutaneous, oral, and intranasal routes of administration [11]. It was concluded that the decreased diameter of particles did not exhibit the desired immunogenicity due to the chances of draining in the lymphatic vessels. Whereas nanovaccines containing a particle diameter of 1000 nm exhibited superb immunoglobulin level or total IgG [12]. Similarly, rod-shaped particles (macrophages and dendritic cells) were swiftly internalized by lysosomes and APCs compared to spherical-shaped particles [13].

A fast endocytosis rate is concerned with spherical particles followed by cubic ones owing to their minimal membrane-bending change [14]. Both anionic and cationic particles exhibited different immune responses. It is reported that neutral and anionic particles get trapped and localized within the lysosomal cells [15]. Particles with excessive molecular weight and hydrophobicity undergo an elevated rate of cellular internalization compared to hydrophilic particles [16]. The mucosal membranes are covering the linings of the ocular (conjunctiva), otic (inner ear), respiratory (aerodigestive), and gastrointestinal tracts; the ducts of all exocrine glands; and the excretory system (urogenital tracts). These mucous membranes are responsible for powerful chemical and mechanical cleansing processes, resulting to the degradation and ejection of most foreign matter from the body. Enormous and extremely specialized adaptive and innate immune systems are endowed with approximately 80% of all immunocytes, and thus shield these mucosal surfaces against potential pathogens. These immunocytes accumulate in or reside between myriad mucosa-associated lymphoid tissues and outline the biggest lymphoid system in mammals [17].

Further, the mammalian mucosal immune system performs important functions including: (i) protection from colonization and infection by devastating microbes; (ii) prevention of internalization of undegraded antigens from food materials and inhaled matters; and (iii) restriction initiation of fatal immune responses. The MALT system encounters a plethora of invaders, guards organs by selecting appropriate chemical or mechanical mechanisms, and avoids further immune exhaustion and tissue damage [18].

11.2 Mucosal Immunity

Mucosa-associated lymphoid tissue is expressed as MALT functions separately. It represents a highly ordered and systematic immunological system. Lymphoid microcompartments include well-defined anatomical structures like mesenteric lymphoid, intestinal solitary follicles, Peyer's patches, and tonsils/adenoids that are present at the mucosal sites and provoke immune responses [19]. A large number of lymphoid cells are diffused and accumulated in the parenchyma of MALT. MALT-containing organs and exocrine glands anchor mucosal effector sites and govern immune responses. Functionally distinct T cells and B cells are highly populated in the MALT and restrict the recirculation of lymphoid cells between these mucosal spots [20]. The entered antigen is directly captured via dendritic cells, lymphocytes, APCs, and macrophages anchored on mucosal inductive sites. These trapped antigens are presented before clusters of differentiation 4 and 8 cells (CD4 and CD8 cells). Both cells fight and kill infections. Other certain intraepithelial T cells and gamma-delta T cells bind with antigens (glycolipid), producing inflammatory reactions of myeloid/lymphoid lineages and thus promoting immune responses. Non-pathogenic antigens, such as food particles, are encountered by T helper cells generated by mucosal APCs and DCs [21].

The mucosal lining around the gut is made up of gut-associated lymphoid tissue commonly known as GALT. A distinct ring (Waldeyer's ring) is surrounded by adenoids and tonsils at the back of the mouth and entrance of gut airways. These areas are decorated with a large aggregation of lymphoid tissues since childhood. Patients with removed adenoids or tonsils have poor immune responses to oral polio vaccines, thus signifying the importance of compartmental mucosal immunity [22]. GALT includes cryptopatches, Peyer's patches, ILFs (isolated lymphoid follicles), cecal patches, gut-draining mesenteric lymph nodes, and colonic patches of lymphoid tissues between the stomach and colon regions [23]. Countless lymphoid aggregates are found in the mammalian gut known as cryptopatches. They are very small in size, found at the base of intestinal villi beneath the crypts, and unequally distributed across gut mucosa. Higher frequency is seen in the ileum and colon region compared to another intestinal part of the duodenum and jejunum. Approximately 1000 small clusters including myeloid cells, lymphocytes, c kit$^+$ IL-7R$^+$, and CD11 are detailing functions of immunity. However, no B and T cells participate here for immune responses. Involved c kit$^+$ IL-7R$^+$ cells are considered innate lymphoid cells (ILC3 cells) that generate interleukins, i.e., IL-17 and IL-22. ILC cells express the RORYt transcription factor, regulate homeostasis, and provoke postnatal production of lymphoid follicles of the intestine [24].

Moreover, isolated lymphoid follicles (ILFs) are a highly complex lymphoid system and occur in almost all areas of the small and large intestines. They contain a sole B cell follicle having a network of follicular dendritic cells. The system is positioned beneath highly specialized M cells that involve the transportation of antigens to the leukocytes [25]. Small intestinal Peyer's patches are another site of mucosal immunity. These are anatomically dome-shaped that stretch across the lumen of the enteric region and impart mucosal immunity. The follicle-associated epithelial region of the Peyer's patches contains both B and M cells. B cells are found around the follicular dendritic cells covering the lining of the Peyer's patches, whereas M cells or microfold cells are less prominent than the enterocytes (microvilli) and deprived of a thick layer of glycocalyx, hence unable to secrete mucus. However, they exhibit adaptive immunity by directly interacting with foreign particles (antigens) found in the gut lumen. These M cells take up antigens or foreign particles via phagocytosis/endocytosis, transported into the interior cell vesicles, and then into the membrane of basal cells. After engulfing foreign particles, these materials are poured out into the extracellular matrix via the process of transcytosis [26].

M cells of GALT are a very important feature of mucosal lymphoid tissues that participate in different cell interactions and signaling via the tumor necrosis factor receptor (TNFR) family. The mucosal lining of the nasal cavity and respiratory tract consists of endogenous microbial communities and is continuously exposed to foreign particles, pollutants, and microbes [27]. Mammalian species have different mucosal lymphoid organs in the upper respiratory tract. Their location and function are also varied according to species. For instance, humans have well-developed diffused nasopharynx and oropharynx, whereas cattle and dogs lack prominent lymphoid tissues in the nasal cavity [28].

Similar to MALT present in the gut, nasal associate lymphoid tissue (NALT) is the major component that expresses CCL20 and CCL9 chemokines. NALT also possesses M cells on the dome-shaped epithelium; efficiently takes up luminal antigens, pollutants, and foreign particles; and lead to the process of transcytosis across the NALT epithelium. Moreover, NALT M cells also express a few markers such as Tnfaip2, GP2, and Spi-B that regulate actin cytoskeleton and promote cell migration in the condition of nasopharyngeal carcinoma [29]. NALT M cell maturation requires interaction with lymphocytes, leucocytes, and chemokine receptor (CCR6) cells. The efficiency of antigen transportation impairs the reduced M cells containing NALT of CCR6 deficit mammalian and leads to poor immune responses against nasally entered antigens. However, B cells directly do not participate in the development of M cells but these cells trigger CD137, a TNFR member that maintains the integrity of M cells, which assists the development of basolateral pockets and thus facilitates effective antigen transportation [30].

NALT, a mucosal lymphoid organ, preferentially releases immunoglobulin A and G (IgA and IgG) after vaccination. Immunological responses are based on the type of adjuvant present in the vaccine. For instance, immunization with cholera toxin generates immunoglobulin A, and protein microparticles release immunoglobulin G response. NALT releases both IgG and IgA against common respiratory tract viruses, i.e., reovirus and influenza. Interestingly, NALT IgA responses more prominently accumulate somatic mutation compared to IgG memory cells due to the interaction with CCR6 eliciting dendritic cells in the epithelial dome region [31].

The bronchus-associated lymphoid tissue, commonly known as BALT, is concerned with the distinct regions of the respiratory tract comprising of trachea, bronchi, and alveoli. These areas are regularly exposed to foreign particles, and inhaled pathogens (i.e., rhinovirus, influenza), and respiratory syncytial virus. Hence, the subordinate respiratory tract must be preserved via robust mucosal immunity. The lung is the most efficient and delicate site where the distance between alveolar epithelium and blood capillary endothelium is minimum and highly susceptible to invaders. The lungs of healthy individuals have relatively fewer mucosal lymphoid areas, i.e., BALT. Whereas adverse lung conditions with chronic obstructive pulmonary abnormalities, pneumonia, and hypersensitivity pneumonitis elicit well-developed BALT that effectively encounters antigens [32]. Figure 11.2 shows the different elements of mucosa-associated lymphoid tissue.

Pulmonary inflammation invites infiltration of leucocytes that often accumulate as inducible bronchus-associated lymphoid tissues or iBALT. Pathogenic infections and inflammatory stimuli provoke the activation of iBALT via a variety of secondary lymphoid organ (SLO) development. These specific tissues contain isolated T-cell and B-cell regions,

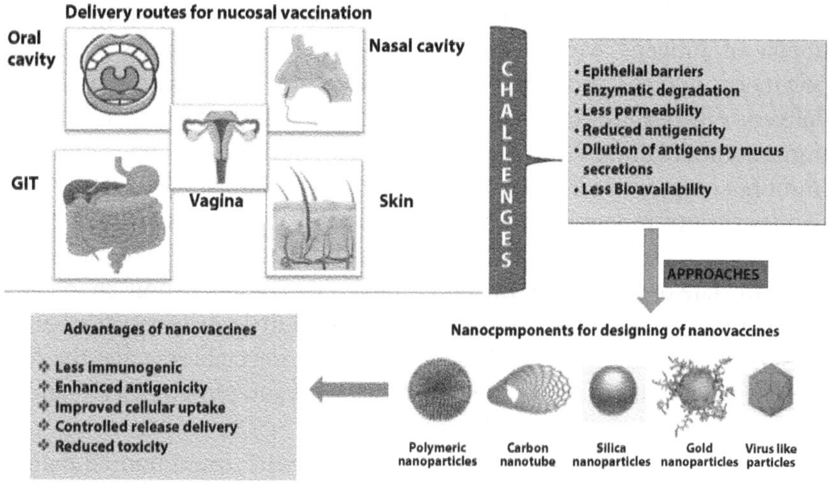

Figure 11.2 Various elements of mucosa-associated lymphoid tissue (MALT).

well-defined stromal cells, and endothelial and lymphatic vessels. iBALT regulates immunity for months even after the cure of pulmonary inflammation. It is located throughout the lungs covering peribronchial, interstitial, and perivascular domains [33].

iBALT is beneficial as it eases faster clearance of invaders without prominent systemic and pulmonary pathology. Several fatal pulmonary infections caused by the SARS virus, *M. tuberculosis*, and *Coxiella burnetii* are successfully confronted by immune responses generated through iBALT. Faster B cell activation and T cell differentiation provide unique iBALT attributes and may be a potential line for the development of nanovaccines that may rapidly initiate antigen-specific cellular and humoral immune responses against pulmonary infections [34].

11.3 Nanovaccine Formulations

Various infectious diseases and disorders including tuberculosis, human immunodeficiency virus (HIV), severe acute respiratory syndrome coronavirus 2 (SARS-CoV), malaria, and other healthcare-associated infections are globally predominating and still not preventable by conventional vaccines, thus requiring a new generation of nanovaccines. SARS-CoV-2 is transmitted via the respiratory route, hence nasal immunization is an ideal and easy method for vaccination. NALT is strategically stretched along the nasal and oropharynx area, and stimulates mucosal and systemic immunity there.

The major function of mucosal vaccines is the promotion of cellular uptake of antigens. Myriad kinds of SARS-CoV nanovaccines including polymeric, lipoidal, proteins, polysaccharides, and nanobiomimetic-based delivery vehicles are reported. Newer technologies containing adjuvants, immunomodulators, cells, and vectors are screened for application on immunocompromised, newborn, and geriatric populations. Figure 11.3 defines various mucosal routes, challenges, and approaches for the design of nanovaccines. Physiological and anatomical sites of the mucosa highly affect the fate of vaccines in the body. Table 11.2 compiles some characteristic parameters that are significant to mucosal drug delivery. The better part of mucosal nanovaccine over conventional injectables is that the former is non-invasive, comfortable (both physical and physiological), and it induces desired immune responses [35].

Nanoscaled composites containing biomimetic immunomodulatory components are desirable carriers or adjuvants for the preparation of nanovaccines. Myriad polymeric, inorganic, and carbon nanotube-based systems are reported for the formulation of nanovaccines owing to their phenomenal characteristics such as antigen high pay loading, *in vitro/in vivo* stability, and temporal targeting. Table 11.3 compiles the merits and demerits of different mucosal routes for mucosal immunization.

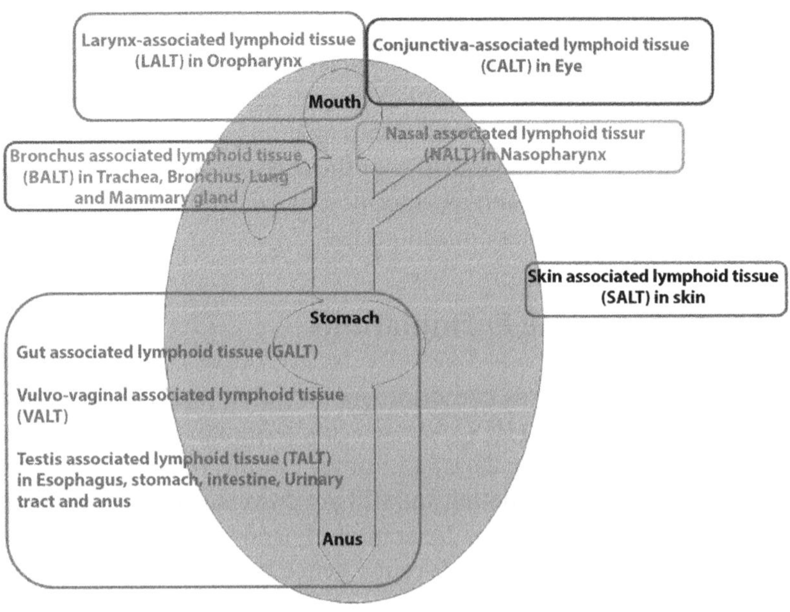

Figure 11.3 Different mucosal immunization routes and approaches for the development of nanovaccines.

Table 11.2 Physiological and anatomical specifications of mucosal sites.

Route of administration	Mucosal thickness	Secretions	Surface area	pH
Oral and buccal	500–800 μm	Approximately 0.5–2 L/day	200 m^2	6.2–7.4
Nasal	800–1000 μm	1–1.5 L/day	160 cm^2	5–6.5
Rectal	200–300 μm	4–5 ml	70–300 cm^2	7.8–8
Ocular	Approximately 500 μm	1.2 μl/min	2 cm^2	7.4

Table 11.3 Comparison between myriad routes for mucosal vaccine delivery.

Mucosal route	Merits	Demerits	Ref.
Oral/buccal route	Safe, easy administration, painless, no experts required, cost-effective, instant absorption	Higher concentration of the vaccine is required for the desired immune response	[36]
Nasal	Easy and self-administration, only a small dose is required due to high vascularization	Poor absorption from the epithelial layer due to immediate removal from the nasal passage, ineffective immune response	[37]
Rectal/vaginal	Stronger immune response at the site, suitable for the avoidance of sexually transmitted pathogens	Poor patient compliance, unacceptability	[38]
Ocular	Suitable for the prevention of Herpes simplex virus	Poor drug permeation and bioavailability	[39]

11.3.1 Polymer-Based Nanovaccines

Mucosal nanovaccines are preferred over parenteral and intramuscular by virtue of their superb patient compliance. Xyloglucan, a tamarind seed polymer, has been employed for intranasal, controlled release delivery systems. Nanovaccine was developed for the production of immunoglobulins against brucellosis, a bacterial infection caused by *B. abortus*. Immunoadjuvant Quillaja saponin and cellular lipopolysaccharide antigen (S19) were conjugated with polymeric xyloglucan nanoparticles via a supercritical method. Prolonged cellular and humoral immunities were observed in female Balb/c mice. Elevated levels of mucosal immunoglobulins, i.e., IgA and IgG, were observed after nasal administration and triggered by the quillaja-immunoadjuvant [40]. The sudden outbreak of COVID-19 has surged the production of mucosal vaccines at an unprecedented rate. Novel immunization techniques involve natural (chitosan, gums, alginates) and synthetic (PLGA) polymers as potential adjuvants and improve the antiviral efficacy of vaccines [41].

Polymeric nanoparticles are widely explored for designing mucosal nanovaccines owing to their potential for limiting antigen dilapidation, and their ability to clear and enhance active uptake via antigen-presenting cells (APCs). Polymeric-based systems are highly versatile and flexible, and can entrap a wide range of antigens/immunomodulators for adaptive immune responses. Antigens can be attached via conjugation, adsorption, and encapsulation on the surface of polymeric nanoparticles. Various biocompatible, biodegradable, and mucoadhesive polymers have been employed for immune and cancer therapy against infectious diseases. Thiomers and thiolated polymers are new-generation mucoadhesive polymers owing to their superb mucosa adhesive properties. The presence of thiol groups of mucins facilitates the formation of stable disulfide covalent bonds with thiolated polymers. Further, they decrease oxidized glutathione in the intestinal epithelial cells by enhancing the local concentration of glutathione, and thus restricting the level of tyrosine phosphate (protein) accountable for the closure of tight junctions in the epithelial region. The result, prolonging the opening of tight junctions, is accompanied by improved permeability of the antigens epithelial cells [42].

11.3.1.1 Polylactide-co-Glycolide (PLGA)-Based Mucosal Nanovaccines

In this series, polylactide-co-glycolide (PLGA) is a well-known and highly desirable polymeric material owing to its biodegradability (metabolized in the body in the form of glycolic acid and lactic acid) and flexibility in design (self-assembled at micro and nano levels). Functionalized PLGA nanogels, nanospheres, micelles, and nanoemulsions are reported for the development of nanovaccines by virtue of the tunable solubility and thermodynamic stability of this polymer [43]. The promising outcomes from preclinical studies of PLGA-based nanovaccines offer researchers a start to clinical trials in humans. Block copolymers composed of PEG or polyetherimide conjugated PLGA can be self-assembled as polymeric micelles that have the potential to entrap hydrophobic proteins or peptide antigens, hence ideal for the development of nanovaccines [44]. Thomas *et al.* formulated porous PLA/PLGA nanoparticles containing nanovaccines for the effective management of hepatitis B in pulmonary disorder. They investigated the effects of hydrophobicity and nanoparticle particle size on mucosal cell-mediated immunity. A certain amount of hepatitis B surface antigen was adjuvanted with the variable preparation of porous PLA/PLGA nanoparticles via double emulsion solvent evaporation method. The levels of IgA and cytokine in the mucosal fluid were estimated in the rat spleen. Nanoparticles greater than 500 nm elicited increased secretion levels of IgA, gamma interferon, and IL-2 compared to nanoparticles of less than 500 nm owing to superb internalization by rat alveolar macrophages. Further, a cell viability study performed on Calu-3 indicated a safe and effective nanovaccine against hepatitis B [45].

P. aeruginosa is one of the most devastating infectious agents that is more prevalent in burn/trauma conditions and immunocompromised individuals. PLGA nanoparticle-based nanovaccines comprising oligopolysaccharide (OPS) and lipopolysaccharide (LPS) antigens have been synthesized via carbodiimide reaction. Studies conducted on BALB/C mice revealed a 10-fold high production of different immunoglobulin responses including IgM, IgG (G1, G2a, and G2b,), and IgA compared to pure antigens. Further, opsonophagocytosic analysis depicted superior performance (immune responses) of antigen (OPS and LPS)-conjugated PLGA nanoparticles against the infection of *P. aeruginosa* [46].

Moreover, cationic lipid PLGA was employed for conjugation with hyaluronic acid, and nanovaccines was developed to augment humoral and cellular responses for the delivery of genes. The conjugate was uptake by dendritic cells (DCs) and major histocompatibility complex (MHCI and MHCII). An *in vivo* study performed on mice revealed preeminent immune responses, i.e., induction of $CD4^+$ and $CD8^+$ T Cells, compared to bare cationic lipid PLGA nanoparticles [47].

11.3.1.2 Chitosan-Based Mucosal Nanovaccines

Natural origin, chitin-derived chitosan is widely pursued for designing mucoadhesive drug delivery approaches as wound dressing materials, skin regenerative, topical film/patches, contact lenses, and for the fabrication of artificial skin. Superb biocompatibility, mucosal adhesiveness, low toxicity, and biodegradability make it an appropriate candidate for the development of mucosal nanovaccines. Soluble chitosan and its derived glycol chitosan have been employed as intranasal adjuvants for enhancing immunity against adenovirus. Bovine herpesvirus I (BoHV-1) glycoprotein D containing intranasal vaccine was examined in cattle calves. Obtained data suggest a great potential mucosal vaccine containing soluble polymeric adjuvants (glycol chitosan) against recombinant viral vectors on intranasal delivery [48].

Glycol chitosan was also explored for mucosal and systemic immunogenicity against hepatitis B after nasal administration. HBsAg, hepatitis surface antigen, was conjugated with glycol chitosan immunoadjuvant nanoparticles (200 nm). The lowest nasal clearance rate and high mucosal uptake were reported in a performed *in vivo* experiment. Immunogenicity was assessed in mice serum by estimating anti-HBsAg antibodies in mice serum. Glycol chitosan nanoparticles provoked higher immune responses (both humoral and mucosal immunities) compared to bare chitosan nanoparticles (only humoral responses) and showed potential as nanovaccine carriers [49].

Chitosan-derived nanoparticles are being widely used to immunize animals via delivering antigens through the mucosal route. The system-targeted macrophages, micro-folded mucosal cells, and dendritic cells resulting in strong initiation of systemic and secretory cell-mediated immunogenicity against pathogens. The bacterial *Salmonella* subunit chitosan-nanoparticle-derived mucosal vaccine was developed by Renu *et al.* in 2020. The system comprised of outer membrane proteins (OMPs) and flagellin protein conjugated with chitosan nanoparticles. Synthesized mucoadhesive nanovaccine was administered to chickens via an oral route that adhered to ileal Peyer's patches. Vaccinated chickens exhibited lymphocytic proliferation, immunoglobulin IgA production, and a high release of toll-like receptors (TLR-2, TLR-4), hence producing cell-mediated and humoral immunities against salmonellosis in poultry.

Lopes *et al.* (2018) encapsulated the infectious bronchitis virus (IBV) vaccine comprising chitosan nanoparticles encoded with -BR-I genotype strain via the ionic gelation method. The system was administered through the oculo-nasal route to the chickens and showed enhanced release of IgA and IgG antibodies. A tracheal ciliostasis study revealed induction of both cell-mediated and humoral immunities against viral replication in the trachea and kidney [50]. Lee *et al.* (2018) examined the slow release potential and mucosal adhesiveness of chitosan/alginate nanoparticles adorned with bee venom (BV). A developed nanovaccine (CH/AL-BV) facilitated systemic immune response and mitigation against porcine reproductive and respiratory syndrome virus (PRRSV). Nasovaccinated pigs exhibited an enhanced population of $CD4^{+}T$ lymphocytes, interleukin -12, and interferon-gamma (IFN-Υ), thus improving immune responses. Further, CH/AL-BV vaccinated groups had a significantly less viral burden in the lungs and bronchial region [51]. Intravaginal hydroxyethylcellulose-based nanovaccines are reported to carry CNP4gp 140 (recombinant trimeric HIV-1 clade -C envelop). The system exhibited enhanced vaginal retention and boosted systemic antibodies in the female genital tract, thus showing potential for the immune response against HIV diseases [52].

11.3.1.3 Alginate-Based Mucosal Vaccines

Grass carp reovirus (GCRV) is a serious threat to the aquaculture industry. Orally administered sodium alginate embedded with VP56-3 antigen and flagellin B adjuvant (SA-VP56-3/FlaB) vaccine were developed to manage GCRV infection in grass carp species. After oral administration, the vaccinated carps exhibited strong immune responses against GCRV [53]. Although chitin-derived chitosan is a highly in demand polymer for the loading of antigens to improve stability and prevention of immediate desorption in gastrointestinal fluid, coating with sodium alginate is suggested. Ovalbumin-embedded chitosan nanoparticles were coated with sodium alginate that provided strong stability, and was easily taken up by gut-associated lymphoid tissues, i.e., M cells of Peyer's patches. The developed nanovaccine had stability in the intestinal region at 37° C [54].

Mosafer *et al.* (2019) have explored negatively charged sodium alginate for the modification of immunostimulatory features of chitosan and trimethyl-derived chitosan nanoparticles. These biopolymers are extensively employed as carriers or adjuvants during the preparation of mucosal nanovaccines. The authors have developed sodium alginate-coated chitosan and trimethyl chitosan embedded with attenuated PR8 influenza virus for effective nasal delivery. The developed system was well-internalized by nasal microfold cells and associated dendritic cells. Further, sodium alginate-coated PR8 chitosan and PR8 trimethyl chitosan elicited improved immune response compared to uncoated PR8 antigens. After nasal delivery in BALB/c mice, a significantly higher ratio of IgG2a/IgG1 was achieved, suggesting the potential application of sodium alginate for the preparation of intranasal mucosal vaccine [55].

Recently, Amin *et al.* have designed surface-modified (sodium alginate) chitosan nanoparticles for intestinal immune system induction. The system was protected from the harsh environment of the gastric region and aimed for M cells present on Peyer's patches. Initially, nanoparticles were embedded with ovalbumin protein, an immunogenic antigen. Their surfaces were modified with sodium alginate and polyethylene glycol (PEG). Improved mucoadhesive, stability, sustained release, and highly immunogenic properties were observed after oral delivery of the vaccine. The added PEG amplified the mucoadhesion power of the polymer and its strong shielding layer made the system stable in the gastrointestinal tract [56].

11.3.1.4 Beta Glucan-Based Mucosal Vaccines

These constituents are carbohydrate-derived components that chiefly occur in the cell walls of bacteria, fungi, yeast, and some cereals (barley and oat). Their sources of origination are accountable for physiological and structural parameters including molecular weight, solubility, and branching. Beta-glucans of fungi and yeast are potential activators of a person's innate immunity as they enhance the functions of microphases, dendritic cells, epithelial cells, and neutrophils [57]. Beta1,3-D-glucan modifies B- and T-cell responses as these moieties bind to dictin-1 present on dendritic cells and macrophages, resulting in the commencement and creation of pro-inflammatory cytokines including IL-2,6,10, 23 and TNF-α [58].

Further, beta-glucan elicits anti-infective activity against several viral, bacterial, fungal, and protozoal diseases. For instance, oat beta-glucan is immunotherapeutic against HIV-1, and lentinan (shiitake mushroom) has a potent action against tuberculosis and pneumonia pathogens [59, 60]. Orally administered beta-glucan-based nanovaccines increase the number of gastrointestinal intraepithelial lymphocytes and trigger cytokines (IFN-Υ). It is a promising anti-infective vaccine adjuvant that stimulates antibody production against *Yersinia ruckeri*, hepatitis B, influenza, coccidioidomycosis, and aspergillosis infections [61]. Another β-glucan 'AFO-202' is a proven potent immune adjuvant that exhibits strong mucosal immunity when administered with avian influenza H5 subtype vaccine [62].

11.3.1.5 Poly-(ε-Caprolactone)/Chitosan-Based Mucosal Vaccines

Developing countries highly require self-administered vaccines for the control of major pathogenic infections including hepatitis B. Nasal mucosal vaccines composed of poly-(ε- caprolactone) PCL/chitosan nanoparticles embedded with attenuated HBsAg (hepatitis B antigen) were found to be advantageous to generate site-specific immune responses owing to mucoadhesive, immunostimulatory, and high-resistance potential against degradation in the biological fluids. These polymeric nanoparticles are efficient when they enter and are retained in pulmonary epithelial cells at a concentration of 5-10 μg [63].

Surface-coated PCL nanoparticles with chitosan were designed for nasal immunization against influenza. Genetically derived influenza A virus protein (A/California/07/2009) was embedded in the polymeric nanoparticles for the stimulation of cellular, humoral, and mucosal immune responses. The system was positively charged, nanoscaled (approximately 125 nm), stable (zeta potential +22.88 mV), and exhibited high antigen entrapment efficiency (74. 84%). Controlled release of antigens for 63 days revealed prolonged immune-stimulating activity in Balb/c mice. After intranasal administration, chitosan-coated PCL nanoparticles exhibited booster immune responses in the lungs owing to an increased number of IgG (IgG1 and IgG2). The outcomes suggested the potential usage of developed nanoparticles as an adjuvant for the preparation of nasal nanovaccine against influenza [64].

A blend of PCL/chitosan polymeric nanoparticles has proven potential for enhanced protein and antigen delivery systems. Positively charged chitosan and the hydrophobicity of PCL facilitate improved protein adsorption and resistant biodegradation, thus efficient and prolonged local immune response occurs. Moreover, this blend polymeric system exhibited synergistic antibody generation and thermal stability with reduced cytotoxic action compared to isolated polymers. This blend does not disturb the secondary structure of adsorbed antigen (ovalbumin), hence no compromised immunological responses are expected on oral administration. The superb stability (6 months, 45°C) of this polymeric combination is quite beneficial to developing countries as it is difficult to maintain cold chain transportation [65]. Table 11.4 compiles several polymers utilized for the formulation of mucosal nanovaccines.

11.3.2 Carbon Nanotube-Based Mucosal Vaccines

The last decade was the era of carbon nanotubes (CNTs) and their derivatives in the field of vaccinology have exhibited a remarkable increase. CNTs are cylindrical structures made up of carbon atoms. Based on the number of walls present, they are classified as single-walled CNTs (0.4–2 nm) and multi-walled CNTs (10–100 nm). CNTs serve as potential carriers for antigens and can efficiently present the antigen to APCs for the instigation of adaptive immune responses [74]. Current research favors the application of CNTs as support materials for the delivery of vaccines [75]. CNTs have numerous characteristic properties that make them a preferred carrier for vaccine delivery. Their inert, non-toxic, and non-immunogenic nature are

Table 11.4 Development of various polymer-based mucosal nanovaccines.

Polymeric structure	Active component	Indication	Inference	Ref.
Poly (lactic-co-glycolic acid) (PLGA) nanoparticles	Recombinant S1-E	SARS-Co-V	Recombinant S1-E protein containing PLGA (Rs1-E-PLGA) nanoparticles exhibited sustained release and good stability. These particles have the potential to induce high titer antibodies, thus surging cellular immune responses.	[66]
PLGA nanoparticles	Lipopolysaccharide O antigen (LPS) and O polysaccharide (OPS)	Brucellosis	Both LPS and OPS nanovaccines exhibited superior performances of antibody production through opsonophagocytosis. Developed nanovaccines were efficient to remove bacterial communities in mice, and thus could be effective in humans against brucellosis. However, the LPS-PLGA composite showed higher immune responses compared to OPS-PLGA conjugate containing nanovaccines.	[67]
Chitosan and sodium alginate	Hepatitis B viral surface antigen	Hepatitis B virus	A higher immunogenic response compared to the control was achieved. The vaccine was stable and cost-effective.	[68]

(Continued)

Table 11.4 Development of various polymer-based mucosal nanovaccines. (*Continued*)

Polymeric structure	Active component	Indication	Inference	Ref.
Poly-ε-caprolactone/ chitosan nanoparticles	Anti-HBsAg	Hepatitis B virus	Mucoadhesive and immunostimulant intranasal nanovaccine was developed that showed highly resistant PCL for biodegradation in biological fluid.	[69]
N-2 hydroxypropyltrimethyl ammonium chloride derivative of chitosan	NDV/La Sota	Newcastle disease virus (NDV)	The developed nanovaccine exhibited stronger cellular and humoral immunity.	[70]
Quaternized chitosan	IBV/H 120, NDV/La Sota	Infectious bronchitis and Newcastle disease	Intranasal animal vaccine containing IBV and NDA enhanced cellular and humoral immunities.	[71]
Trimethyl chitosan nanoparticles	pVAX1-wapA	Mucosal anticaries nanovaccine	Anticaries DNA vaccine was developed to elicit the production of specific IgG and stronger mucosal and systemic responses.	[72]
Alginate nanoparticles	Inactivated influenza virus	Influenza	Improved cellular and humoral responses were observed.	[73]

certain desirable properties that are explored for their role in vaccine delivery. Owing to their hollow structure, nanotubes can entrap diverse molecules, other special properties include high-specific surface area, chemical stability, good shelf life, and *in-vivo* stability. Improved *in-vivo* stability is rendered to the insoluble and particulate nature of CNT; the same feature is responsible for the promotion of engulfment by phagocytic cells that stimulate rapid immune responses [76, 77]. The capability of CNTs to link multiple copies of antigens offers diverse applications of CNTs in vaccine delivery. The nanotubes adjoin bacterial, viral, and protozoal antigens. Functional group modifications in CNT scaffolds like the addition of carboxylic acid groups, amidation, and esterification enhance biological value [78]. The effect of CNTs at the cellular level depends on the recognition of nanotubes by plasma membrane receptors and the activation of immune responses. The uptake of nanomaterials at the cellular level follows energy-dependent and energy-independent mechanisms to target different cellular components [79]. Various CNT-based mucosal vaccines are compiled in Table 11.5.

11.3.3 Inorganic Nanoparticle-Based Mucosal Vaccines

There is a diverse range of inorganic materials that are suitable for the formulation of nanoparticles suitable for vaccine delivery. Inorganic materials-based nanoformulations are primarily used as carriers for vaccine delivery due to therapeutic efficacy, appropriate biodistribution, and pharmacokinetic properties [89]. However, a constraint associated with inorganic materials is that they undergo particle aggregation in biological fluids. Therefore, the nanoparticles are coated with organic materials to overcome the issue of aggregation. Technically, the inorganic nanoparticles are appropriately described as a complex hybrid with an organic outer shell enclosing an inorganic core [90].

11.3.3.1 Silica Based Nanoparticles

Advances in nanoscience play a significant role in vaccine development. Vaccine formulation in the present scenario focuses on the emergence of vaccines that are less immunogenic and have improved antigenic effects [91]. Since the adjuvant options available for human use are fewer in number, the domain of nanotechnology is explored as an alternative delivery vehicle. Silica nanoparticles (Si-Nps) express the ability for utilization as an adjuvant, as well as a vaccine delivery vehicle [92]. Silica is one of the most favorable and promising inorganic components for designing nanovaccines. Silica-based nanoparticles are biocompatible and biodegradable, with low pilot plant scale-up cost. Si-Nps are synthesized with desired

Table 11.5 Development of different CNT-based mucosal vaccines.

Type of CNT	Active molecule	Purpose	Result	Ref.
MWCNTs	Lentinan	To improve immune response	Nanotube could enter dendritic cells and act as effective antigen carriers	[80]
Short CNTs	HIV-1 glycoprotein vaccine	To develop mucosal vaccine for nasal administration	Enhanced local and systemic antibody production in experimental animals	[81]
Single-wall carbon nanotubes (SWCNTs)	Mannosylated nanoparticles	To enhance the effectiveness of immersion vaccine	SWCNTs-based nanovaccine formulations are an effective prophylactic agent for fish viral disease	[82]
Coated CNT	Hepatitis type A virus	To detect the existence of conjugation between the nanotube and HAV	Results revealed the existence of conjugation between the nanotubes and the HAV	[83]
Multi-walled CNTs	Tumor lysate protein	To improve immunotherapy for H22 liver cancer	Marked improvement in tumor cure	[84]

(Continued)

Table 11.5 Development of different CNT-based mucosal vaccines. (*Continued*)

Type of CNT	Active molecule	Purpose	Result	Ref.
Functionalized carbon nano tubes (f-CNT)	*Plasmodium vivax* membrane antigen (AMA-1) peptide	To provide protection and delay in infection	f-CNT offers a promising approach for the development of malarial vaccine formulation	[85]
Carboxylated SWCNTs	Tuberculin purified protein derivative	Potential as a novel nanovehicle	Immunological response was improved	[86]
SWCNTs	Conjugated along with poly(ethylene glycol) (PEG)	Comparison of heme-containing proteins for their capability to degrade PEG-SWCNTs	*In-vivo* studies revealed decreased degradation of CNT by proteins	[87]
SWCNT bundles	Antibodies such as anti-CD3	To develop efficient immunotherapy	A novel example of activation of lymphocytes; beneficial to basic science applications and clinical immunotherapy	[88]

structural properties like size and shape [93]. Appropriate control of particle size and shape of nanoparticles selectively modifies the interaction with the cell. The existence of ample silanol groups on the surface provides features like cell recognition, absorption of biomolecules, improved cell interaction, and increased cellular uptake. Modified versions of Si-Nps include mesoporous silica nanoparticles (MSNs) and hollow Si-NPs; both offer the ability to deliver molecules with diverse molecular weights. As compared to Si-NPs, MSNs have a high loading capacity due to high specific surface area and controlled release related to hollow mesoporous structure [94]. Owing to the insoluble nature of silica and stability in the gastrointestinal environment, Si-NPs efficiently protect proteins, DNA, and RNA loaded in them that are susceptible to degradation in the harsh gastric environment [95]. Table 11.6 details silica nanoparticle-based mucosal vaccines.

11.3.3.2 Gold Nanoparticles (AuNPs)

With progress in the domain of vaccine development for therapeutic efficacy and prophylactic applications, as well as prevention of immunological implications, AuNPs have emerged as a promising tool in the field of vaccine delivery [101]. AuNPs exhibit their application as a delivery system and adjuvant due to their ability to enhance immunological responses and reduce toxicity. In recent years, AuNPs have attracted researchers due to surface functionalization, biocompatibility, tuneable size, surface morphology, and optical characteristics [102]. Size affects the internalization, uptake, and distribution of AuNPs, while surface alteration or functionalization changes the circulation time. Protein-coated AuNPs result in the formation of "bio-corona" [103]. The formation of bio-corona improves targeting potential, binding affinity, and biodistribution [104]. Various gold nanoparticle-based mucosal vaccines are summarized in Table 11.7.

11.3.4 Virus-Like Particles (VLP)-Based Mucosal Vaccines

VLPs offer a novel platform for mucosal vaccine delivery. They are explored as an attractive option for vaccine delivery as compared to conventional live attenuated or inactivated vaccines. VLPs biomimic native viruses in terms of morphology and size. They are made up of multiple copies of viral structure proteins. Further, the absence of genomic information renders them non-competent for replication, which makes them a safer substitute as compared to traditional vaccines [110]. However, VLPs provoke lesser immune responses as compared to attenuated vaccines and is more immunogenic compared to other subunit alternatives. VLPs originate

Table 11.6 Different silica nanoparticle-based mucosal vaccines.

Type of silica nanoparticle	Active molecule	Purpose	Result	Ref.
Three forms of silica (S2, S1, and SBA-15) based nanoparticles	Antigen bovine serum albumin	To design a sustained-release vaccine system	Oral immunization with silica/BSA nanoparticle-boosted humoral and mucosal response	[96]
Mesoporous silica nanoparticles (MSNPs)	Diphtheria toxoid	To provide evidence that inorganic NPs are alternative to currently used adjuvants and could be loaded with toxoids for effective vaccination	Fabricated mesoporous silica nanoparticles are a suitable alternative for currently used adjuvants	[97]
Dendritic mesoporous silica NPs (DMSNs)	Peptide-based subunit vaccines	To assess *in vitro* release profile	*In vitro* results showed that DMSNs depict sustained release in comparison to the reference formulation	[98]
Glutathione-depletion dendritic mesoporous organosilica NPs (GDMONs) adorned with tetrasulfide- framework	Antigen protein (ovalbumin)	To deliver antigen to antigen-presenting cell	GDMONs show potential for use as a novel self-adjuvant for vaccination	[99]
Hollow extra large mesoporous silica NPs	Model antigen	To develop effective immune adjuvant that has the potential to transform the surface charge of the nanoparticles for slow release of antigen	Enhanced activation of the DCs and *in vivo* improved survival rate.	[100]

Table 11.7 Different gold nanoparticle-based mucosal vaccines.

Type of AuNPs	Active molecule	Purpose	Result	Ref.
Gold nanorods	Small molecular-weight polyinosinic-polycytidylic acid	To elicit mucosal immune responses	Gold nanorods offer a promising platform for augmenting the weak adjuvanticity of uPIC (40:400) and the production of low inflammatory cytokine	[105]
Gold nanoparticle 15 nm (sphere)	Two novel proteins, along with Hcp1	To develop glycoconjugate vaccines for protection against *Burkholderia pseudomallei*	*In vivo* results showed that animals treated with AuNP glycoconjugate vaccines developed high specific antibody titers	[106]
Dual-linker gold nanoparticles 18 nm (sphere)	Recombinant trimetric A/Aichi/2/68, hemagglutinin and TLR5 agonist flagellin	To resolve the intrinsic restrictions of the conventional influenza vaccine	Dual-conjugated AuNPs are effective and can promote generation of immune responses	[107]
Gold nanoparticles 12 nm (sphere)	Extracellular region of the matrix 2 protein with or without cytosine-guanine rich oligonucleotide	To formulate a novel vaccine for the management of influenza	Conjugated gold nanoparticles offer a platform for formulating an influenza A vaccine	[108]
Chitosan-reduced gold nanoparticles	Tetanus toxoid (TT) vaccine	To define the efficacy of chitosan-reduced AuNPs over chitosan nanoparticles for effective transmucosal delivery of the vaccine for TT	Subcutaneous administration results in the initiation of a systemic response but without a mucosal response.	[109]

from enveloped and non-enveloped viruses. Utilization of VLPs as a vaccine delivery system provides a defense to the susceptible host against specific viral infections, therefore significant studies have been performed in the last few decades that suggest the efficient nature of VLPs for vaccine delivery [111]. The major concern associated with the application of VLPs is the tedious characterization that includes physical characterization like particle size, shape, density, and homogeneity; the other parameter characterized includes biochemical contents determination like an estimation of total protein content, nucleic acids, and endotoxins. The other obstacles in the path of developing VLP vaccines are regulatory hurdles and additional problems like stability and batch-to-batch uniformity [112]. A few mucosal VLP vaccines and their clinical status are compiled in Table 11.8 [113]. The diverse nature of VLPs as a carrier system allows its exploration to deliver an extensive range of foreign particles/antigens and is suitable for action for mucosal pathogens.

11.4 Future Perspectives

Mucosal delivery of vaccines exhibits potential for effective treatment of diseases and possesses effective prophylactic ability. Although mucosal vaccination offers numerous benefits associated with them like the interaction of antigen with MALT to induce both mucosal and systemic immunity and logistical benefit in terms of ease of delivery, only a few of them are available as licensed commercialized products. Structural barriers of mucosal membranes like harsh physicochemical environments in the stomach and intestinal tract and mucociliary clearance mechanisms for oral and nasal routes, respectively, are primarily responsible for the slow development of mucosal vaccines. Future mucosal vaccine advancements include the application of purified antigen components accompanied by safe, effective adjuvants and delivery systems. Exploration of new biomaterials and research in the field of nanotechnology favor the design of novel mucosal vaccines. However, the economic constraints associated with them require consideration in the near future to prove an edge of mucosal vaccination over pre-existing needle-based vaccines. Mucosal vaccine design exhibits a complex situation, as it involves the proper choice of route, suitable delivery carriers, and adjuvants. In recent years, significant knowledge has been gained but a prominent gap still exists. The proper knowledge of different mucosal surfaces might boost the development of more creative nanotechnology-based licensed products. Table 11.9 gathers registered mucosal nanovaccines containing myriad nanomaterials.

Table 11.8 Examples of VLP vaccines and their clinical status.

Virus	Family	Genome	Particle composition	Vaccine status
Hepatitis type E	Hepeviridae	Single-stranded RNA	Capsid	Preclinical
Hepatitis type C	Flaviviridae	Single-stranded RNA	Core E1, E2	Preclinical
Human immunodeficiency virus (HIV) and Simian immunodeficiency virus (SIV)	Retroviridae	Single-stranded RNA	Gag, env	Preclinical Phase I
Rift Valley Fever	Bunyaviridae	Segmented single-stranded RNA	Capsid, GN, GC	Preclinical
Influenza	Orthomyxoviridae	Segmented single-stranded RNA	HA, (NA), M1/gag	Preclinical Phase I
Papilloma	Papilloviridae	Double-stranded DNA	L1	Licensed
Hepatitis B	Hepadnaviridae	Double-stranded DNA	HBsAg	Licensed

Table 11.9 Approved mucosal nanovaccines manufactured worldwide by pharmaceutical companies.

Nanomaterials	Antigens	Nanovaccine	Company	Registration number
LNP encapsulated in mRNA	mRNA	mRNA-1273LNP	Moderna	Phase III (NCT04470427)
Virosome with inactivated virus particle	Hepatitis A	Epaxal®	Crucell	Phase III NCT01307436
HA (recombinant) protein on Tween 80 nanoparticles	Influenza	NanoFlu™	Novavax	Phase III NCT04120194
LNP with saRNA encapsulated	ssRNA and SARS-CoV-2 S (prefused) protein	LNP-CoV-saRNA	Imperial College, London	NCT04480957
Ceramidecarbamoyl-spermine/cholesterol) liposomal adjuvant	Influenza	VaxiSome™	NasVax	Phase II NCT00915187
LNP with mRNA encapsulated	mRNA -trimerized RBD of coronavirus	mRNA BNT162B2	Pfizer/ BioNTech	
HIV-1 gp41 liposomes	HIV	MPER-656	DAIDS/NIAID/ NIH	Phase I NCT03934541

11.5 Conclusion

Advancement in the field of nanotechnology leads to the formulation of mucosal vaccines that can transit the mucosal barriers and elicit mucosal as well as systemic immunological responses. Mucosal nanovaccine enhances bioavailability, improves patient compliance, and maintains the structural integrity of antigens that are required for the management of devastating pathogens, allergies, and autoimmune diseases. Although most of these mucosal nanovaccines have potential as shown in animal models, their safe and effective applications in humans are yet to be defined. Incorporation of allied sciences including proteomics, genomics, metabolomics, and mathematical models assist to overcome aforesaid issues. Further, genetically diverse humans show significantly different nutritional statuses and physiological and immunological responses to mucosal vaccines. For instance, immune responses of oral polio vaccines, oral cholera, and rotavirus vaccines have less potential in developing countries compared to developed ones. Moreover, the type of adjuvant, targeting site, mucosal cell interaction, and storage criteria are also crucial for the efficacy of attenuated-type mucosal vaccines.

References

1. Vartak, A. and Sucheck, S.J., Recent advances in subunit vaccine carriers. *Vaccines*, 4, 12, 2016.
2. Amico, D., Fontana, C., Cheng, F.R. *et al.*, Development of vaccine formulations: Past, present, and future. *Drug Deliv. Transl. Res.*, 11, 353, 2021.
3. Das, A. and Ali, N., Nanovaccine: An emerging strategy. *Expert Rev. Vaccines.*, 20, 1273, 2021.
4. Nandedkar, T.D., Nanovaccines: Recent developments in vaccination. *J. Biosci.*, 34, 995, 2009.
5. Wagner-Muñiz, D.A., Haughney, S.L., Kelly, S.M., Wannemuehler, M.J., Narasimhan, B., Room temperature stable PspA-based nanovaccine induces protective immunity. *Front. Immunol.*, 9, 325, 2018.
6. Petersen, L.K., Sackett, C.K., Narasimhan, B., High-throughput analysis of protein stability in polyanhydride nanoparticles. *Acta Biomater.*, 6, 3873, 2010.
7. Narasimhan, B., Goodman, J.T., Vela Ramirez, J.E., Rational design of targeted next-generation carriers for drug and vaccine delivery. *Annu. Rev. Biomed. Eng.*, 18, 25, 2016.
8. Thomas, C., Rawat, A., Hope-Weeks, L., Ahsan, F., Aerosolized PLA and PLGA nanoparticles enhance humoral, mucosal and cytokine responses to hepatitis B vaccine. *Mol. Pharm.*, 48, 405, 2011.

9. Ramirez, J.E.V., Boggiatto, P.M., Wannemuehler, M.J., Narasimhan, B., Polyanhydride nanoparticle interactions with host serum proteins and their effects on bone marrow derived macrophage activation. *ACS Biomater. Sci. Eng.*, 3, 160, 2017.
10. Kijanka, G., Bee, J.S., Korman, S.A., Wu, Y., Roskos, L.K., Schenerman, M.A. *et al.*, Submicron size particles of a murine monoclonal antibody are more immunogenic than soluble oligomers or micron size particles upon subcutaneous administration in mice. *J. Pharm. Sci.*, 107, 2847, 2018.
11. Gutierro, I., Hernández, R.M., Igartua, M., Gascón, A.R., Pedraz, J.L., Size dependent immune response after subcutaneous, oral, and intranasal administration of BSA loaded nanospheres. *Vaccine.*, 21, 67, 2002.
12. Reddy, S.T., Van der Vlies, A.J., Simeoni, E., Angeli, V., Randolph, G.J., O'Neil, C.P., Exploiting lymphatic transport and complement activation in nanoparticle vaccines. *Nat. Biotechnol.*, 25, 1159, 2007.
13. Banerjee, A., Qi, J., Gogoi, R., Wong, J., Mitragotri, S., Role of nanoparticle size, shape and surface chemistry in oral drug delivery. *J. Control. Release*, 238, 176, 2016.
14. Li., Y., Kroger., M., Liu, W.K., Shape effect in cellular uptake of PEGylated nanoparticles: Comparison between sphere, rod, cube, and disk. *Nanoscale*, 7, 16631, 2015.
15. Yue, Z.G., Wei, W., Lv, P.P., Yue, H., Wang, L.Y., Su, Z.G. *et al.*, Surface charge affects cellular uptake and intracellular trafficking of chitosan-based nanoparticles. *Biomacromolecules*, 12, 2440, 2011.
16. Goodman, J.T., Vela Ramirez, J.E., Boggiatto, P.M., Roychoudhury, R., Pohl, N.L.B., Wannemuehler, M.J., Nanoparticle chemistry and functionalization differentially regulates dendritic cell-nanoparticle interactions and triggers dendritic cell maturation. *Part. Part. Syst. Char.*, 31, 1269, 2014.
17. Mestecky, J., Lamn, M.N., Ogra, P.L., Strober, W., Bienenstock, J., Mcghee, J., Mayer, L. (Eds.), *Mucosal Immunology*, p. 2064, Academic Press, San Diego, 2005.
18. Holmgren, J. and Czerkinsky, C., Mucosal immunity and vaccines. *Nat. Med.*, 11, 45, 2005.
19. Kiyono, H. and Fukuyama, S., NALT-versus Peyer's-patch-mediated mucosal immunity. *Nat. Rev. Immunol.*, 4, 699, 2004.
20. Guy-Grand, D., Azogui, O., Celli, S., Darche, S., Nussenzweig, M.C., Kourilsky, P., Vassalli, P., Extrathymic T cell lymphopoiesis: Ontogeny and contribution to gut intraepithelial lymphocytes in athymic and euthymic mice. *J. Exp. Med.*, 197, 333–341, 2003.
21. Iwasaki, A. and Kelsall, B.L., Freshly isolated Peyer's patch, but not spleen, dendritic cells produce interleukin 10 and induce the differentiation of T helper type 2 cells. *J. Exp. Med.*, 190, 229, 1999.
22. Janeway Jr., C.A., Travers, P., Walport, M. (Eds.), *Immunobiology: The Immune System in Health and Disease*, Garland Science, New York, 2001.

23. Agace, W.W. and McCoy, K.D., Regionalized development and maintenance of the intestinal adaptive immune landscape. *Immunity*, 46, 548, 2017.
24. Kiss, E.A., Vonarbourg, C., Kopfmann, S., Hobeika, E., Finke, D., Esser, C., Diefenbach, A., Natural aryl hydrocarbon receptor ligands control organogenesis of intestinal lymphoid follicles. *Science*, 334, 1561, 2011.
25. Hamada., H., Hiroi, T., Nishiyama, Y., Takahashi, H., Masunaga, Y., Hachimura, S., Kaminogawa, S., Takahashi, I.H., Iwanaga, T., Kiyono, H., Yamamoto, H., Ishikawa, H., Identification of multiple isolated lymphoid follicles on the antimesenteric wall of the mouse small intestine. *J. Immunol.*, 168, 64, 2002.
26. Vajdy, M., Sethupathi, P., Knight, K.L., Dependence of antibody somatic diversification on gut-associated lymphoid tissue in rabbits. *J. Immunol.*, 160, 2725, 1998.
27. Wang, J., Li, F., Tian, Z., Role of microbiota on lung homeostasis and diseases. *Sci. China. Life. Sci.*, 60, 1407, 2017.
28. Sepahi, A. and Salinas, I., The evolution of nasal immune systems in vertebrates. *Mol. Immunol.*, 69, 131, 2016.
29. Mutoh, M., Kimura, S., Takahashi, I.H., Hisamoto, M., Iwanaga, T., Iida, J., RANKL regulates differentiation of microfold cells in mouse nasopharynx-associated lymphoid tissue (NALT). *Cell. Tissue Res.*, 364, 175, 2016.
30. Hsieh, E.H., Fernandez, X., Wang, J., Hamer, M., Calvillo, S., Croft, M., Kwon, B.S., Lo, D.D., CD137 is required for M cell functional maturation but not lineage commitment. *Am. J. Pathol.*, 177, 666, 2010.
31. Shimoda, M., Nakamura, T., Takahashi, Y., Asanuma, H., Tamura, S., Kurata, T., Mizuochi, T., Azuma, N., Kanno, C., Takemori, T., Isotype-specific selection of high affinity memory B cells in nasal-associated lymphoid tissue. *J. Exp. Med.*, 194, 1597, 2001.
32. Hwang, J.Y., Randall, T.D., Silva, S.A., Inducible Bronchus-Associated Lymphoid Tissue: Taming Inflammation in the Lung. *Front. Immunol.*, 11, 258, 2016.
33. Rangel, M.J., Hartson, L., Navarro, C., Gaxiola, M., Selman, M., Randall, T.D., Inducible bronchus-associated lymphoid tissue (iBALT) in patients with pulmonary complications of rheumatoid arthritis. *J. Clin. Investig.*, 116, 3183, 2006.
34. Richert, L.E., Harmsen, A.L., Rynda-Apple, A., Wiley, J.A., Servid, A.E., Douglas, T., Inducible bronchus-associated lymphoid tissue (iBALT) synergizes with local lymph nodes during antiviral CD4+ T cell responses. *Lymphat. Res. Biol.*, 11, 196, 2013.
35. Dewangan, H.K., Rational application of nanoadjuvant for mucosal vaccine delivery system. *J. Immunol. Methods*, 481, 481–482, 2020.
36. Bhavsar, M.D. and Amiji, M.M., Polymeric nano- and microparticle technologies for oral gene delivery. *Expert Opin. Drug Deliv.*, 4, 197, 2007.
37. Bahadoran, A., Moeini, H., Bejo, M.H., Hussein, M.Z., Omar, A.R., Development of tat-conjugated dendrimer for transdermal DNA vaccine delivery. *J. Pharm. Sci.*, 19, 325, 2016.

38. Lehtinen, M., Söderlund-Strand, A., Vänskä, S., Luostarinen, T., Eriksson, T., Natunen, K. *et al.*, Impact of gender-neutral or girls-only vaccination against human papillomavirus-Results of a community-randomized clinical trial (I). *Int. J. Cancer*, 142, 949, 2018.
39. Moffatt, K., Wang, Y., Raj Singh, T.R., Donnelly, R.F., Microneedles for enhanced transdermal and intraocular drug delivery. *Curr. Opin. Pharmacol.*, 36, 14, 2017.
40. Vyas, S., Dhoble, S., Ghodake, V., Patravale, V., Xyloglucan based mucosal nanovaccine for immunological protection against brucellosis developed by supercritical fluid technology. *Int. J. Pharm. X*, 2, 100053, 2020.
41. Mallakpour, S., Azadi, E., Hussain, C.M., Chitosan, alginate, hyaluronic acid, gums, and β-glucan as potent adjuvants and vaccine delivery systems for viral threats including SARS-CoV-2: A review. *Int. J. Biol. Macromol.*, 182, 1931, 2021.
42. Clausen, A.E., Kast, C.E., Bernkop-Schnürch, A., The role of glutathioneinthe permeation enhancing effect of thiolated polymers. *Pharm. Res.*, 19, 602, 2001.
43. Pippa, N., Gazouli, M., Pispas, S., Recent advances and future perspectives in polymer-based nanovaccines. *Vaccines*, 9, 558, 2021.
44. Liu, Y., Guo, J., Huang, L., Modulation of tumor microenvironment for immunotherapy: Focus on nanomaterial-based strategies. *Theranostics*, 10, 3099, 2020.
45. Thomas, C., Rawat, A., Hope-Weeks, L., Ahsan, F., Aerosolized PLA and PLGA nanoparticles enhance humoral, mucosal and cytokine responses to hepatitis B vaccine. *Mol. Pharm.*, 8, 405, 2011.
46. Maleki, M., Azimi, S., Salouti, M., Protective effect of two new nanovaccines against Pseudomonas aeruginosa based on LPS and OPS: A comparison study. *Immunobiology*, 227, 152278, 2022.
47. Liu, L., Cao, F., Liu, X., Wang, H., Zhang, C., Sun, H., Wang, C., Leng, X., Song, C., Kong, D., Ma, G., Hyaluronic acid-modified cationic lipid-PLGA hybrid nanoparticles as a nanovaccine induce robust humoral and cellular immune responses. *ACS Appl. Mater. Interfaces*, 8, 11969, 2016.
48. Gogev, S., De Fays, K., Versali, M.F., Gautier, S., Thiry, E., Glycol chitosan improves the efficacy of intranasally administrated replication defective human adenovirus type 5 expressing glycoprotein D of bovine herpesvirus 1. *Vaccine*, 22, 1946, 2004.
49. Pawar, D. and Jaganathan, K.S., Mucoadhesive glycol chitosan nanoparticles for intranasal delivery of hepatitis B vaccine: Enhancement of mucosal and systemic immune response. *Drug Deliv.*, 23, 185, 2016.
50. Lopes, P.D., Okino, C.H., Fernando, F.S., Pavani, C., Casagrande, V.M., Lopez, R.F.V. *et al.*, Inactivated infectious bronchitis virus vaccine encapsulated in chitosan nanoparticles induces mucosal immune responses and effective protection against challenge. *Vaccine*, 36, 2630, 2018.
51. Lee, J., Kim, Y.M., Kim, J.H., Cho, C.W., Jeon, J.W., Park, J.K., Nasal delivery of chitosan/alginate nanoparticle encapsulated bee (Apismellifera) venom

promotes antibody production and viral clearance during porcine reproductive and respiratory syndrome virus infection by modulating T cell related responses. *Vet. Immunol. Immunopathol.*, 200, 40, 2018.
52. Donnelly, L., Curran, R.M., Tregoning, J.S., McKay, P.F., Cole, T., Morrow, R.J. *et al.*, Intravaginal immunization using the recombinant HIV-1 clade-C trimeric envelope glycoprotein CN54gp140 formulated within lyophilized solid dosage forms. *Vaccine*, 29, 4512, 2011.
53. Xu, C., Qiao, M., Huo, X., Liao, Z., Su, J., An oral microencapsulated vaccine loaded by sodium alginate effectively enhances protection against GCRV infection in grass carp (*ctenopharyngodonidella*). *Front. Immunol.*, 13, 848958, 2022.
54. Borges, O., Cordeiro-da-Silva, A., Romeijn, S.G., Amidi, M., de Sousa, A., Borchard, G., Junginger, H.E., Uptake studies in rat Peyer's patches, cytotoxicity and release studies of alginate coated chitosan nanoparticles for mucosal vaccination. *J. Control. Release*, 114348, 2006, 2006.
55. Mosafer, J., Sabbaghi, A., Badiee, A., Dehghan, S., Tafaghodi, M., Preparation, characterization and *in vivo* evaluation of alginate-coated chitosan and trimethylchitosan nanoparticles loaded with PR8 influenza virus for nasal immunization. *Asian J. Pharm.*, 14, 216, 2019.
56. Amin, M.K. and Boateng, J.S., Enhancing stability and mucoadhesive properties of chitosan nanoparticles by surface modification with sodium alginate and polyethylene glycol for potential oral mucosa vaccine delivery. *Mar. Drugs*, 20, 156, 2022.
57. Samuelsen, A.B., Schrezenmeir, J., Knutsen, S.H., Effects of orally administered yeast-derived beta-glucans: A review. *Mol. Nutr. Food Res.*, 58, 183, 2014.
58. Leibund Gut-Landmann, S., Gross, O., Robinson, M.J., Osorio, F., Slack, E.C., Tsoni, S.V., Schweighoffer, E., Tybulewicz, V., Brown, G.D., Syk- and CARD9-dependent coupling of innate immunity to the induction of T helper cells that produce interleukin 17. *Nat. Immunol.*, 8, 630, 2007.
59. Wang, S.C., Bligh, S.W.A., Zhu, C.L., Shi, S.S., Wang, Z.T., Hu, Z.B., Crowder, J., Branford-White, C., Vella, C., Sulfated beta-glucan derived from oat bran with potent anti-HIV activity. *J. Agric. Food Chem.*, 56, 2624, 2008.
60. Markova, N., Kussovski, V., Drandarska, I., Nikolaeva, S., Georgieva, N., Radoucheva, T., Protective activity of lentinan in experimental tuberculosis. *Int. Immunopharmacol.*, 3, 1557, 2003.
61. Ikewaki, N., Iwasaki, M., Kurosawa, G., Rao, S., Lakey-Beitia, J., Preethy, S., Abraham, S.J.K., β-glucans: Wide-spectrum immune-balancing food-supplement-based enteric (β-WIFE) vaccine adjuvant approach to COVID-19. *Hum. Vaccin. Immunother.*, 17, 2808, 2021.
62. Le, T., Doan, T.H., Quyen, D., Le, K.X., Pham, V., Nagataki, M., Nomura, H., Ikeue, Y., Watanabe, Y., Agatsuma, T., The adjuvant effect of sophy β-glucan to the antibody response in poultry immunized by the avian influenza A H5N1 and H5N2 vaccines. *J. Microbiol. Biotechnol.*, 21, 405, 2011.

63. Jesus, S., Soares, E., Costa, J., Borchard, G., Borges, O., Immune response elicited by an intranasally delivered HBsAg low-dose adsorbed to poly-ε-caprolactone based nanoparticles. *Int. J. Pharm.*, 17, 59, 2016.
64. Gupta, N.K., Tomar, P., Sharma, V., Dixit, V.K., Development and characterization of chitosan coated poly-(ε-caprolactone) nanoparticulate system for effective immunization against influenza. *Vaccine*, 29, 9026, 2011.
65. Jesus, S., Fragal, E.H., Rubira, A.F., Muniz, E.C., Valente, A.J.M., Borges, O., The inclusion of chitosan in poly-ε-caprolactone nanoparticles: Impact on the delivery system characteristics and on the adsorbed ovalbumin secondary structure. *AAPS Pharm. Sci. Tech.*, 19, 101, 2018.
66. Huang, J., Ding, Y., Yao, J., Peng, K., Deng, K., Zhang, M., Zhang, Y., Zuo, J., The SARS-CoV-2 rS1-E-PLGA nanovaccine and evaluation of its immune effect in BALB/c mice. *Eur. Rev. Med. Pharmacol. Sci.*, 26, 5255, 2022.
67. Hossein, A., Masoud, M., Mojtaba, S., Immunological effects of two new nanovaccines against Brucella based on OPS and LPS antigens conjugated with PLGA nanoparticles. *Eur. Polym. J.*, 139, 110021, 2020.
68. AbdelAllah, N.H., Abdeltawab, N.F., Boseila, A.A., Amin, M.A., Chitosan and sodium alginate combinations are alternative, efficient, and safe natural adjuvant systems for hepatitis B vaccine in mouse model. *Evid.-Based Complement. Altern. Med.*, 2016, 1–8, 7659684, 2016.
69. Jesus, S., Soares, E., Costa, J., Borchard, G., Borges, O., Immune response elicited by an intranasally delivered HBsAg low-dose adsorbed to poly-ε-caprolactone based nanoparticles. *Int. J. Pharm.*, 17, 504, 59–69, 2016.
70. Zhao, K., Li, S., Li, W., Yu, L., Duan, X., Han, J., Wang, X., Jin, Z., Quaternized chitosan nanoparticles loaded with the combined attenuated live vaccine against newcastle disease and infectious bronchitis elicit immune response in chicken after intranasal administration. *Drug Deliv.*, 24, 1574, 2017.
71. Zhao, K., Sun, Y., Chen, G., Rong, G., Kang, H., Jin, Z., Wang, X., Biological evaluation of N-2-hydroxypropyl trimethyl ammonium chloride chitosan as a carrier for the delivery of live newcastle disease vaccine. *Carbohydr. Polym.*, 149, 28, 2016.
72. Li, H., Lu, Y., Xiang, J., Jiang, H., Zhong, Y., Lu, Y., Enhancement of immunogenic response and protection in model rats by CSTM nanoparticles anticaries DNA vaccine. *Nanomedicine*, 11, 1407, 2016.
73. Dehghan, S., Kheiri, M.T., Abnous, K., Eskandari, M., Tafaghodi, M., Preparation, characterization and immunological evaluation of alginate nanoparticles loaded with whole inactivated influenza virus: Dry powder formulation for nasal immunization in rabbits. *Microb. Pathog.*, 115, 74, 2018.
74. Zhou, F., Xing, D., Ou, Z., Wu, B., Resasco, D.E., Chen, W.R., Cancer photothermal therapy in the near-infrared region by using single-walled carbon nanotubes. *J. Biomed. Opt.*, 14, 021009, 2009.
75. de Carvalho Lima, E.N., Diaz, R.S., Justo, J.F., Castilho Piqueira, J.R., Advances and perspectives in the use of carbon nanotubes in vaccine development. *Int. J. Nanomedicine*, 16, 5411, 2021.

76. Kostarelos, K., Lacerda, L., Pastorin, G., Wu, W., Wieckowski, S., Luangsivilay, J., Godefrov, S., Pantarotto, D., Briand, J.-P., Muller, S., Prato, M., Bianco, A., Cellular uptake of functionalized carbon nanotubes is independent of functional group and cell type. *Nat. Nanotechnol.*, 2, 108, 2007.
77. Kam, N.W., O'Connell, M., Wisdom, J.A., Dai, H.J., Carbon nanotubes as multifunctional biological transporters and near-infrared agents for selective cancer cell destruction. *Proc. Natl. Acad. Sci.*, 102, 11600, 2005.
78. Singh, P., Campidelli, S., Giordani, S., Bonifazi, D., Bianco, A., Prato, M., Organic functionalisation and characterisation of single-walled carbon nanotubes. *Chem. Soc. Rev.*, 38, 2214, 2009.
79. Mohammadi, M.R., Nojoomi, A., Mozafari, M., Dubnika, A., Inayathullah, M., Rajadas, J., Nanomaterials engineering for drug delivery: A hybridization approach. *J. Mater. Chem. B*, 5, 3995–4018, 2017.
80. Zheng, S., Luo, L., Huang, Y., Niu, Y., Wang, D., Lentinan-modified carbon nanotubes as an antigen delivery system modulate immune response *in vitro* and *in vivo*. *ACS Appl. Mater. Interfaces*, 8, 19276–19283, 2016.
81. Xu, Y., Jiang, X., Zhou, Z., Ferguson, T., Obliosca, J., Luo, C.C. *et al.*, Mucosal delivery of HIV-1 glycoprotein vaccine candidate enabled by short carbon nanotubes. *Part. Part. Syst. Charact.*, 39, 2200011, 2022.
82. Zhu, B., Zhang, C., Zhao, Z., Wang, X., Targeted delivery of mannosylated nanoparticles improve prophylactic efficacy of immersion vaccine against fish viral disease. *Vaccines*, 8, 87, 2020.
83. Carbone, M., Valentini, F., Caminiti, R., Petrinca, A.R., Donia, D., Divizia, M. *et al.*, Are PEI-coated SWCNTs conjugated with hepatitis a virus? A chemical study with SEM, Z-potential, EDXD and RT-PCR. *Biomed. Mater.*, 5, 35001, 2010.
84. Meng, J., Meng, J., Duan, J., Kong, H., Li, L., Wang, C., Xie, S., Chen, S., Gu, N., Xu, H., Yang, X.-D., Carbon nanotubes conjugated to tumor lysate protein enhance the efficacy of an antitumor immunotherapy. *Small*, 4, 1364, 2008.
85. Yandar, N., Pastorin, G., Prato, M., Bianco, A., Patarroyo, M.E., Lozano, J.M., Immunological profile of a plasmodium vivax AMA-1 N-terminus peptide-carbon nanotube conjugate in an infected Plasmodium berghei mouse model. *Vaccine*, 26, 5864, 2008.
86. Zeinali, M., Jammalan, M., Ardestani, S.K., Mosaveri, N., Immunological and cytotoxicological characterization of tuberculin purified protein derivative (PPD) conjugated to single-walled carbon nanotubes. *Immunol. Lett.*, 126, 48, 2009.
87. Vlasova, I.I., Vakhrusheva, T.V., Sokolov, A.V., Kostevich, V.A., Gusev, A.A., Gusev, S.A. *et al.*, PEGylated single-walled carbon nanotubes activate neutrophils to increase production of hypochlorous acid, the oxidant capable of degrading nanotubes. *Toxicol. Appl. Pharmacol.*, 264, 131, 2012.
88. Fadel, T.R., Steenblock, E.R., Stern, E., Li, N., Wang, X., Haller, G.L., Pfefferle, L.D., Fahmy, T.M., Enhanced cellular activation with single walled carbon nanotube bundles presenting antibody stimuli. *Nano. Lett.*, 8, 2070, 2008.

89. Feliu, N., Docter, D., Heine, M., Del Pino, P., Ashraf, S., Kolosnjaj, J., Macchiarini, P., Nielsen, P., Alloyeau, D., Gazeau, F., Stauber, R.H., Parak, W.J., *In vivo* degeneration and the fate of inorganic nanoparticles. *Chem. Soc. Rev.*, 45, 2440, 2016.
90. Giner-Casares, J.J., Henriksen-Lacey, M., Coronado-Puchau, M., Liz-Marzán, L.M., Inorganic nanoparticles for biomedicine: Where materials scientists meet medical research. *Mater. Today*, 19, 19, 2016.
91. Mamo, T. and Poland, G.A., Nanovaccinology: The next generation of vaccines meets 21st century materials science and engineering. *Vaccine*, 30, 6609, 2012.
92. Navarro-Tovar, G., Palestino, G., Rosales-Mendoza, S., An overview on the role of silica-based materials in vaccine development. *Expert. Rev. Vaccines*, 15, 1449, 2016.
93. Carvalho, L.V., Ruiz, R., Scaramuzzi, K., Marengo, E.B., Matos, J.R., Tambourgi, D.V., Fantini, M.C.A., Sant'Anna, O.A., Immunological parameters related to the adjuvant effect of the ordered mesoporous silica SBA-15. *Vaccine*, 28, 7829, 2010.
94. Cheng, K., El-Boubbou, K., Landry, C.C., Binding of HIV-1 gp120 glycoprotein to silica nanoparticles modified with CD4 glycoprotein and CD4 peptide fragments. *ACS Appl. Mater. Interfaces*, 4, 235, 2012.
95. Ways, T.M.M., Ng, K.W., Lau, W.M., Khutoryanskiy, V.V., Silica nanoparticles in transmucosal drug delivery. *Pharmaceutics*, 12, 751, 2020.
96. Wang, T., Jiang, H., Zhao, Q., Wang, S., Zou, M., Cheng, G., Enhanced mucosal and systemic immune responses obtained by porous silica nanoparticles used as an oral vaccine adjuvant: Effect of silica architecture on immunological properties. *Int. J. Pharm.*, 436, 351, 2012.
97. Huang, X. and Townley, H.E., An assessment of mesoporous silica nanoparticle architectures as antigen carriers. *Pharmaceutics*, 12, 294, 2020.
98. An, W., Defaus, S., Andreu, D., Rivera-Gil, P., *In vivo* sustained release of peptide vaccine mediated by dendritic mesoporous silica nanocarriers. *Front. Immunol.*, 12, 684612, 2021.
99. Lu, Y., Yang, Y., Gu, Z., Zhang, J., Song, H., Xiang, G., Yu, C., Glutathione-depletion mesoporousorgano-silica nanoparticles as a self-adjuvant and co-delivery platform for enhanced cancer immunotherapy. *Biomaterials*, 175, 82, 2018.
100. Lee, J.Y., Kim, M.K., Nguyen, T.L., Kim, J., Hollow mesoporous silica nanoparticles with extra-large mesopores for enhanced cancer vaccine. *ACS Appl. Mater. Interfaces*, 12, 34658, 2020.
101. Kubackova, J., Zbytovska, J., Holas, O., Nanomaterials for direct and indirect immunomodulation: A review of applications. *Eur. J. Pharm. Sci.*, 142, 105139, 2019.
102. Bertoli, F., Garry, D., Monopoli, M.P., Salvati., A., Dawson., K.A., The intracellular destiny of the protein corona: A study on its cellular internalization and evolution. *ACS Nano*, 10, 10471, 2016.

103. Monopoli, M.P., Aberg, C., Salvati, A., Dawson, K.A., Biomolecular coronas provide the biological identity of nanosized materials. *Nat. Nanotechnol.*, 7, 779, 2012.
104. Tazaki, T., Tabata, K., Ainai, A., Ohara, Y., Kobayashi, S., Ninomiya, T., Shape-dependent adjuvanticity of nanoparticle-conjugated RNA adjuvants for intranasal inactivated influenza vaccines. *RSC Adv.*, 8, 16527, 2018.
105. Muruato, L.A., Tapia, D., Hatcher, C.L., Kalita, M., Brett, P.J., Gregory, A.E., Use of reverse vaccinology in the design and construction of nanoglycoconjugate vaccines against Burkholderiapseudomallei. *Clin. Vaccine. Immunol.*, 24, 1, 2017.
106. Wang, C., Zhu, W., Wang, B.Z., Dual-linker gold nanoparticles as adjuvanting carriers for multivalent display of recombinant influenza hemagglutinintrimers and flagellin improve the immunological responses *in vivo* and *in vitro*. *Int. J. Nanomed.*, 12, 4747, 2017.
107. Tao, W., Ziemer, K.S., Gill, H.S., Gold nanoparticle-M2e conjugate coformulated with CpG induces protective immunity against influenza A virus. *Nanomedicine*, 9, 237, 2014.
108. Pokharkar, V., Bhumkar, D., Suresh, K., Shinde, Y., Gairola, S., Jadhav, S.S., Gold nanoparticles as a potential carrier for transmucosal vaccine delivery. *J. Biomed. Nanotechnol.*, 7, 57, 2011.
109. Arnon, R., Overview of vaccine strategies, in: *Vaccine Design Innovative Approaches and Novel Strategies*, R. Rappuoli and F. Bagnoli (Eds.), pp. pp 1–19, Caister Academic Press, Norwolk, UK, 2011.
110. Estes, M.K., Ball, J.M., Guerrero, R.A., Opekun, A.R., Gilger, M.A., Pacheco, S.S., Graham, D.Y., Norwalk virus vaccines: Challenges and progress. *J. Infect. Dis.*, 181, S367, 2000.
111. Buckland, B.C., The process development challenge for a new vaccine. *Nat. Med.*, 11, S16, 2005.
112. Schneider-Ohrum, K. and Ross, T.M., Virus-like particles for antigen delivery at mucosal surfaces. *Curr. Top. Microbiol. Immunol.*, 354, 53, 2012.
113. Azharuddin, M., Zhu, G.H., Sengupta, A., Hinkula, J., Slater, N.K.H., Patra, H.K., Nano toolbox in immune modulation and nanovaccines. *Trends Biotechnol.*, 40, 1195, 2022.

12

Nanovaccine via Intramuscular, Subcutaneous, and Intradermal Routes

Dixa A. Vaghela[1], Maharshi Bhailalbhai Pandya[2], Pooja M. Parmar[2], Sanjay P. Chauhan[2], Akta Vaishnav[1], Rajeshkumar K. Patel[2]*, Dasharath M. Patel[2], Vidhi Ankit Modh[2] and Mihir K. Raval[3]

[1]*Pharmacy Section, L.M College of Pharmacy, Navrangpura Ahmedabad, Gujarat, India*
[2]*Graduate School of Pharmacy, Gujarat Technological University, Gandhinagar, Gujarat, India*
[3]*Department of Pharmacy, S.P. University, Anand, Gujarat, India*

Abstract

A new class of vaccinations known as "nanovaccines" uses nanoparticles (NPs) as adjuvants and/or transporters. Because NPs and viruses are the same sizes, this can effectively activate the immune system, triggering cellular and humoral immunity responses. Intramuscular injection is a method of injecting into the muscle tissue beneath the dermis and subcutaneously. A medication is injected subcutaneously, also known as hypodermically, into the loose connective tissue that lies between the dermis of the skin and the muscle layer. Intradermal injections (ID) are injections administered directly beneath the epidermis into the dermis. It is fascinating to see, how route of administration affects immune response profiles. Nanoparticles help target the delivery system of vaccines, and their efficacy and stable preparation can be obtained using nanoparticles. It can be a vast topic of research in the future.

Keywords: Nanovaccine, nanoparticles, intramuscular, subcutaneous, intradermal

**Corresponding author*: rajeshpatel@gtu.edu.in

Vivek P. Chavda and Vasso Apostolopoulos (eds.) Nanocarrier Vaccines: Biopharmaceutics-Based Fast Track Development, (405–422) © 2024 Scrivener Publishing LLC

12.1 Introduction

The prevalence of sickness has been reduced through vaccination, but this has been done at the cost of off-target reactions and a lack of protection from a variety of pathological disorders. The subunits such as vitamins, hormones, virus particles, proteins, peptides and polysaccharides used in vaccine formulations to increase its effectiveness, safety, control immune response and protection against varieties of pathogenic strains [1]. Nanotechnology has been praised as a potent application for problem-solving. Nanoparticles (NPs), a new class of vaccinations, are used in nanovaccines as transporters and/or adjuvants. The immune system can be effectively stimulated by the NPs' resemblance to viruses in size, resulting in cellular and humoral immunological responses. Because of their outstanding helper role, co-delivering antigen and invariant natural killer T-cell agonists have been shown particularly effective at generating anti-tumor T-cell responses. Long-lasting immunogenic memory and both types of immunity can be produced by nanovaccines. For the prevention of cancer, AIDS, malaria, tuberculosis, influenza, and acquired immunodeficiency syndrome (AIDS), nanovaccines hold enormous promise [2]. Limitations of simple vaccines are, lack of target specificity, lack of modified release, the larger the particles, less will be the penetration and effectiveness will be compromised, and the physiochemical properties of active molecules may not be attained to target the pathogens. To overcome this, nanoparticles with adaptable physicochemical qualities can overcome this limitation by extending circulation periods, bioaccumulating in lymphoid organs, and targeting immune cells. They can also increase the cross-appearance of Antigen Presenting Cells (APCs) and activate the immune system at low quantities. For instance, Poly Lactic Glycolic Acids (PLGA)-based nanoparticles increases ovalbumin cross-presentation through Major Histocompatibility Complex-1 (MHC-I) in bone marrow-derived primary dendritic cells, causing T cells to be stimulated at 1000 times lower concentrations than with soluble antigen [3]. Nanovaccines have been administered via intramuscular (IM), subcutaneous (SC), and transdermal (TD) route to increase the severity and efficacy of vaccines based on their components. According to the study, intranodal and SC administration delivered more antigen to the lymph nodes (LNs), where it stabilized for 4 days, but less antigen to the liver. Subcutaneous administration was less effective at delivering antigen to LN than intranodal treatment, as was to be predicted. After intravenous and subcutaneous doses, hardly any antigen could be found in the spleen, and neither could considerable lymphatic drainage be found after intranodal treatment [4] because they do not have binding sites for cell receptors, and the synthesized nanoparticles communicate with APCs to enhance internalization than functionally.

Nanovaccines can be administered as a preventative or therapeutic measure, before or after the onset of the disease, accordingly. Long-lasting stimulation can be produced by antigens released slowly from nanoparticle depots, negating the need for repeated booster doses [5]. Here, in this chapter, the details have been explained for nanovaccine administration via IM, SC, and TD routes.

12.1.1 What is a Nanovaccine?

The best method for preventing illness today is vaccination. A quick vaccination must be developed to protect people from the spread of infectious disease drug resistance and cancer chemotherapy-related toxicities. The main goal of using nano- and microparticle-based delivery is to extend time that an antigen is retained and the amount of antigen that is taken up by dendritic cells (DCs), which stimulates DCs directly and encourages cross-presentation [6]. Conventional vaccine adjuvant issues can be creatively solved by using microemulsions, liposomes, virosomes, nanogels, micelles, and dendrimers. Nanovaccines may boost targeted delivery, antigen presentation, activation of the body's innate immunity, a potent T-cell response, and safety in the battle against infectious diseases. Additionally, use of nanovaccines could significantly improve the development of potent immunotherapeutic cancer formulations [7].

As nanotechnology has advanced, most scientists are focusing on developing new vaccines using nanomaterials. This is because nanomaterials may be controlled for a variety of immune-stimulating features such as diameter, zetapotential, surface area, shape, and antigen loading efficiency [8, 9].

Nanoparticles (NPs) are utilized as transporters and/or adjuvants in "nanovaccines," a unique type of immunization. Because of the same scale of NPs and viruses, the immune system can be successfully provoked, to induce both types of responses. Additionally, nanovaccines have the potential to provide active targeting, support blood flow stability, extend the shelf life of blood, reduce the requirement for booster doses, maintain the cold chain, and more effectively activate the immune system. When employing just vaccinations, there are extracellular and intracellular hurdles to achieving robust immunity. Examples of extracellular obstacles include the degradation of vaccines by enzymes, interactions with serum proteins, poor cell targetability, and inadequate cellular uptake [10]. The lack of endosomal escape, non-specific intracellular destiny, insufficient vaccine release, and inappropriate nuclear absorption are examples of intracellular obstacles [11]. Following intravenous delivery, the liver and spleen receive the majority of the nanoparticles, whereas local lymph nodes receive the majority of the

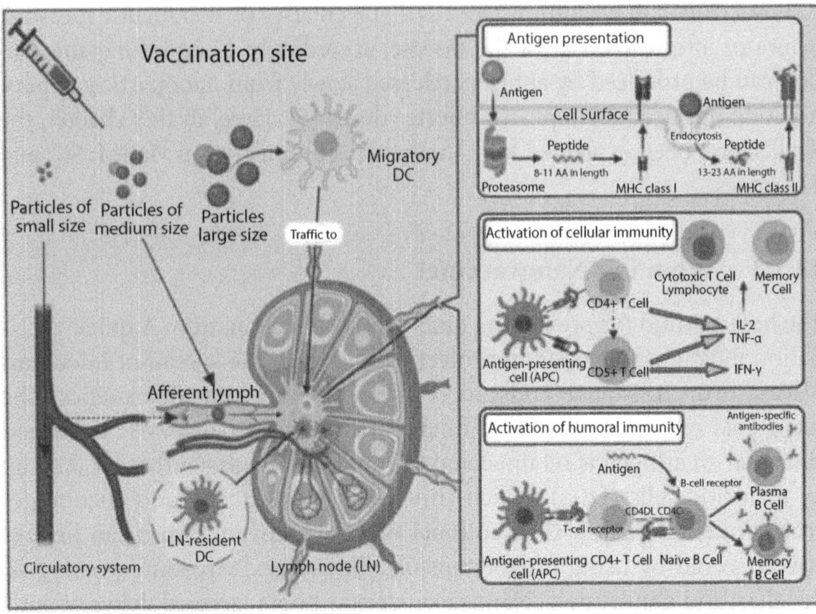

Figure 12.1 After administration of vaccines, small particles in vaccines goes to circulatory systems, a medium size particle goes to the kidney via lymph nodes and dendritic cells moves at adjacent sites. Then antigen presenting cells are activated and it further activates to MHC-class 1 cells to represent antigen, in response to it, CD4+T cells and CD+5 T-cells are activated to produce inflammatory cytokines. Which further activates cytotoxic T cells and memory cells to produce cellular immunity. Activation of humoral immunity by binding of antigen to B-cell receptor, further plasma and memory cells generate antibodies which are specific to antigen and produce humoral immunity [12].

subcutaneous and intranodal injections (Figure 12.1). Malaria, rheumatoid arthritis, acquired immunodeficiency syndrome, and severe autoimmune disorders could all be treated with nanovaccines in the future [4].

12.2 History of Nanovaccination

Historically, vaccinations have been associated with significant gains in overall health. The idea of vaccination was originally noted in Asia, and it was later solidified by research conducted in the 17th and 19th centuries by Jenner and Pasteur that were more documented. The development and production of vaccines later, primarily in the 20th century, were greatly expanded to improve both human and animal health, with a focus on whole dead or attenuated bacteria or viruses. Conventional vaccine designs have aided in the management of various infectious illnesses and the eradication of smallpox. Most individuals have received vaccinations in their lives. Nanosized vaccines have a potential in this research since they may boost the efficacy of subunit vaccines and lead to novel vaccines because of unique features of the nanomaterials. Nanoparticle-based vaccine against the hepatitis B virus and the human papillomavirus, generated in well-established platforms such recombinant yeasts and insect cells, have been made commercially available, thanks to genetic engineering. Innovative recombinant platforms, such as plant cells and algae, are now being developed [46].

Because highly modified viruses are less effective, either the antigenic material of the vaccine must be increased or fresh vaccines must be delivered on an annual basis (for example, influenza virus). The fundamental issue with traditional immunization is that it is ineffective against illnesses whose antigens change after infection (such as human immunodeficiency virus). The primary purpose of traditional vaccines and also the adjuvants included in their mixtures is to generate a humoral, ß cell-mediated immune response. The infection or residency site of the pathogen is frequently intracellular, requiring the usage of cytotoxic T cells to destroy the pathogen or the requirement for helper T cells because B cells cannot effectively combat the pathogen on their own [3].

Vaccines are frequently administered intramuscularly, subcutaneously, or topically. As a result, they frequently need special conditions for storage and transportation. As a result, the formation of a solid powder that will be reconstituted before use is the foundation of the formulation process. The stability of the finished product will affect storage conditions, and the presence of adjuvants, which is crucial in subunit, and toxoid-based vaccines to boost their immunogenicity, are the two key objectives in vaccine formulation development [13].

12.3 Introduction to the Route of Administration

12.3.1 Intramuscular

Intramuscular injection is a way of injecting into the muscle tissue beneath the dermis subcutaneously [14]. Drugs are deeply injected into a few selected muscles via intramuscular injection (IM). The drug is administered quickly, skipping first-pass metabolism, and reaches the systemic circulation before reaching the desired area of action since the large muscles have significant blood flow (Figure 12.2) [15]. The large muscle has a stronger blood flow,

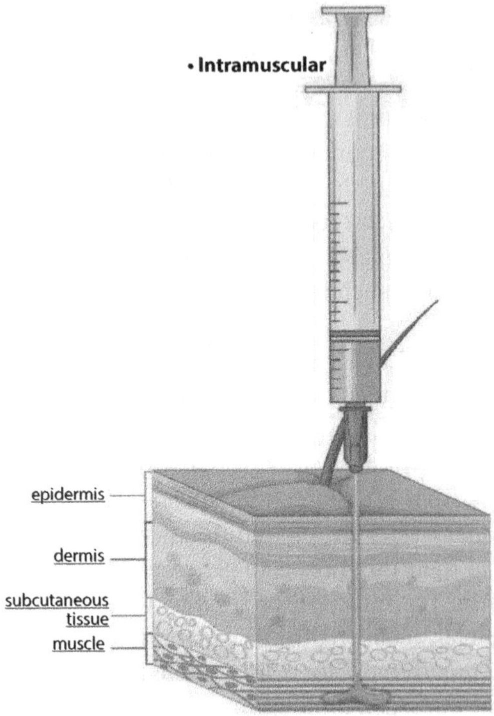

Figure 12.2 Intramuscular needle insertion. Intramuscular injection can be given directly to the muscle by inserting a needle 90° into the skin.

so, the administered drug bypasses the first pass metabolism, and it enters the systemic circulation, and so the targeted action is not achieved [16].

1. Utilize the appropriate syringe.
   - A 1 ml or 3 ml syringe to administer the inoculation
   - A 22- to 25-gauge syringe
   - The appropriate needle length for the patient's gender and weight
   - A 1 to 1.5 inch needle for those above 19 years old.
2. The location of the injection.
   - Site of injection: Deltoid muscle in the upper arm
   - Anatomical markers should be used to find the injection site. The tip of the deltoid muscle's broad, triangular shape is rounded. The bony tip at the end of the shoulder is referred to as the acromion process. Nearly 2 inches

above the axillary fold/armpit and 2 inches below the bone will be the injection site.

3. Appropriately administer the vaccination.
   - Inject the vaccine into the thickest and most central section of the muscle. Insert the needle at a 90° angle and inject the vaccine into the muscle tissue completely.
   - If more than one vaccination is being administered in the same arm, distance the injection sites by 1 inch if feasible.

The deltoid or the anteromedial part of the thigh should be injected with the majority of vaccinations. This increases the vaccine's immunogenicity and lessens injection-site side effects [17]. A common reason for vaccination failure is injecting a vaccine into the layer of subcutaneous fat, where vascularity may slow down the mobilization and processing of the antigen—as in the cases of the hepatitis B, rabies, and influenza vaccines. [18]. In a recent study, 220 adults (presenting for hepatitis B immunization) had the thickness of the fat pad above the deltoid muscle evaluated using high-frequency ultrasonography. The deltoid fat pad varies greatly in thickness, and women have substantially more subcutaneous fat than men do. For genuine deltoid intramuscular injection, a common 16-mm needle would not have penetrated deeply enough in 17% of men and nearly 50% of women in the research sample. It might be safer to use a 25-mm needle for men who weigh 59–118 kg and women who weigh 60–90 kg. A 38-mm needle may be necessary for a woman weighing more than 90 kg [17].

12.3.2 Subcutaneous

Subcutaneous injections are commonly delivered at a 45° angle in the upper outer triceps area for newborns under 12 months and the thigh for those over 12 months (Figure 12.3). If required, a newborn might be given subcutaneous injections in the upper outer triceps area [19, 20].

The depth to which a drug is injected depends on both injection technique and needle size. When utilizing an injection technique, the skin is stretched out flat skin is pinched before the injection, which can call for the use of longer needles. The choice of needle size and injection site should be specific according to their body parameters, for each person in order to ensure that the needle reaches the muscle and that the vaccine does not drip into subcutaneous tissue. The amount of material to be provided, the person's age, and the size of the muscle should also be taken into consideration. Limitations with S.C injections is low volume of drug can be delivered via these routes, several factors can influence the

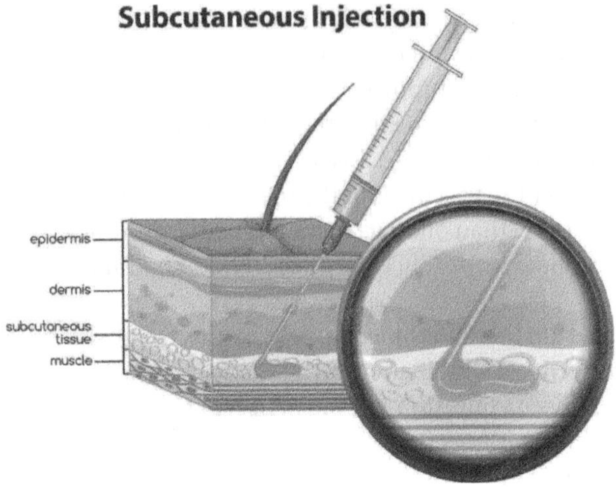

Figure 12.3 Subcutaneous needle insertion. Subcutaneous injection can be given in subcutaneous tissue by inserting a needle 45° into the skin

sensation of pain such as needle features, injection site, volume injected, injection speed, osmolality, viscosity and pH of formulations [47]. Cancer therapeutic nanovaccines are the best instruments for preventing tumor growth and giving the body ongoing immune surveillance that is protective. Conventional subcutaneous (SC) vaccination, however, has low therapeutic efficacy and often only induces a small number of anti-tumor immune responses. Created clay-based nanovaccines and administered them into the spleen via IV injection to create a better anti-tumor immunity with more effectiveness for the treatment of tumors [21].

1. Utilize the proper syringe and needle.
    - Insert a 5/8 inch, 23- to 25-gauge needle into the subcutaneous tissue.
2. The location of the injection.
    - Insert a needle at a 45° angle into the anterolateral thigh fatty tissue. To avoid injection into the muscle, press up on the subcutaneous tissue.
3. Appropriately administer the vaccination.
    - Pinch up on the subcutaneous tissue to avoid injection into the muscle.
    - Place the needle at a 45° angle to the skin.
    - Multiple injections in the same extremities should be spaced by at least 1-inch.

The SC vaccines are generally less preferable than the IM route because upon SC administration of vaccines, seroconversion rates decrease and result to quicker antibody response degradation [22]. Here, we present a simple physical antigen-polymeric nanoparticle mixture used as a minimum nanovaccine that induces a potent cytotoxic T-cell response with minimal systemic cytokine production. The PC7A NP enhances surface presentation while activating type I interferon-stimulated genes, increasing surface presentation by efficiently cytosolically transporting tumor antigens to antigen-presenting cells in draining lymph nodes [23]. STING is necessary for this function but not the Toll-like receptor or the mitochondrial antiviral-signaling protein pathway. The nanovaccine dramatically reduced the growth of tumors in tumor models for melanoma, colon cancer, and human papilloma virus. Together, the PC7A nanovaccine and an anti-PD-1 antibody achieved 100% survival over a 60-day period in a TC-1 cancer model [24]. These tumor-free mice were re-exposed to TC-1 cells, and this resulted in a full suppression of tumor formation, indicating the development of long-lasting antitumor memory [25]. A straightforward, secure, and effective method of enhancing anti-tumor immunity for cancer immunotherapy is provided by the STING-activating nanovaccine [26].

12.3.3 Intradermal

The method of injecting vaccinations beneath the skin's surface is known as intradermal vaccination. Most immunizations are administered intramuscularly (IM) and subcutaneously (SC). Vaccinations are infrequently administered intradermally (ID) (Figure 12.4). To ensure the safety and efficacy of this vaccine delivery method, qualified medical personnel are necessary [27].

ID vaccination was first suggested many years ago. To prevent smallpox, Edward Jenner implanted cowpox virus into the skin in 1796. This vaccine prototype served as the inspiration for later vaccines. ID administration, which uses the Mantoux technique and calls on expertise to produce a wheal signaling successful injection, is now the accepted method for tuberculin skin testing [28]. As a result of improvements in skin immunology and the advantages of ID vaccination, novel technologies for antigen delivery into the skin have been created, and increasing data suggest that ID immunization is advantageous for both adults and children [29]. The technology that are currently on the market, however, were constructed using data on adult skin thickness. Recent research have evaluated skin

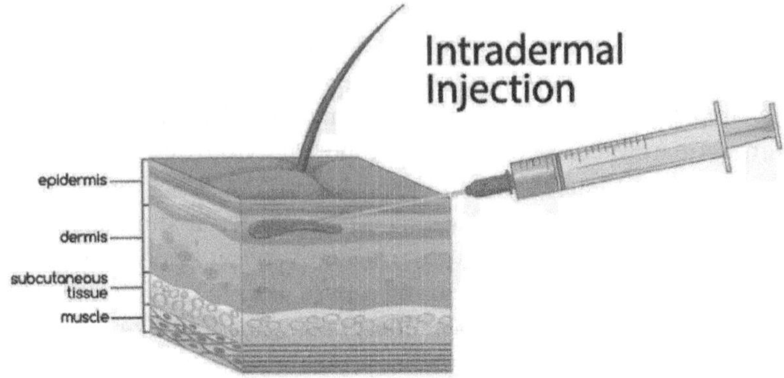

Figure 12.4 Intradermal needle insertion. Intradermal injection can be given in epidermis tissue by inserting a needle into the epidermis layer.

thickness in various pediatric populations to help with the development of effective ID vaccines [30].

1. Utilize the proper syringe and needle.
    - Use a 1-ml insulin syringe with a short (10 mm) 26–27-gauge needle with a short bevel.
2. The location of the injection.
    - Insert a 15° angle needle into the inside surface of the forearm and upper back below the scapula.
3. Appropriately administer the vaccination.
    - Stretch the skin between your fingers and thumb.
    - Bevel facing up, insert the bevel into the dermis to a depth of around 2 mm. The bevel should be visible through the epidermis.

In contrast to conventional immunization techniques, ID vaccination promotes a stronger immune response and necessitates fewer vaccine doses. A commercially viable ID delivery system for adults has been designed to deliver ID vaccines reliably and successfully. Since children receive many immunizations over the course of infancy and childhood, the clinical use of ID vaccines for newborns and children is eagerly anticipated. The evidence currently available is scant and contradictory, and experience with ID vaccinations is limited [31]. The idea of a dermal immune system has been established in recent skin immunology research, and showed that antigen administration into the epidermis and dermis will lead to enhance stimulation in these cells [32]. Therefore, intradermal (ID) vaccination is an effective way to stimulate an effective immune response. In order to get a similar immune response, dose sparing refers to using a dose that is lower than that of intramuscular (IM) and subcutaneous (SC) vaccines [33]. The advantages of ID vaccination may contribute to cost savings, a reduction in vaccine volume, simpler immunization access in areas with limited resources, and better supply during outbreaks of newly or re-emerging infectious diseases [34]. The discovery of new vaccine platforms based on nanoparticle architectures during the past 10 years, including polymer particles, liposomes, and self-assembling protein nanoparticles, has reignited the conversation on route of administration. Many of these platforms are designed to be applied topically to target APCs and drain lymph nodes more successfully (LNs) [35].

12.4 Comparable Adaptive Immune Response After IM, SC, and ID Routes

The vaccination antigen must be delivered to the local lymph nodes (LNs) in order to prime T- and B-cell responses; this was discovered using flow cytometry and positron emission tomography (PET)/computed tomography (CT) [36]. Following IM injection, the vaccine is only transported within the local LNs and not systemically [37]. Based on their proximity to the injection site, LNs were categorized as primary (1°; axillary) or secondary (2°; apical) draining LNs in order to determine which LNs were targeted by immunization. The 1° LNs were shown to be the primary focus of subcutaneous injection, with the 2° LNs showing a lower Env: liposome signal. IM vaccination, on the other hand, nearly exclusively drained the 2° LNs. This is probably because deeper injections target more inwardly localized LNs because of how the lymphatics drain the skin differently from the muscle. The majority of the cells in the evacuating LNs were B cells and Env+ neutrophils [38].

The most popular administration methods are intramuscular (IM) and subcutaneous (SC) injections because of their use and convenience. Contrarily, muscle tissue has a lower density of immune cells and is therefore not considered to be the ideal place for immunization [39]. Therefore, a quick diffusion of vaccination antigens to draining lymph nodes is necessary for the IM route to be effective [40]. Contrarily, APCs are abundant in skin, expected to have an impact on the size, persistence, and direction of immunological memory specific to an antigen. Skin tissue is a potentially preferable vaccine delivery site although receiving less study than conventional administration routes due to practicality. In comparison to the SC and IM methods, intradermal (ID) immunization can reduce the dose by up to 10 times without sacrificing immunogenicity [41]. In mouse models, Mohanan *et al.* [42] showed that the route of administration has a significant impact on the Th1 response, with the IM and ID routes favoring higher IgG2a titers and IFN production over the SC route. The marketed formulation of vaccines is described in Table 12.1.

12.5 Marketed Formulation

Table 12.1 Marketed formulations [43].

Intramuscular	**Subcutaneous**	**Intradermal**
Haemophilus influenzae type b (Hib)	Measles, mumps, and rubella (MMR)	**Influenza**
Hepatitis A (HepA)	Varicella (VAR)	Japanese encephalitis
Hepatitis B (HepB)	Zoster, live (ZVL)	Hepatitis A
Hepatitis A and hepatitis B (HepA-HepB)	**Inactivated polio (IPV)**	Hepatitis B
Human papillomavirus (HPV vaccine)	Pneumococcal polysaccharide (PPSV23)	Human papillomavirus
Influenza vaccine, inactivated (IIV)	Influvac Tetra	Polio
Influenza vaccine, recombinant (RIV4)	Flublok Quadrivalent	Rabies
Meningococcal conjugate (MenACWY)	Menomune	Varicella-zoster
Meningococcal serogroup B (MenB) *f*		Yellow fever
Pneumococcal conjugate (PCV13)		
Pneumococcal polysaccharide (PPSV23)*		
Tetanus and diphtheria toxoid (Td)		
Tetanus toxoid, reduced diphtheria toxoid, and acellular pertussis (Tdap)		
Zoster, recombinant (RZV)		

12.6 Challenges of Vaccine Delivery

Researchers are starting to use vaccines for the treatment of viral diseases and other diseases, including cancer, as their knowledge of the immune system grows. As previously stated, several vaccine versions have been created over time, each with its unique formulation and delivery requirements. Initially, most immunizations contained live organisms that had been attenuated or rendered inactive. This produced an immune reaction, similar to how the initial sickness was battled. The rising danger of serious complications associated with these vaccinations prompted the creation of recombinant antigens with superior safety profiles. Even when an adjuvant system has been added to the formulation to increase immunogenicity, certain proteins, peptides, and nucleic acid antigens typically elicit responses. Aluminum salts such as phosphate and hydroxide are referred to as "alum," which is also utilized as an adjuvant in vaccines. It was the first adjuvant that had been approved for human use. The method by which this adjuvant functions has been widely researched, notwithstanding the initial hypothesis that it established a depot at injection site from which the adsorbed antigen was gradually released into circulation. Increased chemokine and cytokine production for immune cell recruitment, inducing local cell death, and activating innate immunity pathways are a few potential causes for alum's adjuvant effects [44].

The stability of the antigen in vaccines is an essential additional consideration. Temperature, pH, and enzymatic degradation affect contemporary antigens, notably peptides, and proteins. To successfully carry antigen to antigen-presenting cells, delivery methods that shield them from these harmful chemicals (APCs) must be established. The effectiveness of nano and microparticulate carriers, which are particularly applicable for this application, has been proven in the literature.

Antigen targeting of critical tissues with substantial immune cell populations, including lymph nodes (LN) and other lymphatic tissues, severely impedes vaccine delivery. Particle size, surface charge, and chemical changes to the carriers' surfaces have all been found to have a significant influence on how the carriers interact with crucial immune cell populations, particularly APCs, and how they migrate from the administration site to the proper tissues. As a result, these characteristics must be considered while creating novel immunizations against pathogenic viruses. Several studies have compared the delivery of the same vaccination dosage through SC/IM and ID. To determine whether the possibility for dose-sparing is specific to ID administration, more research will be required [45].

12.7 Conclusion

Nanovaccines have generated interest in current years because of distinct physicochemical properties. The potential of nanovaccines as potent vaccines has been investigated in order to increase their therapeutic activity by improving their stability, increasing delivery in response to various biological and environmental stimuli, extending their circulation and site-specific accumulation, and overcoming all physiological obstacles. Antigen stability, improved antigen processing, targeted administration, and avoiding the burst release of antigens and adjuvants as an active immunogenic material to change the immune response are all made possible by nanovaccines. Designing nanovaccines that can elicit strong immune responses to combat tumor immunosuppression requires the use of nanoscale delivery systems. Delivery of nanovaccines via different routes as described in this chapter, is considered based on the nanoparticle usage and the target site of the antigen. Here, it is concluded that among these three routes, intradermal and intramuscular are more preferable than the subcutaneous route of administration. Despite some obstacles and restrictions, the benefits of biomimetic nanovaccines show that they will prevail and open up a range of novel therapeutic modalities for a variety of diseases.

Acknowledgment

We would like to acknowledge the Freepik software, for providing the figures.

References

1. Advancements in prophylactic and therapeutic nanovaccines. PMC n.d. is Elsevier, 21, 22, 2020. https://www.ncbi.nlm.nih.gov/pmc/articles/PMC7163188/ (accessed April 17, 2023).
2. Li, W., Joshi, M.D., Singhania, S., Ramsey, K.H., Murthy, A.K., Peptide vaccine: Progress and challenges. *Vaccines*, 2, 515–36, 2014. https://doi.org/10.3390/vaccines2030515.
3. Shen, H., Ackerman, A.L., Cody, V., Giodini, A., Hinson, E.R., Cresswell, P. *et al.*, Enhanced and prolonged cross-presentation following endosomal escape of exogenous antigens encapsulated in biodegradable nanoparticles. *Immunology*, 117, 11, 2006. https://doi.org/10.1111/j.1365-2567.2005.02268.x.
4. Dölen, Y., Valente, M., Tagit, O., Jäger, E., Van Dinther, E.A.W., van Riessen, N.K. *et al.*, Nanovaccine administration route is critical to obtain pertinent

iNKt cell help for robust anti-tumor T and B cell responses. *Oncoimmunology*, 9, 1738813, 2020. https://doi.org/10.1080/2162402X.2020.1738813.
5. Pati, R., Shevtsov, M., Sonawane, A., Nanoparticle vaccines against infectious diseases. *Front. Immunol.*, 9, 16, 2018.
6. Zhang, X.-Q., Dahle, C.E., Baman, N.K., Rich, N., Weiner, G.J., Salem, A.K., Potent antigen-specific immune responses stimulated by codelivery of CpG ODN and antigens in degradable microparticles. *J. Immunother.*, 30, 469–78, 2007. https://doi.org/10.1097/CJI.0b013e31802fd8c6.
7. Nanovaccine. Expert review vaccines, 10, 20, 2021. https://pubmed.ncbi.nlm.nih.gov/34550859/#:~:text=Nanovaccines can improve targeted delivery, combat infectious diseases and cancers%0A (accessed November 2, 2022).
8. Gheibi Hayat, S.M., Darroudi, M., Nanovaccine: A novel approach in immunization. *J. Cell. Physiol.*, 234, 8, 12530–12536, 2019. doi: 10.1002/jcp.28120. Epub 2019 Jan 11. PMID: 30633361
9. Zhu, G., Zhang, F., Ni, Q., Niu, G., Chen, X., Efficient nanovaccine delivery in cancer immunotherapy. *ACS Nano.*, 11, 3, 2387–2392, 2017.
10. Kheirollahpour, M., Mehrabi, M., Dounighi, N.M., Mohammadi, M., Masoudi, A., Nanoparticles and vaccine development. *Pharm. Nanotechnol.*, 8, 6–21, 2020. https://doi.org/10.2174/2211738507666191024162042.
11. Shtykalova, S., Deviatkin, D., Freund, S., Egorova, A., Kiselev, A., Non-viral carriers for nucleic acids delivery: Fundamentals and current applications. *Life*, 13, 903, 2023.
12. Lu, L., Duong, V.T., Shalash, A.O., Skwarczynski, M., Toth, I., Chemical conjugation strategies for the development of protein-based subunit nanovaccines. *Vaccines*, 9, 563, 2021. https://doi.org/10.3390/vaccines9060563.
13. Amico, C.D., Fontana, F., Cheng, R., Santos, H.A., Development of vaccine formulations: Past, present, and future. *Drug Deliv. Transl. Res.*, 11, 353–72, 2021. https://doi.org/10.1007/s13346-021-00924-7.
14. https://www.nejm.org/doi/pdf/10.1056/NEJM197609022951006
15. Nicoll, L.H. and Hesby, A., Intramuscular injection: An integrative research review and guideline for evidence-based practice. *Appl. Nurs. Res.*, 15, 149–62, 2002.
16. Gutierrez, J.J.P. and Munakomi, S., *Intramuscular Injection*, StatPearls Publishing, StatPearls, Treasure Island (FL, 2023.
17. Zuckerman, J.N., The importance of injecting vaccines into muscle. Different patients need different needle sizes. *BMJ*. 321, 7271, 1237–8, 2000. doi 10.1136/bmj.321.7271.1237. PMID: 11082069; PMCID: PMC1118997
18. Groswasser, J., Kahn, A., Bouche, B., Hanquinet, S., Perlmuter, N., Hessel, L., Needle length and injection technique for efficient intramuscular vaccine delivery in infants and children evaluated through an ultrasonographic determination of subcutaneous and muscle layer thickness. *Pediatrics*, 100, 400–3, 1997. https://doi.org/10.1542/peds.100.3.400.

19. Nanovaccine via Subcutaneous,. Oncoimmunology, 1, 25, 2020. https://www.cdc.gov/vaccines/hcp/acip-recs/general-recs/administration.html#:~:-text=Subcutaneous injections are administered at, of an infant if necessary. (accessed November 2, 2022).
20. How to administer intramuscular and subcutaneous vaccine injections administration by the intramuscular (IM) route. Immunization action coalition, Administration by the subcutaneous (subcut) route. 1–2, 2020.
21. Zhang, L.-X., Jia, Y.-B., Huang, Y.-R., Liu, H.-N., Sun, X.-M., Cai, T. *et al.*, Efficient delivery of clay-based nanovaccines to the mouse spleen promotes potent anti-tumor immunity for both prevention and treatment of lymphoma. *Nano Res.*, 14, 1326–34, 2021. https://doi.org/10.1007/s12274-020-3175-0.
22. Poland, G.A., Borrud, A., Jacobson, R.M., McDermott, K., Wollan, P.C., Brakke, D., Charboneau, J.W., Determination of deltoid fat pad thickness. Implications for needle length in adult immunization. *JAMA*, 277, 21, 1709–11, 1997
23. Rosenberg, S.A. and Restifo, N.P., Adoptive cell transfer as personalized immunotherapy for human cancer. *Science*, 348, 62–8, 2015. https://doi.org/10.1126/science.aaa4967.
24. Reddy, S.T., van der Vlies, A.J., Simeoni, E., Angeli, V., Randolph, G.J., O'Neil, C.P. *et al.*, Exploiting lymphatic transport and complement activation in nanoparticle vaccines. *Nat. Biotechnol.*, 25, 1159–64, 2007. https://doi.org/10.1038/nbt1332.
25. Maldonado, R.A., LaMothe, R.A., Ferrari, J.D., Zhang, A.-H., Rossi, R.J., Kolte, P.N. *et al.*, Polymeric synthetic nanoparticles for the induction of antigen-specific immunological tolerance. *Proc. Natl. Acad. Sci. U. S. A.*, 112, E156–165, 2015. https://doi.org/10.1073/pnas.1408686111.
26. Luo, M., Wang, H., Wang, Z., Cai, H., Lu, Z., Li, Y. *et al.*, A sting-activating nanovaccine for cancer immunotherapy. *Nat. Nanotechnol.*, 12, 648–54, 2017. https://doi.org/10.1038/nnano.2017.52.
27. Hickling, J.K., Jones, K.R., Friede, M., Zehrung, D., Chen, D., Kristensen, D., Intradermal delivery of vaccines: Potential benefits and current challenges. *Bull World Health Organ*, 89, 3, 221–6, 2011.
28. Hata, A., Mano, C., Nakamura, Y., Nishida, H., Kumakura, A., Mizumoto, H. *et al.*, Low response to a monovalent inactivated unadjuvanted influenza A (H1N1) pdm09 vaccine in pediatricians of a general hospital in Japan. *Hum. Vaccin. Immunother.*, 8, 587–91, 2012. https://doi.org/10.4161/hv.19296.
29. Brisse, M., Vrba, S.M., Kirk, N., Liang, Y., Ly, H., Emerging concepts and technologies in vaccine development. *Front. Immunol.*, 11, 583077, 2020. https://doi.org/10.3389/fimmu.2020.583077.
30. Ploin, D., Schwarzenbach, F., Dubray, C., Nicolas, J.-F., Goujon, C., Trong, M.D. *et al.*, Echographic measurement of skin thickness in sites suitable for intradermal vaccine injection in infants and children. *Vaccine*, 29, 8438–42, 2011. https://doi.org/10.1016/j.vaccine.2011.07.111.

31. Saitoh, A. and Aizawa, Y., Intradermal vaccination for infants and children. *Hum. Vaccin. Immunother.*, 12, 2447–55, 2016. https://doi.org/10.1080/21645515.2016.1176652.
32. Nestle, F.O., Di Meglio, P., Qin, J.-Z., Nickoloff, B.J., Skin immune sentinels in health and disease. *Nat. Rev. Immunol.*, 9, 679–91, 2009. https://doi.org/10.1038/nri2622.
33. Zehrung, D., Jarrahian, C., Wales, A., Intradermal delivery for vaccine dose sparing: Overview of current issues. *Vaccine*, 31, 3392–5, 2013. https://doi.org/10.1016/j.vaccine.2012.11.021.
34. Vaccine development for emerging infectious diseases. *Nat. Med.*, 27, 10, 2021. https://www.nature.com/articles/s41591-021-01301-0 (accessed April 19, 2023).
35. Khabbaz, R., Bell, B.P., Schuchat, A., Ostroff, S.M., Moseley, R., Levitt, A. *et al.*, Emerging and reemerging infectious disease threats, in: *Mandell, Douglas, and Bennett's Principles and Practice of Infectious Diseases*, pp. 158–177.e6, 2015. https://doi.org/10.1016/B978-1-4557-4801-3.00014-X.
36. Lindsay, K.E., Bhosle, S.M., Zurla, C., Beyersdorf, J., Rogers, K.A., Vanover, D. *et al.*, Visualization of early events in mRNA vaccine delivery in nonhuman primates via PET-CT and near-infrared imaging. *Nat. Biomed. Eng.*, 3, 371–80, 2019. https://doi.org/10.1038/s41551-019-0378-3.
37. Liang, F., Lindgren, G., Sandgren, K.J., Thompson, E.A., Francica, J.R., Seubert, A. *et al.*, Vaccine priming is restricted to draining lymph nodes and controlled by adjuvant-mediated antigen uptake. *Sci. Transl. Med.*, 9, eaal2094, 2017. https://doi.org/10.1126/scitranslmed.aal2094.
38. Phillips, B., Van Rompay, K.K.A., Rodriguez-Nieves, J., Lorin, C., Koutsoukos, M., Tomai, M. *et al.*, Adjuvant-dependent enhancement of HIV env-specific antibody responses in infant rhesus macaques. *J. Virol.*, 92, e01051–18, 2018.
39. Laurent, P.E., Bonnet, S., Alchas, P., Regolini, P., Mikszta, J.A., Pettis, R. *et al.*, Evaluation of the clinical performance of a new intradermal vaccine administration technique and associated delivery system. *Vaccine*, 25, 8833–42, 2007. https://doi.org/10.1016/j.vaccine.2007.10.020.
40. Neeland, M.R., Shi, W., Collignon, C., Taubenheim, N., Meeusen, E.N.T., Didierlaurent, A.M. *et al.*, The lymphatic immune response induced by the adjuvant AS01: A comparison of intramuscular and subcutaneous immunization routes. *J. Immunol.*, 197, 2704–14, 2016. https://doi.org/10.4049/jimmunol.1600817.
41. Roozbeh, J., Moini, M., Lankarani, K.B., Sagheb, M.M., Shahpoori, S., Bastani, B., Low dose intradermal versus high dose intramuscular hepatitis B vaccination in patients on chronic hemodialysis. *ASAIO J.*, 51, 242–5, 2005. https://doi.org/10.1097/01.MAT.0000161935.66158.97.
42. Mohanan, D., Slütter, B., Henriksen-Lacey, M., Jiskoot, W., Bouwstra, J.A., Perrie, Y. *et al.*, Administration routes affect the quality of immune responses: A cross-sectional evaluation of particulate antigen-delivery

systems. *J. Control. Release*, 147, 342–9, 2010. https://doi.org/10.1016/j.jconrel.2010.08.012.

43. Kim, Y.C., Jarrahian, C., Zehrung, D., Mitragotri, S., Prausnitz, M.R., Delivery systems for intradermal vaccination. *Curr Top. Microbiol. Immunol.*, 351, 77–112, 2012. doi: 10.1007/82_2011_123. PMID: 21472533; PMCID: PMC3173582.
44. Hayat, S.M.G. and Darroudi, M., Nanovaccine: A novel approach in immunization. *J. Cell. Physiol.*, 234, 12530–6, 2019. https://doi.org/10.1002/jcp.28120.
45. Moyer, T.J., Zmolek, A.C., Irvine, D.J., Beyond antigens and adjuvants: Formulating future vaccines. *J. Clin. Invest.*, 126, 3, 799–808, 2016.
46. Bezbaruah, R., Chavda, V.P., Nongrang, L., Alom, S., Deka, K., Kalita, T., Ali, F., Bhattacharjee, B., Vora, L., Nanoparticle-based delivery systems for vaccines. *Vaccines (Basel)*, 10, 11, 1946, 2022. doi: 10.3390/vaccines10111946. PMID: 36423041; PMCID: PMC9694785.

Part 4

APPLICATION AND ADVANCES

13

Nanovaccines for Veterinary Applications

Suneetha Vuppu[1]*, Vivek P. Chavda[2], Toshika Mishra[1], Swati Punetha[1], Nikita Sharma[1], Sathvika Kamaraj[1] and Raj V.[3]

[1]Department of Biotechnology, School of Bio Sciences and Technology, Vellore Institute of Technology, Vellore, India
[2]Department of Pharmaceutics and Pharmaceutical Technology, L M College of Pharmacy, Navrangpura, Ahmedabad, Gujarat, India
[3]Department of Nano and Industrial Engineering, Massachusetts Institute of Technology, Cambridge, MA, USA

Abstract

Disease and death are topics that always interest many scientists. Nanovaccines are unquestionably important techniques for increasing efficiency and combating zoonotic diseases in recent years. However, the greatest challenge is due to persistent infections in animals and rapidly evolving pathogens. The present era is witnessing a rapid rise in the number of contagious diseases and conditions due to the adaptations of pathogenic microbes in all living organisms, including animals. The control of the epidemiology in most zoonotic infectious diseases relies on the utilization of antibiotics, better hygiene to prevent infection, and disease mitigation techniques and vaccinations. Vaccination is the most significant technique for preventing and lowering the prevalence of veterinary contagious diseases. DNA plasmid vaccines are effective against numerous animal infections like avian influenza virus, lymphocytic choriomeningitis virus, rabies virus in dogs and cats, canine parvovirus, bovine viral diarrhea, and feline immunodeficiency virus. At present, edible vaccines are available for veterinary use, but acceptance by the population is missing and controversial. To overcome this challenge, veterinary virologists must create awareness about its advantages. Many of these vaccines are used globally to combat several significant animal pathogens. As an enticing alternative approach to more conventional vaccine platforms, nanoparticle-based vaccine platforms, or "nanovaccines" have recently come to light. Polymer-based nanovaccines for veterinary offer sustained antigen payload release, maintain such

**Corresponding author*: vsuneetha@vit.ac.in

Vivek P. Chavda and Vasso Apostolopoulos (eds.) Nanocarrier Vaccines: Biopharmaceutics-Based Fast Track Development, (425–464) © 2024 Scrivener Publishing LLC

payloads, and stimulate improved cell and antibody-mediated immune responses, both locally and systemically. Recently discovered molecules, including liposomes, biodegradable and biocompatible polymers, virus-like particles, and others, are the foundation of numerous nanovaccine platforms for veterinary use. The cellular and humoral immune systems of animals can be rapidly stimulated by nanovaccines. Veterinary nanovaccines are designed to cross the blood–brain barrier and are simple to administer via various delivery mechanisms, including intravenous, transdermal, and oral. Recent studies have examined the use of pullulan and chitosan for drug and nanovaccines delivery, and these are natural biopolymers screened from agricultural wastes. Rapidly emerging, innovative, and natural mimicking biosystems known as "biomimetic nanoparticles (NPs)" have also been devised and are suitable for a variety of veterinary biomedical applications. Incredible progress has been made by implementing advanced versatile technologies based on chimeric NPs and versatile RNA chaperone functions. Considering the massive benefits of NP-based vaccines, their veterinary applications are indeed very promising. This chapter focuses on the extensive study of the use of different nanovaccines in the veterinary domain, their mode of action, applications, the resultant immune response, regulatory aspects, as well as recent technological advancements.

***Keywords*:** Veterinary, nanovaccines, immunity, polymer-based nanovaccines, dosage

13.1 Introduction

In the area of veterinary science, nanotechnology has demonstrated enormous promise for improving the administration of drugs and vaccinations [1]. As the field of nanoparticles expands, new therapies to treat bacterial or viral infections as well as to speed up the healing of severe wounds will be created. These recently created nanoparticles might also successfully deliver medications to various cells to cure ailments [2]. Nanopharmaceutical products, which offer various benefits in veterinary care, are one of the most enticing and promising areas of nanotechnology [3]. Vaccination is one of the most essential tools for preventing and lowering the prevalence of contagious diseases. Several of these products exhibit differential effectiveness and safety in the field, require additional doses, or are unstable under field circumstances, despite their accessibility and prevalent use to combat numerous significant pathogenic microbes affecting the animals. In the veterinary domain, these are undeniably essential tools for increasing productivity and combating zoonoses [4]. The ongoing requirement for novel vaccine alternatives to regulate outbreaks is highlighted by cases of persistent infections, rapidly emerging pathogens with high variability, or emerging/re-emerging pathogens for which no efficacious vaccines have been devised [4].

As enticing alternative approaches to more conventional vaccine platforms, NP-based vaccine platforms, or "nanovaccines," have recently come to light [5]. Particles in nano-formulations can be either inorganic or organic and have one or more dimensions in the nanoscale (a few nanometers to 1000 nm) [6]. Mainly, polymer-based nanovaccines offer sustained antigen payload release, maintain such payloads, and stimulate improved cell and antibody-mediated immunological reactions, both locally and systemically [5]. In the past 20 years, research on nanovaccines and their carriers has drawn much attention as a platform that has the potential to replace conventional viral vector vaccines in veterinary medicine [7]. Modern nanotechnology allows for the development of vaccines based on nanomaterials with encapsulation capability as well as very enticing characteristics due to their size and surface area, making them useful antigen and immunostimulatory stimulant delivery systems [4]. The creation of veterinary vaccines is similar to that of human vaccines, intending to enhance the general health and welfare of species. The preparation of NP-based veterinary vaccines involves either encapsulating vaccine components inside the NPs or covering the particle surface in viral antigens [8]. Veterinary objectives, on the other hand, tend to focus more on practical large-scale administration techniques and inexpensive to highly beneficial immunization [9]. Over the past few decades, NP-based technologies have drawn enormous interest in the development of unique vaccine candidates due to their innumerable advantages over inactivated viruses or subunit soluble antigens.

13.1.1 Diverse Types of Nanovaccines in the Veterinary Field

In the field of veterinary science, there are several unique categories of nanovaccines available, including polymeric nanoparticles, self-assembling proteins, micelles, liposomes, inorganic nanoparticles, biomimetic nanoparticles, and non-infectious viral particles (VLPs), which are the important systems progressively used in the preparation of nanovaccines in this field [4, 10]. According to their origin, shape, structure, and mode of administration, there are various classifications of nanoparticles in veterinary medicine [11]. Based on their structure and use, they have been categorized as follows:

13.1.1.1 Polymeric Nanoparticles

There are two basic types of polymers: synthetic polymers, like polyethylene glycol (PEG), and patented polymers, like inulin and chitosan, whose structures rely on polysaccharides [12].

13.1.1.2 Liposomes

Liposomes are spherical, PEGylated, biodegradable, and non-destructive particles. The drugs that dissolve in water are illuminated by their fluid beam. Chelated antibodies can be fixed to the exterior surfaces of liposomes. They are highly capable of concurrently stacking water-soluble and water-insoluble materials [13].

13.1.1.3 Fullerenes and Bucky Tubes

The carbon nanoparticles known as fullerenes may successfully interact with cells or pathogens. They may be used as biosensors to identify various components, such as immunoglobulins. Nanotubes can enter cells to cure ailments because of their needle-like shape [14, 15].

13.1.1.4 Microbivores and Respirocytes

Respiratory cells, also known as respirocytes, acquire characteristics from the red and white blood cells. Microbivores prevent infections while productively supplying the tissues with O_2 and avoiding the accumulation of CO_2 by employing extraordinary management sensors [16].

13.1.1.5 Nanoshells

Nanoshells have the advantages of being biocompatible, non-toxic, and free from the negative effects of conventional X-rays [16, 17].

13.1.1.6 Quantum Dots

The structure of the semiconductor components is made of cadmium, zinc, and selenium. In addition to a fluid coating that may be coupled to various biomolecules, they are made up of an inorganic core and shell. The size of the gemstones affects the color of the light produced. These are commonly used for immunodiagnostics and diagnostics [12, 13].

13.1.1.7 Solid Lipid Nanoparticles

Currently, nanoparticles are being used enormously to evaluate the use of fluid lipids. Since lipid nanoparticles have a lipophilic center, they can be utilized to cure cancer. Various hydrophilic drugs or antibodies can be coupled to their hydrophobic exterior [12, 18].

13.1.1.8 Magnetic Iron Oxide Nanoparticles

This group is mostly recognized by an external magnetic field that enables them to be transported to their target cells via circulation. They are more suited for medication administration, thermal treatment, and imaging. Their structure consists of an iron core encased in a silica outer fluorescent coating [18].

13.1.1.9 Dendrimers

Hyper-branched nanomaterials known as dendrimers are remarkably dissolved in water. They are made of polymers, which are far smaller and more numerous than the body's cells. When injected into the blood flow, their small size and composition arrangement keep any unwanted resistance responses at a strategic distance [15].

13.1.1.10 Nanoemulsions

Nanoemulsions provide an excellent opportunity for recovery as bactericidal and virucidal drugs. They are tiny oily droplets in water that are coated with a thin surfactant coating to allow for physical adjustment. However, a few experts warn of the signs of fat drops on sperm cells and red blood cells (RBCs) when administered systemically [19, 20].

13.1.1.11 Nanobubbles

When molecules of liposomal carbon dioxide, hydrogen, and oxygen condense together at room temperature, liposomal nanobubbles are created. They are mostly utilized for medication delivery, particularly drug administration into cancer tissues. Additionally, lipid-based nanobots are used in gene therapy [19].

13.1.1.12 Aluminosilicate Nanoparticles

Short-chain polyphosphate and silica nanoparticles are employed to speed up the body's natural clotting process, which reduces bleeding [21].

13.1.1.13 Polymeric Micelles

Amphiphilic polymers like ε-Caprolactone or Poly lactic-co-glycolic acid (PLGA) are used to create polymeric micelles. They are extremely water soluble because the hydrophobic core is covered in a water-soluble covering. They provide less water-soluble medicines, such as paclitaxel and amphotericin B, employed for focused drug delivery [22, 23].

13.1.1.14 Polymer-Coated Nanocrystals

Polymer-coated nanocrystals help organize a powerful nanosuppression, macrophage-dependent transport to HIV infection and sequestration sites, and inhibit aggregation [22].

13.1.1.15 Polymeric Nanospheres

In vitro, integrin and human epidermal advancement factor receptor cancer cell analysis might also utilize polymeric nanospheres [24].

13.1.1.16 Metallic Nanoparticles

One of the numerous metals employed in the nanosystem is gold, which is primarily used in cancer treatment. Another category of metallic nanoparticles includes silver, manganese, and platinum particles. They all have metallic centers that are covered in protective layers [25].

Based on origin, they have been categorized into three types:

i. Organic nanoparticles: They generally utilize proteins and peptide NPs for their synthesis.
   1.1 Proteins and peptide nanoparticles: Two different types of gelatin, A and B, are obtained by extracting tissues with acids and separating them with bases. Due to their low lethality, ease of production, and outstanding biodegradability, proteins and peptide NPs are primarily used in high-quality delivery methods, such as gelatin [26].
ii. Inorganic nanoparticles: They consist of a variety of inorganic substances such as gold, silver, iron oxide, and calcium phosphate, which are less cytotoxic, biocompatible, and safe. They may be modified during assembly and feature new electrical and optical properties [23, 27].
iii. Hybrid nanoparticles: They are made up of many types of NPs, such as a combination of polymeric NPs and liposomes, which creates a hybrid system made of polymer and lipids [22, 28].

Based on their shape and mode of administration, they are categorized into tubes, spheres, and liquid drops. On the other hand, the nanoparticles can be classified based on their intended administration as a vaccine, for therapeutic use, for diagnostic use, or for nutritional use [22].

The development of several nanovaccine platforms based on liposomes, biodegradable and biocompatible polymers, and virus-like particles in recent years has allowed for the quick and effective stimulation of the cellular and humoral immune systems of animals [5]. In addition, nano-adjuvant vaccinations stimulate the animals' defense system by boosting immunostimulatory characteristics and also in the production of antibodies [29]. Due to their competitive advantages over the use of conventional subunit vaccines, the use of NP-based vaccines, also known as nanovaccines, is emerging as a promising novel approach for vaccination in the veterinary field [30].

13.2 Nanovaccines and Immune Response

Nanotechnology has contributed to the rapid advancements in the production of efficacious vaccines. An essential course of action in nanovaccine production is the use of carriers or delivery system to elicit an immune response [31]. The inclusion of NPs in the vaccine facilitates antigen stability for a longer time, boosts immunogenicity, and enhances antigen presentation time in the target cell by prolonged release. It also provides target-specific delivery, reducing the probability of adverse effects. Both cell-mediated immunity and humoral immunity are elicited by the nanovaccines. These vaccines protect a greater range of pathogens as they can contain multiple antigens in the formulation. This enhanced activity is attributed to the size of NPs further influencing the intensity and type of immune response; however, immunogenicity should be generated specifically by the antigens and not by the NPs [32]. It is condensed into four components: the immune response to microorganisms, which includes the host's development of specialized defense systems through innate and adaptive immunity; the emergence of pathogenic changes; chronic infections; and tissue lesions. Immune system differences can either help or hinder infections, as microbes can elude immune responses or withstand host defense systems.

The vaccine comprises several minor antigens of size <200 nm. On introducing nanovaccines into animals during immunization, the antigens enter the interstitial fluid and reach the lymph nodes. The particulate antigens are confined in the macrophages of the subcapsular sinus, which are located above the B-cell follicles, before the lymphatic fluid leaves through efferent lymphatics close to the medullary sinus. The vaccines within the 5-nm hydrodynamic radius enter the B-cell follicle, and collagen fiber conduits facilitate this entry. In the case of nanovaccines in the size range from 200 to 500 nm, the antigen is lysed by proteins, and the lysed parts are presented via macrophages or conduits. Furthermore, with nanovaccines over the size of 500 nm, the antigen enters the dendritic cells through the Fc (fragment crystallizable) fragment of the FcγRIIb (Fc gamma receptor IIb) receptor, and it is transferred once more to the surface of a dendritic cell to present it to the B cells. This proves that the size of nanovaccines plays a crucial role in determining the route of initiating the immune response. Numerous studies globally aim to understand further details on the mechanism of immune response elicited by nanovaccines. In B cells, the response is generated in the germinal center of the lymph nodes, where the B cells mature by the interaction between the primed B cell and T helper cell. The specific antibodies are generated following somatic hypermutation to select the most fitting antigen-reacting B cell clones. The matured B cells transform into plasma cells for prolonged secretion of antibodies and move out of germinal centers to the bone marrow, providing long-term protection from infection (Figure 13.1). Additionally, the nanoparticles can induce the maturation of dendritic cells and support the cross-presentation of antigens via class I major histocompatibility complex (MHC-I) to CD8+ T cells that consequently result in the generation of innate cytokines that control humoral and cellular immunological responses. This activation of dendritic cells is crucial in integrating innate and adaptive immunity [33]. Pigs from both wild and domesticated populations have had their allele frequency and diversity analyzed. Due to their role in biomedical research, transplantation, regenerative medicine, and vaccine development, MHCs are of interest. Low-resolution Swine Leukocyte Antigen (SLA) haplotyping was utilized by Hammer *et al.* (2021) to characterize European SLA genes [34].

Thus, the immune reaction is triggered by targeting particular immune cells in lymph nodes to produce a wide array of antibodies and select the antibodies most efficient in neutralizing the antigens present in the nanovaccines and providing long-term immunity against the actual infection [35]. Antigen functionalization, drainage to lymph nodes, internalization by presenting cells, maturation of presenting cells, and presentation of the MHC-I peptide immunocomplex to T cells are the five steps that make up the cell complexity for immune response.

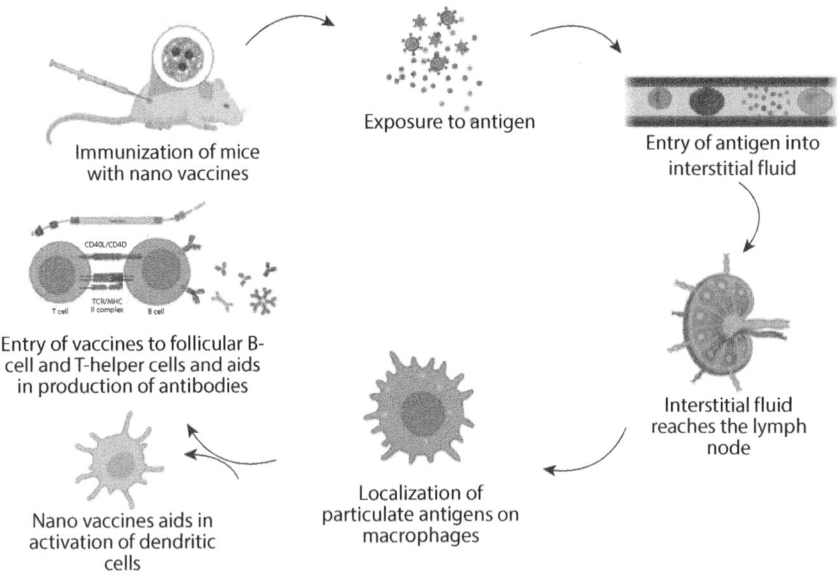

Figure 13.1 Pictorial representation of nanovaccines and immune response (Figure created using biorender.com).

Such unusual antigen presentation results in cellular immunological responses. Nanoparticle-induced endosomal osmotic swelling and rupture expose the antigen, and membrane fusion processes enhance cytosolic export [36].

13.3 Vaccine Production

Nanovaccines have emerged as intriguing alternatives to traditional vaccines [37]. Nanoparticle vaccines, often known as "nanovaccines," are tiny particles of size 20–100 nm that are designed to target the immune system of the host to stimulate the body's defense mechanism against diseases. In the realm of veterinary vaccine preparation, the usage of NPs is increasing. They perform basic immunomodulatory tasks that enhance immune responses. Antigens or adjuvants can be incorporated by encapsulating them inside vesicles or micelles, conjugating them to the surface or core of nanoparticles, or both. Nanoparticles can increase antigen release and circulation time, target antigens to APCs (Antigen Presenting Cells), and also increase the efficacy of these vaccinations. It gives an excellent platform for subunit vaccines. Another typical platform for nanoparticle vaccinations is liposomes. By changing the phospholipids that are employed or by covering the surface with additional compounds like PEG (Polyethylene glycol), liposomes can be easily altered [38].

In veterinary medicine, polymer-based nanovaccines provide sustained antigen payload release, retain payloads, and increase cell-mediated and antibody-mediated immune responses both locally and systemically. Liposomes, biocompatible and biodegradable polymers, virus-like particles, and other recently discovered compounds represent the foundation of several veterinary nanovaccine platforms. However, researchers can accurately adjust the shape, hydrophobicity, surface charge, size, and loading density of antigens and adjuvants by using various materials and production processes [39].

There have been several advancements made in the production of vaccines; various veterinary nanovaccines include- nanopatch TM (for influenza virus), nanoemulsion, gold nanoparticles-based vaccine (for foot-and-mouth disease), PLGA NPs, poly gamma glutamic acid vaccine (for influenza virus), chitosan nanoparticles, nanocapsule (for Newcastle disease), and calcium phosphate nanoparticles vaccines (for herpes simplex virus 2). The nanoemulsion vaccine, like the recombinant influenza virus vaccine and *Bacillus anthracis* spore-based vaccine, induces mucosal immunity after intranasal delivery. The rotavirus capsid vaccine, tetanus toxoid, *Bordetella pertussis* vaccine, and *Helicobacter pylori* vaccine are based on PLGA NPs [40].

Numerous aspects must be taken into account while designing nanovaccines, the first step involves the synthesis of antigens from the microbes. The viruses or microbes can be grown on primary or secondary cell lines, whereas the bacteria against which the vaccines are developed may be cultivated in bioreactors. Followed by the isolation of the antigens, and finally, adjuvants, stabilizers, and preservatives can be added for enhancing the immune response of the antigen, which can be further utilized for stabilizing and preserving dose vials [41]. The final stage of vaccine production involves clinical trials. Clinical veterinary trials are related to research studies that evaluate the safety and efficacy of novel clinical approaches. Clinical trials help us find better, effective ways to diagnose, prevent, and treat diseases. It primarily contrasts a new treatment with an existing one, thereby allowing the identification of the most effective treatments. However, the resulting vaccine yields fewer immune responses, so improvements in vaccine design are required for eliciting the immune system (Figure 13.2).

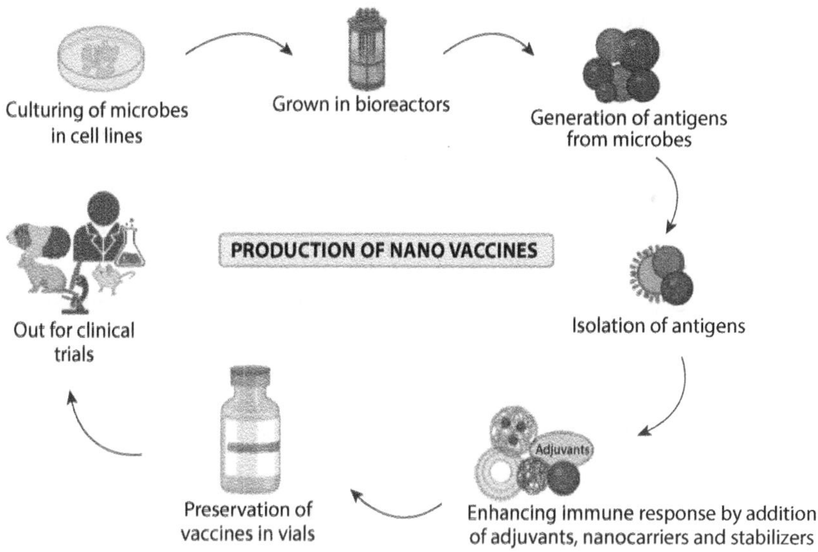

Figure 13.2 Schematic representation of the production of nanovaccines (Figure created using biorender.com).

13.4 Veterinary Applications of Nanovaccines

Vaccination is the most significant technique for preventing and lowering the prevalence of veterinary contagious diseases. The use of vaccines is crucial in preventing and lowering the prevalence of infectious illnesses in animals. Veterinary nanovaccines can be administered easily using several delivery methods, such as intranasal, intravenous, transdermal, and oral, designed to cross the blood-brain barrier [42]. Recent studies have examined the use of pullulan and chitosan for drug and nanovaccine delivery, and these are natural biopolymers screened from agricultural wastes, flower wastes, etc. [43]. DNA plasmid vaccines are effective against numerous animal infections like the avian influenza virus, feline immunodeficiency virus, canine parvovirus, and bovine viral diarrhea. The veterinary field has used nanovaccines in various ways; a few applications of nanovaccines are mentioned below.

13.4.1 Disease Prevention in Cattle

Quick and long-lasting cellular and humoral immunities can be induced by nanovaccines. To facilitate simple uptake by immune cells, nanovaccines can be engineered to mimic the size and structure of pathogens [44]. In mammalian species, the immune system has mostly been conserved. Therefore, humans, rodents, and other animals share many characteristics of innate and adaptive immunological function with cattle. However, several features of the bovine immune system are distinctive and must be considered while developing vaccines and treating infectious illnesses. The innate immune system is mainly activated when it receives signals. These signals might be endogenous, which is the recognition of dead and dying cells, or exogenous, which involves recognizing chemicals produced by microorganisms. Pathogen-associated molecular patterns (PAMPs) are the collective term for exogenous signals, whereas damage-associated molecular patterns (DAMPs) are the term used for endogenous signals [45]. Moreover, the marker vaccines, which are often referred to as "differentiating infected from vaccinated animal" (DIVA) vaccines, are preferred because of their capacity to discriminate between vaccinated and infected animals. Some examples of nanovaccines explored in cattle for viral and bacterial diseases are mentioned in detail.

13.4.1.1 Viral Diseases

Vaccines are crucial for preventing viral infections, which are a vital contributor to a large financial crisis in the global animal husbandry industry. The bovine respiratory disease complex may be influenced by several clinically significant viruses in cattle, including bovine respiratory syncytial virus (BRSV), bovine adenovirus, bovine parainfluenza 3 (BPI3V), bovine herpesvirus-1 (BHV-1) and bovine viral diarrhea virus (BVDV-1). It has been found that n-nanoring vaccination in cattle elicits cell-mediated immune responses and provides moderate protection against experimental BRSV. Additionally, other immunogenic peptides that carry BRSV F and G protein epitopes produce limited protection in calves when exposed to BRSV [46]. Recently, it has been observed that *in vivo* administration of a recombinant BVDV-E2 nanovaccine using hollow MSNs (medium spiny neurons) in mice induced total humoral IgG and cellular virus-specific Interferons (IFN) by T cells [47]. Further, in a calf model, PLGA-based biomaterials nanoparticles have also demonstrated potential when combating BVDV [48].

13.4.1.1.1 Influenza A Virus

The influenza A virus is widely heterogeneous and poses a severe risk to the well-being of humans and animals. The interspecies reassortment of the influenza A virus is mediated mainly by pigs. The virus most likely to cause influenza A infection in pigs is the porcine circovirus type 2 (PCV2). Therefore, it is necessary to develop a vaccine that can combat PCV2 and the swine influenza virus (SIV), simultaneously. Of all influenza subtypes, matrix 2 protein (M2e) has the highest degree of conserved surface protein. Human influenza can be treated with antibodies specific for M2e, and these antibodies can also shield animals from deadly infections by other influenza A viruses. M2e finds an intriguing possibility for developing the global influenza virus vaccine [49]. The steps are mentioned in Figure 13.3.

13.4.1.1.2 Swine Influenza Virus

To find out the effectiveness of the vaccine, nursery pigs were administrated with two intranasal vaccines with either polyanhydride nanoparticle-encapsulated Killed SwIAV H1N2 (δ-lineage) antigens or KAg (KAg nanovaccine) or inactivated SwIAV H1N2 (Swine Influenza A virus- strain H1N2), and the effectiveness against a heterologous zoonotic pathogenic SwIAV H1N1 was assessed. The findings showed that the KAg nanovaccine raised the percentage of T helper cells (CD4+CD8+) and cytotoxic T cells (CD8+) in peripheral blood mononuclear cells and stimulated virus-specific lymphocyte proliferation. Pigs that received KAg nanovaccine were found to be protected against fever and showed fewer viral antigens in their lungs. Various findings showed that intranasal administration of the polyanhydride-based SwIAV nanovaccine enhanced the

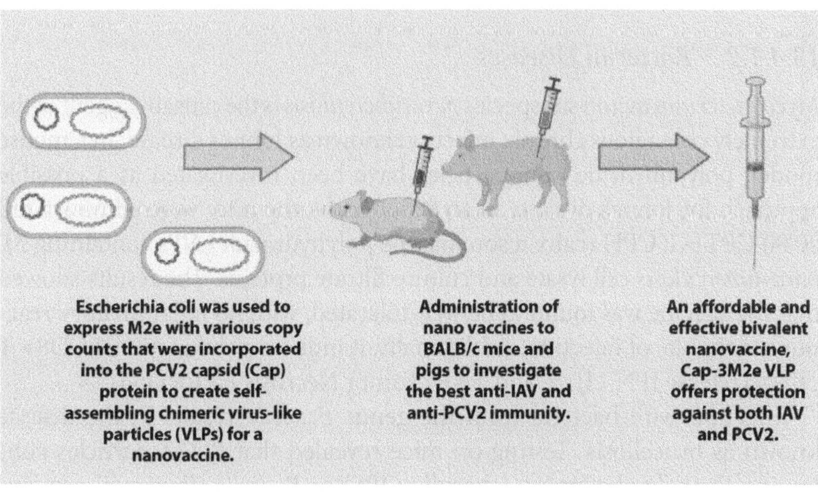

Figure 13.3 Steps involved in developing a vaccine that can combat PCV2 and SIV (Figure created using biorender.com).

cellular immune response to the antigen in pigs and showed the potential for inducing immunity [50].

13.4.1.1.3 Canine Parvovirus 2

The most common factor causing viral enteritis in dogs is canine parvovirus 2 (CPV-2), which continues to be a serious canine infection on a global scale. *In vitro*, the immunogenic activity of W-1 L19 from the VP2 capsid protein of CPV-loaded PLGA nanoparticle was assessed. Results demonstrated that the adjuvant qualities of PLGA nanoparticles improved the ability of W-1 L19 peptide to promote nitric oxide generation *in vitro*. These nanoparticles could be considered canine parvovirus vaccines depending on the outcomes. Shortly, research on PLGA nanoparticle-based delivery systems must be sustained to develop new and more potent nanovaccine formulations [51].

Newcastle Disease Vaccine Tablets. A cutting-edge method for administering the Newcastle disease vaccine to chickens is the development of nano-encapsulated vaccine tablets. Each vaccine tablet contained a nominal oral dose of Newcastle disease virus (NDV) strain I-2 for a single chicken and was made using gelatin, trehalose, and casein as thermostabilized binders. A single oral delivery of these vaccine tablets showed a protective antibody response against virulent Newcastle disease in village chickens. Further, it was predicted that if the method for making vaccine tablets is improved, it will enable the delivery of vaccines to rural areas in tropical countries without relying on cold chains [52].

13.4.1.2 Bacterial Diseases

Mycobacterium avium subspecies *paratuberculosis* is the causative agent of the extremely contagious chronic infection known as Johne's disease. In a mouse model, polyanhydride nanoparticles have been investigated as a possible approach for Johne's disease. In an experiment, the mice were administered 20'80 CPTEG: CPH (calix resorcinarene polyhydrazide) NPs containing *M. para-tuberculosis* cell lysate and culture filtrate proteins. The results showed that the vaccine was found to be well-tolerated, with no inflammatory reactions at the site of injection. Additionally, it induces polyfunctional CD8+ T cells to release IFN-, IL-2, and TNF (Tumor Necrosis Factor) [53].

Infection with bacteria from the genus *Brucella* results in the disease known as brucellosis. Testing on mice revealed that PLGA particles containing *Brucella* oligopolysaccharide (OPS) or *Brucella* ribosomal proteins rL7/L12 elicit humoral and cellular responses [54]. Similar results were also demonstrated using calcium phosphate nanoparticles (CaPNs) coated

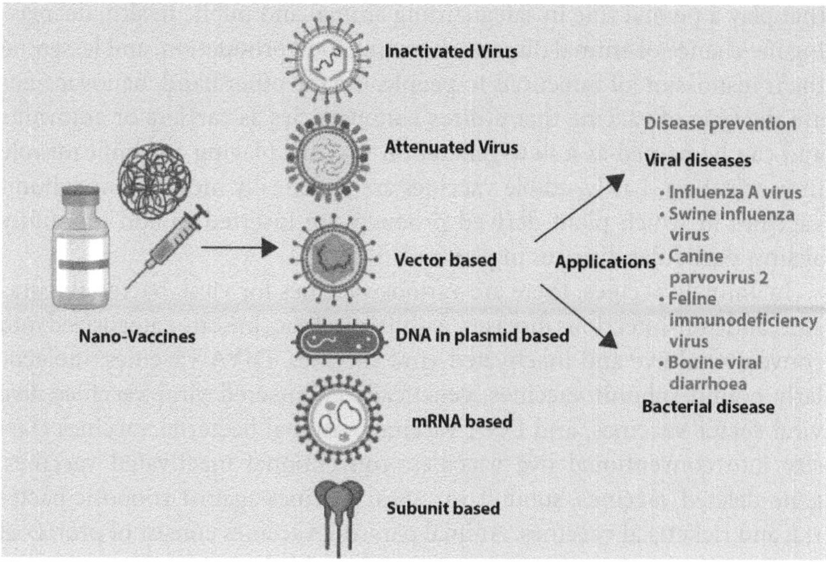

Figure 13.4 Schematic representation of different applications of nanovaccines in viral and bacterial diseases (Figure created using biorender.com).

with three *Brucella* antigens Flagellin (FliC, 7a-HSDH, and BhuA) and two multi-epitopes (poly B and poly T), as well as employing mannosylated chitosan nanoparticles coated with FliC [55]. For longer antigen release in cattle via implants, polyanhydride formulations containing a rough *Brucella abortus* lipopolysaccharide O-antigen-deficient mutant from a virulent field strain have been explored. The findings point to prolonged cellular reactions against *Brucella* antigens without tolerance development [56, 57]. Recently, chitosan-based nanovaccines were tested against avian pathogenic *Escherichia coli*, which has a significant negative influence on poultry production globally. When compared to basic vaccines or the *E. coli* O1 and O78-adjuvanted Montanide ISA 71 R VG adjuvanted vaccine, improved vaccination effectiveness and protection were seen in hens [58, 59]. Wide applications of nanovaccines in veterinary science are illustrated in Figure 13.4.

13.5 Comparative Analysis of Animal Vaccines, Nanovaccines, and Edible Vaccines

There are three major types of prevalent veterinary vaccines, i.e., animal vaccines, nanovaccines, and edible vaccines. Animal vaccines are vaccines that play a pivotal role in safeguarding animal and public health, decreasing the chance of animal diseases, boosting food production, and lessening the transmission of infections to people. On the other hand, nanovaccines are the type of vaccine that utilizes nanoparticles as carriers or adjuvants and can be named as a new-generation vaccine, playing a significant role in vaccination. Lastly, edible vaccines are genetically modified or subunit vaccines in which plant-derived products are inserted to add immunity against particular diseases in plants.

In animal vaccines, there are various vaccines for viral, bacterial, parasitic, and non-infectious diseases. Animal viral vaccines are categorized into conventional live and inactivated viral vaccines, DIVA vaccines, molecularly bound subunit vaccines, genetically engineered viral vaccines, live viral vector vaccines, and DNA vaccines. Animal bacterial vaccines classify into conventional live vaccines, conventional inactivated vaccines, gene-deleted vaccines, subunit vaccines, vaccines against zoonotic bacteria, and rickettsial vaccines. Animal parasitic vaccines consist of protozoal vaccines, live protozoal parasitic vaccines, and helminth and ectoparasitic vaccines. Animal vaccines for non-infectious diseases are allergy vaccines and cancer vaccines. The process of development in animal or veterinary vaccines has both pros and cons as compared with human vaccines. The cost of animal vaccines is very low, and the market size is limited, whereas the host range and pathogens are complex and extensive. In animal viral vaccines, RNA viruses are targeted along with viruses with different serotypes, which shows more variability. According to the market, second-generation vaccines are in much demand [60]. Nanovaccines are stable when stored at room temperature, and it eliminates the requirement for cold chain storage. Nanoparticle platforms can be developed for targeted intranasal, aerosol, or oral administration to promote desired mucosal immunity. In recent times, a wide range of nanovaccine platforms based on liposomes, virus-like particles, and biodegradable and biocompatible polymers have emerged [61, 62]. Though most nanovaccine candidates have not yet advanced past testing in rodent models, an increasing number of them have shown potential in combating contagious illnesses affecting cattle. The application of NP-based vaccinations, also known as nanovaccines, is emerging as a feasible novel approach to vaccination due to their comparative advantages over traditional subunit vaccines. Nanovaccines can prompt cellular and humoral immunities that last long [63–65]. They mimic the size and structure of viruses to facilitate easier immune cell absorption. The particle size, shape, surface charge, and hydrophobicity of NPs can be used to control their self-adjuvant properties, release kinetics, and delivery pathways. Size affects how immune cells internalize and distribute nanovaccines throughout the body. Within a few hours of treatment, particles in the 20–100 nm size range can enter the lymphatic system directly, whereas particles larger than 200 nm must first be absorbed by the antigen-presenting cells [66]. Antigen-presenting cells efficiently integrate particles down to the micron scale, whereas larger particles (>10 μm) have more problems doing so [67, 68]. Rod-shaped particles internalize more quickly by dendritic cells than spherical ones, according to recent research on the impact of particle form. When compared with spherical particles, rod-shaped particles disperse in the blood and gastrointestinal tract for longer periods as revealed by researchers [69].

Edible vaccines primarily consist of live or dead parasite vaccines currently available on the market and have numerous drawbacks. This resulted in a great investment in veterinary transgenic plant vaccines, which turns ongoing research into parasite antigen expression. However, unsuccessful trials use bacterial, yeast, insect, and mammalian expression systems. The extraordinary benefits of plant-based vaccinations make oral delivery particularly alluring. Additionally, veterinary vaccines have a lower regulatory burden than human vaccines. To tackle serious parasite infections such as fasciolosis, schistosomiasis, chicken coccidiosis, swine cysticercosis, and ascariasis, studies using plant-based vaccines have been carried out. In addition, a novel approach to prevent poultry coccidiosis involves delivering antibodies generated in transgenic plants by oral administration. The results of these experiments could soon lead to the creation of promising edible veterinary vaccinations [70]. Consequently, we have discussed various vaccines, their types, modes of action, efficacy, cost-effectiveness, safety, and applications in Table 13.1.

13.6 Regulation of Vaccine Production Process

The products obtained using nanotechnology, like nanovaccines, are distinct from conventional vaccines and approved by the U.S. Food and Drug Administration (FDA), which provides regulation for each type of nanotechnology-based product considering the specific legal standards. FDA supports novelty and innovation in designing and developing nanotechnology-based products. However, the products should prove the safety and efficacy aspects. Regulatory bodies, like FDA, focus on the regulation of products to prevent the chances of harming animals unintentionally. This includes analysis of associated adverse events with the nanovaccines before approval. Also, all standards and guidelines provided by the regulatory authorities must be followed strictly at each step of vaccine production [93].

Table 13.1 List of vaccines, with their characteristic properties and applications.

Vaccines	Types	Mode of action	Efficacy	Cost-effectiveness	Safety	Applications	References
Animal vaccines	• Conventional live and inactivated or killed viral vaccines	They integrate cellular and humoral immunities using non-target hosts, and they need an adjuvant to induce a good humoral immune response.	Extremely dependable and less likely to transition into a virulent phase, despite its inability to target infected cells and excite Tc cells.	Considering the increased production costs and the need for adjuvants during manufacture, these vaccines are more costly.	They are far less effective defensively because they cannot infect cells and activate cytotoxic T cells.	They are only appropriate for clinical studies and not for illnesses since they require strong adjuvants and injections for tolerance.	[71, 72]

(Continued)

Table 13.1 List of vaccines, with their characteristic properties and applications. (*Continued*)

Vaccines	Types	Mode of action	Efficacy	Cost-effectiveness	Safety	Applications	References
	• Differentiating infected from vaccinated animals (DIVA) vaccines	The relevant genes were chosen and subsequently removed from the pathogen.	Chimeric pestiviruses, which substitute the gene encoding either E2 or ERNS with the equivalent gene of another pestivirus species, are the most effective of these vaccinations. The CP7 E2alf vaccine, sold by Zoetis under the name Suvaxyn CSF Marker, is the most efficacious DIVA vaccine presently available.	The next-generation CSF vaccine, C-DIVA, shows much potential as a cost-efficient option.	Due to high specificity, they are less safe for use.	For the testing of several diseases, DIVA vaccines are used for diseases, e.g., infectious bovine rhinotracheitis (IBR), pseudorabies, classical swine fever (CSF), and FMD.	[73–78]

(*Continued*)

Table 13.1 List of vaccines, with their characteristic properties and applications. (*Continued*)

Vaccines	Types	Mode of action	Efficacy	Cost-effectiveness	Safety	Applications	References
	• Molecularly defined subunit vaccines	Repeated adjuvant doses are required. For recombination, they produce viral antigens.	Less efficacy	Less cost-effective	By identifying the protective viral antigens, it may be possible to isolate and/ or create recombinant versions of them, allowing for the administration of vaccines that are safe and does not replicate themselves.	Successful examples of subunit vaccines that can be used as a vaccine for pigs and chickens from viral attacks are as follows: PCV2 and Dow AgroSciences	[79]
	• Genetically engineered viral vaccines	To produce well-defined and reliable attenuated live or inactivated viral vaccines, specific changes or deletions are made to the viral genome.	High efficacy	Cost-effective	It is safer but can be enhanced by deleting multiple genes.	A fascinating development in the field of genetically engineered viral vaccines is the production of chimeric viruses, which combine fragments of two infectious viral genomes, a vaccination with PCV1-2 chimera.	[80]

(*Continued*)

Table 13.1 List of vaccines, with their characteristic properties and applications. (*Continued*)

Vaccines	Types	Mode of action	Efficacy	Cost-effectiveness	Safety	Applications	References
	• Live viral vector vaccines	Exogenous vectors were utilized in these vaccines to transmit vaccine antigens.	Asian H5N1 and the pathogenic H7N7 strain chimera maintains high efficacy, whereas the VP2 gene of IBDV (Infectious Bursal Disease Virus) cannot maintain high efficacy.	Relatively simple and cost-effective	Safe and nonpathogenic	The leading development in vector vaccines is using attenuated viral pathogens as vectors against both the virus that causes Marek's disease and the infectious bursal disease. Chickens with turkey herpesvirus (HVT) against MDV (Marek's disease virus)	[81]
	• DNA vaccines	Naked DNA is employed for the vaccination process because it has the ability to implant cytotoxic cells (Tc cells) into the antigens intracellularly.	The efficacy is high.	Cost-effective	Simple and safe in use	They have a wide range of applications because they naturally produce highly immunogenic virus-like particles, and they are third-generation vaccines which work against melanoma and multiple pathogens. They are used as a marker vaccine as well as to differentiate between immunized and virulent animals.	[82]

(*Continued*)

Table 13.1 List of vaccines, with their characteristic properties and applications. (*Continued*)

Vaccines	Types	Mode of action	Efficacy	Cost-effectiveness	Safety	Applications	References
Animal bacterial vaccines	• Conventional live vaccines	Bacterial antigens are induced to train the immune system by an immune response with the help of certain molecules from the pathogen.	Efficacy is high	Cost-effective	It is safe but sometimes can become lethal	The vaccine, which is delivered by drinking water, will help to increase weight gain and decrease growth variability in pigs. These vaccines can induce both cell-mediated and humoral immunities.	[83]
	• Conventional inactivated vaccines	They contain dead virus strains or bacterins of one or more bacterial species or serotypes, coated in an oil or aluminum hydroxide adjuvant.	These vaccines are now completely approved in New Zealand and conditionally licensed in the United States despite the lack of published effectiveness evidence.	Follow-up booster vaccinations administered by injection are cost-effective, despite being labour- and time-intensive.	Safe in use	Only the certified denatured mucosal vaccines against bacterial infections were used in the preparation of veterinary medicine, which are applicable in novel vaccine technologies.	[83, 84]

(*Continued*)

Table 13.1 List of vaccines, with their characteristic properties and applications. (*Continued*)

Vaccines	Types	Mode of action	Efficacy	Cost-effectiveness	Safety	Applications	References
	• Gene-deleted vaccines	The development of live vaccines with specific deletions of known genes.	Good efficacy with promising results	Cost-effective	Safer in use	A desirable alternative to conventional injectable vaccinations and a mucosal vector for passenger antigens is also available from gene-deleted live bacteria with precise targeted attenuations.	[85]
	• Subunit vaccines	Recombinant proteins were formed against serotypes	Least efficacy	Cost-effective	Simple and safe, with no side effects	The protective impact of anti-ApxI (*Actinobacillus pleuropneumoniae* derived endotoxin 1) and anti-ApxII (*Actinobacillus pleuropneumoniae* derived endotoxin 2) antibodies was countered by the PalA (peptidoglycan-associated outer membrane protein of *A. pleuropneumoniae*) vaccine when combined with RTX (Repeats-in-toxin) toxins.	[86, 87]

(*Continued*)

Table 13.1 List of vaccines, with their characteristic properties and applications. (*Continued*)

Vaccines	Types	Mode of action	Efficacy	Cost-effectiveness	Safety	Applications	References
	• Animal parasitic vaccines	Strong effector leukocytes, mast cells, and eosinophils are activated and transported.	High in efficacy	Cost-effective	Some vaccines are safe, but some have a chance to become virulent, which is unsafe for use.	Several vaccinations like Coccivac B, Coccivac D, Immunocox C1, and Paracox have been developed and made available commercially in different regions of the world.	[88, 89]
	• Animal vaccines for non-infectious diseases	The most frequent therapy for atopic dermatitis is intradermal injection or allergen-specific serum immunoglobulin E tests.	Efficacy ranges from 20%–100%	Cost-effective	Not always safe	Majority of immunizations are based on research for human cancer vaccines, such as DNA immunization with human tyrosinase, or immunizations using canine tumor cell lines transfected with human granulocyte-macrophage colony-stimulating factor or human gp100.	[90]

(*Continued*)

Table 13.1 List of vaccines, with their characteristic properties and applications. (*Continued*)

Vaccines	Types	Mode of action	Efficacy	Cost-effectiveness	Safety	Applications	References
Nanovaccines	• Vaccines for viral diseases • Vaccines for bacterial diseases	Nanoscale-based particles with components to which an immunological response is sought are attached or synthesized.	High efficacy in both viral and bacterial nanovaccines	A reduction in antigen will decrease the cost	Safer in use	A calf model has shown the potential of PLGA-based biomimetic nanoparticles to inhibit BVDV. Nano platforms have been researched for *Anaplasma marginale* vaccines, such as carbon nanotubes carrying the major surface protein and silica vesicles (SV100) containing the type IV secretion system proteins (T4SS) VirB9-1 and VirB9-2, as well as VirB9-1 and VirB10 proteins (MSP) 1a in mice.	[91, 92]
Edible vaccines	• Food vaccine • Oral vaccine • Subunit vaccines	Stimulate the system and mucosal immune systems via a foreign pathogen	Suitable as an immunization delivery method since adjuvants that boost the immune response are not necessary	Edible vaccines are inexpensive to obtain, store, prepare, produce, and ship.	Edible vaccines are easily tolerated, so there is less need for medical professionals and little chance of infection. Therefore, it is safe to use.	Edible vaccines do not include heat-killed pathogens, and they are risk-free since there is no chance that the proteins will reorganize into infectious organisms. Therefore, they are applicable and the most reliable of veterinary vaccines.	[70]

PCV vaccines and swine influenza pandemic H1N1 vaccines are two examples of recently evolving diseases for which veterinary vaccines were recently implemented and approved.

The equine West Nile virus vaccine exemplifies how an animal vaccine can be produced and licensed rapidly in the event of an emergency. Collaborating with the U.S. Department of Agriculture (USDA) Center for Veterinary Biologics, the veterinary vaccine manufacturing sector rapidly came up with a successful vaccine to thwart the disease in horses. The USDA had granted a conditional license for an equine vaccine against the West Nile virus by August 2001. The West Nile virus vaccine has become recognized as one of the most important equine vaccines in the United States [93].

Tha main goal of a statutory governing body such as Pan African Veterinary Vaccine Centre (PANVAC) is to encourage the accessibility of veterinary vaccines that are reliable, efficient, and moderately priced. PANVAC-V and PANVAC-F were considered a priming and boosting vaccine in veterinary science and they were approved by PANVAC.

FDA had approved the theranostic use of iron oxide NPs. These NPs have been proven effective enough to serve as nanovaccines for various infectious conditions. They produced an immunological response specific to the antigen when given intravenously to mice. When intravenously administered in mice, they lead to the production of antigen-specific immune responses [94].

Many of the nanovaccines for veterinary care are in the proof-of-concept study, and it is a long way to go for final approval.

13.7 New Approaches

To deliver individual antigens for vaccinations against various illnesses, vaccine platform technologies use a common backbone or vector. This includes protein-based (for instance, virus-like particles [VLPs]), vector-based (for instance, viruses, bacteria, and protozoa), nucleic-acid-based (for instance, DNA and RNA), and replicon-based (for instance, self-amplifying RNA) as defined by the European Medicines Agency (EMA) Committee for Medicinal Products for Veterinary Use (CMVP). The first prokaryote-expressed veterinary vaccination to protect cats against the retroviral pathogen feline leukemia virus (FeLV) was developed in *E. coli*. *E. coli* was used to produce the FeLVG surface protein, while Quil A and aluminum hydroxide were used to transport it [95–97]. Undoubtedly, the best way to generate recombinant proteins for veterinary virus vaccines is in mammalian or avian cells. Although

Chinese hamster ovary cells have been frequently the preferred mammalian expression cell line, numerous other cell lines have also been used, including Vero cells, human embryonic kidney (HEK) cells, and baby hamster kidney (BHK) cells [98]. Over 30 years ago, the first vector-based vaccines were used in veterinary medicine. Today, various viral, bacterial, and protozoan vector platform technologies are used in vaccine development. The achievement of the vaccine research initiatives started in response to the COVID-19 pandemic has demonstrated the adaptability and efficacy of viral-vector vaccine platforms in therapeutic applications. In the early days of veterinary vaccine development, majority of research was on reproducing major DNA viruses, particularly poxviruses, herpesviruses, and adenoviruses [98, 99]. According to researchers at the University of California, San Diego, the potential for an efficient single-shot vaccine that could be administered intranasally to induce protective mucosal immunity in livestock, such as the BRSV polymeric nanoparticle vaccine, is a very significant advancement in the veterinary vaccine platform technologies [62]. In veterinary medicine, mRNA vaccines against infectious illnesses in animals, especially zoonoses, have significant promise. The primary perceived risks to both industrial livestock production and world health continue to be virus infections. mRNA-based vaccinations are used for combating various illnesses and may potentially help bring reliable and viable vaccines to market. They are strongly applicable in preventing various diseases such as rabies, Zika, and influenza [100]. mRNA vaccines offer the potential for high potency, inexpensive manufacturing, quick development, and secure administration compared to traditional vaccinations. Additionally, since cell cultures and hazardous chemicals are not necessary to produce mRNA vaccines, the end products are not contaminated chemically or biologically [101–104]. Additionally, because mRNA cannot integrate into the host genome, there is no chance of insertional mutagenesis and carcinogenesis. Ultimately, the opportunity to successfully combat emerging and reemerging infectious illnesses is presented by the scaled-up production to commercial levels at a remarkable size and speed. As a result, mRNA vaccinations participate in the initial step of the vaccination process [105].

13.8 Applications of Different Polymer-Based Nanoparticles

Particles or particulate materials having a minimum size in the range of 10–100 nm in one dimension are known as polymeric nanoparticles (PNPs). Due to their small size, high volume-surface area ratio, and tunable pore,

PNPs have been employed in a wide array of implementations in a variety of fields, such as the delivery of drugs, biosensors, stimuli-responsive cargo delivery, nanocomposites, agricultural, and environmental applications. One of the most recent progress in inorganic NP integration into polymers allows for the modification of the polymer's physical and chemical characteristics as well as the improvement features in the polymer system. One of the most popular nanomaterials for use in nanomedicine is PNPs. The possibility of PNPs to change modern medicine is really intriguing. In several restorative applications, biodegradable polymeric nanostructures (BPNs) have demonstrated exceptional reliability [106].

These molecules, which comprise microbial polyesters and microbial polysaccharides, are produced by enzyme hydrolysis (using microbial enzymes). PHB (poly-hydroxy butyrate) is similar to natural polymers in terms of its biodegradability, biocompatibility, piezoelectricity, optical activity, and other unique qualities. Chemical processes are used to create chemically synthesized polymer materials, such as silicone rubber, polyvinyl alcohol, polyurethane (PU), poly (methyl methacrylate), polyester, polyvinylpyrrolidone (PVP), polyurethane (PLGA), and poly (methyl methacrylate) that are used in medical materials. Nanomaterials offer various benefits as delivery vehicles. Our team created Newcastle disease virus (NDV)-encapsulated N-2-hydroxypropyl trimethyl ammonium chloride chitosan nanoparticles (NDV/La Sota-N-2-HACC-NPs). These NDV/La Sota-N-2-HACC-NPs were found to have a high level of safety, extended release, and a very low level of toxicity. They also elicited stronger cellular, humoral, and mucosal immune responses. Starch's hydrophilic microgel is susceptible to environmental changes, including pH, temperature, storage conditions, and salt ions, which can cause varying degrees of swelling and shrinking. The creation of a novel kind of drug carrier called microgel is anticipated. Starch polymers have shown improved bioactivity and capacity to bind to live cells, and regulate medication release by preventing drug degradation [107].

Compared to either AgNPs (silver nanoparticles) or chitosan alone, AgNPs-chitosan NPs were more effective at preventing the development of *E. coli*. To improve AgNPs' ability to bind to proteins found on cell walls, the negatively charged cell surface formed an electrical connection with the positively charged chitosan. It is thought that this results in intracellular components and other protein leaks, which deactivate the bacteria. Chitosan was also employed with AgNPs to inhibit the growth of *E. coli*-specific biofilms and to remove impurities [108].

Pullulan is a crucial, widely distributed, water-soluble, and biocompatible polymer released by *Aureobasidium pullulans*. This polymer has the potential to be an effective antibacterial agent against a variety of microbes, including multidrug-resistant (MDR) pathogens. The use of pullulan-based nanocomposites for wound healing, food packing, and extending the freshness of products are all possible. Nanoparticles made of pullulan are said to have powerful antibacterial properties. The agar well diffusion method was used to assess their antibacterial efficacy against two Gram-positive (*Bacillus subtilis* and *Staphylococcus aureus*) and three Gram-negative (*Escherichia coli* and *Serratia marcescens*) bacterial pathogens [109].

According to reports, pullulan-based nanoparticles effectively combat germs. Using the agar well diffusion method, they were tested for their antibacterial effectiveness against two Gram-positive (*Bacillus subtilis* and *Staphylococcus aureus*) and three Gram-negative (*Escherichia coli* and *Serratia marcescens*) bacterial infections. Depending on their atomic weight, the generated NPs were spherical in form and ranged in size from 10 to 55 nm on average [109].

Pullulan-based NPs could be ideal for future therapeutic uses in transmucosal protein delivery, according to Grenha and Rodrigues. Due to their structural flexibility and propensity to demonstrate biocompatibility and biodegradability, polysaccharides like pullulan are favored to manufacture such carriers [110].

Pullulan-based NPs may be helpful in transmucosal protein distribution, with an emphasis on the respiratory and nasal passages. As a model protein, bovine serum albumin (BSA) was employed. It has been discovered that BSA is released slowly and steadily by these NPs [111].

13.9 Future Prospects

One of the most captivating fields of study is nanomedicines and nanovaccines. Over the past 20 years, there has been a substantial number of research in this area, which has ultimately led to the filing of 1500 patents and the completion of countless clinical trials [112]. Undoubtedly, this trend will continue to be the focus of research and development for the ensuing decades as will the use of nanomedicine and nanodrug delivery systems [112]. If nanovaccines are used extensively in the veterinary industry, there are many factors to take into account in the coming years [5].

It has been noted that both ruminants and chickens have been the focus of studies on nanocarrier-based nanovaccines systems used for nasal immunization, but there is a dearth of information on spray or nebulization vaccine administration. Pulmonary vaccine delivery using nanocarriers in livestock and poultry is anticipated to see explosive growth in both research and application [9].

Numerous research on nanocarrier systems must be conducted shortly, for example, nanoemulsions, since they cannot be administered intravenously. If research on nanoemulsions persists to yield encouraging results, researchers might soon see bactericidal and viricidal products that can be applied topically to animals [113]. Additionally, the field of veterinary medicine can benefit greatly from dendrimers. The identification, management, and eventual eradication of malignant tumors, which frequently afflict the small animal population, will be one of the major contributions of these synthetic nanomaterials in the future. It can be said that nanoshells will be crucial to veterinary medicine in the future, even though more research on this novel optical material needs to be done [113]. Needle-free delivery methods, such as intranasal or oral administration, or application of microprojected patches to the skin, are still in the preliminary phases of research but potentially have a promising future in nanovaccination [114]. To boost vaccine efficiency without triggering inflammatory reactions, scientists are also creating adjuvants made of polystyrene nanobeads coupled to ovalbumin [114]. Worldwide, pigs are adversely affected by viral diseases. The use of existing vaccinations has undoubtedly aided in the achievement of powerful management over various pig viral illnesses, including foot-and-mouth disease, swine fever, transmissible gastroenteritis, and pseudorabies. However, with the aid of cutting-edge nanotechnology, research is constantly being carried out to develop vaccines against porcine viral infections in the future [8].

13.10 Conclusion

In conclusion, NP-based vaccines can function as cutting-edge candidates for immunization against a wide range of viral infections for veterinary usage, potentially enhancing broader protective efficacy in practical settings. When compared to conventional vaccines, using NP platforms as vaccine carriers and/or adjuvants in veterinary medicine has a significant opportunity to enhance vaccines by maintaining stability outside the cold chain, improving the immunogenicity of their payload, engaging both humoral and cellular immune responses, and facilitating long-lasting responses without the requirement for booster doses. These nanovaccine techniques can also be used to create vaccines that are efficacious against viral infections in humans, and the information gained from doing so could be used to boost the potency of current veterinary viral vaccines.

The major constraint for veterinary care is to comprehend and implement these techniques in a way that maximizes the health benefits for animals. The technological prowess will be enhanced by nanotechnology, thereby improving the standard of life and treatment for animals. Further, it will raise the living standards and promote further economic development in the veterinary industry. Ultimately, there are various advantages that nanovaccines have over conventional adjuvants. Numerous research has shown that nanovaccines significantly increase the immunogenicity and efficiency of vaccines in animals. Currently available nano-based medications and vaccines could be repurposed as a useful therapeutic strategy to stop the dissemination and dispersion of infections from animals to other critically endangered species.

Acknowledgments

The authors want to express their gratitude to Dr. G. Viswanathan, Chancellor, VIT University, Vellore, for his constant support and encouragement.

References

1. Bamrungsap, S. *et al.*, Nanotechnology in therapeutics: A focus on nanoparticles as a drug delivery system. *Nanomedicine (Lond.)*, 7, 8, 1253–1271, 2012, doi: 10.2217/NNM.12.87.
2. Hajipour, M.J. *et al.*, Antibacterial properties of nanoparticles. *Trends Biotechnol.*, 30, 10, 499–511, 2012, doi: 10.1016/J.TIBTECH.2012.06.004.
3. Underwood, C. and van Eps, A.W., Nanomedicine and veterinary science: the reality and the practicality. *Vet. J.*, 193, 1, 12–23, 2012, doi: 10.1016/J.TVJL.2012.01.002.
4. Celis-Giraldo, C.T., López-Abán, J., Muro, A., Patarroyo, M.A., Manzano-Román, R., Nanovaccines against animal pathogens: The latest findings. *Vaccines (Basel)*, 9, 9, 988, 2021, doi: 10.3390/vaccines9090988.
5. Maina, T.W., Grego, E.A., Boggiatto, P.M., Sacco, R.E., Narasimhan, B., McGill, J.L., Applications of nanovaccines for disease prevention in cattle. *Front. Bioeng. Biotechnol.*, 8, 608050, 2020, doi: 10.3389/fbioe.2020.608050.
6. Khan, I., Saeed, K., Khan, I., Nanoparticles: Properties, applications and toxicities. *Arab. J. Chem.*, 12, 7, 908–931, 2019, doi: https://doi.org/10.1016/j.arabjc.2017.05.011.
7. Fawzy, M., Khairy, G.M., Hesham, A., Rabaan, A.A., El-Shamy, A.G., Nagy, A., Nanoparticles as a novel and promising antiviral platform in

veterinary medicine. *Arch. Virol.*, 166, 10, 2673–2682, 2021, doi: 10.1007/s00705-021-05177-w.

8. Dhakal, S. and Renukaradhya, G.J., Nanoparticle-based vaccine development and evaluation against viral infections in pigs. *Vet. Res.*, 50, 1, 90, 2019, doi: 10.1186/s13567-019-0712-5.
9. Calderon-Nieva, D., Goonewardene, K.B., Gomis, S., Foldvari, M., Veterinary vaccine nanotechnology: Pulmonary and nasal delivery in livestock animals. *Drug Deliv. Transl. Res.*, 7, 4, 558—570, 2017, doi: 10.1007/s13346-017-0400-9.
10. Feng, C. *et al.*, Emerging vaccine nanotechnology: From defense against infection to sniping cancer. *Acta Pharm. Sin. B*, 12, 5, 2206–2223, 2022, doi: https://doi.org/10.1016/j.apsb.2021.12.021.
11. Youssef, F.S., El-Banna, H.A., Elzorba, H.Y., Galal, A.M., Application of some nanoparticles in the field of veterinary medicine. *Int. J. Vet. Sci. Med.*, 7, 1, 78–93, 2019, doi: 10.1080/23144599.2019.1691379.
12. Mohanty, N., Palai, T., Prusty, B., Mohapatra, J., An overview of nanomedicine in veterinary science. *Vet. Res. Int.*, 2, 4, 90–95, 2014. scholar.archive.org. [Online] Available: https://scholar.archive.org/work/etfkij5ksfe3ndgv56ej7qj66a/access/wayback/http://jakraya.com/journal/pdf/6-vriArticle_2.pdf.
13. Torres-Sangiao, E., Holban, A.M., Gestal, M.C., Andronescu, E., Grumezescu, A.M., Advanced nanobiomaterials: Vaccines, diagnosis and treatment of infectious diseases. *Molecules*, 21, 7, 867, 2016, doi: 10.3390/MOLECULES21070867.
14. Meena, N.S., Sahni, Y.P., Thakur, D., Singh, R.P., Applications of nanotechnology in veterinary therapeutics. *J. Entomol. Zool. Stud.*, 6, 2, 167–175, 2018.
15. Jurj, A., Braicu, C., Pop, L.A., Tomuleasa, C., Gherman, C.D., Berindan-Neagoe, I., The new era of nanotechnology, an alternative to change cancer treatment. *Drug Des. Devel. Ther.*, 11, 2871–2890, 2017, doi: 10.2147/DDDT.S142337.
16. Freitas Jr., R.A., Microbivores: Artificial mechanical phagocytes using digest and discharge protocol. *J. Evol. Technol.*, 14, 1, 54–106, 2005. Accessed: Jan. 30, 2023. [Online]. Available: https://jetpress.org/volume14/freitas.pdf?utm_medium=email&utm_source=transaction.
17. Manuja, A., Kumar, B., Singh, R.K., Nanotechnology developments: Opportunities for animal health and production. *Nanotechnol. Dev.*, 2, 1, 4, 2012, doi: 10.4081/nd.2012.e4.
18. Elgqvist, J. and Stephan, C., Nanoparticles as theranostic vehicles in experimental and clinical applications—focus on prostate and breast cancer. *Int. J. Mol. Sci.*, 18, 5, 1102 2017.
19. Rapoport, N., Gao, Z., Kennedy, A., Multifunctional nanoparticles for combining ultrasonic tumor imaging and targeted chemotherapy. *J. Natl. Cancer Inst.*, 99, 14, 1095–1106, 2007, doi: 10.1093/JNCI/DJM043.

20. Rodríguez-Burneo, N., Busquets, M.A., Estelrich, J., Magnetic nanoemulsions: Comparison between nanoemulsions formed by ultrasonication and by spontaneous emulsification. *Nanomaterials*, 7, 190, 2017, doi: 10.3390/NANO7070190.
21. Kudela, D. *et al.*, Clotting activity of polyphosphate-functionalized silica nanoparticles. *Angew. Chem. Int. Ed.*, 54, 13, 4018–4022, 2015, doi: 10.1002/ANIE.201409639.
22. El-Sayed, A. and Kamel, M., Advanced applications of nanotechnology in veterinary medicine. *Environ. Sci. Pollut. Res.*, 27, 16, 19073–19086, 2020, doi: 10.1007/S11356-018-3913-Y.
23. Erathodiyil, N. and Ying, J.Y., Functionalization of inorganic nanoparticles for bioimaging applications. *Acc. Chem. Res.*, 44, 10, 925–935, 2011, doi: 10.1021/AR2000327/ASSET/IMAGES/MEDIUM/AR-2011-000327_0013.GIF.
24. McMillan, J., Batrakova, E., Gendelman, H.E., Cell delivery of therapeutic nanoparticles, in: *Progress in Molecular Biology and Translational Science*, vol. 104, pp. 563–601, 2011.
25. Landers, J.J. *et al.*, Prevention of influenza pneumonitis by sialic acid–conjugated dendritic polymers. *J. Infect. Dis.*, 186, 9, 1222–1230, 2002, doi: 10.1086/344316.
26. Jahanshahi, M. and Babaei, Z., Protein nanoparticle: A unique system as drug delivery vehicles. *Afr. J. Biotechnol.*, 7, 25, 4926–4934, 2008.
27. Lee, S.J., Yhee, J.Y., Kim, S.H., Kwon, I.C., Kim, K., Biocompatible gelatin nanoparticles for tumor-targeted delivery of polymerized siRNA in tumor-bearing mice. *J. Control. Release*, 172, 1, 358–366, 2013, doi: 10.1016/J.JCONREL.2013.09.002.
28. Prabhu, R.H., Patravale, V.B., Joshi, M.D., Polymeric nanoparticles for targeted treatment in oncology: Current insights. *Int. J. Nanomedicine*, 10, 1001–1018, 2015, doi: 10.2147/IJN.S56932.
29. Krishnan, S. *et al.*, Nanotechnology-based therapeutic formulations in the battle against animal coronaviruses: An update. *J. Nanopart. Res.*, 23, 10, 229, 2021, doi: 10.1007/s11051-021-05341-y.
30. Olivera-Ugarte, S.-M. *et al.*, A nanoparticle-based COVID-19 vaccine candidate elicits broad neutralizing antibodies and protects against SARS-CoV-2 infection. *Nanomedicine*, 44, 102584, 2022, doi: 10.1016/j.nano.2022.102584.
31. Sengupta, A., Azharuddin, M., Al-Otaibi, N., Hinkula, J., Efficacy and immune response elicited by gold nanoparticle-based nanovaccines against infectious diseases. *Vaccines*, 10, 505, 2022, doi: 10.3390/VACCINES10040505.
32. Azharuddin, M., Zhu, G.H., Sengupta, A., Hinkula, J., Slater, N.K.H., Patra, H.K., Nano toolbox in immune modulation and nanovaccines. *Trends Biotechnol.*, 40, 10, 1195–1212, 2022, doi: 10.1016/J.TIBTECH.2022.03.011.
33. Dhakal, S. and Renukaradhya, G.J., Nanoparticle-based vaccine development and evaluation against viral infections in pigs. *Vet. Res.*, 50, 1, 1–14, 2019, doi: 10.1186/S13567-019-0712-5.

34. Hammer, S.E. *et al.*, Comparative analysis of swine leukocyte antigen gene diversity in european farmed pigs. *Anim Genet.*, 52, 4, 523–531, 2021, doi: 10.1111/AGE.13090.
35. Singh, A., Eliciting B cell immunity against infectious diseases using nanovaccines. *Nat. Nanotechnol.*, 16, 1, 16–24, 2020, doi: 10.1038/s41565-020-00790-3.
36. Monrad, J.T., Sandbrink, J.B., Cherian, N.G., Promoting versatile vaccine development for emerging pandemics. *NPJ Vaccines*, 6, 1, 2021, doi: 10.1038/S41541-021-00290-Y.
37. Bhardwaj, P., Bhatia, E., Sharma, S., Ahamad, N., Banerjee, R., Advancements in prophylactic and therapeutic nanovaccines. *Acta Biomater.*, 108, 1–21, 2020, doi: 10.1016/J.ACTBIO.2020.03.020.
38. Shen, Y., Hao, T., Ou, S., Hu, C., Chen, L., Applications and perspectives of nanomaterials in novel vaccine development. *MedChemComm*, 9, 2, 226–238, 2018, doi: 10.1039/c7md00158d.
39. Zhao, L. *et al.*, Nanoparticle vaccines. *Vaccine*, 32, 3, 327–337, 2014, doi: 10.1016/J.VACCINE.2013.11.069.
40. El-Sayed, A. and Kamel, M., Advanced applications of nanotechnology in veterinary medicine. *Environ. Sci. Pollut. Res. Int.*, 27, 16, 19073–19086, 2020, doi: 10.1007/S11356-018-3913-Y.
41. Kelly, H.G., Kent, S.J., Wheatley, A.K., Immunological basis for enhanced immunity of nanoparticle vaccines. *Expert Rev. Vaccines*, 18, 3, 269–280, 2019, doi: 10.1080/14760584.2019.1578216.
42. Chenthamara, D. *et al.*, Therapeutic efficacy of nanoparticles and routes of administration. *Biomater. Res.*, 23, 1, 2019, doi: 10.1186/S40824-019-0166-X.
43. Suneetha, V. and Punetha, S., Microbial exploitation of biofertilizer, pullulan and biochar from floral waste. *Austin Biochem.*, 6, 1, 2021.
44. Narasimhan, B., Goodman, J.T., Vela Ramirez, J.E., Rational design of targeted next-generation carriers for drug and vaccine delivery. *Annu. Rev. Biomed. Eng.*, 18, 25–49, 2016, doi: 10.1146/ANNUREV-BIOENG-082615-030519.
45. Kalina, W.V., Woolums, A.R., Berghaus, R.D., Gershwin, L.J., Formalin-inactivated bovine RSV vaccine enhances a Th2 mediated immune response in infected cattle. *undefined.*, 22, 11–12, 1465–1474, 2004, doi: 10.1016/J.VACCINE.2003.10.024.
46. Blodörn, K. *et al.*, Vaccine safety and efficacy evaluation of a recombinant bovine respiratory syncytial virus (BRSV) with deletion of the SH gene and subunit vaccines based on recombinant human RSV proteins: N-nanorings, P and M2-1, in calves with maternal antibodies. *PloS One*, 9, 6, e100392–e100392, 2014, doi: 10.1371/JOURNAL.PONE.0100392.
47. Mahony, D. *et al.*, Immunisation of sheep with bovine viral diarrhoea virus, E2 protein using a freeze-dried hollow silica mesoporous nanoparticle formulation. *PloS One*, 10, 11, e0141870, 2015, doi: 10.1371/JOURNAL.PONE.0141870.

48. Riitho, V. *et al.*, Design and evaluation of the immunogenicity and efficacy of a biomimetic particulate formulation of viral antigens. *Sci. Rep.*, 7, 1, 13743–13743, 2017, doi: 10.1038/S41598-017-13915-X.
49. Ding, P. *et al.*, Nanovaccine confers dual protection against influenza a virus and porcine circovirus type 2. *Int. J. Nanomedicine*, 14, 7533–7548, 2019, doi: 10.2147/IJN.S218057.
50. Dhakal, S. *et al.*, Polyanhydride nanovaccine against swine influenza virus in pigs. *Vaccine*, 35, 8, 1124–1131, 2017, doi: 10.1016/J.VACCINE.2017.01.019.
51. Derman, S., Mustafaeva, Z.A., Abamor, E.S., Bagirova, M., Allahverdiyev, A., Preparation, characterization and immunological evaluation: Canine parvovirus synthetic peptide loaded PLGA nanoparticles. *J. BioMed. Sci.*, 22, 1, 2015, doi: 10.1186/S12929-015-0195-2.
52. Wambura, P.N., Formulation of novel nano-encapsulated Newcastle disease vaccine tablets for vaccination of village chickens. *Trop. Anim. Health Prod.*, 43, 1, 165–169, 2011, doi: 10.1007/S11250-010-9669-0.
53. Thukral, A. *et al.*, A single dose polyanhydride-based nanovaccine against paratuberculosis infection. *NPJ Vaccines*, 5, 1, 15–15, 2020, doi: 10.1038/S41541-020-0164-Y.
54. Maleki, M., Salouti, M., Ardestani, M.S., Talebzadeh, A., Preparation of a nanovaccine against Brucella melitensis M16 based on PLGA nanoparticles and oligopolysaccharide antigen. *Artif. Cells Nanomed. Biotechnol.*, 47, 1, 4248–4256, 2019, doi: 10.1080/21691401.2019.1687490.
55. Sadeghi, Z., Fasihi-Ramandi, M., Bouzari, S., Nanoparticle-based vaccines for brucellosis: Calcium phosphate nanoparticles-adsorbed antigens induce cross protective response in mice. *Int. J. Nanomedicine*, 15, 3877–3886, 2020, doi: 10.2147/IJN.S249942.
56. Maleki, M., Salouti, M., Ardestani, M.S., Talebzadeh, A., Preparation of a nanovaccine against brucella melitensis M16 based on PLGA nanoparticles and oligopolysaccharide antigen. *Artif. Cells Nanomed. Biotechnol.*, 47, 1, 4248–4256, 2019, doi: 10.1080/21691401.2019.1687490.
57. Celis-Giraldo, C.T., López-Abán, J., Muro, A., Patarroyo, M.A., Manzano-Román, R., Nanovaccines against animal pathogens: The latest findings. *Vaccines (Basel)*, 9, 9, 988, 2021, doi: 10.3390/VACCINES9090988.
58. Wang, F., Chen, G., Zhao, Y., Biomimetic nanoparticles as universal influenza vaccine. *Smart Mater Med.*, 1, 21–23, 2020, doi: 10.1016/J.SMAIM.2020.03.001.
59. Mohammed, G.M., ElZorkany, H.E.S., Farroh, K.Y., Abd El-Aziz, W.R., Elshoky, H.A., Potential improvement of the immune response of chickens against E. coli vaccine by using two forms of chitosan nanoparticles. *Int. J. Biol. Macromol.*, 167, 395–404, 2021, doi: 10.1016/J.IJBIOMAC.2020.11.200.
60. Meeusen, E.N.T., Walker, J., Peters, A., Pastoret, P.P., Jungersen, G., Current status of veterinary vaccines. *Clin. Microbiol. Rev.*, 20, 3, 489, 2007, doi: 10.1128/CMR.00005-07.

61. Chenthamara, D. Subramaniam, S., Ramakrishnan, S.G. *et al.*, Therapeutic efficacy of nanoparticles and routes of administration. *Biomater. Res.*, 23, 1, 20, 2019, doi: 10.1186/S40824-019-0166-X.
62. McGill, J.L. *et al.*, Efficacy of mucosal polyanhydride nanovaccine against respiratory syncytial virus infection in the neonatal calf. *Sci. Rep.*, 8, 1, 3021, 2018, doi: 10.1038/S41598-018-21292-2.
63. Bhardwaj, P., Bhatia, E., Sharma, S., Ahamad, N., Banerjee, R., Advancements in prophylactic and therapeutic nanovaccines. *Acta Biomater.*, 108, 1–21, 2020, doi: 10.1016/J.ACTBIO.2020.03.020.
64. Wagner, D.A. *et al.*, Single-dose combination nanovaccine induces both rapid and long-lived protection against pneumonic plague. *Acta Biomater.*, 100, 326–337, 2019, doi: 10.1016/J.ACTBIO.2019.10.016.
65. Saluja, V. and Sekhon, B.S., Biosimilars: An overview. *Biosimilars*, 1, 1–11, 2011, doi: 10.2147/BS.S16120.
66. Bachmann, M.F. and Jennings, G.T., Vaccine delivery: A matter of size, geometry, kinetics and molecular patterns. *Nat. Rev. Immunol.*, 10, 11, 787–796, 2010, doi: 10.1038/NRI2868.
67. Thomas, C., Rawat, A., Hope-Weeks, L., Ahsan, F., Aerosolized, P.L.A., and PLGA nanoparticles enhance humoral, mucosal and cytokine responses to hepatitis B vaccine. *Mol. Pharm.*, 8, 2, 405–415, 2011, doi: 10.1021/MP100255C.
68. Vela Ramirez, J.E., Boggiatto, P.M., Wannemuehler, M.J., Narasimhan, B., Polyanhydride nanoparticle interactions with host serum proteins and their effects on bone marrow derived macrophage activation. *ACS Biomater. Sci. Eng.*, 3, 2, 160–168, 2017, doi: 10.1021/ACSBIOMATERIALS.6B00394/ASSET/IMAGES/MEDIUM/AB-2016-00394U_0007.GIF.
69. Zhao, Y. *et al.*, A comparison between sphere and rod nanoparticles regarding their *in vivo* biological behavior and pharmacokinetics. *Sci. Rep.*, 7, 1, 2017, doi: 10.1038/S41598-017-03834-2.
70. Jacob, S.S., Cherian, S., Sumithra, T.G., Raina, O.K., Sankar, M., Edible vaccines against veterinary parasitic diseases—current status and future prospects. *Vaccine*, 31, 15, 1879–1885, 2013, doi: 10.1016/J.VACCINE.2013.02.022.
71. Minke, J.M., Audonnet, J.C., Fischer, L., Equine viral vaccines: The past, present and future. *Vet. Res.*, 35, 4, 425–443, 2004, doi: 10.1051/VETRES:2004019.
72. Day, M.J., Vaccine side effects: Dact and fiction. *Vet. Microbiol.*, 117, 1, 51–58, 2006, doi: 10.1016/J.VETMIC.2006.04.017.
73. van Oirschot, J.T., Kaashoek, M.J., Rijsewijk, F.A.M., Advances in the development and evaluation of bovine herpesvirus 1 vaccines. *Vet. Microbiol.*, 53, 1–2, 43–54, 1996, doi: 10.1016/S0378-1135(96)01233-3.
74. Reimann, I., Depner, K., Trapp, S., Beer, M., An avirulent chimeric Pestivirus with altered cell tropism protects pigs against lethal infection with classical swine fever virus. *Virology*, 322, 1, 143–157, 2004, doi: 10.1016/j.virol.2004.01.028.
75. Leifer, I. *et al.*, Modified live marker vaccine candidate CP7_E2alf provides early onset of protection against lethal challenge infection with classical

swine fever virus after both intramuscular and oral immunization. *Vaccine*, 27, 47, 6522–6529, 2009, doi: 10.1016/J.VACCINE.2009.08.057.

76. van Gennip, H.G.P., van Rijn, P.A., Widjojoatmodjo, M.N., de Smit, A.J., Moormann, R.J.M., Chimeric classical swine fever viruses containing envelope protein E(RNS) or E2 of bovine viral diarrhoea virus protect pigs against challenge with CSFV and induce a distinguishable antibody response. *Vaccine*, 19, 4–5, 447–459, 2000, doi: 10.1016/S0264-410X(00)00198-5.
77. Blome, S., Wernike, K., Reimann, I., König, P., Moß, C., Beer, M., A decade of research into classical swine fever marker vaccine CP7_E2alf (suvaxyn® CSF marker): A review of vaccine properties. *Vet. Res.*, 48, 1, 51, 1–10 2017, doi: 10.1186/S13567-017-0457-Y.
78. Pannhorst, K., Fröhlich, A., Staubach, C., Meyer, D., Blome, S., Becher, P., Evaluation of an Erns-based enzyme-linked immunosorbent assay to distinguish Classical swine fever virus-infected pigs from pigs vaccinated with CP7_E2alf. *J. Vet. Diagn. Invest.*, 27, 4, 449–460, 2015, doi: 10.1177/1040638715592446.
79. Allan, G.M. *et al.*, PMWS: Experimental model and co-infections. *Vet. Microbiol.*, 98, 2, 165–168, 2004, doi: 10.1016/j.vetmic.2003.10.009.
80. Fenaux, M., Opriessnig, T., Halbur, P.G., Elvinger, F., Meng, X.J., A chimeric porcine circovirus (PCV) with the immunogenic capsid gene of the pathogenic PCV type 2 (PCV2) cloned into the genomic backbone of the nonpathogenic PCV1 induces protective immunity against PCV2 infection in pigs. *J. Virol.*, 78, 12, 6297–6303, 2004, doi: 10.1128/JVI.78.12.6297-6303.2004.
81. da Silva, V.O. *et al.*, A phase III trial of efficacy of the FML-vaccine against canine kala-azar in an endemic area of Brazil (São Gonçalo do Amaranto, RN). *Vaccine*, 19, 9–10, 1082–1092, 2000, doi: 10.1016/S0264-410X(00)00339-X.
82. Seregin, A. and virus infectious, D.N.A. *et al.*, and its noninfectious derivatives. *Virology*, 356, 1–2, 115–125, 2006, doi: 10.1016/J.VIROL.2006.07.038.
83. McOrist, S., Jasni, S., Mackie, R.A., MacIntyre, N., Neef, N., Lawson, G.H.K., Reproduction of porcine proliferative enteropathy with pure cultures of ileal symbiont intracellularis. *Infect. Immun.*, 61, 10, 4286–4292, 1993, doi: 10.1128/IAI.61.10.4286-4292.1993.
84. Jorge, S. and Dellagostin, O.A., The development of veterinary vaccines: A review of traditional methods and modern biotechnology approaches. *Biotechnol. Res. Innovation*, 1, 1, 6–13, 2017, doi: 10.1016/J.BIORI.2017.10.001.
85. Walker, J.A. and Timoney, J.F., Construction of a stable non-mucoid deletion mutant of the Streptococcus equi Pinnacle vaccine strain. *Vet. Microbiol.*, 89, 4, 311–321, 2002, doi: 10.1016/S0378-1135(02)00205-5.
86. Frey, J., Virulence in actinobacillus pleuropneumoniae and RTX toxins. *Trends Microbiol.*, 3, 7, 257–261, 1995. doi: 10.1016/S0966-842X(00)88939-8.
87. van den Bosch, H. and Frey, J., Interference of outer membrane protein PalA with protective immunity against actinobacillus pleuropneumoniae infections in vaccinated pigs. *Vaccine*, 21, 25–26, 3601–3607, 2003, doi: 10.1016/S0264-410X(03)00410-9.

88. Maizels, R.M., Balic, A., Gomez-Escobar, N., Nair, M., Taylor, M.D., Allen, J.E., Helminth parasites–masters of regulation. *Immunol. Rev.*, 201, 89–116, 2004, doi: 10.1111/J.0105-2896.2004.00191.X.
89. Balic, A. and Meeusen, E.N.T., The immunobiology of gastrointestinal nematode infections in ruminants. *Adv. Parasitol.*, 45, 181–241, 2000, doi: 10.1016/S0065-308X(00)45005-0.
90. Bergman, P.J. *et al.*, Development of a xenogeneic DNA vaccine program for canine malignant melanoma at the Animal Medical Center. *Vaccine*, 24, 21, 4582–4585, 2006, doi: 10.1016/j.vaccine.2005.08.027.
91. Zhao, L. *et al.*, A partially purified outer membrane protein VirB9-1 for low-cost nanovaccines against anaplasma marginale. *Vaccine*, 35, 1, 77–83, 2017, doi: 10.1016/J.VACCINE.2016.11.037.
92. Zhao, L. *et al.*, Immunogenicity of outer membrane proteins VirB9-1 and VirB9-2, a novel nanovaccine against anaplasma marginale. *PloS One*, 11, 4, e0154295, 2016, doi: 10.1371/JOURNAL.PONE.0154295.
93. Saiz, J.C., Animal and human vaccines against West Nile virus. Pathogens. 9, 12, 1073, 2020. doi: 10.3390/pathogens9121073.
94. Cordeiro, A.S., Patil-Sen, Y., Shivkumar, M., Patel, R., Khedr, A., Elsawy, M.A., Nanovaccine delivery approaches and advanced delivery systems for the prevention of viral infections: From development to clinical application. *Pharmaceutics*, 12, 2091, 2021, doi: 10.3390/PHARMACEUTICS13122091.
95. Entrican, G. and Francis, M.J., Applications of platform technologies in veterinary vaccinology and the benefits for one health. *Vaccine*, 40, 20, 2833–2840, 2022, doi: 10.1016/J.VACCINE.2022.03.059.
96. Berg, P. *et al.*, Potential biohazards of recombinant DNA molecules. *Science*, 185, 4148, 303, 1974. doi: 10.1126/SCIENCE.185.4148.303.
97. Marciani, D., Kensil, C., Beltz, G., Hung, C., Cronier, J., Aubert, A., Genetically-engineered subunit vaccine against feline leukaemia virus: Protective immune response in cats. *Vaccine Elsevier*, 9, 2, 89–96, 1991. Accessed: Dec. 11, 2022. [Online]. Available: https://www.sciencedirect.com/science/article/pii/0264410X91902625.
98. Francis, M.J., Recent advances in vaccine technologies. *Vet. Clin. North Am. Small Anim. Pract.*, 48, 2, 231–241, 2018. vetsmall.theclinics.com, Accessed: Dec. 11, 2022. [Online]. Available: https://www.vetsmall.theclinics.com/article/S0195-5616(17)30119-5/abstract.
99. Baron, M., Iqbal, M., Nair, V., Recent advances in viral vectors in veterinary vaccinology. *Curr. Opin. Virol. Elsevier*, 29, 1–7, 2018. Accessed: Dec. 12, 2022. [Online]. Available: https://www.sciencedirect.com/science/article/pii/S1879625718300178.
100. Le, T., Sun, C., Chang, J., Zhang, G., Yin, X., mRNA vaccine development for emerging animal and zoonotic diseases. *Viruses*, 14, 2, 401, 1–19, 2022, doi: 10.3390/V14020401.

101. Anand, P. and Stahel, V.P., Correction to: The safety of Covid-19 mRNA vaccines: A review. *Patient Saf. Surg.*, 15, 1, 20, 1–9, 2021, doi: 10.1186/S13037-021-00296-4.
102. Pardi, N., Hogan, M.J., Porter, F.W., Weissman, D., mRNA vaccines-a new era in vaccinology. *Nat. Rev. Drug Discovery*, 17, 4, 261–279, 2018, doi: 10.1038/NRD.2017.243.
103. Chaudhary, N., Weissman, D., Whitehead, K.A., mRNA vaccines for infectious diseases: Principles, delivery and clinical translation. *Nat. Rev. Drug Discovery*, 20, 11, 817–838, 2021, doi: 10.1038/S41573-021-00283-5.
104. Teijaro, J.R. and Farber, D.L., COVID-19 vaccines: Modes of immune activation and future challenges. *Nat. Rev. Immunol.*, 21, 4, 195–197, 2021, doi: 10.1038/S41577-021-00526-X.
105. Wang, Y., Zhang, Z., Luo, J., Han, X., Wei, Y., Wei, X., mRNA vaccine: A potential therapeutic strategy. *Mol. Cancer*, 20, 1, 33, 2021, doi: 10.1186/S12943-021-01311-Z.
106. Adhikari, C., Polymer nanoparticles-preparations, applications and future insights: A concise review. *Polym.-Plast. Technol. Mater.*, 60, 18, 1996–2024, 2021, doi: 10.1080/25740881.2021.1939715.
107. Han, J., Zhao, D., Li, D., Wang, X., Jin, Z., Zhao, K., Polymer-based nanomaterials and applications for vaccines and drugs. *Polym. (Basel)*, 10, 1, 31, 1–14, 2018, doi: 10.3390/POLYM10010031.
108. Sanpui, P., Murugadoss, A., Prasad, P.V.D., Ghosh, S.S., Chattopadhyay, A., The antibacterial properties of a novel chitosan-Ag-nanoparticle composite. *Int. J. Food Microbiol.*, 24, 2, 142–146, 2008, doi: 10.1016/J.IJFOODMICRO.2008.03.004.
109. Ganduri, V.S.R.K., Mangamuri, U., Muvva, V., Poda, S., Pullulan-stabilized silver nanoparticles-their synthesis, characterization and application as bactericidal agents. *J. Appl. Pharm. Sci.*, 6, 7, 027–037, 2016, doi: 10.7324/JAPS.2016.60704.
110. Grenha, A. and Rodrigues, S., Pullulan-based nanoparticles: Future therapeutic applications in transmucosal protein delivery. *Ther. Deliv.*, 4, 11, 1339–1341, 2013, doi: 10.4155/TDE.13.99.
111. Dionísio, M., Cordeiro, C., Remuñán-López, C., Seijo, B., da Costa, A.M.R., Grenha, A., Pullulan-based nanoparticles as carriers for transmucosal protein delivery. *Eur. J. Pharm. Sci.*, 50, 1, 102–113, 2013, doi: 10.1016/J.EJPS.2013.04.018.
112. Patra, J.K. *et al.*, Nano based drug delivery systems: Recent developments and future prospects. *J. Nanobiotechnology*, 16, 1, 1–33, 2018, doi: 10.1186/s12951-018-0392-8.

113. Chakravarthi, P.V. and Balaji, S.N., Applications of nanotechnology in veterinary medicine. *Vet. World*, 3, 10, 477–480, 2010, doi: 10.5455/vetworld.2010.477-480.
114. Manuja, A., Kumar, B., Singh, R.K., Nanotechnology developments: Opportunities for animal health and production. *Nanotechnol. Dev.*, 2, 1, 4, 2012, doi: 10.4081/nd.2012.e4.

14

Regulatory Pathways for Nanocarrier Vaccine

Niva Rani Gogoi[1], Rajashri Bezbaruah[1,2*], Vishwa Patel[3], Riyansi Satasia[3], Bedanta Bhattacharjee[4] and Bhaskar Mazumder[1]

[1]Department of Pharmaceutical Sciences, Faculty of Science and Engineering, Dibrugarh University, Dibrugarh, Assam, India
[2]Institute of Pharmacy, Assam Medical College and Hospital, Dibrugarh, Assam, India
[3]L. M. College of Pharmacy, Ahmedabad, Gujarat, India
[4]Department of Pharmacology, Girijananda Chowdhary Institute of Pharmaceutical Sciences, Tezpur, Sonitpur, Assam, India

Abstract

Although vaccines have been proven to possess remarkable effectiveness over the last centuries, the conventional technique is solely based on attenuated microorganisms. It has evolved from live attenuated microorganisms to toxoids, each with its own set of benefits and limitations. Nano-vaccination is currently the most efficient strategy to eradicate infections. In the next few decades, nanotechnology-based vaccination is projected to have a tremendous influence on diagnosis, mitigation, and treatment. Also, since 2020, a tremendous increase in the number of authorized nanocarrier vaccines in the market have been noticed. The nanovaccine (NV) market is majorly dominated by polymeric nanoparticles, lipid-based nanoparticles, virosomes, proteosomes, nanoemulsions, nanocrystals, polymeric-protein conjugates, and nanoconjugates. Although there is a lot of anticipation in the field of NV and its impact on the pharmaceutical industries, there exists very minimal regulatory counseling in this field despite a continuous appeal from the research community. The global approval of nanocarrier vaccines is strictly regulated following applicable literature, legislation, and guidelines to guarantee that suitable procedures are followed to assure product safety. This chapter aims to explore the existing regulatory requirements and the regulatory flow pathways for the development of nanocarrier vaccines, requirements and

**Corresponding author*: rajashribezbaruah@dibru.ac.in

Vivek P. Chavda and Vasso Apostolopoulos (eds.) Nanocarrier Vaccines: Biopharmaceutics-Based Fast Track Development, (465–486) © 2024 Scrivener Publishing LLC

challenges faced in regulating NVs, and the present regulation employed in various countries for their clinical and marketing approval.

***Keywords*:** Nanocarrier vaccine, nanovaccine, vaccine delivery, regulatory framework, regulatory challenges

14.1 Introduction

In the 21st century, infectious diseases have become severe health hazard to millions of people globally and have accounted for thousands of fatalities posing a massive devastating impact on socio-economic and healthcare sectors [1]. A well-designed vaccine delivery system and the selection of an appropriate administration route, are critical for boosting immunization efficacy and vaccine stability and can aid in regulating dose frequency and patient convenience [2]. Today's vaccination influence is extensive and far-reaching, since it is a critical public health tool with enormous societal advantages, also, heavily influencing the global economy and even influencing the biodefense tactics [2, 3]. Therefore, the development of nanocarrier vaccine is among the most successful approaches that have contributed to mitigating numerous diseases.

Moreover, in terms of efficiency, nanocarrier vaccine outperforms conventional vaccine. Undoubtedly, NV is a promising tool in terms of the effective immunity it may provide; yet, it is critical to examine the risks associated with its development [4, 5]. Also, the precise method of interaction of NVs with biological systems is yet undetermined. Concerns about the biosafety of NVs have resulted in the formation of the discipline of nanotoxicology and approaches for investigating the nanotoxicity of NVs. Similarly, assessing the double-use potential of NVs is critical for biosafety and biosecurity [6]. As a result, assessing the risks and impacts of NVs becomes important.

Many assessments have analyzed the applicability of present regulatory regulations for nanotechnology-based vaccines in general and their use in healthcare; nevertheless, no nanoformulation-specific legislation has been adopted as of now [7]. But current regulatory practice requires further guidance to adequately encompass the specificities of NVs. A nanotechnology regulatory framework for nanocarrier vaccine might serve as a blueprint for NV governance [8]. It also poses unusual challenges and necessitates the establishment of a particular governance structure since it is intrinsically associated with human health. Moreover, present drug and medical device laws are neither NV-specialized nor sufficient to handle the complex concerns of NVs [9]. This contributes to the regulatory uncertainties impacting NV research and development as well as its commercialization. As a result, having a distinct regulatory route is crucial, and addressing issues at the appropriate time is critical. The universal concerns related to NV include biosafety, toxicity, and accessibility; however, there exists a few nation-specific challenges as well. Since the law is territorial, a nation-specific regulatory structure is highly required. The United States (US), the United Kingdom (UK), Russia, Canada, Australia, and Europe are all currently engaged with the framing of NV regulatory and risk assessment guidelines [10]. The US Food and Drug Administration (FDA) and the European Medicines Agency (EMA) have established guidelines for firms developing nanocarrier vaccines [11]. It is worth noting that Asian nations such as India, Taiwan, Thailand, China, and Japan are defining appropriate governance frameworks and developing regulatory laws to handle concerns associated with NV and its uses [12].

Keeping everything in mind, the authors have attempted to investigate the regulatory initiatives and gaps in the existing regulatory system globally. The current study recommends the development of a region-specific regulatory framework incorporating numerous stakeholders worldwide that plays a vital role in the long-term development of this novel concept of nanocarrier vaccination. Several approaches for the regulation of nanocarrier vaccines are covered in the following sections.

14.2 The Need for a Regulatory Framework

Since the advent of NV, the use of various nanomaterials has proliferated in a full range of immunization. Several characteristics of NV regarding high payload, low dose, and targeting ability make them attractive candidates for efficient immunization, but an important aspect to be considered clinically is that these NVs remain immune toxic [13]. For *in vivo* applications, the interface between NV and biological agents plays a fundamental role in the fate of nanomaterials in effective delivery, clearance, and accumulation.

Moreover, despite the recent advancements in the emerging era of nanocarrier vaccines, loopholes persist in the regulatory guidelines for the same. The initial contact of NVs with biological fluids exposes them to various proteins, forming a crown effect commonly referred to as protein corona particles [14]. Other biomolecules such as DNA, RNA, and ribose sugars also interact with NVs, but they are underrepresented in the literature and the resulting scientific investigation. Interactions of NVs with plasma proteins and other biomolecules alter their fate, as various biochemical changes alter the bioactivity of NVs [15]. Most of the NVs function by directly interacting with the genetic makeup of the host or with the biomolecules essential for normal genomic function, all of which have the potential of causing mutagenicity and genotoxicity [16]. Such toxicity arbitrates to inflammatory responses that lead to the extensive production of reactive nitrogen and oxygen species, which ultimately causes nitrosative and oxidative stress. The deposition of such radicals in the body can cause extensive damage.

Adverse interactions between the immune system and NVs have often been reported for immunostimulatory or immunosuppressive purposes, leading to inflammatory or autoimmune diseases, thereby triggering the receptor body [17, 18]. Also, there are just too many unknown factors that may represent potential hazards for NVs that become difficult to trace after administration. But, without uniform regulatory guidelines in this domain, little is likely to change. In the past, regulatory agencies were required to be attentive after market approval. An inactivated vaccine with adjuvant enterotoxin was withdrawn from the market after it considerably caused Bell's palsy [19–21]. Such instances signify the need for well-documented regulatory norms following safety review procedures throughout the NV development method, as shown in Figure 14.1. Another element to consider regarding the impact of NVs is their potential environmental impact after usage, disposal, and manufacturing. It is accepted that conventional vaccines are eventually

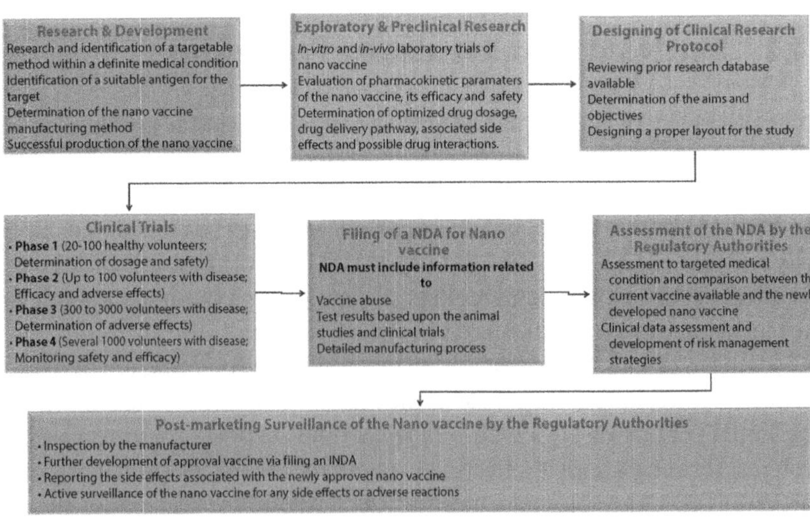

Figure 14.1 Flow diagram depicting the authorities involved in the approval process for the nanocarrier vaccine; starting from the development of the nanocarrier vaccine to the final product in the market to its post-market surveillance.

recovered in the environment, and it is believed that NVs would behave similarly, raising the possibility of NVs negatively impacting the environment. However, the regulatory authorities cite a lack of evidence to establish the biosafety of NVs both to humans and to the environment. This has a detrimental influence on NV development and inhibits the future usage of nanoscale medications since uncertainty affects future financing, research and development, and also public acceptability. All these factors demand a need for properly framed regulatory guidelines for NVs.

14.3 Regulatory Requirements for the Manufacturing of NVs

There are many quality, safety, and efficacy-related regulatory requirements for the production of NVs, beginning with the early phases of development and concluding with storage and commercialization. In this section, we shall focus on these requirements, primarily for downstream and upstream manufacturing of NV research and development.

14.3.1 Upstream Manufacturing Process

For upstream processing, the production method of NVs is similar to that of the traditional vaccine production method [22]. Once the immunogen synthesis process has been determined, the antigen for the NV may be developed and thoroughly tested to assure purity and safety [23].

Depending on the starting material, the virus can be grown on primary cell lines, continuous cell lines, or chicken eggs [24]. Each of these manufacturing systems has advantages and disadvantages. Consequently, the following factors should be examined during development and selection: yields, impacts of post-translational modifications, cost, contamination concerns, scalability, complexity, manufacturing timeline, glycosylation profiles, and regulatory approval frameworks [25, 26].

Further modification of the immunogen along with encapsulation of the immunogen into a nanocarrier system can be contemplated in the downstream processing.

14.3.2 Downstream Manufacturing Process

Numerous additional ways for producing and incorporating the immunogen into a nanocarrier system can be utilized; however, factors like consistency, scalability, environmental factors, and the use of organic solvents should be addressed throughout the manufacturing process [27]. When the immunogen has been loaded into the nanocarrier, size modification is frequently performed by a method like homogenization, which employs shear pressures to ensure a uniform size distribution [28].

Depending on the findings of their stability tests, intermediate products can now be kept at low temperatures [29]. As a result, the thermostability of the nanocarrier vaccines during different periods should be considered throughout the validation and certification of the NV and the process of manufacturing [27, 30]. Stability tests are essential for physicochemical parameters and biological testing on both the intermediate and the final product [31]. A stability protocol's implementation should be based on thorough information regarding the tests performed.

14.4 Clinically Approved Nanocarrier Vaccines

In general, vaccine research and development must undergo multiple rounds of preclinical and clinical screening before being approved for manufacture and sale. The preclinical stage of the NV development process is where the infectious agent is intensively examined for the immunogenic antigen that can elicit immune responses in the recipient [1, 32]. The outcome of the preclinical investigations is evaluated by the regulators, who authorize and allow the developer to carry out clinical trials only if the advantages of the NV surpass the risks and unwanted adverse reactions [33]. Clinical trials entail examining the impact of NVs in development on human participants in three stages [34]. Phase I entails testing the NV on a limited population of healthy volunteers to determine that the developed product is free of serious safety problems, as well as evaluating the dose range and the evoked immune response. In Phase II trials, a pilot effectiveness study for a larger population volunteers are conducted to validate safety. If the vaccine under study shows effectiveness and a minimal risk of previously reported toxicity, it will proceed towards the third phase of clinical testing. Phase III trials involve a larger population group, often hundreds or thousands of participants from different regions with increased viral transmission rates, elderly people, and people who have medical histories, in order to confirm safety, efficacy, and the effective dosage levels of vaccine in these groups. Successful NVs from Phase 3 can apply for marketing authorization from the regulatory bodies for mass manufacturing and distribution. Once the NVs are approved for marketing, they enter the Phase IV of clinical trials where they will be regularly and meticulously assessed for safety and any kind of adverse drug reactions [35]. In emergency pandemic situations like COVID-19, full approval of a vaccine or an NV is not done but is authorized to manufacture under "emergency use authorization" to expedite its accessibility and usage for immunization, even if clinical studies are still ongoing [36]. For COVID, currently, eight vaccines are undergoing Phase IV clinical trials, including mRNA-1273 and BNT162b2 of Moderna and Pfizer, respectively, both of which are lipid nanoparticle-based nanocarrier vaccines [37, 38].

Despite the limitations associated with conventional vaccines, which include tedious and time-consuming manufacturing processes, lesser effectivity, and higher toxicity, there exist only a few fully approved vaccines that use nanocarriers for their manufacturing and delivery [39]. Traditionally, virosome nanocarriers have demonstrated clinical effectiveness with HBV and HAV vaccines, in addition to influenza and lately for preventive HPV vaccination [40]. Nevertheless, Shingrix®, a zoster vaccine produced by the company GlaxoSmithKline, is among the few clinically authorized NV that employs liposomes to deliver the antigen [41–43]. As a result, there exists a lot of work to be done to develop clinically approved NVs that could meet the demanding quality and safety standards of the regulatory authorities. The recent development of the lipid-based nanoparticle for SARS-CoV-2 in Phase IV of clinical trials has opened up new avenues of enhanced nanotechnological techniques for the development of more NVs [44].

Along with various nanotechnology-based methods for the development of NV, microneedle-based methods for immunization following the transdermal route are also now under clinical trials [45–47]. Earlier studies have also established the clinical effectiveness of microneedle-based vaccination, notably of an influenza vaccine encased with a microneedle array developed by the Prausnitz group of Georgia Institute of Technology [48].

The implementation of nanotechnologies is a potential approach for the effective and targeted vaccine delivery, that is currently being thoroughly tested for SARS-CoV-2 but could be effectively applied for the development of other NVs in the coming years. Nevertheless, a regulatory framework for clinical and marketing authorization challenges for NVs is being developed, which must be considered from the beginning of research and development (Table 14.1) [49].

Table 14.1 List of NVs approved for clinical use by the regulatory authorities.

NV name/type	Target disease/ virus	Developer company	Sponsor	Clinical trial phase	NCT number
VLP plant	Influenza	Medicago	-	I, II	NCT04622592
VLP plant	SARS-CoV-2	Medicago	-	II, III	NCT04636697
VLP recombinant	Encephalitis	SRI International	-	I	NCT03776994
VLP recombinant	Chikungunya	Emergent Biosolutions	-	II, III	NCT05065983, NCT05072080
OMV	Gonorrhea	NIAID	-	II	NCT04722003, NCT04350138
Nanocage	Influenza	NIAID	-	I	NCT04579250
Nanocage	Epstein-Barr virus	NIAID	-	I	NCT04645147
Nanocage	SARS-CoV -2	U.S. Army	-	I	NCT04784767
-	Melanoma	-	University of Florida	I	NCT05264974

(Continued)

Table 14.1 List of NVs approved for clinical use by the regulatory authorities. *(Continued)*

NV name/type	Target disease/virus	Developer company	Sponsor	Clinical trial phase	NCT number
-	COVID-19 (healthy volunteers)	-	Korea University of Hong Kong	II	NCT05175950
-	SARS-CoV -2	-	The University of Hong Kong	IV	NCT05057182
mRNA-123	SARS-CoV -2	Moderna	-	Emergency use authorization	-
-	Immune response of 25 ug of ChulaCov19-BNA159 mRNA vaccine	Chulalongkorn University Chula Clinical Research Center	Faculty of Medicine, Chulalongkorn University	I	NCT05231369
-	SARS-CoV -2 (Booster)	Murdoch Children's Research Institute Coalition for Epidemic Preparedness Innovations	The Peter Doherty Institute for Infection and Immunity	III	NCT05658523

14.5 Regulatory Challenges

An NV's lifespan may be split into seven primary stages, namely conceptualization, scientific research, NV development and optimization, intellectual property protection, clinical trials (safety evaluations), commercialization, and post-market vigilance [50, 51]. However, even though it has been more than 15 years since the FDA authorized the first albumin-based nano medication, Abraxane®, for usage in 2005, lately, NV continues to pose regulatory challenges as shown in Figure 14.2 [52, 53]. Both NVs and nanomedicines are regulated within the existing regulatory guidelines for small drug molecules and medical devices, but it may necessitate further safety and quality evaluations due to the unique characteristics of nanocarrier vaccines. Identification and universal agreement on regulatory standards relevant to NV evaluation are, therefore, a prerequisite for a smooth approval procedure of an NV [17, 50]. The primary purpose of the regulatory bodies is to ensure efficacy, safety, and quality following a well-established regulatory framework; however, as scientific breakthroughs and market expectations grow, setting explicit rules becomes increasingly challenging.

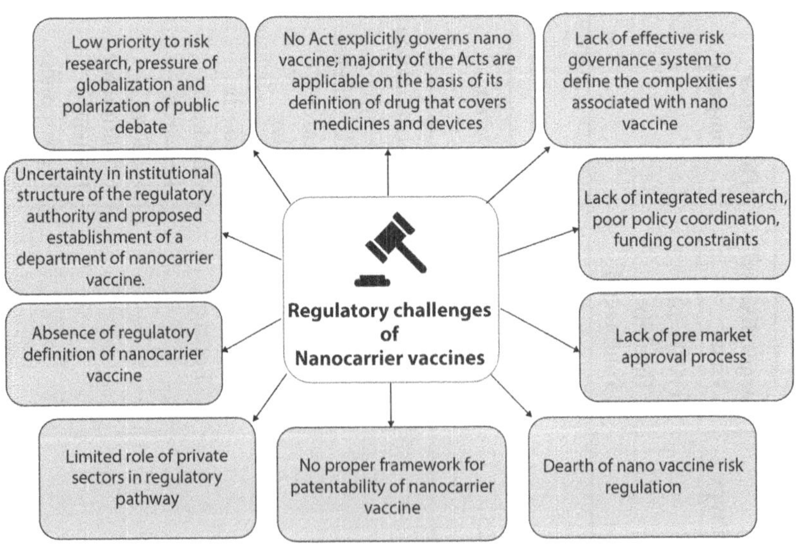

Figure 14.2 The major regulatory challenges faced throughout the process of research and development of nanocarrier vaccine to its marketing, followed by post-marketing vigilance.

14.5.1 Existing Regulatory Guidelines are not Adequate for the Regulation of Nanocarrier Vaccines

Presently, very little guidance is provided due to a lack of comprehensive datasets that allow for a definite conclusion on the information required for regulatory decisions. Moreover, the rising complexity and diversity of the next generation of NVs may necessitate a more specialized direction. The lack of rules and standardization from various regulatory agencies creates significant ambiguity for product developers, impeding the development and commercialization of breakthrough NVs [54, 55]. This scenario creates significant ambiguity for product developers, and the lack of such guidelines to demonstrate regulatory compliance may impede the development and commercialization of NV [56, 57]. At the same time, high-quality data on safety, effectiveness, and quality for NV, which are required for improved regulatory definition, will not be developed.

To ensure successful approval of NVs, the identification and agreement on the regulatory requirements relevant to the NV are hence a precondition. In general, regulatory authorities seek the drug release and stability profile from the pilot batches that were manufactured following the same techniques as that of the NV intended for marketing. Additionally, they also need detailed information on all the excipients used, with a clear explanation of the manufacturing process employed, along with a justification for the same. A rationale for the NV used in the preclinical studies from animal models assessing the immune response for every individual component of the NV and also for that of the final NV should be provided. Also, toxicological findings from clinical studies would be required to prove safety. These findings should include complete data from preclinical studies on local tolerance and repeated dosage toxicity.

14.5.2 Lack of an Appropriate Risk Governance Organization

Proper characterization of nanocarrier vaccine for risk and biosafety assessment necessitates the provision of dependable and well-grounded procedures to address the regulatory demands of nanocarrier vaccine [58]. However, due to the distinctive properties of NV, not all conventional methods can be applied to provide reliable results for risk assessment [59]. Since standardized methods do not apply to the governance of risk assessment, there is a necessity to use extremely specific and complex methodologies that are still under development.

Furthermore, the applicability of approaches specific to NVs might vary based on the excipient used, the antigen incorporated, and the dispersion characteristics [60]. Inadequate procedures might result in an inaccurate outcome that demands risk governance and regulatory intervention.

Moreover, particle contamination with microbial endotoxin is difficult to quantify and has a significant influence on toxicological study outcomes. As a result, current techniques should be assessed for their applicability to test NVs.

14.5.3 No Act is Nanovaccine-Specific Framed

Numerous legislations, rules, and regulatory frameworks available are for small drug molecules, therapeutic agents, and medical devices, no law or act is specifically designed for NVs [61, 62]. All acts are applied to medications that include active pharmaceutical ingredients and technologies used for diagnosis, mitigation, and treatment.

14.5.4 Lack of Proper Regulatory Framework for Patentability

A precise definition of NV does not exist and this has posed challenges to scientists in patent offices when filing new patents for NV [63]. The absence of a uniform definition has consequences for patent search and categorization, as well as for tracking the NV patenting trends [64]. This increases the likelihood that an NV patent will be invalidated and that similar or conflicting patents will be issued [65, 66]. The situation is further aggravated by the inclusion of vague or imprecise terminology, which is confusing, thus making it challenging for the patent examiners to judge how an intervention is different and novel from the previous patents.

14.6 Global Strategies for Clinical Approval

Regulatory authorities develop specialized interdisciplinary knowledge for evaluating NVs based on the latest risk–benefit analysis approaches. A few such strategies for the clinical approval of NVs are discussed in the following section.

14.6.1 The United States of America

A competitive organization on nanotechnology was established by the US Center for Drug Evaluations and Research (CDER) in 2007 to promote the use of nanomaterials in biotherapeutics, diagnosis, and food items [67].

Since then, the task force has released three significant guidelines that are centered on the evaluation and approval of nanomedicine. Yet, all the procedures created for the production of nanostructures, pharmaceutical development, clinical development, and post-marketing evaluations are predicated on the supposition that various toxicity studies in vaccines based on nanocarriers will be evaluated to accomplish a toxicology study [68]. In 2017, the FDA and the Nanotechnology Task Force released their report guideline outlining the regulatory relevant safety, effectiveness, and risk quality concerns for nanocarrier vaccines [69]. The final guideline was published earlier this year after asking for stakeholder feedback on the original proposal. The recently released guideline includes numerous noteworthy new recommendations [70]. The FDA has listed several significant definitions and classifications pertaining to the development of industry awareness of key concepts in nanotechnology [71]. In-depth manufacturing, non-clinical testing, clinical development, and environmental effect based on nanomaterials are also covered in the document. The fact that this guideline includes a thorough risk assessment framework is particularly noteworthy. It recognizes how even slight changes to the structure, size, and route of administration of drugs based on nanocarriers can have a significant impact on their bioequivalence, pharmacokinetics, and toxicology. Also, the FDA categorized nanocarrier drugs as high, medium, or low risk, with associated risk reduction plans for each class [72]. The FDA guideline also describes the efficacy and safety factors for generic and biosimilar nanocarrier medications. To accomplish this, the regulations outline how manufacturers can show bioequivalence by comparing the *in vivo* pharmacokinetics and *in vitro* physiochemical properties of the generic medication and its equivalent reference mentioned to the nanocarrier vaccination.

14.6.2 The European Union

Even if the nanocarrier material employed for NVs has already received approval, registration with the appropriate authority for NVs necessitates a complete set of preclinical and clinical investigations because they are all regarded as novel entities [73]. This is justified by the fact that NVs may be utilized to alter a drug's bioavailability, which may then alter its pharmacokinetic and pharmacodynamic profiles and, ultimately, its safety. Antigen-loaded NVs are regarded as non-biological complicated medicines (NBCDs) because of their complex structure [71]. The effectiveness and safety of this drug class depend on the interaction between the medicine and the nano biomaterial, which together constitute the active pharmacological ingredient. Similar to their biological counterparts (such as therapeutic proteins), NBCDs cannot be fully characterized; as a result, it appears difficult to produce and register "follow-on" drug nanomedicines as generics [74]. Although these follow-on nanomedicine products have previously received marketing authorizations through the generic pathway, discussions are still ongoing among stakeholders about establishing a regulatory strategy for "nano similar" products that would be similar to the biosimilars for complex biological drugs. Future follow-on nanomedicine development and commercialization will have an extra challenge as a result of the need for studies that go beyond the demonstration of bioequivalence between the original drug and the planned "nano similar" product [75].

14.6.3 The United Kingdom

The Medicines and Healthcare Products Regulatory Authority (MHRA) in the United Kingdom (UK) oversees the regulation of all medications, including NVs [76, 77]. The UK lacks particular regulations for items based on nanotechnology, much like in the EU. The MHRA handles the approval of NVs on a case-by-case basis, and researchers are urged to get in touch with them for assistance at any stage of the development process [78, 79].

14.6.4 Canada

Based on the current recommendations made by the Organization for Economic Co-operation and Development (OECD) Council for the safety assessment of nanomaterials, Canada regulates products based on nanotechnology. Regarding the EU and the USA, Health Canada evaluates the risks and characteristics of NVs to assist and guide manufacturers throughout the development process. The Canadian Health Portfolio Nanotechnology Working Group was created to assist regulatory organizations (such as Health Canada and the Canadian Institutes of Health Research) in discussing and classifying concerns related to nanotechnology. To date, Health Canada has released preliminary guidelines for meals and health products based on nanotechnology.

14.6.5 Japan

The Ministry of Health, Labour and Welfare (MHLW) and the Pharmaceuticals and Medical Devices Agency are in charge of this procedure in Japan, and in 2016, they published detailed rules for the regulation of therapeutic products based on liposome [80, 81]. Similar to the USA and EU, Japan's Pharmaceutical Affairs Law framework closely collaborates with the EMA to govern NVs on a case-by-case basis.

14.6.6 Other Countries

Regulations pertaining to NVs are now being developed in India, China, and Thailand. A working group for regulating and developing rules for products based on nanotechnology has been established in India by the government and the Department of Science and Technology [82, 83]. The development of new medications and their comparison with already existing entities were included in the first nanomedicine regulation guideline published by the Indian government in 2019.

14.7 Conclusion and Future Prospects

Regardless of the absence of precise regulatory guidelines, nearly 15 NVs have entered the market, and this number is continuously increasing. A global concern is the lack of official regulation of NV synthesis for infectious diseases. Because of discrepancies among regulatory bodies, certain NVs are classified as simply vaccines while others are classified as nanomedicines. To date, the establishment of functional regulatory groups has not resulted in action; nanotechnology is new, and the need and urgency for NVs for specific infectious diseases cannot be provided under the current regulatory system.

While some attempts have been made by academic groups and government organizations to establish Nanotechnology Characterization Laboratory, more specific and strict vigilance and guidelines are still required from major regulating authorities such as the FDA and MHRA. Many illnesses do not discriminate based on race or geography, hence a worldwide collaboration for NV regulation should be established to advance these objectives and provide official direction to the research community. During the last decade, millions of dollars have been invested in NV development; but, without clear guidelines from regulatory agencies, these initiatives would not result in NVs reaching the market, and future growth will be directed elsewhere.

References

1. Bloom, D.E. and Cadarette, D., Infectious disease threats in the twenty-first century: Strengthening the global response. *Front. Immunol.*, 10, 549, Mar. 2019. doi: 10.3389/fimmu.2019.00549.
2. Lemoine, C. *et al.*, Technological approaches for improving vaccination compliance and coverage. *Vaccines (Basel)*, 8, 2, 304, Jun. 2020. doi: 10.3390/vaccines8020304.

3. Rodrigues, C.M.C. and Plotkin, S.A., Impact of vaccines; health, economic and social perspectives. *Front. Microbiol.*, 11, 1526, Jul. 2020. doi: 10.3389/fmicb.2020.01526.
4. Feng, C. *et al.*, Emerging vaccine nanotechnology: From defense against infection to sniping cancer. *Acta Pharm. Sin. B*, 12, 5, 2206–2223, May 2022. doi: 10.1016/j.apsb.2021.12.021.
5. Celis-Giraldo, C.T., López-Abán, J., Muro, A., Patarroyo, M.A., Manzano-Román, R., Nanovaccines against animal pathogens: The latest findings. *Vaccines (Basel)*, 9, 9, 988, Sep. 2021. doi: 10.3390/vaccines9090988.
6. Khoury, D.S. *et al.*, Neutralizing antibody levels are highly predictive of immune protection from symptomatic SARS-CoV-2 infection. *Nat. Med.*, 27, 7, Art. no. 7, 1205–1211, Jul. 2021. doi: 10.1038/s41591-021-01377-8.
7. Mishra, A.K., Das, R., Sahoo, S., Biswal, B., Chapter ten - global regulations and legislations on nanoparticles usage and application in diverse horizons, in: *Comprehensive Analytical Chemistry, Environmental Nanotechnology: Implications and Applications*, vol. 99, N.B. Turan, G.O. Engin, M.S. Bilgili (Eds.), pp. 261–290, Elsevier, Edinburgh, 2022. doi: 10.1016/bs.coac.2021.12.004.
8. Allan, J. *et al.*, Regulatory landscape of nanotechnology and nanoplastics from a global perspective. *Regul. Toxicol. Pharm.*, 122, 104885, Jun. 2021. doi: 10.1016/j.yrtph.2021.104885.
9. Baran, A., Nanotechnology: Legal and ethical issues. *Ekonomia i Zarzadzanie*, 8, 1, 47–54, Mar. 2016. doi: 10.1515/emj-2016-0005.
10. Falkner, R. and Jaspers, N., Regulating nanotechnologies: Risk, uncertainty and the global governance gap. *Glob. Environ. Polit.*, 12, 1, 30–55, 2012.
11. O. of the Commissioner, Nanotechnology programs at FDA, FDA, Feb. 23, 2021, https://www.fda.gov/science-research/science-and-research-special-topics/nanotechnology-programs-fda (accessed Apr. 09, 2023).
12. Tsai, T.F., Rao, R.D.S.V., Xu, Z.Y., Immunization in the Asia-Pacific Region, in: *Plotkin's Vaccines*, pp. 1466–1485.e2, 2018. doi: 10.1016/B978-0-323-35761-6.00075-4.
13. Smith, D.M., Simon, J.K., Baker, J.R., Applications of nanotechnology for immunology. *Nat. Rev. Immunol.*, 13, 8, 592–605, Aug. 2013. doi: 10.1038/nri3488.
14. Park, Y.-M. *et al.*, Nanoparticle-based vaccine delivery for cancer immunotherapy. *Immune Netw.*, 13, 5, 177, 2013. doi: 10.4110/in.2013.13.5.177.
15. Azharuddin, M., Zhu, G.H., Sengupta, A., Hinkula, J., Slater, N.K.H., Patra, H.K., Nano toolbox in immune modulation and nanovaccines. *Trends Biotechnol.*, 40, 10, 1195–1212, Oct. 2022. doi: 10.1016/j.tibtech.2022.03.011.
16. Gulla, S.K. *et al.*, *In vivo* targeting of DNA vaccines to dendritic cells using functionalized gold nanoparticles. *Biomater. Sci.*, 7, 3, 773–788, Feb. 2019. doi: 10.1039/C8BM01272E.
17. Milane, L. and Amiji, M., Clinical approval of nanotechnology-based SARS-CoV-2 mRNA vaccines: Impact on translational nanomedicine. *Drug Deliv. Transl. Res.*, 11, 4, 1309–1315, Aug. 2021. doi: 10.1007/s13346-021-00911-y.

18. Friedrichs, S. and Bowman, D.M., COVID-19 may become nanomedicine's finest hour yet. *Nat. Nanotechnol.*, 16, 4, 362–364, Apr. 2021. doi: 10.1038/s41565-021-00901-8.
19. Bell's Palsy following intranasal vaccination, https://www.who.int/groups/global-advisory-committee-on-vaccine-safety/topics/bell-s-palsy-following-intranasal-vaccination (accessed Apr. 09, 2023).
20. Valli, E. *et al.*, LTA1 is a safe, intranasal enterotoxin-based adjuvant that improves vaccine protection against influenza in young, old and B-cell-depleted (μMT) mice. *Sci. Rep.*, 9, 1, 15128, Art. no. 1, Oct. 2019. doi: 10.1038/s41598-019-51356-w.
21. Stephenson, I. *et al.*, Phase I evaluation of intranasal trivalent inactivated influenza vaccine with nontoxigenic escherichia coli enterotoxin and novel biovector as mucosal adjuvants, using adult volunteers. *J. Virol.*, 80, 10, 4962–4970, May 2006. doi: 10.1128/JVI.80.10.4962-4970.2006.
22. Cid, R. and Bolívar, J., Platforms for production of protein-based vaccines: From classical to next-generation strategies. *Biomolecules*, 11, 8, 1072, Jul. 2021. doi: 10.3390/biom11081072.
23. Charlton Hume, H.K. and Lua, L.H.L., Platform technologies for modern vaccine manufacturing. *Vaccine*, 35, 35, 4480–4485, Aug. 2017. doi: 10.1016/j.vaccine.2017.02.069.
24. Fang, Z. *et al.*, Application of bioreactor technology for cell culture-based viral vaccine production: Present status and future prospects. *Front. Bioeng. Biotechnol.*, 10, 921755, 2022, Accessed: Apr. 09, 2023. [Online]. Available: https://www.frontiersin.org/articles/10.3389/fbioe.2022.921755.
25. Nooraei, S. *et al.*, Virus-like particles: Preparation, immunogenicity and their roles as nanovaccines and drug nanocarriers. *J. Nanobiotechnol.*, 19, 1, 59, Feb. 2021. doi: 10.1186/s12951-021-00806-7.
26. Tenchov, R., Bird, R., Curtze, A.E., Zhou, Q., Lipid nanoparticles–from liposomes to mRNA vaccine delivery, a landscape of research diversity and advancement. *ACS Nano*, 15, 11, 16982–17015, Nov. 2021. doi: 10.1021/acsnano.1c04996.
27. Suk, J.S., Xu, Q., Kim, N., Hanes, J., Ensign, L.M., PEGylation as a strategy for improving nanoparticle-based drug and gene delivery. *Adv. Drug Delivery Rev.*, 99, Pt A, 28, Apr. 2016. doi: 10.1016/j.addr.2015.09.012.
28. Pati, R., Shevtsov, M., Sonawane, A., Nanoparticle vaccines against infectious diseases. *Front. Immunol.*, 9, 2224, 2018, Accessed: Apr. 09, 2023. [Online]. Available: https://www.frontiersin.org/articles/10.3389/fimmu.2018.02224.
29. *Guidance for the validation of analytical methodology and calibration of equipment used for testing of illicit drugs in seized materials and biological specimens.* United Nations Publications, New York, 2009.
30. Solaro, R., Chiellini, F., Battisti, A., Targeted delivery of protein drugs by nanocarriers. *Mater. (Basel)*, 3, 3, 1928–1980, Mar. 2010. doi: 10.3390/ma3031928.

31. Herdiana, Y., Wathoni, N., Shamsuddin, S., Muchtaridi, M., Scale-up polymeric-based nanoparticles drug delivery systems: Development and challenges. *OpenNano*, 7, 100048, Jul. 2022. doi: 10.1016/j.onano.2022.100048.
32. Fifis, T. *et al.*, Size-dependent immunogenicity: Therapeutic and protective properties of nano-vaccines against tumors. *J. Immunol.*, 173, 5, 3148–3154, Sep. 2004. doi: 10.4049/jimmunol.173.5.3148.
33. Tian, Y. *et al.*, The novel complex combination of alum, CpG ODN and HH2 as adjuvant in cancer vaccine effectively suppresses tumor growth *in vivo*. *Oncotarget*, 8, 28, 45951–45964, Jul. 2017. doi: 10.18632/oncotarget.17504.
34. Clinical evaluation of vaccines, https://www.who.int/teams/health-product-policy-and-standards/standards-and-specifications/vaccine-standardization/clinical-evaluation-of-vaccines (accessed Apr. 09, 2023).
35. How vaccines are developed and approved for use, CDC, Mar. 30, 2023, https://www.cdc.gov/vaccines/basics/test-approve.html (accessed Apr. 09, 2023).
36. Health Institutes of Turkey, Randomized, double-blind, placebo-controlled phase III clinical trial for evaluation of efficacy and safety of SARS-CoV-2 vaccine (vero cell), inactivated, clinicaltrials.gov, Clinical trial registration NCT04582344, Aug. 2022, Accessed: Apr. 06, 2023. [Online]. Available: https://clinicaltrials.gov/ct2/show/NCT04582344.
37. The Moderna COVID-19 (mRNA-1273) vaccine: What you need to know, https://www.who.int/news-room/feature-stories/detail/the-moderna-covid-19-mrna-1273-vaccine-what-you-need-to-know (accessed Apr. 09, 2023).
38. Simnani, F.Z., Singh, D., Kaur, R., COVID-19 phase 4 vaccine candidates, effectiveness on SARS-CoV-2 variants, neutralizing antibody, rare side effects, traditional and nano-based vaccine platforms: A review. *3 Biotech.*, 12, 1, 15, Jan. 2022. doi: 10.1007/s13205-021-03076-0.
39. Ogholikhan, S. and Schwarz, K.B., Hepatitis vaccines. *Vaccines*, 4, 1, 6, Art. no. 1, Mar. 2016. doi: 10.3390/vaccines4010006.
40. Qian, C. *et al.*, Recent progress on the versatility of virus-like particles. *Vaccines*, 8, 1, 139, Art. no. 1, Mar. 2020. doi: 10.3390/vaccines8010139.
41. Shingrix approved in the US for prevention of shingles in adults aged 50 and over, GSK, Oct. 23, 2017, https://www.gsk.com/en-gb/media/press-releases/shingrix-approved-in-the-us-for-prevention-of-shingles-in-adults-aged-50-and-over/ (accessed Apr. 09, 2023).
42. Irvine, D.J., Aung, A., Silva, M., Controlling timing and location in vaccines. *Adv. Drug Deliv. Rev.*, 158, 91–115, Jan. 2020. doi: 10.1016/j.addr.2020.06.019.
43. AboulFotouh, K., Xu, H., Williams, R.O., Cui, Z., Development of dry powder formulations of AS01B containing vaccines using thin-film freeze-drying. *bioRxiv*, 2022.02.27.482135, 622, 121825, Mar. 01, 2022. doi: 10.1101/2022.02.27.482135.
44. Wilson, B. and Geetha, K.M., Lipid nanoparticles in the development of mRNA vaccines for COVID-19. *J. Drug Deliv. Sci. Technol.*, 74, 103553, Aug. 2022. doi: 10.1016/j.jddst.2022.103553.

45. He, X., Sun, J., Zhuang, J., Xu, H., Liu, Y., Wu, D., Microneedle system for transdermal drug and vaccine delivery: Devices, safety, and prospects. *Dose Response*, 17, 4, 1559325819878585, Oct. 2019. doi: 10.1177/1559325819878585.
46. Alimardani, V. *et al.*, Microneedle arrays combined with nanomedicine approaches for transdermal delivery of therapeutics. *J. Clin. Med.*, 10, 2, 181, Jan. 2021. doi: 10.3390/jcm10020181.
47. D'Amico, C., Fontana, F., Cheng, R., Santos, H.A., Development of vaccine formulations: Past, present, and future. *Drug Deliv. Transl. Res.*, 11, 2, 353–372, Apr. 2021. doi: 10.1007/s13346-021-00924-7.
48. Esser, E.S. *et al.*, Microneedle patch delivery of influenza vaccine during pregnancy enhances maternal immune responses promoting survival and long-lasting passive immunity to offspring. *Sci. Rep.*, 7, 1, 5705, Art. no. 1, Jul. 2017. doi: 10.1038/s41598-017-05940-7.
49. Home - ClinicalTrials.gov, https://clinicaltrials.gov/ (accessed Apr. 09, 2023).
50. Ahmed, S. *et al.*, Vaccine development against COVID-19: Study from pre-clinical phases to clinical trials and global use. *Vaccines*, 9, 8, 836, Art. no. 8, Aug. 2021. doi: 10.3390/vaccines9080836.
51. Farlow, A. *et al.*, The future of epidemic and pandemic vaccines to serve global public health needs. *Vaccines*, 11, 3, 690, Art. no. 3, Mar. 2023. doi: 10.3390/vaccines11030690.
52. Kianfar, E., Protein nanoparticles in drug delivery: Animal protein, plant proteins and protein cages, albumin nanoparticles. *J. Nanobiotechnol.*, 19, 1, 159, Art. no. 1, Dec. 2021. doi: 10.1186/s12951-021-00896-3.
53. Gupta, N., Hatoum, H., Dy, G.K., First line treatment of advanced non-small-cell lung cancer - specific focus on albumin bound paclitaxel. *Int. J. Nanomed.*, 9, 209–221, 2014. doi: 10.2147/IJN.S41770.
54. Stern, A.D., Innovation under regulatory uncertainty: Evidence from medical technology. *J. Public Econ.*, 145, 181–200, Jan. 2017. doi: 10.1016/j.jpubeco.2016.11.010.
55. Kabir, E.R., Moreino, S.S., Sharif Siam, M.K., The breakthrough of biosimilars: A twist in the narrative of biological therapy. *Biomolecules*, 9, 9, 410, Aug. 2019. doi: 10.3390/biom9090410.
56. O. of the Commissioner, FDA's approach to regulation of nanotechnology products, FDA, Mar. 2019, Accessed: Apr. 09, 2023. [Online]. Available: https://www.fda.gov/science-research/nanotechnology-programs-fda/fdas-approach-regulation-nanotechnology-products.
57. World Trade Organization and World Health Organization, and World Intellectual Property Organization, *Promoting access to medical technologies and innovation: Intersections between public health, intellectual property and trade*, WTO, Geneva 21, Switzerland, 2013. doi: 10.30875/63a4aa65-en.
58. Ahmad, A., Imran, M., Sharma, N., Precision nanotoxicology in drug development: Current trends and challenges in safety and toxicity implications of customized multifunctional nanocarriers for drug-delivery

applications. *Pharmaceutics*, 14, 11, 2463, Art. no. 11, Nov. 2022. doi: 10.3390/pharmaceutics14112463.

59. Valdivia-Olivares, R.Y., Rodriguez-Fernandez, M., Álvarez-Figueroa, M.J., Kalergis, A.M., González-Aramundiz, J.V., The importance of nanocarrier design and composition for an efficient nanoparticle-mediated transdermal vaccination. *Vaccines*, 9, 12, 1420, Art. no. 12, Dec. 2021. doi: 10.3390/vaccines9121420.
60. Farjadian, F., Ghasemi, A., Gohari, O., Roointan, A., Karimi, M., Hamblin, M.R., Nanopharmaceuticals and nanomedicines currently on the market: Challenges and opportunities. *Nanomedicine (Lond)*, 14, 1, 93–126, Jan. 2019. doi: 10.2217/nnm-2018-0120.
61. Nelson, B.C., Minelli, C., Doak, S.H., Roesslein, M., Emerging standards and analytical science for nanoenabled medical products. *Annu. Rev. Anal. Chem. (Palo Alto Calif.)*, 13, 1, 431–452, Jun. 2020. doi: 10.1146/annurev-anchem-091619-102216.
62. Medical device & diagnostics, https://cdsco.gov.in/opencms/opencms/en/Medical-Device-Diagnostics/Medical-Device-Diagnostics/ (accessed Apr. 09, 2023).
63. Wu, X., Innovation and patenting activities of COVID-19 vaccines in WTO members: Analytical review of Medicines Patent Pool (MPP) COVID-19 Vaccines Patent Landscape (VAXPAL), World Trade Organization.
64. Singh & Associates and Rastogi, P., Patenting of nanotechnology invention: Issues and challenges, Lexology, Oct. 31, 2012, https://www.lexology.com/library/detail.aspx?g=7025304d-89d2-4927-8c2e-c38a70164475 (accessed Apr. 09, 2023).
65. The impact of effective patents on future innovations in nanomedicine, Pharmaceutical Patent Analyst, https://www.future-science.com/doi/abs/10.4155/ppa.11.4?journalCode=ppa (accessed Apr. 09, 2023).
66. Bawa, R., Patents and nanomedicine. *Nanomedicine*, 2, 3, 351–374, Jun. 2007. doi: 10.2217/17435889.2.3.351.
67. Marques, M.R.C., Choo, Q., Ashtikar, M., Rocha, T.C., Bremer-Hoffmann, S., Wacker, M.G., Nanomedicines - tiny particles and big challenges. *Adv. Drug Deliv. Rev.*, 151–152, 23–43, 2019. doi: 10.1016/j.addr.2019.06.003.
68. Mühlebach, S., Regulatory challenges of nanomedicines and their follow-on versions: A generic or similar approach? *Adv. Drug Deliv. Rev.*, 131, 122–131, Jun. 2018. doi: 10.1016/j.addr.2018.06.024.
69. Duvall, M.N., FDA regulation of nanotechnology.
70. de Vlieger, J.S.B. *et al.*, Report of the AAPS guidance forum on the FDA draft guidance for industry: Drug products, including biological products, that contain nanomaterials. *AAPS J.*, 21, 4, 56, Apr. 2019. doi: 10.1208/s12248-019-0329-7.
71. Emily, M., Ioanna, N., Scott, B., Beat, F., Reflections on FDA draft guidance for products containing nanomaterials: Is the abbreviated new drug

application (ANDA) a suitable pathway for nanomedicines. *AAPS J.*, 20, 5, 92, Aug. 2018. doi: 10.1208/s12248-018-0255-0.
72. A. FDA/CDER/"Yeaton, Drug products, including biological products, that contain nanomaterials - guidance for industry, 2017.
73. Annual report of the EFSA scientific network of risk assessment of nanotechnologies in food and feed for 2020, EFSA, Mar. 04, 2021. https://www.efsa.europa.eu/en/supporting/pub/en-6502 (accessed Apr. 09, 2023).
74. Sweeney, F., EU's regulatory process for evaluation and approval of vaccines.
75. Schmutz, M. *et al.*, A methodological safe-by-design approach for the development of nanomedicines. *Front. Bioeng. Biotechnol.*, 8, - 258, 2020, Accessed: Apr. 09, 2023. [Online]. Available: https://www.frontiersin.org/articles/10.3389/fbioe.2020.00258.
76. Vagnoni, C., Rough, E., Bunn, S., UK vaccination policy, Sep. 2023, Accessed: Apr. 09, 2023. [Online]. Available: https://commonslibrary.parliament.uk/research-briefings/cbp-9076/.
77. Richards, N. and Hudson, I., UK medicines regulation: Responding to current challenges. *Br. J. Clin. Pharmacol.*, 82, 6, 1471–1476, Dec. 2016. doi: 10.1111/bcp.13077.
78. The medicines and healthcare products regulatory agency annual report and accounts 2021/22, Gov.UK, https://www.gov.uk/government/publications/medicines-and-healthcare-products-regulatory-agency-annual-report-and-accounts-2021-to-2022/the-medicines-and-healthcare-products-regulatory-agency-annual-report-and-accounts-202122 (accessed Apr. 09, 2023).
79. Regulatory approval of Pfizer/BioNTech vaccine for COVID-19, Gov.UK, Mar. 17, 2023. https://www.gov.uk/government/publications/regulatory-approval-of-pfizer-biontech-vaccine-for-covid-19 (accessed Apr. 09, 2023).
80. MHLW pharmaceuticals and medical devices safety information, Pharmaceuticals and Medical Devices Agency, https://www.pmda.go.jp/english/safety/info-services/drugs/medical-safety-information/0002.html (accessed Apr. 09, 2023).
81. C. H. P. B. M. 7315 W. A. S. 400W Bethesda *et al.*, Japan medical device & pharmaceutical regulations - PMDA, MHLW, Pacific Bridge Medical, Aug. 12, 2018. https://www.pacificbridgemedical.com/regulation/japan-medical-device-pharmaceutical-regulations/ (accessed Apr. 09, 2023).
82. Mission on Nano Science and Technology (Nano mission), Department Of Science & Technology, https://dst.gov.in/scientific-programmes/mission-nano-science-and-technology-nano-mission (accessed Apr. 09, 2023).
83. A multilevel governance framework for regulation of nanomedicine in India, https://www.degruyter.com/document/doi/10.1515/ntrev-2016-0083/html?lang=en (accessed Apr. 09, 2023).

Index

Printed and bound by CPI Group (UK) Ltd, Croydon, CR0 4YY

17/09/2026

14964279-0006